An Introduction to
Data Structures and Algorithms

Consulting Editor
John C. Cherniavsky, National Science Foundation

J.A. Storer

An Introduction to
Data Structures and Algorithms

BIRKHÄUSER   •   SPRINGER
BOSTON   NEW YORK

James A. Storer
Department of Computer Science
Brandeis University
Waltham, MA 02454
U.S.A.

Library of Congress Cataloging-in-Publication Data

Storer, James A. (James Andrew), 1953–
   An introduction to data structures and algorithms / J.A. Storer.
     p. cm.
   Includes bibliographical references and index.
   ISBN 0-8176-4253-6 (alk. paper) – ISBN 3-7643-4253-6 (alk. paper)
    1. Data structures (Computer science) 2. Computer algorithms. I. Title.

   QA76.9.D33 S73 2001
   005.7'3–dc21   2001043503

AMS Subject Classifications: Primary—68P05, 68W40, 68W10; Secondary—68Q25, 68P10, 68W20, 68W25

Printed on acid-free paper
©2002 J.A. Storer

ISBN 0-8176-4253-6   SPIN 10841937
ISBN 3-7643-4253-6

Typeset by the author
Cover design by Mary Burgess, Cambridge, MA
Printed and bound by Hamilton Printing, Rensselaer, NY
Printed in the United States of America

9 8 7 6 5 4 3 2 1

Preface

This book is for college-level students who are attending accompanying lectures on data structures and algorithms. Except for the introduction, exercises, and notes for each chapter, page breaks have been put in manually and the format corresponds to a lecture style with ideas presented in "digestible" page-size quantities.

It is assumed that the reader's background includes some basic mathematics and computer programming. Algorithms are presented with "pseudo-code", not programs from a specific language. It works well to complement a course based on this book with a programming "lab" that uses a manual for a specific programming language and assigns projects relating to the material being taught.

Chapters 1 through 4 go at a slower pace than the remainder of the book, and include sample exercises with solutions; some of the sections in these chapters are starred to indicate that it may be appropriate to skip them on a first reading. A first course on data structures and algorithms for undergraduates may be based on Chapters 1 through 4 (skipping most of the starred sections), along with portions of Chapter 5 (algorithm design), the first half of Chapter 6 (hashing), the first half of Chapter 12 (graphs), and possibly an overview of Chapter 13 (parallel models of computation). Although Chapters 1 through 4 cover more elementary material, and at a slower pace, the concise style of this book makes it important that the instructor provide motivation, discuss sample exercises, and assign homework to complement lectures. For upper class undergraduates or first-year graduate students who have already had a course on elementary data structures, Chapters 1 through 4 can be quickly reviewed and the course can essentially start with the algorithm design techniques of Chapter 5.

There is no chapter on sorting. Instead, sorting is used in many examples, which include bubble sort, merge sort, tree sort, heap sort, quick sort, and several parallel sorting algorithms. Lower bounds on sorting by comparisons are included in the chapter on heaps, in the context of lower bounds for comparison based structures. There is no chapter on NP-completeness (although a section of the chapter on graph algorithms does overview the notion). Formal treatment of complexity issues like this is left to a course on the theory of computation.

Although traditional serial algorithms are taught, the last chapter presents the PRAM model for parallel computation, discusses PRAM simulation, considers how the EREW PRAM, HC/CCC/Butterfly, and mesh models can be used for PRAM simulation, and concludes with a discussion of hardware area-time tradeoffs as a way of comparing the relative merits of simulation models. Although it is not clear what parallel computing models will prevail in practice in the future, the basic concepts presented in the chapter can provide a foundation for further study on specific practical architectures. Also, from the point of view of studying algorithms, it is instructive to see how basic principles taught in the previous chapters can be adapted to other models.

Acknowledgements

Many people provided helpful comments during the writing of this book. In particular I would like to thank (in alphabetical order): R. Alterman, B. Carpentieri, J. Cherniavsky, M. Cohn, I. Gessel, A. Kostant, G. Motta, T. Hickey, J. Reif, F. Rizzo, J. E. Storer, and I. Witten.

J.A.S. 2001

Contents

6. Hashing 203

10. Graphs 289

1. RAM Model

Introduction

Efficient data structures and algorithms can be the key to designing practical programs. Most of this book addresses the traditional model of a computer as pictured below; some memory together with a single processor that can perform basic arithmetic operations (e.g., add, subtract, multiply, divide), make logical tests (e.g., test if a value is less than zero), and follow flow of control accordingly (e.g., *goto* instructions). The processor may include a few special *registers* that can be manipulated faster than normal memory, but the bulk of memory is uniform in the sense that access to all locations of memory is equally fast, from which comes the name *Random Access Memory* (RAM). Of course, computers in practice are much more complicated, with cache memory, secondary storage, multiple processors, etc. However, design methodology based on the RAM model has traditionally served well in more complex models.

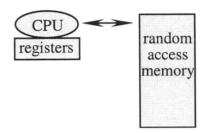

A key simplifying assumption is that operations are *unit-cost*. Each basic instruction takes one "unit" of time, and we count the number of basic instructions executed to measure time. Furthermore, the size of a number in an operation is not taken into account. For example, it is one time unit to add two numbers, independent of their size. Although these assumptions are not true in practice, if we do not "abuse" the unit-cost RAM model, our designs will be sound and provide the basis for practical algorithms on a variety of architectures. For example, if we have enough memory in practice to store n integers, then it is reasonable to assume that basic machine instructions can manipulate integers in the range 0 to n.

On the other hand, consider the following program that repeatedly multiplies by 10:

1. Set the variable x equal to 10.
2. Repeat for n steps the operation of multiplying x by 10.

Since each multiplication adds another digit to x, this program computes a value with over n digits. When n is large, it may not be realistic to assume that numbers of n digits can be stored in a single machine register and multiplied with a single machine instruction. A more "dramatic" example repeatedly squares a value:

1. Set the variable x equal to 10.

2. Repeat for n steps the operation of multiplying the variable x times itself.

Since each multiplication in this program about doubles the number of base 10 digits to represent x (i.e., x is a 1 followed by a number of 0's that doubles with each multiplication), this program computes a value with roughly 2^n digits. For example, even when n is only 25, this is more than 30 million digits.

It is possible to strengthen the RAM model by charging a cost per operation that depends on the size of the numbers involved. However, the extra complication is really not necessary if we are willing to exercise good judgement to understand when it would not be practical to manipulate in a fixed time unit numbers that are involved in an algorithm.

The only data structure that we consider to be part of the basic RAM model is the standard one dimensional array, or equivalently the ability to perform indirect addressing. For a fixed integer x, not only are there machine instructions to read or write to memory location x, but we can also read or write to memory location y, where y is the value stored in memory location x. Indirect addressing corresponds exactly to the notion of a one dimensional array. For example, setting a variable x to the value of $A[i]$ does not place the contents of the memory location i into x, but rather, it copies the contents of a memory location that is determined by the value of i. In fact, we shall see how indirect addressing can be used to implement arrays of any dimension.

In practice, one generally avoids the complexity and tedium of programs written directly with machine instructions and relies on compilers to translate higher level programming languages into "assembly language" that can be more or less directly mapped to the actual machine instructions encoded with 0's and 1's. Since we are concerned here only with basic data structures and algorithm design, and not with all of the "messy stuff" that is needed in practice (input-output, graphics, systems calls, file storage, etc.), we express algorithms with simple "pseudo-code" using a mixture of English and constructs that are available, in some form, in most languages (*if-then-else*, etc.). Hopefully, the syntax of pseudo-code used in this book is self-explanatory to anyone who has had some programming background.

The first example of this chapter computes the factorial function; that is, $n!$ is the product of the integers 1 through n (define $n!=1$ for $n<1$). Factorial makes a nice first example of a program because it employs no data structures and simply performs a sequence of n multiplications. However, like the computation of n multiplications of a number by itself, this is a good example of a program for which the unit-cost model of arithmetic operations is not realistic; the value of $n!$ gets very large very quickly (although not as extreme as repeatedly squaring a value — see the sample exercises). The first program for $n!$ is pseudo-code that sets a variable to 1 and then successively multiplies it by the values 2 through n. The second program shows what $n!$ might look like in assembly language. In fact, the assembly language code has been commented to follow the flow of control of the pseudo-code (if a compiler made the translation from the first program, it probably would not look this neat, but would work equivalently); this is just "pseudo-assembly" language for a very simple single register machine (all computations are performed with respect to a single register called the *accumulator*). We will not use assembly language again, but it is always important to have in mind the basic RAM model, and an understanding of what can be done in a fixed amount of time.

Central to this chapter is the *O* notation that will be used to measure the *asymptotic* time and space complexity of algorithms. We will not attempt to pin down how fast an algorithm runs on a particular machine on an input of a particular size. Rather, we consider a machine independent notion of how, using the RAM model, the time or space used by the algorithm increases as a function of the input size. So for example, for a given problem of size n (e.g., given n bank account files, compute their monthly interest), an algorithm that executes a number of basic steps that is proportional to n will be considered better than one that uses a number of steps that is proportional to n^2, even if the constant of proportionality for the n^2 algorithm is much lower. It is the large inputs that concern us most, and as n gets large, an extra factor of n dominates the savings derived from a smaller constant of proportionality.

Unless otherwise stated, we measure the *worst case* time or space used by an algorithm; that is, as a function of n the most time or space that can be used for an input of size n. Occasionally we consider the *expected* time or space used. That is, under some assumption about the probability distribution of the inputs of size n (e.g., all inputs are equally likely) or about the expected effect of randomized decisions made by the algorithm, we will be able to analyze the average time or space used over all inputs (which is sometimes better than the worst case).

In practice, presentation of a clean description (e.g., pseudo-code) of an algorithm that performs well for large inputs is only the first step in what may be a substantial development process that leads to a practical computer program (or possibly computer hardware). This chapter ends by listing a host of other issues that may arise, but are outside of the basic data structures and algorithm design that are the subject of this book. In addition, because a problem is not complex in the sense that it does not require much time to be executed does not mean that it is not a difficult one to program in practice. Many complex operations in practice use only a small fixed amount of time, and are trivial from a theoretical point of view, but require complicated programming that may not be easily described with pseudo-code. Data structures and algorithm design is a valuable tool for what is often a much larger task.

The examples presented in this chapter do not employ any data structures besides standard 1-dimensional arrays. In later chapters, we will be able to design algorithms for more complex and varied applications by using appropriate data structures.

Pseudo-Code

Idea: Express algorithms with reasonably self-explanatory "pseudo-code" that mixes English with syntax that is equivalent to instructions available in most programming languages. We assume that all conditions and expressions are evaluated from left to right.

Examples:

$a := b$
>Assignment statement that sets a to the value of b.

if *condition* **then** statement 1 **else** statement 2
>If *condition* is true then do statement 1, otherwise do statement 2.

for $i := a$ **to** b **do** statement
>Repeatedly set $i := a$, $i := a+1$, $i := a+2$, ... $i := b$ and perform the statement.

for $i := a$ **to** b **by** c **do** statement
>Repeatedly set $i := a$, $i := a+c$, $i := a+2c$, ... and perform the statement until $i>b$.

for $i := a$ **downto** b **do** statement
>Repeatedly set $i := a$, $i := a-1$, $i := a-2$, ... $i := b$ and perform the statement.

for $1 \leq i \leq n$ **do** statement
>Execute the statement for all values of i in this range, the order does not matter; e.g., this is equivalent to: **for** $i := 1$ **to** n **do** statement

while *condition* **do** statement
>Repeatedly test *condition* and perform the statement until *condition* is false.

repeat statements **until** *condition*
>Repeatedly perform the statements and test *condition* until *condition* is true.

begin statement ... statement **end**
>Group these statements together so they can be placed anywhere that you would normally place a single statement.

function NAME(*arguments*)
>⋮
>
>**end**
>
>Statements inside a function are performed when it is called. Arguments are evaluated left to right and the result stored in a temporary location so that when a function returns, original values of the arguments are not affected (although an array name can be passed and its contents modified). A function is a **procedure** when it does not return a value (it may still modify global variables or an array whose name was passed as an argument). The statement **return** *value* immediately exits the function and returns the *value*. For a procedure, **return** exits immediately (without a **return** statement, the procedure returns when its end is reached).

Array Arguments: When passing arguments to functions, we sometimes use the expression $A[i]...A[j]$ as *notation* to mean the three parameters i, j, the location in memory of A; so only space for three variables is used to specify this portion of A.

Example: *n*! — Pseudo-Code Versus Machine Code

Recall that for an integer $n \geq 1$, $n!$ denotes the product of the integers from 1 through n (define $n! = 1$ for $n \leq 1$).

Idea: The pseudo-code below initializes a variable to 1 and then successively multiplies it by the integers 2 through n (or does nothing if $n \leq 1$). "Higher-level" languages are typically translated into a "low-level" language that corresponds closely to the basic instructions on the machine being used. The "assembly language" code presented here uses instructions of a generic type that would be available (in some equivalent syntax) on most any computer today. The *accumulator* refers to a special register used by the processor to perform computations; in practice machines may have a number of registers.

Pseudo-code for *n*!:

> **read** n
> $x := 1$
> **for** $i := 2$ **to** n **do** $x := x * i$
> **write** x

Assembly type code for *n*! on a single register machine: We follow the flow of the pseudo-code, where n, x, and i are stored in memory locations 1, 2, and 3.

	read	read the input n into the accumulator
	store 1	store accumulator into memory location 1 (save the value of n)
	load =1	load the constant 1 into the accumulator
	store 2	store accumulator into memory location 2 (initialize $x=1$)
	store 3	store accumulator into memory location 3 (initialize $i=1$)
loop:	load 3	load memory location 3 into accumulator (load value of i)
	subtract 1	subtract memory location 1 from accumulator (subtract n)
	goto(≥ 0) *done*	go to *done* if the accumulator is ≥ 0 (done if $i \geq n$)
	load 3	load memory location 3 into accumulator (load the value of i)
	add =1	add the constant 1 to the accumulator (compute $i+1$)
	store 3	store accumulator in memory location 3 (save the value of i)
	multiply 2	multiply accumulator by memory location 2 (compute $x*i$)
	store 2	store accumulator in memory location 2 (save new value of x)
	goto *loop*	repeat the main loop
done:	load 2	place memory location 2 in accumulator (load value of x)
	write	write contents of accumulator to output device (write x)

We will not directly consider assembly language again in this book, but when considering the complexity of an algorithm that is expressed with pseudo-code, it will always be important to have a basic understanding of the underlying RAM model and what operations are or are not realistic in a fixed amount of time.

Motivation for Asymptotic Time Complexity

Problem: How can we make a meaningful statement about the complexity of a task, especially for large inputs, that is reasonably machine independent? Consider the basic *sorting* problem of arranging a list of integers into increasing order:

Sorting 100 integers can be done quickly.

> *What do you mean by quickly?*

Sorting 100 integers can be done in 100 seconds.

> *On what computer?*

Sorting 100 integers can be done in 100 seconds on Brand X computer.

> *What about sorting 200 integers?*

Sorting n integers can be done in $10n^2$ seconds on Brand X computer.

> *What about a different computer?*

Sorting n integers can be done in time proportional to n^2 on any computer.

> *What do you mean by any computer?*

Sorting n integers can be done in time proportional to n^2 on any computer of the standard random access type built today.

> *What do you mean by time; if there is a slowdown because the disk drive needs to be powered up, does this count?*

Sorting n integers can be done in a number of basic steps proportional to n^2 on any computer of the standard random access type built today.

> *Although this statement is somewhat vague about the definition of basic steps and the definition of a computer, at least it is machine independent.*

> *A typical way to show that such a statement is true is to exhibit a particular computer program.*

> *There are still many unanswered questions, such as whether it is possible to do better than n^2.*

Note: This discussion assumes that the n data items can fit into memory; see the chapter notes for references about sorting large data files in secondary memory such as a disk or tape drive. Assuming that the data does fit into memory, we shall see a number of practical ways to sort n integers in time proportional to $n\log(n)$.

Worst case vs. expected case: Unless otherwise stated, we address the *worst case* performance of an algorithm; that is, the *most* amount of time that can be spent on *any* input of a given size. Occasionally we also consider how fast the algorithm runs on the average under some assumption about the probability distribution of possible inputs. Usually, the assumption is that all inputs are equally likely.

CHAPTER 1

Analyzing Algorithms with Asymptotic Notation

Idea: Without limiting ourselves to a particular machine or programming language it is difficult to bound the precise constants in an expression denoting the time (or space) used by an algorithm. However, we can at least get a crude measure of *asymptotically* how much time or space an algorithm uses by identifying how the time or space it uses grows as a function of some parameter n, which usually denotes the size of the input (but sometimes, as in the case of $n!$, may be the value of the input or some natural parameter associated with it). We define a precise notion of when, ignoring constant factors, one function $f(n)$ is "smaller" (grows no faster) than another function $g(n)$. Typically, $f(n)$ will denote the time or space used by a particular algorithm and $g(n)$ will be a "well-behaved" function (n, $n\log(n)$, n^2, n^3, etc.), but the definition itself simply relates mathematically two functions f and g.

The O Notation

Let f and g be two functions from the non-negative integers to the non-negative integers:

O: A function $f(n)$ is $O(g(n))$ if there exists real numbers $a,b > 0$ such that for all integers $n \geq a$, $f(n) \leq b*g(n)$.

Ω: A function $f(n)$ is $\Omega(g(n))$ if there exists a real number $c > 0$ such that for infinitely many integers n, $f(n) \geq c*g(n)$.

Θ: A function $f(n)$ is $\Theta(g(n))$ if it is both $O(g(n))$ and $\Omega(g(n))$.

Intuition: If constants are ignored, O is like $\leq$ between two functions, Ω is like $\geq$ between two functions, and θ is like $=$ between two functions. Saying that $f(n)$ is $O(1)$ says that $f(n)$ is bounded above by a constant. Saying that $f(n)$ is $\Omega(1)$ does not really say anything (unless functions that can have values less than one — see the exercises).

Examples of asymptotic relationships (see the sample exercises):

A. 100 is $\Theta(1)$

B. $2n$ is $\Theta(n)$

C. n is $O(n^2)$ but not $\Omega(n^2)$

D. $n^2/2 - n/2$ is $\Theta(n^2)$.

E. $n\sqrt{n}$ is $O(n^2)$ but not $\Omega(n^2)$

F. n^2 is $\Omega(1000n)$ but not $O(1000n)$

G. $n^2 + n\sqrt{n} + 25n + 100$ is $\Theta(n^2)$

Note: In practice, when the running time of an algorithm is $\Theta(f(n))$, we will talk about the algorithm being $O(f(n))$ in discussing its efficiency and $\Omega(f(n))$ in discussing its limitations. We use the Θ notation occasionally when it is appropriate to emphasize that the stated asymptotic running time is "tight". The definition of Ω is intentionally not symmetric with that of O; see the sample exercises.

Example: Bubble Sort

Problem: Rearrange the array elements $A[1] \ldots A[n]$ into increasing order.

Idea: Repeatedly "percolate" the next largest element up to its correct position.

Bubble sort algorithm:
> **for** $i := n-1$ **downto** 1 **do**
> > **for** $j := 1$ **to** i **do**
> > > **if** $A[j] > A[j+1]$ **then** exchange $A[j]$ and $A[j+1]$

Time: Bubble sort is very slow when n is large. It makes the same number of comparisons no matter what the ordering of the input. For all inputs of length n the *if* statement is executed exactly the same number of times (once for every iteration of the inner loop) and consumes $O(1)$ time. That is, it always does the work of testing if $A[j] > A[j+1]$ even if an exchange is not necessary. Hence, the same time analysis can be used for both O and Ω. Since the inner loop goes from 1 to i, and i is going from $n-1$ down to 1 with each iteration of the outer loop, the number of times the *if* statement is executed is:

$$= (n-1) + (n-2) + \cdots + 1$$
$$= 1 + 2 + 3 + \cdots + (n-1)$$
$$= (\text{number of terms})(\text{average value of a term})$$
$$= (\text{number of terms})\left(\frac{\text{smallest term } + \text{ largest term}}{2} \right)$$
$$= \tfrac{1}{2}n^2 - \tfrac{1}{2}n$$

Since this expression is less than n^2 for all n, it follows that it is $O(n^2)$. We also saw in the example on the previous page that it is $\Omega(n^2)$, where the mathematics is shown in the sample exercises. Hence it is $\Theta(n^2)$.

Space: Although the time is poor, bubble sort uses only $O(1)$ space in addition to the $O(n)$ space used by the input array A.

Note: We will see faster ways to sort in later chapters.

Example: Run-Length Codes

Idea: Given a string that has many *runs* (blocks of repeated characters), it may take less space to represent by the lengths of each run. We limit our attention to binary strings (see the exercises for generalization to larger alphabets). For example, the binary string 0 1 1 1 1 0 0 1 1 1 0 0 0 0 0 1 0 1 0 1 1 could be encoded as 1 4 2 3 5 1 1 1 1 2.

Implementation issues: Assume that the string starts with 0 (if not, begin by encoding a run of zero 0's) and that the encoder and decoder have agreed in advance on the number of bits k to be used to represent each integer in the encoding. Since k bits represent the integers 0 through 2^k-1, when 2^k-1 is received by the decoder, it expects that the next run will be more of the same bit; in particular, when the count is exactly 2^k-1 (or some multiple of it), it is followed by a zero count for the same bit (unless it is the last count). See the exercises for practical improvements in compression over what is presented here.

Binary run-length encoding algorithm:
> $MaxCount := 2^k\text{-}1$
> $count := 0$
> $PrevBit := \text{"0"}$
> **while** input remains **do begin**
>> $CurBit :=$ the next input bit
>> **if** $PrevBit=CurBit$ and $count<MaxCount$ **then** $count := count+1$
>> **else begin**
>>> Output $count$ using k bits.
>>> **if** $count=MaxCount$ and $PrevBit{\neq}CurBit$ **then** Output 0 using k bits.
>>> $count := 1$
>>> **end**
>>
>> $PrevBit := CurBit$
>> **end**
>
> Output $count$ using k bits.

Binary run-length decoding algorithm:
> $MaxCount := 2^k\text{-}1$
> $parity := 0$
> **while** input remains **do begin**
>> Read k bits from the input stream to get the integer x.
>> **if** $parity=0$ **then** output x "0" bits **else** output x "1" bits
>> **if** $x<MaxCount$ **then** $parity := 1 - parity$
>> **end**

Complexity: Space is $O(1)$ for both encoding and decoding. Assuming that input or output of k bits is $O(k)$ time, for a string of n bits, encoding and decoding is $O(kn)$ in the worst case (e.g., an alternating sequence of 0's and 1's), which is $O(n)$ assuming that k is constant with respect to n.

Example: Horner's Method for Polynomial Evaluation

Problem: Given the $n \geq 1$ coefficients $A[0]$ ... $A[n-1]$, evaluate:

$$A[n-1]x^{n-1} + A[n-2]x^{n-2} + ... + A[1]x + A[0]$$

"Brute force" solution: Multiply x times itself i times to compute x^i, multiply x^i times $A[i]$, and add this to a running total, working from $i=(n-1)$ down to $i=0$ (where the last step adds $A[0]$).

> **input** x
> $value := 0$
> **for** $i := (n-1)$ **downto** 0 **do begin**
> $temp := 1$
>   **for** $j := 1$ **to** i **do** $temp := temp*x$
> $value := value + A[i]*temp$
>   **end**
> **output** $value$

Time: Since the body of the outer *for* loop is $O(1)$ time excluding the time used by the inner *for* loop, the time is proportional to the total number of multiplications performed by the inner loop, which is always:

$$= (n-1) + (n-2) + \cdots + 1 + 0$$

This is exactly the same expression we had for bubble sort, and it follows that the algorithm is $\Theta(n^2)$.

Space: $O(1)$ space in addition to the $O(n)$ space used by the input array A.

Horner's method: We again work from $i=(n-1)$ down to $i=0$, but by multiplying the entire running total by x, and then adding $A[i]$:

> compute $A[n-1]$
> compute $A[n-1]x + A[n-2]$
> compute $(A[n-1]x + A[n-2])x + A[n-3]$
> compute $((A[n-1]x + A[n-2])x + A[n-3])x + A[n-4]$
> and so on ...

The algorithm is now a single loop rather than a nest pair of loops:

> **input** x
> $value := 0$
> **for** $i := (n-1)$ **downto** 0 **do** $value := value*x + A[i]$
> **output** $value$

Time: $O(n)$, since the loop body is $O(1)$ and the loop is executed n times.

Space: $O(1)$ space in addition to the $O(n)$ space used by the input array A.

Example: Matrix Multiplication

Definition: The product of two "square" n by n matrices is defined as:

$$\begin{pmatrix} A[1,1] & \cdots & A[1,n] \\ \vdots & \vdots & \vdots \\ A[n,1] & \cdots & A[n,n] \end{pmatrix} \begin{pmatrix} B[1,1] & \cdots & B[1,n] \\ \vdots & \vdots & \vdots \\ B[n,1] & \cdots & B[n,n] \end{pmatrix} = \begin{pmatrix} C[1,1] & \cdots & C[1,n] \\ \vdots & \vdots & \vdots \\ C[n,1] & \cdots & C[n,n] \end{pmatrix}$$

$$C[i,j] = (\text{row } i \text{ of } A)(\text{column } j \text{ of } B)$$
$$= A[i,1]B[1,j] + A[i,2]B[2,j] + \cdots + A[i,n]B[n,j]$$

For example:

$$\begin{pmatrix} 1 & 2 \\ 3 & 4 \end{pmatrix} \begin{pmatrix} 5 & 6 \\ 7 & 8 \end{pmatrix} = \begin{pmatrix} 1*5+2*7 & 1*6+2*8 \\ 3*5+4*7 & 3*6+4*8 \end{pmatrix} = \begin{pmatrix} 19 & 22 \\ 43 & 50 \end{pmatrix}$$

This definition generalizes naturally to multiplying a m by w matrix A with a w by n matrix B to form a n by m matrix C. That is, computing an entry of C by multiplying a row of A by a column of B is well defined as long as the length of a row in A is the same as the height of a column in B. The multiplication of a row by a column is called a *dot product* or an *inner product*; in general, for $1 \leq i \leq n$ and $1 \leq j \leq m$:

$$C[i,j] = A[i,1]B[1,j] + A[i,2]B[2,j] + \cdots + A[i,w]B[w,j]$$

For example:

$$\begin{pmatrix} 1 & 2 \\ 3 & 4 \\ 5 & 6 \\ 7 & 8 \end{pmatrix} \begin{pmatrix} 1 \\ 2 \end{pmatrix} = \begin{pmatrix} 1*1+2*2 \\ 3*1+4*2 \\ 5*1+6*2 \\ 7*1+8*2 \end{pmatrix} = \begin{pmatrix} 5 \\ 11 \\ 17 \\ 21 \end{pmatrix}$$

Standard matrix multiplication algorithm:

> **for** $1 \leq i \leq m,\ 1 \leq j \leq n$ **do begin**
> $C[i,j] := 0$
> **for** $k=1$ **to** w **do** $C[i,j] = C[i,j] + A[i,k]B[k,j]$
> **end**

Time: The inner *for* loop is $O(w)$, and it is executed by the outer *for* loop for all mn possible values of i,j, for a total of $O(mwn)$ time. For the case of square matrices where $m=w=n$, the time is $O(n^3)$.

Space: $O(1)$ in addition to the space used by A, B, and C.

Example: Pascal's Triangle of Binomial Coefficients

The number of different ways to choose m items out of a total of n items (without repetitions), $n \geq 1$ and $0 \leq m \leq n$, is given by the *binomial coefficient*:

$$\binom{n}{m} = \frac{n!}{(n-m)!\,m!}$$

That is, there are n ways to choose the first item, $n-1$ ways to choose the second, etc. This is like $n!$ except we do not want the terms below $n-m$ (giving $n!/(n-m)!$); in addition, we must divide by the $m!$ different ways to reorder the m elements chosen.

Pascal's triangle: The infinite table of all binomial coefficients is "Pascal's triangle". Below are rows 0 through 5. The triangle has one row for each value of n, where each row shows from left to right the values of the binomial coefficients going from $m=0$ to $m=n$. On the right is the equivalent rectangular array.

$n=0$:					1					1	0	0	0	0	0
$n=1$:				1		1				1	1	0	0	0	0
$n=2$:			1		2		1			1	2	1	0	0	0
$n=3$:		1		3		3		1		1	3	3	1	0	0
$n=4$:	1		4		6		4		1	1	4	6	4	1	0
$n=5$:	1	5		10		10	5		1	1	5	10	10	5	1

Straightforward computation: For the row $n=i$, first compute a table of the values of j factorial, $0 \leq j \leq i$, and then compute each term of the row with three arithmetic operations, for a total of $O(k^2)$ multiplications to compute rows $n=0$ through $n=k$.

Simpler computation: Rows $n=0$ through $n=k$ can be computed in $O(n^2)$ time with one addition for each internal entry. An entry is 1 if $m=0$ or $m=n$, and in general:

$$\binom{n}{m} = \frac{n!}{(n-m)!\,m!}$$

$$= \frac{m(n-1)!}{(n-m)!\,m!} + \frac{(n-m)(n-1)!}{(n-m)!\,m!}$$

$$= \frac{(n-1)!}{((n-1)-(m-1))!\,(m-1)!} + \frac{(n-1)!}{(n-m-1)!\,m!}$$

$$= \binom{n-1}{m-1} + \binom{n-1}{m}$$

Compute each row from entries of the previous row:

> **for** $n := 0$ **to** k **do**
>> **for** $m := 0$ **to** n **do**
>>> **if** $m=0$ or $m=n$ **then** $A[i,j] := 1$ **else** $A[n,m] := A[n-1,m-1]+A[n-1,m]$

Binomial Sums

The classic *binomial sum* is the expansion of the expression $(a+b)^n$.

Theorem:

$$(a+b)^n = \sum_{m=0}^{n} \binom{n}{m} a^{n-m} b^m$$

Proof: If we consider the process of computing $(a+b)^n = (a+b)(a+b)...(a+b)$ as forming the 2^n products that result from the n choices of selecting an a or b from each of the n clauses $(a+b)$, then each of these 2^n products is one of the $n+1$ values $a^{n-m}b^m$, $0 \leq m \leq n$. The number of copies of $a^{n-m}b^m$ is the number of ways of selecting which m of the n choices are b (the remaining choices must be a). Hence, the coefficient of the term $a^{n-m}b^m$ is the corresponding binomial coefficient.

Corollary: Since the binomial sum is symmetric, we also have:

$$(a+b)^n = \sum_{m=0}^{n} \left(\binom{n}{m} a^m b^{n-m} \right)$$

When both *a* and *b* are 1: It is interesting to note that when $a=b=1$, 2^n is just a sum of binomial coefficients. For example, for $n=5$:

$$2^5 = \sum_{m=0}^{5} \binom{5}{m}$$

$$= \binom{5}{0} + \binom{5}{1} + \binom{5}{2} + \binom{5}{3} + \binom{5}{4} + \binom{5}{5}$$

$$= \frac{5!}{5!0!} + \frac{5!}{4!1!} + \frac{5!}{3!2!} + \frac{5!}{2!3!} + \frac{5!}{1!4!} + \frac{5!}{0!5!}$$

$$= \frac{(5!/5!)}{0!} + \frac{(5!/4!)}{1!} + \frac{(5!/3!)}{2!} + \frac{(5!/2!)}{3!} + \frac{(5!/1!)}{4!} + \frac{(5!/0!)}{5!}$$

$$= 1 + 5 + \frac{5*4}{2} + \frac{5*4*3}{3*2} + \frac{5*4*3*2}{4*3*2} + 1$$

$$= 1 + 5 + 10 + 10 + 5 + 1$$

$$= 32$$

Note: Another way to see that the case $a = b = 1$ yields powers of 2 is to examine the figure on the preceding page that shows Pascal's triangle; the sum of the ith row is 2^i.

(*) Example: Solving Sets of Linear Equations

Problem: For $1 \leq i \leq n$, determine values for X that satisfy all n equations of the form:

$$A[i,1]X[1] + A[i,2]X[2] + \cdots + A[i,n]X[n] = A[i,n+1]$$

Idea: Subtract multiples of one equation from another to put them in "triangular form", and then work "backwards" (there are faster methods — see the chapter notes).

Example: Suppose $n=4$, denote $X[1]...X[4]$ by w, x, y, and z, and denote the four rows of A by a, b, c, and d, where:

a:	w	$+x$	$+2y$	$+2z$	$=$	17
b:	$2w$	$+x$	$+y$	$+3z$	$=$	19
c:	w	$+3x$	$+y$	$+2z$	$=$	18
d:	$3w$	$+x$	$+y$	$+z$	$=$	12

Phase 1, Put the equations in triangular form:

Clear the first column by subtracting a multiple of row a from rows b, c, and d:

a:	w	$+x$	$+2y$	$+2z$	$=$	17	
b:		$-x$	$-3y$	$-z$	$=$	-15	$(b := b - 2a)$
c:		$2x$	$-y$	$+0z$	$=$	1	$(c := c - a)$
d:		$-2x$	$-5y$	$-5z$	$=$	-39	$(d := d - 3a)$

Clear the second column by subtracting a multiple of row b from rows c and d:

a:	w	$+x$	$+2y$	$+2z$	$=$	17	
b:		$-x$	$-3y$	$-z$	$=$	-15	
c:			$-7y$	$-2z$	$=$	-29	$(c := c - (-2)b)$
d:			y	$-3z$	$=$	-9	$(d := d - 2b)$

Clear the third column by subtracting a multiple of row c from row d:

a:	w	$+x$	$+2y$	$+2z$	$=$	17	
b:		$-x$	$-3y$	$-z$	$=$	-15	
c:			$-7y$	$-2z$	$=$	-29	
d:				$-(23/7)z$	$=$	$-92/7$	$(d := d - (-1/7)c)$

Phase 2, Back-substitute:

Substitute $z = (-92/7)/(-23/7) = 92/23 = 4$:

a:	w	$+x$	$+2y$	$=$	9
b:		$-x$	$-3y$	$=$	-11
c:			$-7y$	$=$	-21

Substitute $y = (-21)/(-7) = 3$:

a:	w	$+x$	$=$	3
b:		$-x$	$=$	-2

Substitute $x = (-2)/(-1) = 2$:

a:	w	$=$	1

Substitute $w = (1)/(1) = 1$.

The final solution is: $w = 1, x = 2, y = 3, z = 4$

Back-Substitution Algorithm

In *Phase* 1 for $i=1$ to $n-1$ subtract multiples of row i from higher numbered rows so the only remaining values of A that can be non-zero are:

$$
\begin{array}{rcl}
A[1,1]X[1] \;+\; A[1,2]X[2] \;+\; A[1,3]X[3] \;+\cdots+\; A[1,n]X[n] &=& A[1,n+1] \\
A[2,2]X[2] \;+\; A[2,3]X[3] \;+\cdots+\; A[2,n]X[n] &=& A[2,n+1] \\
A[3,3]X[3] \;+\cdots+\; A[3,n]X[n] &=& A[3,n+1] \\
\vdots \\
A[n,n]X[n] &=& A[n,n+1]
\end{array}
$$

Once the equations are in this "triangular form", in *Phase* 2 "back-substitute" to determine the values for X; that is, we know $X[n]=A[n,n+1]/A[n,n]$, and once we know $X[n]$, we can determine $X[n-1]$, and once we know $X[n-1]$ and $X[n-2]$ we can determine $X[n-3]$, and so on. In Phase 1 rearrange the equations at the kth step so that $A[k,k]\neq0$ and Phase 2 will not divide by zero (the equations do not have a unique solution if they cannot be so arranged — see the exercises); if $A[k,k]=0$, exchange this equation with a higher numbered equation that has a non-zero value in the kth column (for simplicity, instead of exchanging, add the higher numbered equation to the kth one).

Phase 1, *put the equations in triangular form:*
for $k := 1$ **to** $n-1$ **do**
    if $A[k,k]=0$ **then begin** (add a higher numbered equation with non-zero first term)
        $x := k$
        while $x \leq n$ and $A[x,k]=0$ **do** $x := x+1$
        if $x>n$ **then** ERROR — equations do not have a unique solution
        else for $y := k$ **to** $n+1$ **do** $A[k,y] := A[k,y] + A[x,y]$
        end
    for $i := (k+1)$ **to** n **do**
        for $j := k$ **to** $n+1$ **do**
$$
A[i,j] = A[i,j] - \left(\frac{A[i,k]}{A[k,k]} \right) A[k,j]
$$

Phase 2, *back-substitute to compute* $X[1]..X[n]$:
    for $i := n$ **downto** 1 **do begin**
$$
X[i] = \frac{A[i,n+1]}{A[i,i]}
$$
        for $j := (i-1)$ **downto** 1 **do** $A[j,n+1] := A[j,n+1] - A[j,i]X[i]$
        end

Complexity: The time is dominated by three nested loops ranging over at most n values, for a total of $O(n^3)$ time. $O(1)$ space is used in addition to the space used for A.

(*) Example: Lagrange Interpolation of Polynomials

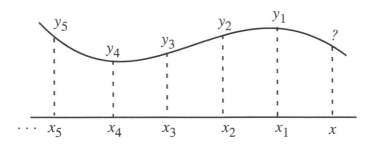

Idea: If y_1, y_2, ..., y_n denote the values of a function at distinct points x_1, x_2, ..., x_n, then a unique polynomial of degree at most $(n-1)^{th}$ passes through these points (exactly degree $n-1$ unless there is a lower order dependence among the values — see the exercises). It can be used to "predict" the value of the function at a new point x. The figure shows the points equally spaced in sequential order with x to the right; in general they may be in any order with arbitrary spacing, as long as they are distinct.

Solution by formulation as a set of linear equations: The solution to a set of k linear equations gives the coefficients of the corresponding $(n-1)^{th}$ degree polynomial.

Example: If $(x_1,y_1)=(1,15)$, $(x_2,y_2)=(2,32)$, and $(x_3,y_3)=(3,57)$, then the three equations

$$a(1)^2 \quad +b(1) \quad +c \quad = \quad 15$$
$$a(2)^2 \quad +b(2) \quad +c \quad = \quad 32$$
$$a(3)^2 \quad +b(3) \quad +c \quad = \quad 57$$

simplify to:

$$a \quad +b \quad +c \quad = \quad 15$$
$$4a \quad +2b \quad +c \quad = \quad 32$$
$$9a \quad +3b \quad +c \quad = \quad 57$$

Subtracting 4 times the 1st row from the 2nd and 9 times the 1st from the 3rd gives:

$$a \quad +b \quad +c \quad = \quad 15$$
$$-2b \quad -3c \quad = \quad -28$$
$$-6b \quad -8c \quad = \quad -78$$

Finally, subtracting 3 times the 2nd row from the 3rd gives

$$a \quad +b \quad +c \quad = \quad 15$$
$$-2b \quad -3c \quad = \quad -28$$
$$c \quad = \quad 6$$

from which $c=6$ follows, and then back-substituting $c=6$ into the second equation gives $b=5$, and finally back-substituting $b=5$, $c=6$ into the first equation gives $a=4$. Hence, the polynomial that passes through the three (x,y) coordinates $(1,15)$, $(2,32)$, and $(3,57)$ is $P(x) = 4x^2+5x+6$.

The Lagrange Interpolation Formula

Although we could solve a set of linear equations each time we needed to compute the n coefficients of a polynomial given its values at n distinct points $x_1...x_n$, the polynomial can always be expressed in the following regular form:

$$P(x) = \sum_{1 \le i \le n} \left(\left[\prod_{i \ne j} \frac{(x - x_j)}{(x_i - x_j)} \right] y_i \right)$$

$$= \frac{(x - x_2)(x - x_3)\cdots(x - x_k)}{(x_1 - x_2)(x_1 - x_3)\cdots(x_1 - x_k)} y_1 + \frac{(x - x_1)(x - x_3)\cdots(x - x_k)}{(x_2 - x_1)(x_2 - x_3)\cdots(x_2 - x_k)} y_2$$

$$+ \frac{(x - x_1)\cdots(x - x_{k-2})(x - x_k)}{(x_{k-1} - x_1)\cdots(x_{k-1} - x_{k-2})(x_{k-1} - x_k)} y_{k-1} + \cdots + \frac{(x - x_1)(x - x_2)\cdots(x - x_{k-1})}{(x_k - x_1)(x_k - x_1)\cdots(x_k - x_{k-1})} y_k$$

Since a unique polynomial of degree at most $n-1$ passes through n distinct points, correctness of the Lagrange interpolation formula follows by observing:

- $P(x)$ a polynomial of degree at most $n-1$.
- Suppose that $x = x_i$ for some $1 \le i \le n$. For the fraction multiplying y_i, the numerator (top portion) is the same as the denominator (bottom portion), and hence y_i is multiplied by 1. On the other hand for $j \ne i$, the product in the numerator contains the term $(x - x_i) = (x_i - x_i) = 0$, and hence this term is 0. Thus, for any of the points x_i, $1 \le i \le n$, $P(x_i) = y_i$.

Complexity: Each of the n terms can be computed in $O(n)$ time, for a total of $O(n^2)$ time. This improves upon the $O(n^3)$ time used by solving a set of linear equations.

Examples:

$n=2$: $\quad P(x) = \frac{(x - x_2)}{(x_1 - x_2)} y_1 + \frac{(x - x_1)}{(x_2 - x_1)} y_2$

$n=3$: $\quad P(x) = \frac{(x - x_2)(x - x_3)}{(x_1 - x_2)(x_1 - x_3)} y_1 + \frac{(x - x_1)(x - x_3)}{(x_2 - x_1)(x_2 - x_3)} y_2 + \frac{(x - x_1)(x - x_2)}{(x_3 - x_1)(x_3 - x_2)} y_3$

$n=4$:
$$P(x) = \frac{(x - x_2)(x - x_3)(x - x_4)}{(x_1 - x_2)(x_1 - x_3)(x_1 - x_4)} y_1 + \frac{(x - x_1)(x - x_3)(x - x_4)}{(x_2 - x_1)(x_2 - x_3)(x_2 - x_4)} y_2$$
$$+ \frac{(x - x_1)(x - x_2)(x - x_4)}{(x_3 - x_1)(x_3 - x_2)(x_3 - x_4)} y_3 + \frac{(x - x_1)(x - x_2)(x - x_3)}{(x_4 - x_1)(x_4 - x_2)(x_4 - x_3)} y_4$$

Note: Stability of these expressions in the presence of round-off errors can be a concern in practice; see the chapter notes.

Linear Time Lagrange Interpolation for Unit Spacing

Idea: Sample n points of an unknown function so that a polynomial $P(x)$ can be constructed to predict the value of the function at new points. Often, the choice of sample points is arbitrary, and so we can choose them to be equally spaced by 1 unit apart. By convention we assume that they are listed from right to left.

$O(n)$ time algorithm: We already know that $P(x_i)=y_i$; so we may assume that $x \neq x_i$, $1 \leq i \leq n$. If we define $Q(n,x)$ to be the product over $1 \leq i \leq n$ of the terms $(x-x_i)$, then the numerator (top portions) of the fraction for y_i in the Lagrange interpolation formula is just $Q(n,x)/(x-x_i)$. For the denominators (bottom portions), the ith has a product of negative differences $(x_i-x_1)(x_i-x_2)...(x_i-x_{i-1}) = (-1)^{i-1}(i-1)!$ times a product of positive differences $(x_i-x_{i+1})(x_i-x_{i+2})...(x_i-x_n) = (n-i)!$, where $0!$ is defined to be 1. Hence:

$$P(x) = \sum_{1 \leq i \leq n}\left(\frac{Q(n,x)}{L[n,i](x-x_i)}\, y_i \right) \quad \text{where} \quad Q(n,x) = \prod_{1 \leq i \leq n}(x-x_i) \quad \text{and} \quad L[n,i] = (-1)^{i-1}(i-1)!(n-i)!$$

A table of the values of $i!$, $0 \leq i \leq n$, can be computed in $O(n)$ time, and from it, a table of the constants $L[n,i]$ can be computed in $O(n)$ time (in practice, we may get these values from a fixed table that has been computed once for all computations). Given x, $Q(n,x)$ can be computed in $O(n)$ time with n subtractions and $n-1$ multiplications, and using $Q(n,x)$ and the constants $L[n,i]$, $P(x)$ can be computed in $O(n)$ time.

The special case $x_0 = x_1+1$: If $x_0 = x_1+1$, then $(x_0-x_i)=i$, $Q(k,x_0)=n!$, and:

$$P(x_0) = \sum_{1 \leq i \leq n}\left(\frac{Q(n,x)}{L[n,i](x-x_i)}\, y_i \right)$$

$$= \sum_{1 \leq i \leq n}\left(\frac{n!}{(-1)^{i-1}(i-1)!(n-i)!\,i}\, y_i \right)$$

$$= \sum_{1 \leq i \leq n}\left((-1)^{i-1}\frac{n!}{i!(n-i)!}\, y_i \right)$$

$$= \sum_{1 \leq i \leq n}\left((-1)^{i-1}\binom{n}{i} y_i \right) \quad - \text{ where } \binom{n}{i} \text{ is a binomial coefficient}$$

A practical example is when the next point in a trajectory of equally spaced samples is approximated locally by a polynomial of the last few sample points (e.g., one of the coordinates of computer "mouse" as a function of time). The binomial coefficients can be stored once and for all, and the computation is just a linear combination; for example:

$n=2$: $P(x_0) = 2y_1 - y_2$
$n=3$: $P(x_0) = 3y_1 - 3y_2 + y_3$
$n=4$: $P(x_0) = 4y_1 - 6y_2 + 4y_3 - y_4$
$n=5$: $P(x_0) = 5y_1 - 10y_2 + 10y_3 - 5y_4 + y_5$

Logarithms and Exponentials

Idea: For most of what we do, it suffices to know that $\log_a(r)$ is a real number such that

$$\lfloor \log_a(r) \rfloor \le \log_a(r) \le \lceil \log_a(r) \rceil$$

and we can simply take the closest integers to $\log_a(r)$ as definitions of integer valued functions that share essentially the same properties.

Logarithms with integer values: For $a > 1$ and $r \ge 1$ real numbers define:

$\lceil \log_a(r) \rceil =$ Starting with 1, the minimum number of times that we must multiply by a to get an integer $\ge r$; for example, $\lceil \log_2(9) \rceil = \lceil \log_2(16) \rceil = 4$.

$\lfloor \log_a(r) \rfloor =$ Starting with 1, the maximum number of times that we can multiply by a to get an integer $\le r$; for example, $\lfloor \log_2(8) \rfloor = \lfloor \log_2(15) \rfloor = 3$.

Note: $\lceil \log_a(1) \rceil = \lfloor \log_a(1) \rfloor = 0$

Exponentials with integer exponents: For $a > 1$ a real number and $i \ge 0$ an integer:

$a^i = a*a*a*a...*a$ (i times); that is, $a^0 = 1$ and for $i > 0$, $a^i = a*a^{i-1}$

Facts about integer logarithms and exponentials (see the exercises):

A. For any real number $a > 1$ and integer $i \ge 0$: $\left\lceil \log_a\left(a^i\right) \right\rceil = \left\lfloor \log_a\left(a^i\right) \right\rfloor = i$

B. For any real number $a > 1$ and integer $i \ge 0$: $a^{\left\lfloor \log_a(i) \right\rfloor} \le i \le a^{\left\lceil \log_a(i) \right\rceil}$

C. For any real number $a > 1$ and integers $i, j \ge 0$:

$$\left\lfloor \log_a(i) \right\rfloor + \left\lfloor \log_a(j) \right\rfloor \le \left\lfloor \log_a(ij) \right\rfloor \le \left\lceil \log_a(ij) \right\rceil \le \left\lceil \log_a(j) \right\rceil + \left\lceil \log_a(j) \right\rceil$$

$$j \left\lfloor \log_a(i) \right\rfloor \le \left\lfloor \log_a\left(i^j\right) \right\rfloor \le \left\lceil \log_a\left(i^j\right) \right\rceil \le j \left\lceil \log_a(i) \right\rceil$$

D. For any real number $a > 1$ and integer $i \ge 0$:

$$\frac{\left\lfloor \log_b(i) \right\rfloor}{\left\lceil \log_b(a) \right\rceil} \le \left\lfloor \log_a(i) \right\rfloor \le \left\lceil \log_a(i) \right\rceil \le \frac{\left\lceil \log_b(i) \right\rceil}{\left\lfloor \log_b(a) \right\rfloor}$$

E. For any real number $a > 1$ and integers $i, j \ge 0$:

$$a^{(i+j)} = a^i a^j$$

$$a^{(ij)} = \left(a^i\right)^j$$

$$a^{(ij)} \ne a^{\left(i^j\right)} \text{ except when } j = 1 \text{ or } i = j = 2$$

(*) Non-Integer Logarithms and Exponentials

Idea: We could generalize to rational exponents (e.g., $a^{i/j} = \sqrt[j]{a^i}$ and $a^{-i} = 1/a^i$) and then for a real number r define a^r as a limit using successively closer rational numbers to r (see the exercises). Simple calculus provides another way.

Logarithms and exponentials with real values: Let $x>0$ be a real number:

For any real number $x>0$: $\ln(x) = \int_1^x \frac{1}{t} dt$ (the "natural" logarithm of x)

$e = 2.71821828459... =$ that real number such that $\ln(e)=1$ (the "natural" logarithm)

For any real number x, e^x is the real number such that $\ln(e^x)=x$.

For any real numbers $a,x>0$ such that $a\neq1$: $\log_a(x) = \ln(x)/\ln(a)$

For any real numbers $a>0$ and x: $a^x = e^{x\ln(a)}$

Facts about logarithms and exponentials (see the exercises): The facts listed earlier for integer logarithms and exponentials remain true; in addition:

A. For any real numbers $a,x>0$ such that $a\neq1$:

$$\frac{d\ln(x)}{dx} = \frac{1}{x} \text{ and } \frac{d\log_a(x)}{dx} = \left(\frac{1}{\ln(a)}\right)\left(\frac{1}{x}\right)$$

B. For any real numbers $a>0$ and x: $\dfrac{de^x}{dx} = e^x$ and $\dfrac{da^x}{dx} = \ln(a)(a^x)$

   Note: Because e^x is its own derivative, e is in some sense a "natural" logarithm base. However, in computer science, it is often convenient to use base 2.

C. For any real numbers $a>0$ and x such that $a\neq1$: $\log_a(a^x) = a^{\log_a(x)} = x$

D. For any real numbers $a,x,y>0$ such that $a\neq1$:

   for $x \neq 0$: $\log_a(1/x) = -\log_a(x)$
   $\log_a(xy) = \log_a(x) + \log_a(y)$
   $\log_a(x^y) = y\log_a(x)$

E. For any real numbers $a,b,x>0$ such that $a,b\neq1$:

$$\log_a(x) = \frac{\log_b(x)}{\log_b(a)} \text{ and } a^x = b^{x/\log_a(b)}$$

F. For any real numbers $a>0$ and x:

$$a^{(-x)} = \frac{1}{a^x}$$
$$a^{(x+y)} = a^x a^y$$
$$a^{(xy)} = \left(a^x\right)^y$$

G. For any real numbers $a>0$ and x, and any integer $i>0$: $a^{(x/i)} = \sqrt[i]{a^x} = \left(\sqrt[i]{a}\right)^x$

Logarithms and Exponentials Versus Polynomials

Logarithm base does not change asymptotic complexity.

Theorem: For any integer $k \geq 2$, $\log_k(n)$ is $\Theta(\log_2(n))$.

Proof: Since $\log_k(n) \leq \log_2(n)$, $\log_k(n)$ is $O(\log_2(n))$. Since $\log_k(n) = \log_2(n)/\log_k(2)$, $\log_k(n)$ is $\Omega(\log_2(n))$ – choose $c = 1/\log_k(2)$ in the definition of Ω.

> * Because logarithm base does not affect asymptotic complexity, we often omit it; for example, we may say a function is $O(n\log(n))$ rather than $O(n\log_2(n))$.

Logarithms are smaller than roots.

Theorem: For any real number $r > 0$, $\log_2(n)$ is $O(n^r)$.

Proof: Since $\log_2(n) \leq n$ for all $n \geq 1$, $\log_2(n) = (1/r)\log_2(n^r) \leq (1/r)n^r$. Hence, $a = 1$ and $b = (1/r)$ satisfies the definition of O.

A polynomial is O of its first term.

Theorem: For any integer $k \geq 0$, k^{th} degree polynomial $P(n)$ is $O(n^k)$.

Proof: Let m be the absolute value of the largest coefficient; then each of the $k+1$ terms of $P(n)$ is less than mn^k, and so any $a \geq 0$ and $b \geq (k+1)m$ satisfies the definition of O.

Note: In fact, for any $\varepsilon > 0$, there is a constant a such that for all $n > a$, $P(n) < (1+\varepsilon)hn^k$, where h is the coefficient of the high order term (see the sample exercises).

Exponentials are larger than polynomials.

Theorem: For any integer $k \geq 0$, 2^n is $\Omega(n^k)$.

Proof: $2^n = 2^{n\log_2(n)/\log_2(n)} = \left(2^{\log_2(n)}\right)^{n/\log_2(n)} = n^{n/\log_2(n)}$

Hence, any $c > 0$ satisfies the definition of Ω; for example, if we let $c = 1$ then $2^n \geq n^k$ for all n such that $n/\log_2(n) \geq k$.

The constant used for an exponential affects asymptotic complexity.

Theorem: For any real numbers $0 < x < y$, y^n is not $O(x^n)$.

Proof: Since

$$y^n = \left(\frac{y}{x}x\right)^n = \left(\frac{y}{x}\right)^n x^n$$

there can be no constants a and b that satisfy the definition of O, because $y/x > 1$ and hence $(y/x)^n$ is greater than any constant for sufficiently large n.

Example: Binary Search

Problem: Given part of an array $A[a] \ldots A[b]$ that is already arranged in increasing order, and a value x, find an i, $a \leq i \leq b$, such that $x=A[i]$, or determine that no such i exists.

Binary search algorithm: Compute the midpoint m between a and b, check whether $x \leq A[m]$, and repeat this process on the appropriate half of the array.

> **while** $a<b$ **do begin**
> $\quad m := \lfloor (a+b)/2 \rfloor$
> $\quad$ **if** $x \leq A[m]$ **then** $b := m$ **else** $a := m+1$
> **end**
> **if** $x=A[a]$ **then** $\{x$ is at position $a\}$ **else** $\{x$ is not in $A\}$

Correctness: We verify correctness by considering the quantity $(b-a)$.

The case $(b-a)=0$ occurs when only one element is being searched and $a=b$; the *while* loop is not executed and the *if* statement after it correctly tests if $x=A[a]=A[b]$.

For $(b-a)>0$, each iteration of the *while* loop computes $m := \lfloor (a+b)/2 \rfloor$. Since $a<b$ and the computation of m rounds $(a+b)/2$ down, it must be that $a \leq m < b$, and hence both halves defined by m are smaller than the original; that is, $(m-a) < (b-a)$ and $(b-(m+1)) < (b-a)$.

Since A is sorted in increasing order, if $x \leq A[m]$, it cannot be that $x=A[i]$ for some $m < i \leq b$, unless $x=A[m]$ (i.e., it could be that $A[m] = A[m+1] = \ldots = A[i]$), and hence it is ok to restrict the search to $A[a]\ldots A[m]$. Similarly, if $x>A[m]$ it cannot be that $x=A[i]$ for some $a \leq i \leq m$, and hence it is ok to restrict the search to $A[m+1]\ldots A[b]$.

Thus, since each iteration of the *while* loop correctly reduces the problem (so $(b-a)$ is smaller), it eventually gets down to $a=b$, and exits to the final *if* statement.

Time: Let n denote the number of elements. To simplify our analysis, let N be the smallest power of 2 that is $\geq n$ and imagine that the input is padded with copies of a special $+\infty$ value so that the number of elements is exactly N (i.e., the initial value of $b-a+1$ is exactly N). Then since each iteration of the main loop halves $b-a+1$, the number of iterations is $\leq \log_2(N) = \lceil \log_2(n) \rceil$, and hence the algorithm is $O(\log(n))$ since the time for each iteration is $O(1)$.

Space: $O(1)$ space in addition to the $O(n)$ space used by A.

Observation: Binary search is similar to how one typically finds a word in a dictionary or a phone book. Instead of starting at the first page and flipping forward one page at a time, one opens the book in the middle, if past the word, opens the book somewhere in the left half, or if before the word, opens the book somewhere in the right half, and so on. In practice, people typically guess where to open the book based on whether the word is near the start or end of the alphabet (binary search could be similarly adapted).

Example: Binary Numbers

Idea: Binary integers (i.e., base 2 integers) use only the digits ("bits") 0 and 1; for example the integers zero through ten are written in binary as:

$$0, \quad 1, \quad 10, \quad 11, \quad 100, \quad 101, \quad 110, \quad 111, \quad 1000, \quad 1001, \quad 1010$$

Base ten versus binary: Base ten integers express a quantity as a sum of each digit times the corresponding power of ten, that is if the digits of a number x written in base ten are $t_k t_{k-1} ... t_1 t_0$, then:

$$x = t_k 10^k + t_{k-1} 10^{k-1} + t_{k-2} 10^{k-2} + \cdots + t_1 10 + t_0$$

Similarly, if the digits (i.e., bits) in the binary representation of x are $b_k b_{k-1} ... b_1 b_0$, then:

$$x = b_k 2^k + b_{k-1} 2^{k-1} + b_{k-2} 2^{k-2} + \cdots + b_1 2 + b_0$$

For real numbers, to the right of the decimal point, base ten sums powers of 1/10 whereas base 2 (binary) sums powers of 1/2. For example:

$1/2 = .5$ base ten $= .1$ binary

$1/8 = .125$ base ten $= .001$ binary

$11/16 = .6875$ base ten $= .1011$ binary

A real number represented in binary with a finite number of bits can be represented in base 10 with a finite number of digits; the reverse is not always true (see the exercises).

Arithmetic with binary numbers: Work in the same way as base 10. For example, to add two binary numbers, just add each column and if the result is a two bit number (10 or 11), then carry the leading 1 to the next column; in this example, the carry bits are written in italics on top:

$$27 + 25 = 52 \quad \longrightarrow \quad
\begin{array}{ccccc}
\textit{1} & & \textit{1} & \textit{1} & \\
1 & 1 & 0 & 1 & 1 \\
1 & 1 & 0 & 0 & 1 \\
\hline
1 \quad 1 & 0 & 1 & 0 & 0
\end{array}$$

The number of bits in a binary number: Assuming that the leftmost bit must be 1, it can be shown (see the exercises):

- The number of bits needed for n distinct numbers is $\lceil \log_2(n) \rceil$; that is, for n a power of two, $0, 1, ... , n-1$ can use $\log_2(n)$ bits, but we need $\log_2(n)+1$ bits for n.
- For $n \geq 1$, the least power of two that is $\geq n$ uses $\lceil \log_2(n+1) \rceil$ bits.
- The number of bits to represent a binary integer $n \geq 0$ is $\lfloor \log_2(n) \rfloor + 1$. For example, it takes 3 bits to represent 7 (111) and 4 bits to represent 8 (1000).
- If $n > 1$ is a binary integer of i bits, then $2n$ has $i+1$ bits and $\lfloor n/2 \rfloor$ has $i-1$ bits.
- If $n > 1$ is a binary integer of i bits and $m > 1$ is a binary integer of j bits, then $n*m$ has at least $(i*j)-1$ bits and at most $i*j$ bits.

(*) Representing Arrays

We usually think of an array reference like $A[1,2]$ as a fundamental operation and assume that the time for such a reference is $O(1)$. However, in practice, the compiler must translate this reference to code that finds the appropriate position in memory. This can be done in time proportional to the dimension of the array (the number of indices), which is typically a constant. Although it is rare in practice to use arrays of dimension greater than three, we consider a general method for any dimension.

Notation:

$$base = \text{starting address for } A$$
$$esize = \text{size of each element of } A \text{ (in basic units of memory – usually bytes)}$$
$$(l_1, u_1)...(l_d, u_d) = \text{the lower and upper limits of each dimension}$$
$$r_i = u_i - l_i + 1 = \text{the number of positions in the } i^{th} \text{ dimension}$$

Idea: Think of a d-dimensional array as a 1-dimensional array, where each element is a $(d-1)$-dimensional array. Process a reference $A[i_1...i_d]$ by first skipping over (i_1-l_1) dimension $(d-1)$ elements and then proceeding as with a dimension $(d-1)$ reference.

1. $A[i_1] \rightarrow base + esize * (i_1 - l_1)$

2. $A[i_1, i_2] \rightarrow base + esize * ((i_1 - l_1)r_2 + (i_2 - l_2))$

3. $A[i_1, i_2, i_3] \rightarrow base + esize * ((i_1 - l_1)r_2 r_3 + (i_2 - l_2)r_3 + (i_3 - l_3))$

$\vdots$

d. $A[i_1 \cdots i_d] \rightarrow base + esize * \sum_{j=1}^{d} \left((i_j - l_j) \prod_{k=j+1}^{d} r_k \right)$

Computation in $O(d)$ time: Similar to Horner's Method for polynomial evaluation.

$offset := 0$

for $j := 1$ **to** d **do** $offset := r_j * offset + (i_j - l_j)$

$A[i_1...i_d] \rightarrow base + esize * offset$

Practical considerations: Array dimensions are typically statically defined (they do not change as the program runs), and we can compute at compile time

$$R_j = esize \prod_{k=j+1}^{d} r_k$$

$$B = base - \sum_{j=1}^{d} l_j R_j$$

so that references at run time can be computed in $O(d)$ time by doing:

$$A[i_1 \cdots i_d] \rightarrow B + R_1 i_1 + \cdots + R_d i_d$$

The Significance of Asymptotic Complexity

Idea: With poor asymptotic complexity, things can often become hopeless as n gets large. Suppose that your computer executes 1000 operations per second (it's probably much faster, but that does not change the relative comparisons made here). Consider six different hypothetical programs, P_1 through P_6, to solve a given problem of size n (where P_1 might be a very complex linear time algorithm with a large constant and P_6 might be an extremely simple program that tries all possibilities in exponential time):

program	running time $f(n)$	maximum size that can be solved in 1 second (1000 steps)	maximum size that can be solved in 10 seconds (10,000 steps)	maximum size that can be solved in 1 hour $(3.6*10^6$ steps)
P_1	$100n$	10	100	36,000
P_2	$20n\lceil \log_2(n) \rceil$	12	71	12,857
P_3	$5n^2$	14	44	850
P_4	$2n^3$	7	17	121
P_5	$1.5n^4$	5	9	39
P_6	2^n	9	13	21

Now suppose you are considering buying a new computer that is 10 times faster and compute how much larger a problem size that you could solve in a given time T:

program	running time $f(n)$	The real number r such that $f(r)=T$; only sizes $\leq \lfloor r \rfloor$ can be solved in time T.	Maximum problem size that can be solved on the new computer in time T.
P_1	$100n$	r_1	$\lfloor 10r_1 \rfloor$
P_2	$20n\lceil \log_2(n) \rceil$	r_2	$\lfloor \approx 10r_2 \rfloor$ — see sample exercises
P_3	$5n^2$	r_3	$\lfloor \sqrt{10}r_3 \rfloor < \lfloor 3.2r_3 \rfloor$
P_4	$n^3/2$	r_4	$\lfloor \sqrt[3]{10}r_4 \rfloor < \lfloor 2.2r_4 \rfloor$
P_5	$1.5n^4$	r_5	$\lfloor \sqrt[4]{10}r_5 \rfloor < \lfloor 1.8r_5 \rfloor$
P_6	2^n	r_6	$\lfloor r_6 + \log_2(10) \rfloor < \lfloor r_6 + 3.4 \rfloor$

Basic Approach to Algorithm Design

Idea:

First develop pseudo-code with the asymptotic time and space as small as possible.

Then, based on the pseudo-code, develop a program in a specific language with the constant on the asymptotic bound as small as possible.

Note: Sometimes "hybrid" algorithms are appropriate; e.g.:

> **if** $n < 25$
> > **then** simple but asymptotically poor algorithm
> > **else** complex but asymptotically fast algorithm

Other considerations:

- Program will be used only a few times.

- Program will be used on only small inputs.

- How easy it is to develop the code.

- How easy it is to maintain and modify the code.

- Is the code likely to have other applications.

- The possibility of incorporating already existing code.

- Numerical stability, round-off errors.

- Time-space tradeoffs.

- Instruction set of the particular machine being used.

- Features available in different programming languages.

- "Portability" of the code to other systems.

- System and hardware issues for secondary storage.

- Special purpose hardware that may be available.

- Different model of computation (e.g., parallel machine).

- Use of patented or proprietary methods.

Sample Exercises

1. Just like the pseudo-code for $n!$, the assembly-like code for $n!$ counted from $i=1$ up to n, successively multiplying x by i. It is actually a bit simpler to count down from $i=n$ to 1. Write a version that does this (use essentially the same set of instructions with obvious variations such as subtract instead of add). You may assume that $n \geq 1$.

Solution:

	read	read n into accumulator
	store 1	store accumulator into memory location 1
loop:	subtract =1	subtract 1 from accumulator
	goto(≤0) done	go to done if accumulator is ≤ 0
	store 2	store accumulator in memory location 2
	multiply 1	multiply accumulator by memory location 1
	store 1	store accumulator in memory location 1
	load 2	place memory location 2 in accumulator
	goto loop	go to loop
done:	load 1	place memory location 1 in accumulator
	write	write contents of accumulator to output

2. Give an algorithm with a running time of $\theta(n^4 \log(n))$.

Solution: Although it is not very interesting, a simple way to construct such an algorithm is to place a "dummy" procedure that uses $\theta(\log(n))$ time inside four nested loops that index from 1 to n. The dummy procedure could be anything that consumes $O(\log(n))$ time, such as a loop that starts with 1 and repeatedly multiplies by 2 until a value $\geq n$ is obtained. That is:

```
for i := 1 to n do
    for j := 1 to n do
        for k := 1 to n do
            for l := 1 to n do begin
                m := 1
                while m<n do m := 2m
            end
```

3. Prove each by applying directly the definitions of O, Ω, and Θ:

A. 100 is $\Theta(1)$

B. $2n$ is $\Theta(n)$

C. n is $O(n^2)$ but not $\Omega(n^2)$

D. $n^2/2 - n/2$ is $\Theta(n^2)$.

E. $n\sqrt{n}$ is $O(n^2)$ but not $\Omega(n^2)$

F. n^2 is $\Omega(1000n)$ but not $O(1000n)$

G. $n^2 + n\sqrt{n} + 25n + 100$ is $\Theta(n^2)$

Solution:

A. For all integers $n \geq 0$ (in fact n does not make any difference here!), $100 = 100*1$; so $a=1$ and any $b \geq 100$ suffices for the definition of O and any $c \leq 1/100$ suffices for the definition of Ω. Since 100 is both $O(1)$ and $\Omega(1)$, it is $\Theta(1)$.

B. For all integers $n \geq 0$, $2n \leq 2n$; so $a=1$ and $b \geq 2$ suffices for the definition of O. For all integers $n \geq 0$, $2n \geq n$; so any $0 < c \leq 2$ suffices for the definition of Ω. Since $2n$ is both $O(n)$ and $\Omega(n)$, it is $\Theta(n)$.

C. For all integers $n \geq 0$, $n \leq n^2$; so $a=1$ and $b \geq 1$ suffices for the definition of O. However, n cannot be $\Omega(n^2)$ since no matter how small a value $c>0$ is chosen, for all integers $n > 1/c$:
$$n = c(1/c)n < cnn = cn^2$$

D. For all integers $n \geq 0$, $n^2/2 - n/2 \leq n^2$; so $a=1$ and $b \geq 1$ suffices for the definition of O. Since $n^2/2 - n/2 = n^2/4 + (n^2/4 - n/2)$, for all $n \geq 2$, $n^2/2 - n/2 \geq n^2/4$, so any $0 < c \leq 1/4$ suffices for the definition of Ω. Hence, $n^2/2 - n/2 = \Theta(n^2)$.

In fact, any $0 < c < 1/2$ works for the definition of Ω, since for all $n \geq 1/(1-2c)$:
$$\tfrac{1}{2}n^2 - \tfrac{1}{2}n \geq \frac{1}{2(1-2c)^2} - \frac{1}{2(1-2c)}$$
$$= \frac{c}{(1-2c)^2}$$
$$= cn^2$$

E. For all integers $n \geq 0$, $n\sqrt{n} \leq n*n$; so $a=1$ and $b \geq 1$ suffices for the definition of O. However, $n\sqrt{n}$ cannot be $\Omega(n^2)$ since no matter how small a value $c>0$ is chosen, for all integers $n > 1/(c^2)$:
$$n\sqrt{n} = c\left(\sqrt{1/c^2}\right)n\sqrt{n} < cn\sqrt{n}\sqrt{n} = cnn = cn^2$$

F. Any $c>0$ suffices for the definition of Ω since for all integers $n \geq 1000c$, $n^2 = nn \geq 1000cn$. However, n^2 cannot be $O(1000n)$ since no matter how large a value b is chosen, for all integers $n > 1000b$:
$$n^2 = nn > 1000bn$$

G. Since for all integers $n \geq 25$, each of the four terms is $\leq n^2$, $a=25$ and any $b \geq 4$ suffices for the definition of O. Since removing the last three terms only makes the expression smaller, any $0 < c \leq 1$ suffices for the definition of Ω. Since the expression is $O(n^2)$ and $\Omega(n^2)$, it is $\Theta(n^2)$.

4. Prove each by applying directly the definitions of O and Ω:

A. $100n^2$ is $O(n^3)$

B. n^2 is $O(2^n)$

C. n^{100} is $O(2^n)$

D. n^{100} is $\Omega(100n)$

E. $n^2 + \log_2(n)^3$ is $O(n^2)$

F. $n(\log_2(n))^2$ is $O(n^2)$

Solution:

A. For all $n \geq 100$, $100n^2 \leq n*n^2 = n^3$. So $a=100$ and $b=1$ suffices for the definition of O.

B. Since $2^n = 2^{n\log_2(n)/\log_2(n)} = n^{n/\log_2(n)}$ and for $n \geq 4$, $n/\log_2(n) \geq 2$, we can use $a=4$ and $b=1$ in the definition of O.

C. Since $2^n = 2^{n\log_2(n)/\log_2(n)} = n^{n/\log_2(n)}$ and for $n \geq 1024$, $n/\log_2(n) \geq 100$, we can use $a=1024$ and $b=1$ in the definition of O.

D. Since for any $n \geq 2$, $n^{99} > 100$, we can use $c=1$ in the definition of Ω.

E. Since for any $n \geq 8$, $\log_2(n)^3 < n^2$, we can use $a=8$ and $b=2$ in the definition of O.

F. We know that both $\log_2(n)$ and $n^{1/2}$ are both smooth functions, so if it is true that for a particular value $n=a$ that $\log_2(a) \leq a^{1/2}$, then it will stay true for all $n \geq a$. So for example, $\log_2(100) < 7$ but $100^{1/2} = 10$. So for all $n \geq 100$:

$$n\log_2(n)^2 < nn^{1/2} < nn^{1/2}n^{1/2} = n^2$$

Hence, we can use $a=100$ and $b=1$ in the definition of O.

5. Recall the example that was given in the introduction of squaring a number n times, which we can write as the following function:

function BIG(n)

    Set x equal to 10.

    Multiply x times itself n times.

    return x

    end

Compare the number of digits, as a function of n, in $BIG(n)$ versus $n!$.

Solution: First let us consider the number of digits in $BIG(n)$. Since x starts at 10, each time we multiply x by itself, the number of zero's in the number doubles (e.g., $10*10 = 100$, $100*100 = 10,000$, $10,000*10,000 = 10^8$, $10^8*10^8 = 10^{16}$, etc.). Thus, $BIG(n)$ has $2^n+1 = \Theta(2^n)$ digits (a one followed by 2^n zeros). Now consider the number of digits in $n!$. Since each of the values 1 through n can be represented by at most $\lfloor\log_{10}(n)\rfloor+1$ base-10 digits, and multiplying two numbers produces a number that has at most as many digits as the sum of the digits in the two numbers, $n!$ cannot have more than $n(\lfloor\log_{10}(n)\rfloor+1) = O(n\log(n))$ digits. On the other hand, if we just consider the portion of $n!$ consisting of $(n/2)(n/2+1)...(n-1)n$, that is $n/2$ values of at least $\lfloor\log_{10}(n)-1\rfloor+1 = \lfloor\log_{10}(n)\rfloor$ digits each, for a total of at least $(n/2)\lfloor\log_{10}(n)\rfloor = \Omega(n\log(n))$ digits. And hence $n!$ uses $\Theta(n\log(n))$ digits.

6. In our discussion of the significance of asymptotic complexity, in the second row of the second table it was stated that the function $f(n)=20n\lceil \log_2(n) \rceil$ scales up approximately linearly when we go to larger n. Explain why, for a given time T, if r_2 is the real number such that $f(r_2)=T$, then the new machine which is 10 times faster can solve problems that are almost as large as $10r_2$ when T is large.

Solution: Since solving a problem in time T on the new machine is equivalent to solving a problem in time $10T$ on the old machine, it suffices to show that $f(10n)$ is not much larger than $10f(n)$ for large n. That is, it suffices to show that for any real number $\varepsilon>0$, there is an integer N such that for all $n \geq N, f(10n) \leq (10+\varepsilon)f(n)$:

$$f(10n) = 20(10n)\lceil \log_2(10n) \rceil$$

$$= 20(10n)\lceil \log_2(n) + \log_2(10) \rceil$$

$$\leq 10\left(20n\lceil \log_2(n) \rceil\right) + 200\lceil \log_2(10) \rceil n$$

$$< 10\left(20n\lceil \log_2(n) \rceil\right) + 200(3.4)n$$

$$= 10\left(20n\lceil \log_2(n) \rceil\right) + 680n$$

$$\leq (10 + \varepsilon)\left(20n\lceil \log_2(n) \rceil\right) \quad \text{for all } n \text{ such that } \varepsilon 20\lceil \log_2(n) \rceil \geq 680$$

$$= (10 + \varepsilon)\left(20n\lceil \log_2(n) \rceil\right) \quad \text{for all } n \text{ such that } \lceil \log_2(n) \rceil \geq \frac{34}{\varepsilon}$$

$$\leq (10 + \varepsilon)\left(20n\lceil \log_2(n) \rceil\right) \quad \text{for all } n \text{ such that } \log_2(n) + 1 \geq \frac{34}{\varepsilon}$$

$$= (10 + \varepsilon)\left(20n\lceil \log_2(n) \rceil\right) \quad \text{for all } n \geq 2^{(34/\varepsilon)-1}$$

$$= (10 + \varepsilon)f(n) \quad \text{for all } n \geq 2^{(34/\varepsilon)-1}$$

7. Show that:

$$\sum_{i=0}^{n}\left(\frac{n}{2^i}\right) \text{ is } O(n)$$

Solution: Since

$$\sum_{i=0}^{n}\left(\frac{n}{2^i}\right) = n\sum_{i=0}^{n}\left(\frac{1}{2^i}\right) = n\sum_{i=0}^{n}(r^i), \text{ where } r = \frac{1}{2}$$

we have n times the geometric sum $1+2+4+... < 2$, and hence (see the appendix), we can use $a=0$ and $b=2$ in the definition of O. To prove this from "scratch", if we let S denote the sum and $(1/2)S$ denote the same sum with every term multiplied by $1/2$, then if we subtract $(1/2)S$ from S, all terms cancel except the first and last, and

$$S-(1/2)S = n-n/(2^{n+1})$$

which means that $S = 2n - n/(2^n) \leq 2n$.

8. We have already seen that a polynomial $P(n)$ of degree k is $O(n^k)$. However, the proof was "crude" because a much larger constant was used than necessary. Show that for any $\varepsilon > 0$, there is a constant a such that for all $n > a$, $P(n) < (1+\varepsilon)hn^k$, where h is the coefficient of the high order term.

Solution: Let:

$$P(n) = A[k]nk + \cdots + A[1]n + A[0]$$

For $0 \leq i < k$, let $M[i]$ be the smallest integer such that:

$$A[i]M[i]^i < (\varepsilon/k)A[k]M[i]^k$$

(To see that $M[i]$ exists, divide each side by $M[i]^i$ and observe that the left side becomes a constant, but because $i<k$, the right side has $M[i]$ raised to a positive power and can thus be made as large as needed.)

Now let a be the maximum of the $M[i]$, $0 \leq i < k$.

Then for all $n > a$, each of the k lower order terms satisfies:

$$A[i]n^i < (\varepsilon/k)A[k]n^k , \quad 0 \leq i < k$$

Hence for all $n > a$:

$$P(n) < (1+\varepsilon)A[k]n^k$$

9. We have already seen that logs are asymptotically smaller than roots. Show that for any integer $k \geq 1$, $\log_2(n)$ is not only $O(n^{1/k})$, but for sufficiently large a, it suffices to use $b=1$ in the definition of O.

Solution: To simplify notation, all logs in this solution are assumed to be base 2. It suffices to show that there exists an integer a such that for all $n \geq a$, $\log(n)^k \leq n$. Since

$$n = 2^{\log(n)} = \log(n)^{\log(n)/\log\log(n)}$$

it suffices to show that for all $n \geq a$, $k \leq \log(n)/\log\log(n)$. Since k is fixed but $\log(n)/\log\log(n)$ gets larger as n gets larger, this is always true for sufficiently large a. For example, if we choose $a=k^{2k}$, then any $n \geq a$ can be written as $n=x^{2k}$ for some real number $x \geq k$, and:

$$\frac{\log(n)}{\log\log(n)} = \frac{2k\log(x)}{\log(k)+\log\log(x)+1}$$

$$\geq \frac{2k\log(x)}{\log(x)+\log\log(x)+1} \quad\text{— since } x \geq k$$

$$\geq k \quad\text{— since } x \geq k \geq 2 \text{ implies } (\log(x)\text{-}1) \geq \log\log(x)$$

10. Explain why if $f_1(n)$ and $f_2(n)$ are both $O(g(n))$ then $f_1(n)+f_2(n)$ is $O(g(n))$.

Solution: If a_{max} denotes the maximum of the values of a needed for the proofs that $f_1(n)$ and $f_2(n)$ are $O(g(n))$ and b_{max} denotes the maximum of the values of b needed for the proofs that $f_1(n)$ and $f_2(n)$ are $O(g(n))$, then $a = a_{max}$ and $b = 2b_{max}$ suffice to verify that $f_1(n)+f_2(n)$ is $O(g(n))$. That is, for all integers $n \geq a_{max}$:

$$f_1(n)+f_2(n) \leq b_{max}g(n) + b_{max}g(n) \leq 2_{bmax}g(n) = bg(n)$$

11. Sometimes in the literature a definition of Ω that is symmetric with the definition of O is used; we call this the *stronger definition* of Ω:

> A function $f(n)$ is $\Omega(g(n))$ if there exists real numbers $a,b>0$ such that for all integers $n \geq a$, $f(n) \geq b*g(n)$.

Clearly, if a function $f(n)$ is $\Omega(g(n))$ under the stronger definition, then it also is under the standard definition. Although the reverse is not true, for virtually all algorithms addressed in this book, if the running time is $\Omega(f(n))$ under the standard definition, then it is also under the stronger definition. The standard definition can capture situations such as when infinitely often the algorithm takes only $O(1)$ time to indicate that the input is illegal for some reason but on the other hand infinitely often uses $f(n)$ time. The stronger definition has nice mathematical symmetry with the definition of O and, as we will observe in this problem, simple mathematical identities work out more cleanly with the stronger definition.

Let $f(n)$ and $g(n)$ be any non-negative functions and let $F(n)$ and $G(n)$ denote functions that are $\Omega(f(n))$ and $\Omega(g(n))$ respectively.

A. Using the stronger definition of Ω, prove that for $n>0$ the following identities are always true:

  1. $f(n)$ is $\Omega(g(n))$ if and only if $g(n)$ is $O(f(n))$.

  2. $F(n)+G(n)$ is $\Omega(f(n)+g(n))$.

  3. $F(n)*G(n)$ is $\Omega(f(n)*g(n))$.

B. Using the standard definition of Ω, give examples that make the identities of Part A false.

Solution:

A. The symmetry of the definition of O and the stronger version of the definition of Ω makes it easy to "put proofs together":

  1. If the proof that $f(n)$ is $\Omega(g(n))$ uses the constants a and b, then the proof that $g(n)$ is $O(f(n))$ can use the constants a and $1/b$ (since if for $n \geq a$ $f(n) \geq bg(n)$, then for $n \geq a$ $g(n) \leq (1/b)f(n)$).

  2. Let a_F, b_F be the constants in a proof that $F(n)$ is $\Omega(f(n))$ and let a_G, b_G be the constants in a proof that $G(n)$ is $\Omega(g(n))$. Then we can use the constants $MAXIMUM(a_F,a_G)$ and $MINIMUM(b_F,b_G)$ to prove that $F(n)+G(n)$ is $\Omega(f(n)+g(n))$. This is because if $n \geq a_F$ and $n \geq a_G$, then we know $F(n) \geq b_F f(n)$ and $G(n) \geq b_G g(n)$, and hence:

  $$F(n)+G(n) \geq b_F f(n)+b_G g(n) \geq MINIMUM(b_F,b_G)(f(n)+g(n))$$

3. Let a_F, b_F be the constants in a proof that $F(n)$ is $\Omega(f(n))$ and let a_G, b_G be the constants in a proof that $G(n)$ is $\Omega(g(n))$. Then we can use the constants $MAXIMUM(a_F,a_G)$ and b_F*b_G to prove that $F(n)*G(n)$ is $\Omega(f(n)*g(n))$. This is because if $n \geq a_F$ and $n \geq a_G$, then we know $F(n) \geq b_F f(n)$ and $G(n) \geq b_G g(n)$, and hence:

$$F(n)*G(n) \geq b_F f(n)*b_G g(n) \geq (b_F *b_G)(f(n)*g(n))$$

B. For example, let:

$$f(n) = \begin{cases} 1 \text{ if } n \text{ is even} \\ n \text{ if } n \text{ is odd} \end{cases}$$

$$g(n) = \begin{cases} n \text{ if } n \text{ is even} \\ 1 \text{ if } n \text{ is odd} \end{cases}$$

$$F(n)=G(n)=1.$$

1. Since infinitely often (when n is odd) $f(n) \geq g(n)$, it follows that $f(n)$ is $\Omega(g(n))$. However, $g(n)$ cannot be $O(f(n))$ since $g(n)=n$ and $f(n)=1$ for all even values of n; that is, no matter how large a value for b you choose, for *all* even integers $n > b$, $g(n)>bf(n)$.

2. $F(n)+G(n) = 2$ but for all integers $n \geq 1$, $f(n)+g(n) = n+1$, and hence $F(n)+G(n)$ cannot be $\Omega(f(n)+g(n))$ because a constant cannot be Ω of a function like n. That is, no matter how small a value for b you choose, it cannot be that $F(n)+G(n) > b(f(n)+g(n))$ for infinitely many integers n, because for *all* integers $n > 2/b$, $b(f(n)+g(n)) = b(n+1) > bn > 2$.

3. Similar reasoning to Part 2 since $F(n)*G(n) = 1$ but $f(n)*g(n) = n$.

Note: Examples like this are useful for a mathematical understanding of the definitions of O, Ω, and Θ. However, in practice we shall measure the running times of programs with "well-behaved" functions and it is natural to consider the running time of a program to be "as bad as $g(n)$ in the worst case" if its running time is that bad infinitely often.

12. Consider the arithmetic sum:

$$\sum_{i=0}^{n} i^k$$

A. The appendix shows that:

$$\sum_{i=0}^{n} i^k = \tfrac{1}{k+1}n^{k+1} + \tfrac{1}{2}n^k + O\left(n^{k-1}\right)$$

Explain why this means that this sum is $\Theta(n^{k+1})$.

B. Using only simple algebra show that this sum is $\Theta(n^{k+1})$.

Solution:

A. We can first verify that each of the three terms is $O(n^{k+1})$; in the definition of O, we can use $a=1$ and $b=1/(k+1)$ for the first term, $a=1$ and $b=1/2$ for the second term, and the third term is $O(n^{k+1})$ because $n^{k-1} \leq n^{k+1}$ (and so the same values of a and b can be used that show it is $O(n^{k-1})$). If a_{max} denotes the maximum of the values of a needed for the O proofs for the three terms and b_{max} denotes the maximum of the values of b needed for the O proofs of the three terms, then $a = a_{max}$ and $b = 3b_{max}$ suffice to verify that the entire expression is $O(n^{k+1})$.

B. Since each term is at most n^k, we see that:

$$\sum_{i=0}^{n} i^k = \sum_{i=1}^{n} i^k \leq \sum_{i=1}^{n} n^k = nn^k = n^{k+1} = O\left(n^k\right)$$

On the other hand, by simply throwing out terms we see that:

$$\sum_{i=0}^{n} i^k \geq \sum_{i=(n/2)}^{n} i^k \geq \sum_{i=(n/2)}^{n} \left(\tfrac{n}{2}\right)^k = \left(\tfrac{n}{2}\right)\left(\tfrac{n}{2}\right)^k = \frac{1}{2^{k+1}} n^{k+1} = \Omega\left(n^k\right)$$

13. Give an exact expression for the number of additions performed in the computation of Pascal's triangle.

Solution: No additions are used for Rows 0 and 1, one is used for Row 2, two are used for Row 3, ... and $n-1$ are used for Row n. Hence, the number of additions is given by the arithmetic sum:

$$1+2+3+\cdots+(n-1) = (\text{number of terms})(\text{average value of a term})$$
$$= (n-1)\left(\frac{(n-1)+1}{2}\right)$$
$$= \frac{n^2 - n}{2}$$

14. Prove that if the first n rows of Pascal's triangle are constructed by computing each term independently with three separate factorial computations, the time is $\Omega(n^3)$.

Solution: Computing the three factorials independently for each entry on the ith row takes $O(i)$ time for a total of $O(i^2)$ time; we can bound the time for n rows as follows (see the Appendix for bounds with a better constant):

$$\sum_{i=0}^{n} i^2 \geq \sum_{i=\lceil\frac{n}{2}\rceil}^{n} i^2 \geq \sum_{i=\lceil\frac{n}{2}\rceil}^{n} \lceil\tfrac{n}{2}\rceil^2 \geq \left(\tfrac{n}{2}\right)\left(\tfrac{n}{2}\right)^2 = \frac{1}{8} n^3 = \Omega(n^3)$$

15. Give an expression for the value computed by the following function (that takes a single integer input) and determine its asymptotic running time as a function of n, the length of the input when written in binary. You may make the usual assumption that all arithmetic operations take constant time and all variables consume constant space.

```
function WOOD(k)
    index := 0
    power := 1
    while power<k do begin
        power := power*2
        index := index+1
    end
    return index
end
```

Solution: For an input of length $n > 1$, WOOD computes how many times k can be divided by 2 before it becomes ≤ 1; that is, $\lceil \log_2(k) \rceil$. WOOD uses $O(1)$ space and has time complexity $O(\log_2(k)) = O(bits\ to\ represent\ k\ in\ binary) = O(n)$.

16. Define the *parity* of an integer n to be 0 if the binary representation of n has an even number of 1's and 1 if the binary representation of n has an odd number of 1's. Equivalently, the parity of an integer n is the logical *exclusive or* (XOR) of its bits, where the XOR of two bits is 1 if they are different and 0 if they are the same. The order in which XOR of the bits is performed together does not matter; for example, computing the parity of 27 from right to left gives:

parity of 25 = parity of 11001 = 1 XOR (1 XOR (0 XOR (0 XOR 1))) = 1

Give pseudo-code to compute the parity of an integer n that is represented by m bits.

Solution: The straightforward solution initializes a variable *parity* to 0 and then for $i := 1$ to m sets *parity* to $1 - parity$ if the ith bit of n is 1. If we do not have arithmetic operations that can access individual bits of n, we can repeatedly shift the bits of n to the right and test whether n is odd or even. We define:

ShiftRight(x, 1) = shift the bits of x to the right by 1 position and place a 0 in the leftmost position that was vacated (i.e., $x = \lfloor x/2 \rfloor$)

We now can proceed from right to left through the bits of n:

```
parity := 0
for i := 1 to m do begin
    if n is odd then parity := 1-parity
    n := ShiftRight(n, 1)
end
```

When n is a power of 2, we can compute the parity more quickly by shifting by amounts that double at each step. That is, we first compute the parity of each block of two bits (and store it in the rightmost bit of each block), then compute the parity of each block of 4 bits (and store it in the rightmost bit of each block), and so on, until the parity of all n bits is stored in the rightmost bit. We make use of the following two basic operations:

x XOR y = bitwise logical XOR (XOR of two bits = 0 if they are the same, 1 otherwise)

$$ShiftRight(x,i) = \quad \text{shift the bits of } x \text{ to the right by } i \text{ positions and place 0's in the leftmost } i \text{ positions that have been vacated (i.e., } x = \lfloor x/2^i \rfloor)$$

We start with $k=1$. Repeatedly, the *while* loop XOR's bits that are distance k from each other (so, in particular, the rightmost bits of two adjacent blocks of k bits are combined with a XOR), and then doubles k.

function parity(n)
    $k := 1$
    while $k<m$ **do begin**
        $n := n$ XOR ShiftRight(n,k)
        $k := 2*k$
        end
    if n is even **then return** 0 **else return** 1
    end

For example, if $m=32$, in C we could unwind the *while* loop to compute the parity of a 32-bit integer n in only 5 arithmetic operations, where $>>$ is how ShiftRight(n,k) is written in C and $\text{\textasciicircum}=$ is a C assignment statement that XOR's n with the right-hand side. At the end, the return statement uses the AND operator $\&$ to return the rightmost bit of n.

```
int parity_32(int n) {
    n ^= n >> 1;
    n ^= n >> 2;
    n ^= n >> 4;
    n ^= n >> 8;
    n ^= n >> 16;
    return(n & 1);
}
```

This doubling approach uses only $\log_2(m)$ shift operations and $\log_2(m)$ XOR operations, as compared to the straightforward approach that used $O(m)$ ShiftRight($n,1$) operations and $O(m)$ operations that adjusted the parity bit. Assuming that ShiftRight(n,k) takes $O(1)$ time, the doubling approach is exponentially faster asymptotically.

17. Suppose that we wish to write data to an external storage device such as a disk and at a later time read it back in, and that the external storage device requires that data be written or read a byte at a time. Suppose further that we wish to write and read variable numbers of bits, which are not necessarily multiples of 8 (and we do not want to waste any space). For $m \geq 1$, give a procedure WRITEBITS(i,k) that writes the rightmost k bits of the unsigned integer m-bit integer variable i, where it can be assumed that all the other bits of i are 0. Also give a function READBITS(k) that reads k bits and returns them as the rightmost bits of an unsigned m-bit integer variable (where the rest of the bits are made 0). An unsigned m-bit integer is a variable that has values in the range 0 to 2^m-1 and is represented by m bits. You may assume that the following logical operations on unsigned m-bit integers are $O(1)$ time:

$$x \text{ OR } y = \quad \text{bitwise logical } OR \text{ (} OR \text{ of two bits} = 1 \text{ if either is } 1, 0 \text{ otherwise)}$$

$$x \text{ AND } y = \text{bitwise logical } AND \text{ } (AND \text{ of two bits} = 1 \text{ if both are } 1, 0 \text{ otherwise})$$

$ShiftLeft(x,i) = $ shift the bits of x to the left by i positions and place 0's in the rightmost i positions that have been vacated (i.e., $x = x*2^i$)

$ShiftRight(x,i) = $ shift the bits of x to the right by i positions and place 0's in the leftmost i positions that have been vacated (i.e., $x = \lfloor x/2^i \rfloor$)

Solution: For convenience, assume that $k \leq m\text{-}7$ (e.g., if m=32, then $k \leq 25$). Also assume only reading or only writing, but not both, so we can use a single set of global variables:

BitBuffer: An unsigned m-bit integer that holds the bits output from the encoder but yet written or the bits input to the decoder but yet processed; bits enter from the left (high order side) and leave from the right (low order side).

Size: The current number of bits in *BitBuffer*.

MASK[i]: *MASK*[1] .. *MASK*[m] are m-bit unsigned integers where the rightmost i bits of *MASK*[i] are all ones, and the rest of *MASK*[i] is all zeros. These masks are used for getting bits from *BitBuffer*.

If we are simultaneously reading bits from one file and writing bits to another, two copies of *BitBuffer* and *Size* (with different names) can be used. MASK never changes and can be initialized once before any reading or writing, by successively shifting in a 1 from the right to an unsigned m-bit integer that is initially 0:

```
procedure BuildMasks
    k := 0
    for i := 1 to m do begin
        ShiftLeft(k,1)
        k := k OR 1
        MASK[k] := k
        end
    end
```

WRITEBITS works by adding the new bits with a logical OR and then removing bytes with a logical AND until the number of bits left in *BitBuffer* is less than eight:

```
procedure WRITEBITS(i,k):
    ShiftLeft(i,Size)
    BitBuffer := BitBuffer OR i
    Size := Size +k
    while Size ≥ 8 do begin
        i := BitBuffer AND MASK[8]
        ShiftRight(BitBuffer,8)
        Size := Size−8
        Output the value of i as an unsigned byte.
        end
    end
```

READBITS works by adding in new bytes with a logical OR until there are enough bits in *BitBuffer* to remove k bits with a logical AND:

> **function** READBITS(k)
> > **while** *Size*<k **do begin**
> > > Read another byte from the input stream and put it in the unsigned integer i.
> > > ShiftLeft($i,Size$)
> > > *BitBufer* := *BitBuffer* OR i
> > > *Size* := *Size*+8
> > > **end**
> >
> > i := *BitBuffer* AND *MASK*[k]
> > ShiftRight(*BitBuffer,k*)
> > *Size* := *Size*−k
> > **return** i
> > **end**

18. Pseudo-code is a highly effective way to present an *algorithm*. In this book, we usually consider algorithms where the *programming*, although often time consuming, is relatively straightforward. However, it can often be that the particular programming language in question has a great effect on the practical aspects of implementation and optimization, and it sometimes can be useful to go beyond pseudo-code to explain an algorithm, or at least a particular part of its implementation. Consider the silly but instructive example of the problem of writing a program that prints itself. It is well known (see the chapter notes) that this can be done in any full programming language. The general idea can be described with pseudo-code. If we let START denote whatever is necessary to start a program in the language in question, and END whatever is needed to end a program, then generic self-printing pseudo-code might be:

> START
>
> x[1] := "START"
>
> x[2] := "the remainder of this program after this closing quote"
>
> print x[1]
>
> **for** i := 1 **to** 2 **do** print the string x[i] := ", the string x[i], and the string "
>
> print x[2]
>
> END

However, to do this in a reasonably compact fashion, the details can be quite language specific. For example, in Pascal, a reasonably straightforward implementation might avoid special characters (e.g., quotes inside quotes), by using the Pascal *chr* function to refer to characters by their ASCII codes. That is, for an integer $0 \le x < 127$, the Pascal function $chr(x)$ returns the character that corresponds to the ASCII code x. A full table of the ASCII codes is presented later in the exercises; here we make use of the following codes:

ASCII code	corresponding character
39	'
58	:
59	;
61	=
91	[
93	]
120	x

We use an array of length 9 to store the pieces of the program:

```
{Self-Printing Pascal Program}
program self(input,output);
var i: integer; x: array[1..9] of packed array[1..62] of char;
begin
x[1]:='{Self-Printing Pascal Program}                                ';
x[2]:='program self(input,output);                                   ';
x[3]:='var i: integer; x: array[1..9] of packed array[1..62] of char;';
x[4]:='begin                                                         ';
x[5]:='for i:=1 to 4 do writeln(x[i]);                               ';
x[6]:='for i:=1 to 9 do writeln(chr(120),chr(91),i:1,chr(93),        ';
x[7]:='    chr(58),chr(61),chr(39),x[i],chr(39),chr(59));            ';
x[8]:='for i:=5 to 9 do writeln(x[i]);                               ';
x[9]:='end.                                                          ';
for i:=1 to 4 do writeln(x[i]);
for i:=1 to 9 do writeln(chr(120),chr(91),i:1,chr(93),
    chr(58),chr(61),chr(39),x[i],chr(39),chr(59));
for i:=5 to 9 do writeln(x[i]);
end.
```

Here, there are 4 lines of START saved in $x[1]...x[4]$, 4 lines of code saved in $x[5]...x[8]$, and one line of END saved in $x[9]$. With a little more work, this program can be rewritten without the *chr* function; for example:

```
{Self-Printing Pascal Program - with no chr function}
program self(input,output);
var i: integer; x: array[1..16] of packed array[1..63] of char;
begin
x[ 1]:='{Self-Printing Pascal Program - with no chr function}        ';
x[ 2]:='program self(input,output);                                  ';
x[ 3]:='var i: integer; x: array[1..16] of packed array[1..63] of char;';
x[ 4]:='begin                                                        ';
x[ 5]:='for i:=1 to 4 do writeln(x[i]);                              ';
x[ 6]:='for i:=1 to 15 do writeln(x[12]:2,i:2,x[13]:3,x[16]:1,       ';
x[ 7]:='                          x[i],x[16]:1,x[14]:1);             ';
x[ 8]:='writeln(x[12]:2,16:2,x[13]:3,x[16]:1,x[16]:1,                ';
x[ 9]:='          x[16]:1,x[15]:61,x[16]:1,x[14]:1);                 ';
x[10]:='for i:=5 to 11 do writeln(x[i]);                             ';
```

```
x[11]:='end.                                              ';
x[12]:='x[                                                ';
x[13]:=']:=                                               ';
x[14]:=';                                                 ';
x[15]:='                                                  ';
x[16]:='''                                                ';
for i:=1 to 4 do writeln(x[i]);
for i:=1 to 15 do writeln(x[12]:2,i:2,x[13]:3,x[16]:1,
                          x[i],x[16]:1,x[14]:1);
writeln(x[12]:2,16:2,x[13]:3,x[16]:1,x[16]:1,
        x[16]:1,x[15]:61,x[16]:1,x[14]:1);
for i:=5 to 11 do writeln(x[i]);
end.
```

Write a self-printing C-program.

Solution: It is possible to "hack" very short self printing C-programs (there are also short ones in Pascal — see the Chapter notes), such as:

char *a="char *a=%c%s%c;main(){printf(a,34,a,34);}";main(){printf(a,34,a,34);}

However, it takes familiarity with C to figure out exactly how this program works. Also, although many compilers will give you a warning message, this program causes some compilers to give an error message because it leaves out the standard statement #include <stdio.h>.

The following program essentially copies the Pascal program, except formatting is put in x[0] and the second *for* statement fits on one line:

```
/*Self-Printing C Program*/
#include <stdio.h>
int i; char *x[10];
main () {
x[0]="%c%c%c%c%c%c%c%s%c%c%c";
x[1]="/*Self-Printing C Program*/";
x[2]="#include <stdio.h>";
x[3]="int i; char *x[10];";
x[4]="main () {";
x[5]="for(i=1;i<=4;i++) puts(x[i]);";
x[6]="for(i=0;i<=8;i++) printf(x[0],120,91,48+i,93,61,34,x[i],34,59,10);";
x[7]="for(i=5;i<=8;i++) puts(x[i]);";
x[8]="return 0;}";
for(i=1;i<=4;i++) puts(x[i]);
for(i=0;i<=8;i++) printf(x[0],120,91,48+i,93,61,34,x[i],34,59,10);
for(i=5;i<=8;i++) puts(x[i]);
return 0;}
```

If we want this program to look even more like the Pascal program, we can modify it to not pre-store the formatting information. To make it more readable (by a human), we include a macro that defines the single character *P* to mean *putchar*; this makes shorter and easier to read lines.

```
/*Self-Printing C Program*/
#include <stdio.h>
#define P putchar
main () {int i; char *x[10];
x[1]="/*Self-Printing C Program*/";
x[2]="#include <stdio.h>";
x[3]="#define P putchar";
x[4]="main () {int i; char *x[10];";
x[5]="for (i=1;i<=4;i++) puts(x[i]);";
x[6]="for (i=1;i<=9;i++) {P(120); P(91); P(48+i); P(93); P(61);";
x[7]="                    P(34); printf(x[i]); P(34); P(59); P(10);}";
x[8]="for (i=5; i<=9; i++) puts(x[i]);";
x[9]="return 0;}";
for (i=1;i<=4;i++) puts(x[i]);
for (i=1;i<=9;i++) {P(120); P(91); P(48+i); P(93); P(61);
                    P(34); printf(x[i]); P(34); P(59); P(10);}
for (i=5; i<=9; i++) puts(x[i]);
return 0;}
```

Exercises

19. Integer round-off notation occurs often in algorithms and algorithm analysis:

$\lceil i \rceil$ = The least integer $\geq i$.

$\lfloor i \rfloor$ = The greatest integer $\leq i$.

Prove that:

A. For any real number x:

$$x-1 < \lfloor x \rfloor \leq x \leq \lceil x \rceil < x+1$$

B. For any two integers i and j such that $j \neq 0$:

$$\left\lfloor \frac{i}{j} \right\rfloor + \left\lceil i - \frac{i}{j} \right\rceil = i$$

(for example, $\lceil i/2 \rceil + \lfloor i/2 \rfloor = i$)

C. For any integers i, j, k such that $j \neq 0$ and $k \neq 0$:

$$\left\lceil \frac{\lceil i/j \rceil}{k} \right\rceil = \left\lceil \frac{i}{jk} \right\rceil$$

$$\left\lfloor \frac{\lfloor i/j \rfloor}{k} \right\rfloor = \left\lfloor \frac{i}{jk} \right\rfloor$$

D. For any positive integers i, j, k such that $j \geq k$:

$$i \left\lceil \frac{j}{k} \right\rceil \geq \left\lceil \frac{ij}{k} \right\rceil$$

$$i \left\lfloor \frac{j}{k} \right\rfloor \leq \left\lfloor \frac{ij}{k} \right\rfloor$$

In addition, for both inequalities, give a class of examples where the difference can be arbitrarily close to a factor of 2.

20. It was pointed out in the presentation of O and Ω notation that $O(1)$ is a useful concept but $\Omega(1)$ is not. Suppose that we generalized our definitions of O and Ω to functions from the non-negative integers to the non-negative real numbers (e.g., $f(n) = 1/n$). How does this change the status of $\Omega(1)$? What if we allow functions to be from the real numbers to the real numbers (possibly negative)?

21. We have limited ourselves mainly to the O and Ω notation, with occasional use of Θ. Some authors also use the "little o" notation when they wish to emphasize that f is not just bounded by a constant times g, but truly grows more slowly than g. Formally:

little o: A function $f(n)$ is $o(g(n))$ if for every real number $b > 0$ there exists an integer $a > 0$ such that for all integers $n \geq a$, $f(n) < b*g(n)$.

Informally, if one thinks of O as $\leq$ between functions, little o can be thought of as $<$ between functions. That is, no matter how small you make b, $f(n)$ is still smaller than $bg(n)$ for all but a finite number of values.

A. Prove that:

n is not $o(2n)$ but n is $o(n^2)$

$\log_{10}(n)$ is $O(\log_2(n))$ but not $o(\log_2(n))$

for real numbers $x, y > 1$, $\log_x(n)$ is $o(n^y)$

B. Prove that if the condition $f(n) < b*g(n))$ is replaced by the condition $f(n) \leq b*g(n)$, the meaning of the little o definition is unchanged.

C. Prove that $f(n)$ is $o(g(n))$ if and only if $f(n)$ is not $\Omega(g(n))$.

D. Recall from the sample exercises the stronger definition of Ω:

Stronger definition of Ω: A function $f(n)$ is $\Omega(g(n))$ if there exists real numbers $a, b > 0$ such that for all integers $n \geq a$, $f(n) \geq b*g(n)$.

Prove that if $f(n)$ is $o(g(n))$ then it is not $\Omega(g(n))$ for either the standard or stronger definition of Ω. However, also prove that if $f(n)$ is not $\Omega(g(n))$ under the stronger definition of Ω, it is not necessarily true that $f(n)$ is $o(g(n))$, and hence Part C is not true for the stronger definition of Ω.

22. *Reverse bubble sort* is like bubble sort, but instead of repeatedly percolating the next largest element up, it repeatedly percolates the next smallest element down:

for $i=2$ **to** n **do**
    for $j=n$ **downto** i **do**
        if $A[j] < A[j-1]$ **then** EXCHANGE($A[j], A[j-1]$)

Give an exact expression for the number of times the *if* statement is executed (i.e., the number of times the test $A[j] < A[j-1]$ is made) and a formal proof that this expression is $\Theta(n^2)$.

23. *Odd-even bubble sort* has two inner *for* loops; one that checks $A[1]$, $A[3]$, $A[5]$, ... and one that checks $A[2]$, $A[4]$, $A[6]$, ... Although odd-even bubble sort is similar in structure to normal bubble sort, it does not have the same percolation property; in one iteration of the outer *for* loop, an element can move at most two positions up in the array. In addition, unlike normal bubble sort, the inner *for* loops always go all the way to n:

repeat
    flag := 0
    for $j := 1$ **to** $n-1$ **by** 2 **if** $A[j] > A[j+1]$ **then begin**
        Exchange $A[j]$ and $A[j+1]$.
        flag := 1
    end
    for $j := 2$ **to** $n-1$ **by** 2 **if** $A[j] > A[j+1]$ **then begin**
        Exchange $A[j]$ and $A[j+1]$.

```
    flag := 1
    end
until flag=1
```

A. Prove that odd-even bubble sort correctly sorts A.

B. Determine an upper for the maximum number of iterations of the repeat loop, as a function of n.

24. *Position sort* copies an array $A[1]...A[n]$ into an array $B[1]...B[n]$ in sorted order by, for each element $A[i]$, computing its position in the sorted order by counting the number of elements that must come before it.

```
for 1 ≤ i ≤ n do begin
    count[i] := 0
    for j := 1 to n do
        if A[j]<A[i] or (j<i and A[j]=A[i]) then count[i] := count[i]+1
    end
for i := 1 to n do B[1+count[i]] := A[i]
```

A. Analyze the time and space used.

B. Prove that position sort places the elements of A into B such that:

1. B is in sorted order; that is, if $i<j$, then $B[i] \le B[j]$.

2. Duplicates in A appear in the same relative order in B; that is, if $i<j$, $A[i]=A[j]$, $A[i]$ is copied to $B[x]$, and $A[j]$ is copied to $B[y]$, then $x<y$.

25. *Displacement sort* copies an array $A[1]...A[n]$ into an array $B[1]...B[n]$ in sorted order by, for each element $A[i]$, counting the number of elements that are out of order with respect to $A[i]$, and then displacing elements up or down according to their counts:

```
for 1 ≤ i ≤ n do begin
    count[i] := 0
    for j := 1 to i−1 do if A[j]>A[i] then count[i] := count[i]−1
    for j := i+1 to n do if A[j]<A[i] then count[i] := count[i]+1
    end
for i := 1 to n do B[i+count[i]] := A[i]
```

A. Analyze the time and space used.

B. Prove that displacement sort places the elements of A into B such that:

1. B is in sorted order; that is, if $i<j$, then $B[i] \le B[j]$.

2. Duplicates in A appear in the same relative order in B; that is, if $i<j$, $A[i]=A[j]$, $A[i]$ is copied to $B[x]$, and $A[j]$ is copied to $B[y]$, then $x<y$.

26. We analyzed the time and space of run-length decoding as a function of the number of bits output. Discuss its complexity as a function of the number of input bits.

27. Consider again the binary run-length encoding and decoding algorithms.

  A. Zero counts are used in only two ways.

   - A count of zero is sent if the string starts with a 1.

   - After sending a sequence of 2^k-1 counts, the encoder has to send a zero count (since after receiving a count of 2^k-1, the decoder expects an additional count for that bit).

  Suppose that we modify the run-length method as follows:

   - Adopt the convention that the run length coding starts with a single bit that is 0 if the first run is 0's and 1 if it is 1's.

   - Change the meaning of a count i to mean $i+1$ bits (so now counts can represent runs in the range 1 to 2^k). The only exception is when a count of 2^k is sent; in this case, all subsequent counts for that run denote a value in the range 0 to 2^k-1 as before.

  For only the "price" on one additional bit at the beginning, this modified method can represent values that are one larger with a "normal" count, and is no worse off when additional counts are used for values that are too large to fit into a single count.

  Give pseudo-code for this modified run-length coding method.

  B. Augment your encoding pseudo-code of Part A to scan the string in advance and determine the value of k that makes the shortest encoding, and send this value of k at the beginning of the encoding (you need a convention for the decoder to know when it has received the bits for k and the encoding is starting). What is the complexity of your algorithm?

  C. Suppose that you modified your algorithm of Part A to handle runs of length 2^k and longer as follows:

  > If a run is 2^k or longer, send k bits for 2^k and then $2k$ bits. If that is not enough, then send $4k$ bits, then $8k$ bits, and so on, until the run is encoded.

  Modify your pseudo-code of Part A for this variation.

  D. Discuss how to generalize your answer to Part B to work for Part C.

28. Consider run-length codes for alphabets of any size ≥ 2.

  A. Give pseudo-code to generalize the binary run-length encoding and decoding algorithms to where a run is preceded by index of the character it represents.

  B. Discuss improved compression that could be achieved with two scans of the input by the encoder, where the first scan gathers statistics on the runs present.

  C. Discuss how to implement the ideas of Part B in a single pass.

29. Prove each by applying directly the definitions of O, Ω, and Θ:

A. $53n$ is $\Theta(n)$

B. $n^2\sqrt{n}$ is not $\Omega(n^3)$

C. n^2 is $\Omega(20n)$

D. $2n^4 - 3n^2 + 32n\sqrt{n} - 5n + 60$ is $\Theta(n^4)$

E. $100n^2$ is $O(n^3)$

F. $2n^3$ is $O(2^n)$

G. $10n^{10}$ is $O(2^n)$

H. n^{10} is $\Omega(1000n)$

I. $n^3 + \log_2(n)^{10}$ is $O(n^3)$

J. $n\sqrt{n}\log_2(n)$ is $O(n^2)$

30. The ASCII character code is the most commonly used standard to represent characters in a machine. The code goes from 0 to 127 and includes the lower and uppercase letters of the alphabet, the digits 0 through 9, punctuation marks, and special characters. Although 7 bits suffice to store each character, it is typical to store characters one per byte.

000 nul	001 soh	002 stx	003 etx	004 eot	005 enq	006 ack	007 bel
008 bs	009 ht	010 nl	011 vt	012 np	013 cr	014 so	015 si
016 dle	017 dc1	018 dc2	019 dc3	020 dc4	021 nak	022 syn	023 etb
024 can	025 em	026 sub	027 esc	028 fs	029 gs	030 rs	031 us
032 sp	033 !	034 "	035 #	036 $	037 %	038 &	039 '
040 (	041)	042 *	043 +	044 ,	045 -	046 .	047 /
048 0	049 1	050 2	051 3	052 4	053 5	054 6	055 7
056 8	057 9	058 :	059 ;	060 <	061 =	062 >	063 ?
064 @	065 A	066 B	067 C	068 D	069 E	070 F	071 G
072 H	073 I	074 J	075 K	076 L	077 M	078 N	079 O
080 P	081 Q	082 R	083 S	084 T	085 U	086 V	087 W
088 X	089 Y	090 Z	091 [	092 \	093]	094 ^	095 _
096 `	097 a	098 b	099 c	100 d	101 e	102 f	103 g
104 h	105 i	106 j	107 k	108 l	109 m	110 n	111 o
112 p	113 q	114 r	115 s	116 t	117 u	118 v	119 w
120 x	121 y	122 z	123 {	124		125 }	126 ~	127 del

ASCII Character Code

Suppose that we define the following ordering:

- The digits 0 through 9 have the usual order with respect to each other but are all smaller than any lower case letter.

- The lower case letters *a* through *z* have the usual order with respect to each other but are all smaller than any upper case letter.

- The upper case letters *A* through *Z* have the usual order with respect to each other but are all smaller than any special character.

- Special characters (all characters other than digits, lower case letters, or upper case letters) are ordered according to their ASCII values.

Give pseudo-code to compare two character strings and decide which is smaller.

31. A *perfect integer square partition* is a square where the interior area is partitioned into distinct square regions, in which the dimensions of all of these regions are integers. For example, as shown in the figure below, a square of size 112 by 112 can be partitioned into distinct square regions where the smallest two have size 2 by 2 and 4 by 4 and the largest has size 50 by 50:

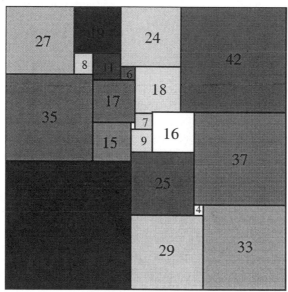

Describe an algorithm and give pseudo-code which, for an integer *n*, computes a perfect integer square partition of the *n* by *n* square or determines that none is possible. Analyze the time and space used by your algorithm.

32. To compute Pascal's triangle by computing each row from the previous row, we proved the identity:

$$\binom{n}{m} = \binom{n-1}{m-1} + \binom{n-1}{m}$$

Give a physical interpretation to this identity in terms of the process of choosing m out of n items.

Hint: Consider the two possibilities of whether or not the first element is chosen.

33. Discuss how large numbers can get during the computation of Pascal's triangle using brute force computation (i.e., just plugging into the formula) versus computing the table entries by summing according to the recurrence formula.

34. In the back-substitution algorithm presented for solving a set of linear equations, the error check in the first phase prevented a divide by zero in the second phase.

A. Assuming that there is no error reported, explain why there is always a unique solution.

B. Explain why there cannot be a unique solution if an error is reported.

C. If there is not a unique solution, explain under what circumstances it is because there is more than one solution and when it is because there is no solution.

D. If there is more than one solution, explain why there must be an infinite number of solutions and modify the algorithm to produce one of them.

E. When the back substitution algorithm is used to determine the coefficients of a polynomial of degree at most $(k-1)$ that passes through $k \geq 2$ distinct points, explain under what conditions a polynomial of degree exactly $k-1$ results, and when one of lower degree results. Also, explain why there must always be exactly one such polynomial.

F. Recall the Lagrange interpolation formula of a set of points y_1, y_2, ..., y_k. Explain why it must yield a polynomial of degree at most $k-1$, and that for any of the points x_i, $1 \leq i \leq k$, it evaluates to y_i. Also, explain how it is possible for it to yield a polynomial of degree less than k.

35. Consider the definitions of logarithms and exponentials presented:

A. Starting only with the definitions of integer logarithms and exponentials, give proofs for the facts presented about them.

B. Starting only with the calculus definitions for logarithms and exponentials, give proofs for the facts presented about them.

C. Generalize the definitions of integer logarithms and exponentials to apply to rational numbers so that $a^{i/j} = \sqrt[j]{a^i}$ and $a^{-i/j} = 1/a^{i/j}$, then define for any real number r, a^r to be a limit of approximations by successively closer rational numbers, then give alternate proofs of the facts about logarithms and exponentials given for the calculus definitions.

36. Recall the facts that were presented about the number of bits in a binary number.

A. Prove each of them.

B. Generalize each of them to base b, for any integer $b \geq 2$. For example, show that the number of bits to represent an integer $n \geq 0$ in base b is $\lfloor \log_b(n) \rfloor + 1$.

37. The quantity $k\log_k(n)$ arises sometimes in algorithm analysis. For example, to make a crude estimate of the amount of hardware needed to build a register to hold a number n in base k, we charge cost k to build a hardware module that can represent k states and then we need $\log_k(n)$ of these modules for the register.

A. Prove that for integers $k \geq 2$, this expression takes its minimum at $k=3$.

Hint: Since $k\log_k(n) = \ln(n)k/\ln(k)$, where ln denotes the natural logarithm, we can view n as a scaling parameter on the function $k/\ln(k)$. The following plot shows k on the x-axis and the function $k/\ln(k)$ on the y-axis. For real numbers $k \geq 2$, show that for any $n>1$, this function takes its minimum at natural logarithm base e, and then use the fact that 3 is the closest integer to e.

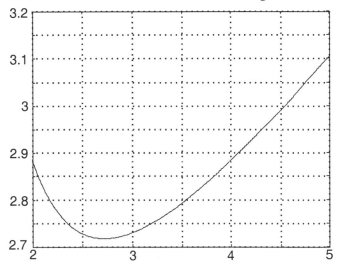

Note: Although $e=2.71...$ is closer to 3 than 2, it is not far from being an equal distance between 2 and 3, and binary has many other advantages for practical hardware.

B. Although it looks similar, prove that for any $n>1$, and considering integer values of $k \geq 2$, the expression $\lceil (k-1)\log_k(n) \rceil$ is minimum at $k=2$ (in fact, for real values of $k>1$ it increases monotonically as a function of k).

Hint: Since $(k-1)\log_k(n) = k\log_k(n) - \log_k(n)$, this function is like $k\log_k(n)$ except that an increasing amount is subtracted as k gets larger, so it suffices to check the $k=2$ and $k=3$ cases to see which is less.

C. Show that for integer values of k, the function $(k+1)\log_k(n)$ takes its minimum at $k=4$.

D. For real numbers $k,n>1$ and any real number d (possibly negative), give an expression for the minimum of the function $(k+d)\log_k(n)$.

38. Recall the sample exercise that showed how to write the procedure WRITEBITS(i,k) and the function READBITS(k). Rewrite these procedures under the assumption that logical operations are not available and you only have addition, subtraction, multiplication, and integer division (i.e., x divided by y is defined to be $\lfloor x/y \rfloor$).

39. We have proved that binary search of n elements is $O(\log(n))$ time. Prove that it is also $\Omega(n\log(n))$ time.

40. Binary search worked by dividing the problem in half.

A. Give pseudo-code for *ternary search* that divides the problem into three parts. That is, it makes at most two comparisons and works on a problem of size $n/3$.

B. Generalize your solution to Part A to a *k-ary search* algorithm that divides the problem into k parts. That is, it makes at most $k-1$ comparisons and works on a problem of size n/k.

C. For any fixed value of k, explain why your solution to Part B is $O(\log(n))$ time.

D. Explain why the number of inequality tests made by the algorithm of Part B is at most $\lceil (k-1)\log_k(n) \rceil$ in the worst case.

41. Consider the problem of converting a number written in base 10 to binary.

A. Give pseudo-code to convert a sequence of bits $B_n B_{n-1}...B_0$ representing an integer $i \geq 0$ in binary to a sequence of digits $T_x T_{x-1}...T_0$ that represent i in base 10.

B. Conversion of real numbers from one base to another can require an approximation in order to prevent an infinite number of digits to the right of the decimal point. For example, in base 3, $1/3 = .1$, but in base 10, it is .33333... Show that this can be a problem when converting base 2 to base 10 by showing that $1/5 = .2$ in base 10 but .001100110011... in binary.

C. Show that it is always possible to convert a real number that uses a finite number of digits in binary notation to a finite number of digits in base 10 notation.

   Hint: Show how to convert $1/2^i$ to a base 10 fraction of i digits and then sum up the digits in a binary fraction.

D. When does a fraction $1/i$ have a finite representation in a given base b?

E. Generalize your solution to Part A to convert a sequence of bits

$$B_n B_{n-1}...B_0.b_1 b_2...b_m$$

representing a real number $r \geq 0$ in binary to a sequence of digits

$$T_x T_{x-1}...T_0.t_1 t_2...t_y$$

that represent i in base 10.

Also, given a precision parameter k that limits the number of bits to the right of the decimal point to at most k, give pseudo-code to go in the reverse direction (from base 10 to binary).

42. Referring to the appendix, another way of viewing the linear weighted geometric sum for the case $r<1$ is as an infinite table of values where the sum first adds up the values in each column and then adds up the columns:

$$\sum_{i=1}^{n} i r^i = \begin{cases} r & +r^2 & +r^3 & +r^4 & +r^5 & +\cdots \\ & +r^2 & +r^3 & +r^4 & +r^5 & +\cdots \\ & & +r^3 & +r^4 & +r^5 & +\cdots \\ & & & +r^4 & +r^5 & +\cdots \\ & & & & +r^5 & +\cdots \end{cases}$$

A. Show how to compute the bound for the linear weighted geometric sum for $r<1$ by first expressing each row as a power of r times a simple geometric sum and then solving the resulting simple geometric sum:

$$\sum_{i=1}^{n} i r^i = \begin{cases} r\left(1 + r^2 + r^3 + r^4 + r^5 + \cdots\right) \\ r^2\left(1 + r^2 + r^3 + r^4 + r^5 + \cdots\right) \\ r^3\left(1 + r^2 + r^3 + r^4 + r^5 + \cdots\right) \\ r^4\left(1 + r^2 + r^3 + r^4 + r^5 + \cdots\right) \\ r^5\left(1 + r^2 + r^3 + r^4 + r^5 + \cdots\right) \end{cases}$$

B. Generalize your answer to Part A for the case $r>1$.

C. Generalize your answer to Part A for the quadratic weighted geometric sum.

D. Generalize your answer to Part A for arbitrary weighted geometric sums.

43. When we examined how a compiler can represent arrays in memory and compute a d-dimensional array reference in $O(d)$ time, we assumed that the basic RAM model includes the ability to perform *indirect addressing*. That is, for a fixed integer x, not only are there machine instructions to read or write to memory location x, but in $O(1)$ time we can also read or write to the memory location y, where y is the value stored in memory location x. Indirect addressing corresponds exactly to the notion of a one dimensional array. For example, an assignment statement like $x := A[i]$ does not place the contents of the memory location i into x; rather, it copies the contents of a memory location that is determined by the value of i. Assume that computer programs are encoded in binary and placed in one portion of memory, and another portion of memory is used for the data that is manipulated by the program.

A. Suppose that programs do not have indirect addressing but are allowed to modify memory locations in the area where the program is stored. Show how such programs can effectively perform indirect addressing in $O(1)$ time by modifying arguments to their own instructions.

B. Suppose that programs do have indirect addressing but are not allowed to modify memory locations in the area where the program is stored. Describe how the actions of a program that can modify itself can be simulated in $O(1)$ time per instruction.

C. Discuss the practical merits of the models of Parts A and B, and whether it is useful to have the features of both.

44. Consider the self-printing Pascal program presented in the sample exercises that did not make use of the chr function to make a character from an ASCII value. Write a C program that works along the same lines.

45. Consider the self-printing Pascal program presented in the sample exercises. It padded all the lines with spaces (although you cannot see them) so that they all have the length for which the array x is declared. Rewrite it so that there is no space padding.

Hint: Define a length array L where $L[i]$ holds the length of $x[i]$, and then you can format the *writeln* statements properly.

46. Consider again the one-line self-printing C program in the sample exercises.

    char *a="char *a=%c%s%c;main(){printf(a,34,a,34);}";main(){printf(a,34,a,34);}

We noted that although many compilers will give you a warning message, this program causes some compilers to give an error message, because it leaves out:

    #include <stdio.h>

Write a self-printing C program that works along these lines but has this include statement. Because of the language specific nature of this problem, you should enter this program electronically, compile it, run it, and use a difference checker to verify that the output is identical to the source code.

47. *Cube and peg puzzles* consist of k^3 unit-size cubes ($k \geq 2$), each of which has holes and pegs that must be assembled to form a single cube of k units on a side.

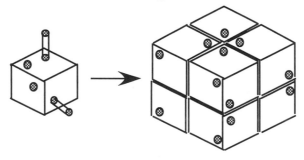

In the standard version, pegs are unit length, holes go completely through cubes (and are slightly bigger in diameter than the pegs), and all holes and pegs are positioned on only one of four standard positions (e.g., at each of the corners). The most simple case is *k*=2, as shown in the previous figure.

A. Describe an algorithm to compute the number of distinct standard puzzles for *k*=2, where in order to count, it must be possible to disassemble the puzzle.

B. Give an algorithm to solve the standard for *k*=2. That is, your algorithm takes as input a description of the hole positions and outputs a solution (or determines that there is no solution).

C. Generalize your solution to Part B to any *k* and analyze the time and space used by your algorithm.

D. Generalize your solution to Part C to non-standard puzzles (arbitrary peg lengths, arbitrary hole depths, and arbitrary hole positions).

48. The *Kev's Kubes* puzzle has twenty seven cubes threaded together with an elastic cord so that they can be "folded up" by rotating adjacent cubes with respect to each other; the object is to form a 3 by 3 by 3 cube. Given an algorithm to solve a problem of *n* cubes and analyze its running time.

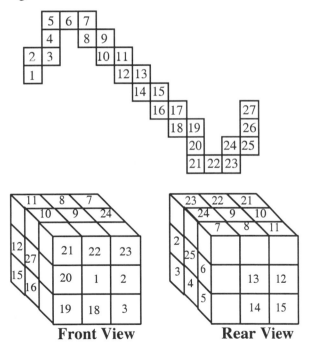

Front View   **Rear View**

Chapter Notes

For additional reading on the general subject of data structures and algorithms, the reader may refer to a host of books, including: Cormen, Leiserson, Rivest, and Stein [2001], Goodrich and Tamassia [2001], Miller and Boxer [2001], Sedgewick and Flajolet [1998], Adamson [1996], Neapolitan and Naimipour [1996], Brassard and Bratley [1996], Goodrich and Tamassia [1995], Beidler [1995], Nievergelt and Hinrichs [1993], Kozen [1992], Banachowski, Kreczmar, and Rytter [1991], Gonnet and Baeza-Yates 1991], Lewis and Denenberg [1991], Moret and Shapiro [1991], Kingston [1990], Manber [1989], Baase [1988], Brassard and Bratley [1988], Purdom and Brown [1985], Mehlhorn [1984], Tremblay and Sorenson [1984], Reingold and Hansen [1983], Aho, Hopcroft and Ullman [1983], Standish [1980], Gotlieb and Gotlieb [1978], Horowitz and Sahni [1998], Kronsjo [1979], Aho, Hopcroft and Ullman [1974], Knuth [1969]. The three volume text of Knuth [1969] and the text of Aho, Hopcroft, and Ullman [1974] were early in the field and remain classic texts. The book of Cormen, Leiserson, Rivest, and Stein [2001] includes a wide range of topics and provides an excellent reference. Atallah [1999] and Skiena [1998] are dictionary-like references.

The unit-cost model for the cost of storing integers and performing arithmetic operations is fine if not "abused". It is possible, however, to embed complex computations into arithmetic operations that operate on integers with a number of bits that depends on the input size. For this reason, especially when considering lower bounds, more precise models can be employed that charge for operations in proportion to the number of bits involved. See, for example, Aho, Hopcroft, and Ullman [1974] for a presentation of the "log-cost" model.

Horner's method is presented in many texts in both computer science and mathematics; see, for example, texts on numerical analysis such as Henrici [1964] or Dahlquist and Bjorck [1974].

Sorting is used for examples throughout this book and is a problem that underlies the complexity of many problems. Bubble sort is often one of the first sorting algorithms one encounters because it is so simple. Virtually all textbooks on the general subject of data structures and algorithms contain material on sorting. See for example Knuth [1969] for detailed analysis of the constants involved in classic sorting methods.

A number of sorting methods will be presented as examples through this book (e.g., merge sort and quick sort); in particular, see the chapter on algorithm design techniques and the chapter on parallel algorithms.

The book of Pratt [1979] is devoted to *shell sort*, an interesting method (not necessarily the best, but challenging to analyze) that sorts by a number of passes on increasingly closer spaced sub-sequences; see also the paper of Jiang, Li, and Vitanyi [2000] for lower bounds on the average-case complexity of shell sort and more recent references.

The standard matrix multiplication algorithm presented in this chapter directly reflects the definition and has a low constant factor on the running time in practice (it's as about a simple as an $O(n^3)$ algorithm can get). However, asymptotically faster methods are known. One such method (Strassen's algorithm) will be presented later as an example of "divide and conquer" strategies, and chapter notes for that chapter contain additional references.

The binomial coefficients on which Pascal's triangle is based are a common subject of texts on algorithms and discrete mathematics; see for example Ross and Wright [1999], Graham, Knuth, and Patashnik [1989], Liu [1985], Sahni [1985], Preparata and Yeh [1973], and Knuth [1969]. Binomial coefficients $\binom{n}{m}$ can be generalized to allow m to be a real number; for example, this allows the analysis of the arithmetic sum given in the appendix to be generalized to when the exponents are arbitrary real numbers. The book of Collani and Drager [2001] is devoted to binomial distributions.

The basic mathematics of solving sets of linear equations (conditions for when the solution is unique, etc.) is contained in standard texts on linear algebra such as Hoffman and Kunze [1971], Noble and Daniel [1977], and O'Nan [1976]. Asymptotically faster methods than the back-substitution method (often called Gaussian elimination) presented here can be obtained by employing fast matrix multiplication; see, for example, the presentation in Aho, Hopcroft, and Ullman [1974].

Lagrange interpolation can be found in may basic mathematics texts, including reference books like Weisstein [1999]. The book of Knott [2000] is devoted to interpolating cubic splines.

For a more formal discussion of logarithms and exponentials and for proofs of the facts cited see a calculus text such as Thomas [1968]; see also the presentation of the constant e in the chapter on hashing and the exercises to that chapter.

For background on discrete mathematics, the reader may refer to a host of text books, including: Cooke and Bez [1984], Grassmann and Tremblay [1996], Green and Knuth [1982], Kolman and Busby [1984], Levy [1980], Liu [1985], Mott, Kandel, and Baker [1986], Preparata and Yeh [1973], Ross and Wright [1999], Sahni [1981], Stanat and McAllister [1977].

The presentation of the significance of asymptotic complexity by considering tables that consider times for hypothetical problems of different sizes is based on the presentation in Chapter 1 of Aho, Hopcroft, and Ullman [1974].

The self-printing Pascal and C program that were presented in the sample exercises were written by the author; the single line self-printing C program that was also presented in the sample exercises is one of many self-printing programs that have been passed anonymously around the internet The fact that self-printing programs can be written in any full programming language follows from the *recursion theorem*, due to Kleene [1952]. In its basic form, the recursion theorem guarantees that any computable function from the non-negative integers to the non-negative integers has a fixed point in the following sense.

By writing all inputs and outputs in binary, we can view program inputs and outputs as single integers (that represent the corresponding binary string). In addition, programs are just character strings that can also be written in binary, and so each program can be represented by a single integer. An integer that does not correspond to a program can be taken to be a new program that for all inputs produces no output. Thus, we can imagine an infinite list of all programs P_0, P_1, P_2, ... , each of which takes an integer input and produces an integer output (or possibly runs forever and produces no output). The recursion theorem says:

For any computable function f, there is an integer k such that $P_k = P_{f(k)}$.

(For all inputs, program k and program $f(k)$ produce the same output.)

The existence of self-printing programs can now be argued as follows:

Let $f(i)$ be an integer such that $P_{f(i)}$ always ignores its input and produces the output i. Since it is easy to write such a program, $f(i)$ is computable. By the recursion theorem, there is an integer k, such that the program k is equivalent to the program $f(k)$; that is, program k must produce output k, and hence is a self-printing program.

A proof of this basic form of the recursion theorem (which is presented in most texts on the theory of computation such as Hopcroft and Ullman [1979], Hopcroft, Motwani, and Ullman [2001], or Sipser [1997]) goes as follows:

Let $g(i)$ be a function such that $P_{g(i)}$ is a program which for every input x, ignores x, computes $P_i(i)$, and if the computation terminates with a value y, computes $P_y(x)$. That is, for all i and x, either

$$P_{g(i)}(x) = P_{P_i(i)}(x)$$

or both computations are undefined. Now let j be an integer such that P_j is a program which for all inputs x computes $f(g(x))$. Then we have

$$P_{g(j)}(x) = P_{P_j(j)}(x) = P_{f(g(j))}(x)$$

and hence the value $k=g(j)$ satisfies the recursion theorem.

In fact, the recursion theorem guarantees that any full programming language has infinitely many self-printing programs (i.e., since there are infinitely many equivalent programs to a given program by just adding dummy statements, we could remove program k from the language and still have a full language from which the recursion theorem could find another self printing program). However, the recursion theorem says nothing about how small or how elegant the self-printing program may be. Writing compact self-printing programs has traditionally been a challenge for recreational programmers (see for example, Hofstadter [1985], Hay [1980], Burger, Brill, and Machi

[1980], and Bratley and Millo [1972]); the web site of gthompso [2001] has a listing of self-printing programs in many programming languages.

The parity exercise was suggested by Motta [2000], and is essentially a simpler version of the method presented in exercises to the chapter on strings to count the number of 1's in the binary representation of an integer.

The perfect integer square exercise was suggested by Hickey [1997], who constructed the figure included with the exercise to depict the smallest solution.

2. Lists

Introduction

Lists are perhaps the most simple and most commonly used data structure. We use them in everyday activities (e.g., a grocery list). Formally, a *list* is an ordered sequence of *vertices* where associated with each vertex is a *data* item and a *next* vertex.

For some applications, a simple *array implementation* (items are stored in sequential locations of an array) suffices to achieve $O(1)$ time for basic operations. The most common examples are lists where data is inserted and deleted only at the ends. A *stack* is a list with only the operations of inserting and deleting data items at one end (the "top"). A *queue* is a list with only the operations of inserting at one end (the "rear") and removing from the other end (the "front"). Stacks model a "last-in, first-out" process, whereas queues model a "first-in, first-out" process (like a line at the grocery store).

Although we shall see that arrays work well to represent stacks and queues, if we wish to perform more general operations on the list such as inserting and deleting elements from a position in the middle of the list, the use of an array becomes problematic. For example, in an array implementation, deleting an item may require sliding all elements in higher numbered positions of the array down a position to close up the "hole" left by the deleted item. To make insertion, deletion, and related operations possible in $O(1)$ time, we employ *pointers* to represent lists. A pointer is a variable whose value denotes an address in memory. For example, consider the problem of inserting an item somewhere in the middle of a list. If each element of a list is represented by a vertex that contains the data and a pointer to the next element, then we can simply modify pointer fields to "hook" the new item in. The figure below depicts how a new vertex containing a data item d can be hooked in after a vertex v in a list L.

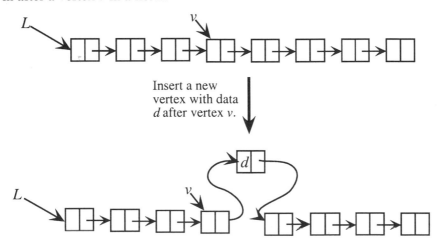

Most compilers provide pointer variables (we shall also see how they can be implemented "manually"). Pointers only increase the memory needed for the list by a constant factor. In the example above, if each element of the list is a simple integer, then the use of pointers doubles the space used by an array representation, but in exchange, allows for many operations to be done in $O(1)$ time. It could be that the items stored in a list are records of many bytes (e.g., names and addresses), and the additional memory for the pointers is far less than a factor of two.

An advantage of pointers is that it is not necessary to choose a maximum list size in advance, as we do when we declare an array to hold a stack or a queue (although it is possible to re-allocate memory when a stack or queue becomes full — see the exercises). Most programming languages allow one to create new list vertices as you need them. When vertices are no longer used (such as when deleted from a list), some languages allow one to explicitly reclaim their memory for future use, and others do automatic "garbage collection" of vertices that are no longer referenced by any active pointers.

With a *singly linked* list (as in the preceding figure), each vertex has a *data* field that contains the data for that vertex and a *next* field that contains the address of the next item in the list (or a special *nil* value in the case of the last item). In a *doubly linked* list, each vertex also has a *previous* field that contains the address of the previous item in the list (or *nil* in the case of the first item). Since it only increases space by a constant factor to have double links, our basic definition of linked lists will include them. In practice, when operations that need links going in both directions to be fast are not used or can be avoided, single links can be used to improve by a constant factor the space used by vertices and the time to manipulate them.

Although for some applications lists can be implemented to allow the "sharing" of list tails, we restrict our attention to implementations where each vertex has a unique predecessor, and hence the previous field is uniquely defined.

Array Implementation of Stacks

Definition: A *stack* is a special type of list where the only operations are inserting and deleting items from the same end.

Idea: A variable *top* is the index of the first unused location. If there are currently k items, locations 0 through $k-1$ store them, $top=k$, and locations k through $n-1$ are unused.

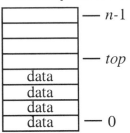

Specification of the array implementation of a stack:

An array $S[0] \ldots S[n-1]$.

An integer variable *top* that is initially 0.

PUSH places a data item d on the stack S:

    procedure PUSH(d,S)
        if $top \geq n$ **then** {stack overflow}
        else begin
          $S[top] := d$
          $top := top+1$
        end
    end

POP removes (and returns) the data item on top of the stack S:

    function POP(S)
        if $top \leq 0$ **then** {stack underflow}
        else begin
          $top := top-1$
          return $S[top]$
        end
    end

TOP simply returns data item on top of the stack S:

    function TOP(S) **if** $top \leq 0$ **then** {error} **else return** $S[top-1]$ **end**

Simplified Notation: When S is understood, we may omit it as an argument. Also, we sometimes use POP as a procedure (so it decrements *top* and returns nothing).

Example: Using a Stack to Reverse an Array

Problem: Rearrange $A[1]...A[n]$ so that the elements are in reverse order.

Idea: A stack works like plates stacked on a shelf in the kitchen; a plate is added or removed from the top of the pile. Of course, here, instead of plates, we shall push elements of the array. Since the last item into a stack is the first item out, if everything is first pushed into a stack, then popping everything back out will produce the array in reverse order.

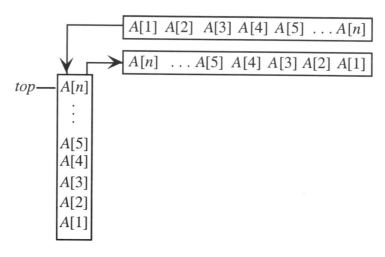

Algorithm:

> Initialize an empty stack capable of holding n elements.
> **for** i=1 **to** n **do** PUSH($A[i]$)
> **for** i=1 **to** n **do** A[i]=POP

Time: $O(n)$ since both PUSH and POP are $O(1)$ time and they are each executed n times.

Space: $O(n)$ space for the stack, in addition to the $O(n)$ space used by A (although it is also possible to reverse A in place using only $O(1)$ space in addition to the space used by A — see the sample exercises).

(*) Example: Evaluating a Postfix Expression with a Stack

Infix notation for arithmetic expressions: Standard *infix* notation has the form *operand operator operand* (e.g., $a+b$), where an operand is a value or a sub-expression. Although assuming that multiplication and division take precedence over addition and subtraction allows parentheses to be omitted in many cases, they are necessary in general; for example, $a+(b \bullet c) \neq (a+b) \bullet c$.

Postfix notation: *Postfix* notation has the form *operand operand operator* (e.g., $ab+$). An advantage is that parentheses are not needed and evaluation is always unique. For example, the infix expression $(a+(b \bullet c)) \bullet d$ is written as $dcb \bullet a+ \bullet$ in postfix notation. When the order of operations is not specified in an infix expression (e.g., $a+b+c$), there are more than one equivalent postfix expressions (e.g., $ab+c+$ and $abc++$).

Evaluating a postfix expression: For this example, we assume that the postfix expression is in an array $P[1]...P[n]$, where for $1 \leq i \leq n$, $P[i]$ is a *value* or a *binary operator* (i.e., operators like $+$, $-$, $*$, and $/$ that take two operands). To evaluate P, we use a stack, which is initially empty, where values can be pushed when they are encountered and popped when the appropriate operator needs them. The result of an operation is pushed back onto the stack to become an operand for a later operation.

> **function** postEVAL
>> **for** $i := 1$ **to** n **do begin**
>>> **if** $P[i]$ is a value **then** PUSH($P[i]$)
>>>
>>> **else if** $P[i]$ is a binary operator **then begin**
>>>> **if** S is empty **then** *ERROR — missing value*
>>>> $b := $ POP
>>>> **if** S is empty **then** *ERROR — missing value*
>>>> $a := $ POP
>>>> PUSH(the result of applying $P[i]$ to a,b)
>>>> **end**
>>>
>>> **else** *ERROR — illegal input*
>>
>> **end**
>>
>> **if** stack has only one item **then return** POP **else** *ERROR — missing operator*
>
> **end**

Unary operators: This algorithm can be generalized to handle operators that take only one operand (e.g., leading minus, a percent, or an exponent); see the sample exercises.

Correctness: A proof of correctness for this algorithm can also be used to argue that the evaluation of a postfix expression is always unique; see the sample exercises.

Array Implementation of Queues ("Circular Queues")

Definition: A *queue* is a special type of list where the only operations allowed are inserting items at one end and deleting items from the other end.

Idea: Over time, as items are inserted (and *rear* is incremented) and items are deleted (and *front* is incremented), the queue data migrates to the right. To avoid a jam at the right end, allow the data to wrap back to position 0.

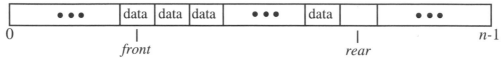

Specification of the array implementation of a queue:

Array $Q[0]$ ... $Q[n-1]$ and variables *front*, *rear,* and *size* initially 0.

The ENQUEUE operation to enter d at the rear of the queue Q:

> **procedure** ENQUEUE(d,Q)
> > **if** *size* $\geq n$ **then** {queue overflow}
> > **else begin**
> > > $Q[rear] := d$
> > > *rear* := *rear*+1; **if** *rear*=*n* **then** *rear* := 0
> > > *size* := *size*+1
> > > **end**
> > **end**

The DEQUEUE operation to remove from the front of the queue Q:

> **function** DEQUEUE(Q)
> > **if** *size*=0 **then** {queue underflow}
> > **else begin**
> > > *temp* := $Q[front]$
> > > *front* := *front*+1; **if** *front*=*n* **then** *front* := 0
> > > *size* := *size*−1
> > > **return** *temp*
> > > **end**
> > **end**

FRONT and REAR simply return the front and rear data items of the queue Q:

> **function** FRONT(Q) **if** *size*=0 **then** {error} **else return** $Q[front]$ **end**
> **function** REAR(Q)
> > **if** *size*=0 **then** {error}
> > **else if** *rear*=0 **then return** $Q[n-1]$
> > **else return** $Q[rear-1]$
> > **end**

Simplified Notation: When Q is understood, we may omit it as an argument. Also, we sometimes use DEQUEUE as a procedure (so it deletes and returns nothing).

Example: Using a Queue to Separate an Array

Problem: Rearrange $A[1]...A[n]$ so that the odd numbered elements are in positions 1 through $\lceil n/2 \rceil$ and the even numbered elements are in positions $\lceil n/2 \rceil + 1$ through n.

Idea: A queue is like a line at the post office; people enter the line at the rear and are served (deleted from the line) at the front. Allocate two queues, $Q1$ and $Q2$, of size $\lceil n/2 \rceil$. Enqueue the odd elements into $Q1$ and the even elements into $Q2$ (this preserves the relative order of elements). Empty $Q1$ into the first half of A and $Q2$ into the second half.

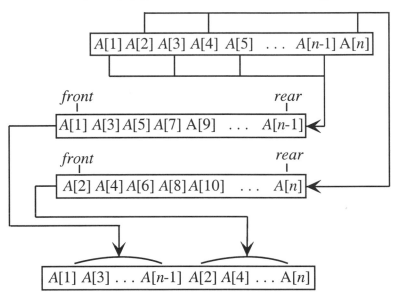

Algorithm:

> Initialize an empty queue $Q1$ capable of holding at least $\lceil n/2 \rceil$ elements.
> Initialize an empty queue $Q2$ capable of holding at least $\lfloor n/2 \rfloor$ elements.
> **for** $i=1$ **to** n **do**
>   **if** i is odd **then** ENQUEUE($A[i], Q1$)
>   **else** ENQUEUE($A[i], Q2$)
>   **end**
> **for** $i=1$ **to** $\lceil n/2 \rceil$ **do** $A[i]$=DEQUEUE($Q1$)
> **for** $i=\lceil n/2 \rceil + 1$ **to** n **do** $A[i]$=DEQUEUE($Q2$)

Time: $O(n)$ since both ENQUEUE and DEQUEUE are $O(1)$ time and they are each executed n times.

Space: $O(n)$ space for the queues, in addition to the $O(n)$ space used by A. (although it is possible to use less additional space — see the exercises).

General Purpose Lists

Definition: A *list* is an ordered sequence of *vertices* where associated with each vertex is a *data* item, a *previous* vertex, and a *next* vertex; the two exceptions being the vertices previous to the first vertex and next from the last vertex, which are defined to be *nil*.

Basic operations:

CREATE:	Create and return an empty list.
DESTROY(L):	Reclaim any remaining memory used by list L (assumes that all of its vertices have already been deleted).
EMPTY(L):	Return *true* if list L is empty, *false* otherwise.
SIZE(L):	Return the current number of vertices in list L.
FIRST(L):	Return the first vertex of list L (*nil* if L is empty).
LAST(L):	Return the last vertex of list L (*nil* if L is empty).
DATA(v,L):	Return the data stored at vertex v in list L.
NEXT(v,L):	Return the vertex that follows vertex v in list L (or *nil* if v is the last vertex in L); NEXT(*nil*) is defined to be *nil*.
PREV(v,L):	Return the vertex that precedes vertex v in list L (or *nil* if v is the first vertex in L); PREV(*nil*) is defined to be *nil*.
INSERT(d,v,L):	Create a new vertex to the right of vertex v in list L and initialize it to contain data d (if $v=nil$, then insert at the beginning of L).
DELETE(v,L):	Delete vertex v from list L, reclaim the memory it consumed, and return the data it contained.
SPLICE($L1,v,L2$):	Make the vertices of list $L1$ follow vertex v in list $L2$ (if $L1$ is empty then $L2$ is unchanged, if $v=nil$ then insert the vertices of $L1$ at the beginning of $L2$).
CUT($v1,v2,L$):	Remove and return as a new list the portion of list L that starts with vertex $v1$ and ends with vertex $v2$.

Note: SPLICE and CUT could be implemented as sequences of INSERT or DELETE operations but we shall view them as "basic" operations because in many implementations they can be performed in constant time.

Simplified notation: When L is understood, we may omit it as an argument to the operations DATA, NEXT, and PREV.

Example: Using the Basic List Operations

Output a list L:

    procedure PRINT(L)
        v := FIRST(L)
        while $v{\neq}nil$ **do begin**
            output DATA(v)
            v := NEXT(v)
            end
        end

Make a copy of a list L:

    function COPY(L)
        M := CREATE
        v := FIRST(L)
        while $v{\neq}nil$ **do begin**
            INSERT(DATA(v,L),LAST(M),M)
            v := NEXT(v,L)
            end
        return M
        end

Search for a data item d in a list L (return nil if d is not in L):

    function SEARCH(d,L)
        v := FIRST(L)
        while $v{\neq}nil$ **and** $DATA(v){\neq}d$ **do** v := NEXT(v)
        return v
        end

Add up the data in vertices v through w of L (assumes L stores numbers):

    function ADD(v,w,L)
        sum := DATA(w)
        while $v{\neq}w$ **do begin**
            sum := $sum+DATA(v)$
            v := NEXT(v)
            end
        end

Stack operations:

    PUSH(d,S) is another name for INSERT(d,nil,S)
    POP(S) is another name for DELETE(FIRST(S),S)

Queue operations:

    ENQUEUE(d,Q) is another name for INSERT(d,LAST(Q),Q)
    DEQUEUE(Q) is another name for DELETE(FIRST(Q),Q)

Representing Lists With Pointers

Problem: A list can be stored in an array. However, except for special types of lists (such as stacks and queues), operations such as INSERT can be time consuming because elements may have to be "shifted" to make room for the new element.

Idea: Associate with each list vertex fields that contain *pointers* to the addresses in memory of the previous and next elements in the list. For example, a pointer that has value 323 does not denote the 323rd element of the list (except possibly by pure chance); rather, it simply denotes a position in memory. When starting at the first vertex of the list and following next pointers to the second item, third item, and so on, one is in general skipping around between arbitrary positions in memory. A list *header* can store global information about a list such as the list size and pointers to the first and last vertices, as well as any information specific to the application in question.

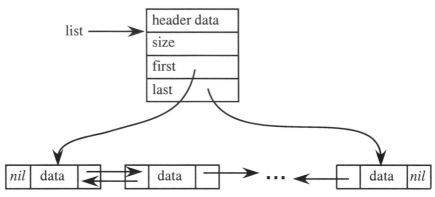

Headers: Headers are often not used because they can make it more cumbersome to manipulate lists; in this case, needed global information about the list must be explicitly stored in additional variables. When headers are used, they may contain data (such as the current size of the list) that must be updated when operations are performed.

Reclaiming Memory: Memory must be allocated when creating new vertices and in many applications should be reclaimed when they are no longer needed (the DESTROY operation may have to reclaim the memory used by the header).

Singly Linked Lists: Often lists are *singly linked* to reduce the space and the time to manipulate list vertices. Each list vertex has a data field and next field, but no field containing a pointer to the previous vertex. PREV(v,L) can now be expensive (e.g., if no information about the preceding vertex has been explicitly saved, then it is necessary to traverse the list from the beginning until a vertex with its next field pointing to v is found). However, most other operations are unaffected or can be easily modified to still work in $O(1)$ time. For example, DELETE(v,L) can be redefined to delete the vertex that follows v (see the exercises).

CHAPTER 2

Pointer Variables

Idea: In most programming languages, a variable can be declared that contains the address in memory of a data item, and operators are provided to go between a pointer and the data that is pointed to.

Example: In Pascal, $^\wedge x$ is a pointer to the variable x and $x^\wedge$ is the value pointed to by x; so the sequence of statements $p := {}^\wedge y$; $x := p^\wedge$ is equivalent to doing $x := y$. Declaration of a list with integer data and a header that contains a character string might look like:

```
TypeData = integer;

TypeHData = packed array [1..80] of char;

TypeVertexBody = record
        data: TypeData;
        left: TypeVertex;
        right: TypeVertex
        end;

TypeVertex = ^ TypeVertexBody;

TypeHeader = record
        hdata: TypeHData;
        size: integer;
        first: TypeVertex;
        last: TypeVertex
        end;

TypeList = ^ TypeHeader;
```

Example: In C, $\&x$ is a pointer to the variable x and $*x$ is the value pointed to by x; so the sequence of statements $p=\&y$; $x=*p$ is equivalent to doing $x=y$. Declaration of a list with integer data and a header that contains a character string might look like:

```
typedef int TypeData;

typedef char TypeHData[80];

typedef struct vbody {
        TypeData data;
        struct vbody *left;
        struct vbody *right;
        } TypeVertexBody;

typedef TypeVertexBody *TypeVertex;

typedef struct header {
        TypeHData hdata;
        int size;
        struct vbody *first;
        struct vbody *last;
        } TypeHeader;

typedef TypeHeader *TypeList;
```

(*) Implementing Pointers

Idea: When pointer variables are not available, or possibly for time-critical applications, pointers can be implemented "manually" with array fields. In the context of learning about pointers, it is instructive to see how this is done (and at a simplified level, how a typical compiler implements pointer variables).

Direct implementation of doubly linked lists: To store a collection of lists with up to a total of n vertices, maintain three arrays *data*, *prev*, and *next* that range from 1 to n.

A pointer is an index between 1 and n.

Each list can be represented by the index of its first element. In addition, a special list *FREE* can store all unused locations. *FREE* initially goes sequentially from 1 to n, and we return free vertices to the front of the *FREE* list.

The illegal address 0 can be used for the *nil* pointer.

Example: Suppose $n=8$ and we have two lists $L1=(a,b,c)$ and $L2=(x,y)$. If we insert the items in the order a, b, c, x, y then $L1$ and $L2$ end up sequentially in the array. But in general, operations will scatter vertices everywhere. For example, the figure below shows the result of continuing with *INSERT(d,2,L1)*, *DELETE(5,L2)*, DELETE(5,L1):

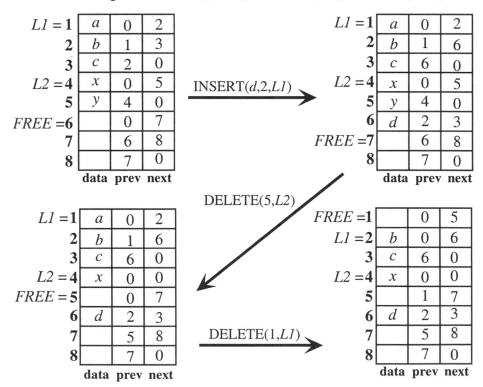

CHAPTER 2

Implementation of the Basic List Operations in $O(1)$ Time

Idea: With a list header being a pointer to a structure containing *first* and *last* fields and a list vertex being a pointer to a structure containing *data*, *next*, and *previous* fields, all of the basic operations are $O(1)$ time:

CREATE: Define a new header, set the first and last fields to *nil*, set the size field to 0.
DESTROY(*L*): Reclaim the memory used by the list header.

EMPTY(*L*): Test if the header size field is 0 (or if the first or last field is *nil*).
SIZE(*L*): Return the size field of the header.

FIRST(*L*), LAST(*L*): Return the appropriate field from the header of *L*.

DATA(*v,L*), NEXT(*v,L*), PREV(*v,L*): Return the appropriate field of *v*.

INSERT(*d,v,L*), SPLICE(*L1,v,L2*) — "hook in" (and update the header size field):

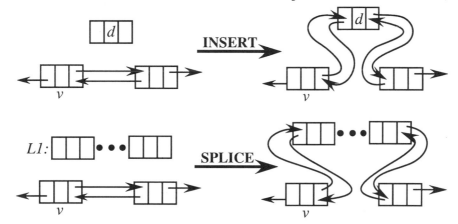

DELETE(*v,L*), CUT(*v1,v2,L*) — "bypass" (and update the header size field):

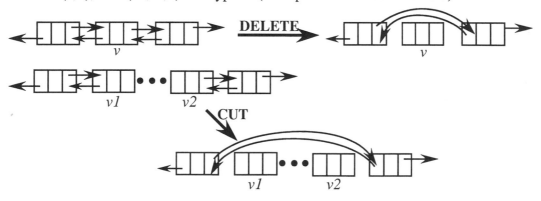

Example: Details of the INSERT Operation

Notation for "low-level" operations: If p is a pointer to a list vertex and L a pointer to a list header, we use the following notation:

$$
\begin{aligned}
p.data &= \text{data field} \\
p.next &= \text{next field} \\
p.prev &= \text{previous field} \\
L.size &= \text{size field} \\
L.first &= \text{first field} \\
L.last &= \text{last field}
\end{aligned}
$$

procedure INSERT(d,v,L):

    Allocate memory for a new vertex x.

    $x.data := d$

    $L.size := L.size+1$

    if $L.first=nil$ **then begin** (L is an empty list)
        $x.prev := nil$
        $x.next := nil$
        $L.first := x$
        $L.last := x$
        end

    else if $v=nil$ **then begin** (x goes at the front of L)
        $x.prev := nil$
        $x.next := L.first$
        $(L.first).prev := x$
        $L.first := x$
        end

    else if $v=L.last$ **then begin** (x goes at the end of L)
        $x.prev := v$
        $x.next := nil$
        $v.next := x$
        $L.last := x$
        end

    else begin (x goes in the middle of L)
        $x.prev := v$
        $x.next := v.next$
        $(v.next).prev := x$
        $v.next := x$
        end

    end

Example: Details of the DELETE Operation

Idea: Basically the reverse of the INSERT. We use the same notation for low level operations:

procedure DELETE(v,L):

    $d := \text{DATA}(v,L)$

    $L.size := L.size-1$

    if $L.first=L.last$ **then begin** (v is the only vertex in L)
            $L.first := nil$
            $L.last := nil$
        end

    else if $v=L.first$ **then begin** (v is the first item of L)
            $L.first := v.next$
            $(v.next).prev := nil$
        end

    else if $v=L.last$ **then begin** (v is the last item of L)
            $L.last := v.prev$
            $(v.prev).next := nil$
        end

    else begin (v is in the middle of L)
            $(v.prev).next := v.next$
            $(v.next).prev := v.prev$
        end

    Reclaim the memory that was used by v.

    return d

    end

Illegal arguments: Like the code for INSERT, this code includes no check for illegal arguments. Many but not all illegal inputs can be checked without changing the asymptotic time of the operation (see the exercises).

Example: Reversing a Singly Linked List

Problem: Given a singly linked list L of n elements that has a header containing pointers to its first and last elements, and where each vertex has only a *next* pointer field, reverse the order of the elements.

Method 1 (makes a new list, uses basic operations): Copy L to a second list R one item at a time. Because items are removed from the front of L and placed at the front of R, the effect is like pushing them on a stack, and R ends up in the reverse of the order of L.

> R := CREATE
> **while** (**not** EMPTY(L)) **do** INSERT(DELETE(FIRST(L),L),*nil*,R)
> DESTROY(L)
> L := R

Complexity: $O(n)$ time and $O(1)$ space in addition to the space consumed by the list vertices (since all vertices placed in R are first deleted from L).

Method 2 (in place, uses "low-level" operations): Traverse L to reverse each *next* pointer and then finish by exchanging the *first* and *last* pointers of the header. Although some of the steps can be expressed using the basic list operations, "low-level" assignments are needed to modify the *next* field and the *first* and *last* fields of the header; we use the same notation for low level operations as the previous examples.

> *previous* := *nil*
> *current* := *L.first*
> **while** (*current* $\neq$ *nil*) **do begin**
>   *temp* := *current.next*
>   *current.next* := *previous*
>   *previous* := *current*
>   *current* := *temp*
>   **end**
> *L.last* := *L.first*
> *L.first* := *previous*

Complexity: $O(n)$ time and $O(1)$ space in addition to the space consumed by the list vertices.

CHAPTER 2

(*) Example: The POP(*i*) Stack Operation

Idea: Although pointer based list implementations usually save time by allowing a larger range of operations that can be done efficiently, sometimes there are "cheap" tricks that can be done with array implementations. Fortunately, in some cases, the total time of the corresponding work done in the pointer implementation for a sequence of operations is not so bad. The POP(*i*) operation is an example of this.

Definition: The operation POP(*i*) removes $i-1$ items from the stack and then does a normal POP operation (if the stack has less than *i* items, there is an underflow error).

Problem: With an array implementation, in $O(1)$ time we can simply subtract *i* from *top* and return *stack*[*top*], but with pointers we must do:

> **procedure** POP(*i*):
>   **for** $j := 1$ **to** $i-1$ **do** POP
>   **return** POP
>   **end**
>
> *Note:* Here we have used POP as both a procedure that simply discards the element on top of the stack and as a function that deletes and returns it.

Idea: An individual POP(*i*) operation could be as bad as $\Omega(n)$. However, *n* operations are $O(n)$ because the cost of a POP(*i*) can be charged against the PUSH operations that placed the *i* elements there in the first place.

Theorem: Starting with an empty stack, a sequence of *n* PUSH and POP(*i*) operations takes $O(n)$ time.

Proof: Give each PUSH two "credits" and each POP(i) one credit, where a credit "pays" for $O(1)$ computing time.

When a PUSH is performed, one credit pays for the operation and the other credit is associated with the item.

When a POP(*i*) is performed, the credit associated with each item that is popped can be used to pay for that iteration of the *while* loop, and the credit with the POP(*i*) pays for the remaining $O(1)$ time for the operation.

Hence, the total credits used (which is proportional to total time used) is linear in the number of operations.

Amortized time analysis: This analysis of the time used by the POP(i) operation is a simple example of an *amortized* time bound. To show that a sequence of *n* PUSH and POP(*i*) operations are $O(n)$, we did not show that each POP(*i*) operation was $O(1)$ in the worst case; in fact, an individual POP(*i*) can be $\Omega(n)$.

Sample Exercises

49. We have already seen how to reverse an array using a stack. However, if we are more careful, the stack is not necessary. Given an array $A[1]$ ... $A[n]$, describe (in English and with pseudo-code) how to reverse the order of the elements in A in $O(n)$ time and using only $O(1)$ space in addition to the space used by A.

Solution: Work from the ends towards the middle exchanging elements:

> $left := 1$
> $right := n$
> **while** $left < right$ **do begin**
>   Exchange the contents of $A[left]$ and $A[right]$.
> $left := left+1$
> $right := right-1$
> **end**

The algorithm is $O(n)$ time since the *while* loop is executed $\lfloor n/2 \rfloor$ times and the body of the *while* loop takes $O(1)$ time. It uses $O(1)$ space in addition to the space for A.

50. Recall the algorithm presented to evaluate a postfix expression (containing only binary operators) with a stack.

    A. Prove that a postfix expression (containing only binary operators) can always be evaluated correctly by working from left to right. That is, by repeatedly replacing the leftmost operator and the two values to its left by a value.

    B. Prove that the algorithm correctly evaluates a legal expression (that contains only binary operators).

    C. Prove that the algorithm always reports an error for an illegal expression.

    D. Generalize the algorithm to allow *unary operators* (operators that take only one operand, such as a leading minus, a percent, or an exponent). You may assume that the same symbol is not used for both a unary and binary operator. So for example, an expression such as $a - (-b)$ would have to be represented in postfix by either subtracting from 0 (i.e., $a\ 0\ b - -$), or by using a different symbol for a unary minus sign (e.g., if _ denotes unary minus, then we would have $a\ b\ \_ -$).

Solution:

    A. By definition, a binary postfix operator takes the two operands to its left. If an operator x is the leftmost operator in the postfix expression and occurs at position j, then there are no other operators to the left of x (i.e., no operators in positions 1 through $j-1$) that could cause two operands to be combined into a single one. Hence, the only possibility for x in a legal evaluation of the expression is that $j \geq 3$ and x is applied to the values at positions $j-1$ and $j-2$. Applying x to these two values has the effect of replacing x and these two values by a single value, resulting is a shorter legal postfix expression. This reasoning can be repeated until we are left with a single value equal to the value of the expression.

B. Since the *for* loop proceeds from left to right in P and works by pushing all values and applying all operators immediately to the top two items of the stack, it performs a left to right evaluation of the expression, and hence by Part A correctly evaluates a legal expression.

C. If for some i, $P[i]$ is not a binary operator or an operand, then the *if* and *else if* statements will fail, and the final *else* statement will report an error. Now suppose that all entries of P are legal binary operators or values. Since by Part B we know that the algorithm evaluates the expression from left to right, it will always be able to successfully evaluate the expression, except when an operator is encountered but there are not two items on the stack to which it can be applied; this is detected by the two checks for an empty stack in the *else if* statement. If evaluation is successful, but there is more than one operand left on the stack (the expression did not have enough operators), then this is detected by the final check before returning.

D. We can insert the following *else if* statement between the current *if* and *else if* statements:

> **else if** $P[i]$ is a unary operator **then begin**
>> **if** S is empty **then** *ERROR — missing value*
>> $a := \text{POP}$
>> PUSH(the result of applying $P[i]$ to a)
> **end**

This generalization preserves the left to right evaluation (and Parts A, B, and C can be generalized).

51. In the array implementation of queues presented, a variable *size* kept track of the current number of elements in the queue. This can be handy when the size has to be checked often for some reason, however, it is in some sense redundant because the current number of elements can always be computed from front and rear. Rewrite the pseudo-code to eliminate the variable *size*.

Solution: To eliminate the problem of distinguishing an empty queue from a completely full one, we "waste" one position and only allow a maximum of $n-1$ elements in the queue. To simplify presentation, we employ the MOD notation where for $0 \leq i < n$, $i+1$ MOD n is 0 if $i = n-1$ and $i+1$ otherwise, and $i-1$ MOD n is $n-1$ if $i=0$ and $i-1$ otherwise.

Array $queue[0] \dots queue[n-1]$ and variables *front* and *rear* initially 0.

The ENQUEUE operation to enter d at the rear of the queue Q:

> **procedure** ENQUEUE(d,Q)
>> **if** $rear=((front-1)$ MOD $n)$ **then** {queue overflow}
>> **else begin**

$$queue[rear] := d$$
$$rear := rear+1 \text{ MOD } n$$
end

      end

The DEQUEUE operation to remove from the front of the queue Q:

function DEQUEUE(Q)

    if *front=rear* **then** {queue underflow}

    else begin
$$temp := queue[front]$$
$$front := front+1 \text{ MOD } n$$
      return *temp*
      end

    end

52. Suppose singly linked lists are represented without headers so that:

- A list is just a pointer to its first element.
- Each vertex contains a data field and a next field.
- The data field is a single pointer to whatever data is associated with that vertex.
- The data field of the last element of the list is always a pointer to a structure containing the header information for whatever is appropriate to the application (so an empty list contains exactly one vertex).

Given a pointer p to a vertex in such a list (and nothing else, so you do not have a pointer to the vertex preceding the one pointed to by p), describe how to remove an element in constant time. Your algorithm must maintain the property that each vertex of a list, except the last one, is associated with exactly one data item of the list.

Solution:

procedure DELETE(p):

    Replace the data of p by DATA[NEXT[p]].

    Change the next field of p to be NEXT[NEXT[p]].

    end

53. Suppose lists of integers are represented with single links (so each node has a data field and a next field) and headers are not used (so a list is just a pointer to its first element).

A. Describe in English and give pseudo-code for a procedure listMERGE that merges two sorted lists, *list1* and *list2*, into a single sorted list, *list3*. Work at the low level list notation of *p.next* to denote the next field of the vertex pointed to by p. Be sure your code works when one or both lists are empty.

B. Using your solution to Part A, describe how to do "bottom-up" (non-recursive) merge sorting; that is, given a singly linked list L of n integers, rearrange L so that its elements are in non-decreasing order.

C. Analyze the time and space used by your solution to Part B.

Solution:

A. Repeatedly compare the first elements of *list1* and *list2* and place the smaller at the end of *list3*; when one of the lists runs out, the other can be appended to the end of *list3*; care must be taken when lists are initially empty or become empty. The variable *last* maintains a pointer to the last vertex of *list3*. Even though the next field of the vertex pointed to by *last* may not be *nil* during the execution of the program, the program always terminates by appending the remainder of *list1* or *list2* to *list3*, thus properly ending *list3* with the next field of the last vertex *nil*.

function listMERGE(*list1,list2*)

> *list3* := *last* := *nil*
>
> **while** (*list1≠nil*) **and** (*list2≠nil*) **do begin**
>
>> **if** (*list1=nil*) **then begin**
>>> **if** (*list3=nil*) **then** *list3* := *list2* **else** *last.next* := *list2*
>>> *list2* := *nil*
>>> **end**
>>
>> **else if** (*list2=nil*) **then begin**
>>> \<symmetric to the previous case\>
>>> **end**
>>
>> **else if** (DATA(*list1*)<DATA(*list2*)) **then begin**
>>> **if** *list3=nil* **then** *list3* := *list1* **else** *last.next* := *list1*
>>> *last* := *list1*
>>> *list1* := *list1.next*
>>> **end**
>>
>> **else begin**
>>> \<symmetric to the previous case\>
>>> **end**
>>
>> **end**
>
> **return** *list3*
>
> **end**

B. We can make a single pass through L to convert L to a list of lists where each list in L contains exactly one of the elements originally in L. Then, we can repeatedly make passes through L that merges pairs of lists; that is the first and second list in L are merged, the third and fourth lists are merged, etc. If L has an odd number of lists, then after the third to last is merged with the second to last list, one additional merge can merge this list with the last list in L. We stop when L contains only one list, at which time we can change L to be this list.

C. Each pass through L takes time proportional to the sum of the lengths of each list in L, which is $O(n)$. Since the minimum size of a list in L doubles with each pass, there are at most $\lfloor \log_2(n) \rfloor$ passes, and hence the algorithm is $O(n \log(n))$.

Exercises

54. In the code for the basic stack operations, to check for an overflow or underflow error, PUSH could test if $top=n$ and POP / TOP could test if $top=0$. Clearly it does no harm to make the tests $top \geq n$ and $top \leq 0$. Discuss circumstances in practice where these tests might provide an extra margin of "safety" to detect an error that might result, for example, from a mistake in programming changing the value of top outside of the procedures PUSH and POP.

55. In the sample exercises we saw how to eliminate the variable *size* from the pseudo-code for an array implementation of a queue (although it is not clear that it is advantageous to do so). The solution "wasted" one position to avoid the problem of distinguishing an empty queue from a completely full one, and never stored more than $n-1$ elements. Suppose that instead of having *rear* be the index of the next available location, it is the index of the item at the rear of the queue. Give pseudo-code for ENQUEUE and DEQUEUE that does not use the variable *size* and can store n elements in the queue, being careful with how you handle the case that the queue contains only 0 or 1 elements. Can you suggest other approaches that do not use *size* and can store n elements?

56. A *doubly ended queue* is a list where items may be inserted or deleted from either end. In a way analogous to what was presented for queues, define key variables (front, rear, etc.) and present pseudo-code for the following operations:

FrontInsert(i): Inserts i at the front (or does nothing and detects an error if the array is full).

FrontDelete: Deletes and returns the front element (or does nothing and detects an error if no elements are currently stored in the array).

RearInsert(i): Inserts i at the rear (or does nothing and detects an error if the array is full).

RearDelete: Deletes and returns the rear element (or does nothing and detects an error if no elements are currently stored in the array).

Hint: Think of *RearInsert* and *FrontDelete* as a normal queue, and now you just need to add two symmetric operations.

57. We have seen how to separate the odd and even elements of an array using two queues.

A. As presented, the code used two *for* loops to fill the queues and then two *for* loops to empty them. Rewrite the code with a single *for* loop that makes a single pass through A to fill the queues and then a single *for* loop that makes a single pass through A to empty the queues.

B. If we are more careful, two queues are not necessary. Given an array A[1] ... A[n], describe (in English and with pseudo-code) how to, in $O(n)$ time and using only $O(1)$ space in addition to the space that is used by A and the space consumed by a single queue capable of holding $\lceil n/2 \rceil$ elements, move the odd numbered elements of A to positions 1 through $\lceil n/2 \rceil$ (in the same relative order) and the even numbered elements of A to positions $\lceil n/2 \rceil + 1$ through n (in the same relative order).

C. Suppose that we do not care about the order of elements, as long as all odd elements end up in the left half of A (but not necessarily in the same relative order) and all even elements end up in the right half of A. For example, it would be ok to transform the sequence 1, 2, 3, 4, 5, 6, 7, 8, 9, 10 into 1, 9, 3, 7, 5, 6, 4, 8, 2, 10. Give an algorithm to separate an array using only $O(1)$ space in addition to the space used by A, by working from the outside in performing swaps on every other element in a fashion similar to the way the sample exercise reversed an array.

Note: We shall see when we study in-place permutations that no queues are necessary, and we can do this in place (preserving the correct order of elements), using only $O(1)$ space in addition to that used by A.

58. Suppose that you do not know in advance how large a stack will get, but are using a language that allows you to dynamically create new arrays and destroy ones that are no longer needed (and reclaim the associated memory). One possible strategy is to start with some standard initial size of the stack (e.g., 10), and then, whenever the stack fills, declare a new stack array of twice the size, push all the elements into the new stack, and destroy the old stack.

A. Explain why n stack operations (PUSH, POP, or TOP) are $O(n)$ time.

Hint: Make use of the fact that geometric sums are the order of their last term; that is, $1 + 2 + 4 + \cdots + n = 2n - 1$.

B. Suppose we generalized this technique to move to a stack of half the size whenever the size of the stack became half its current maximum size. Generalize your answer to Part A to show that the time and space remain $O(n)$.

C. Explain how to modify your answers to Parts A and B for queues.

59. In the sample exercises we saw how to generalize the algorithm to evaluate a postfix expression using a stack to allow unary operators, as long as the same symbol was not used for both a unary and binary operator. Suppose that we have only two operators. minus and plus, where we allow the minus symbol to be both unary and binary.

A. Explain why the postfix expression

$$a\, b - c\, d\ ef + + + + +$$

can be evaluated in only one way:

$$a + ((-b) + (c + (d + (e + f))))$$

B. Explain why the postfix expression

   $a\ b - -$

   can be evaluated in two different ways that yield different answers:

   $-(a-b) = -a+b$ or $a-(-b) = a+b$

C. Explain why the postfix expression

   $a\ b - c\ d - + +$

   can be evaluated in two different ways that yield the same answer:

   $(a-b) + (c+(-d)) = a - b + c - d$

   or

   $a+((-b)+(c-d)) = a-b+c-d$

D. Give an algorithm that determines if such a postfix expression has exactly one legal evaluation.

E. Give an algorithm that produces a legal evaluation if one or more legal evaluations is possible (it may arbitrarily choose which of the legal evaluations to produce).

60. Consider the function LSEARCH(d,L) that returns the last vertex in L that contains data d (or *nil* if d is not in L); that is, if there is more than one copy of d in L, then it returns the last one encountered when traversing the list from the first to the last item.

A. Give pseudo-code for LSEARCH that works from the first to the last element.

B. Give pseudo-code for LSEARCH that works from the last to the first element.

61. The pseudo-code presented for INSERT(d,v,L) and DELETE(v,L) did not check for illegal input (e.g., L is a *nil* pointer). Discuss the ways in which the input could be illegal, describe how to augment the pseudo-code to check for them, and discuss the complexity of your error checking (in particular, what things can be checked in $O(1)$ time and what errors take more time).

62. With singly linked lists, we do not have a previous field with each vertex from which the PREV operation can be implemented in $O(1)$ time.

A. Give pseudo-code for the PREV(v,L) operation for singly linked lists that traverses the list from the front to find the vertex whose next field points to v.

B. With singly linked lists, the lack of a previous field with each vertex makes the DELETE operation expensive (we have to use Part A to find the vertex to the left before the list can be patched up) or awkward (loops that traverse lists must keep track of both the current position and the previous position, etc.). Suppose that we redefine DELETE(v,L) for singly linked lists to delete the vertex that follows v. Rewrite the low-level pseudo-code that was presented in the example showing the details of the DELETE(v,L) operation.

63. Present pseudo-code, in the same style as that presented for the INSERT operation, for the SPLICE operation.

64. Present pseudo-code, in the same style as that presented for the DELETE operation, for the CUT operation.

65. A doubly linked list can be reversed in constant time if the meaning of "previous" and "next" is reversed. Discuss how this idea might be implemented in practice.

66. Suppose that list headers were not used and a list is simply defined as a pointer to its first element. Discuss how this affects the definitions of the basic list functions and how one implements operations that make use of information in the header.

67. Give pseudo-code for bottom-up merge sorting that was described in the sample exercises that works at the level list notation of *p.next* to denote the next field of the vertex pointed to by *p*.

68. Describe how the implementation of the basic list operations must be changed (and which operations must be eliminated or modified) to allow lists to share tails, as depicted for three lists in the following figure:

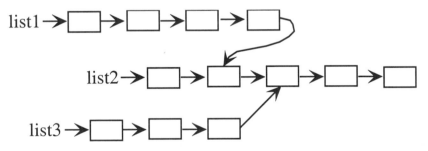

69. We have seen how pointers for doubly linked lists can be implemented "manually" with an array of dimension *n* by 3, where *n* is the maximum number of list vertices present in all lists and the three columns of the array are used to store the *data*, *prev*, and *next* fields of the vertex corresponding to that row; a special list *FREE* was used to hold all unused vertices. Give detailed pseudo-code (that explicitly handles the array) for each of the basic linked list operations.

Chapter Notes

Stacks, queues, and linked lists are presented in virtually all texts on data structures and algorithms; the reader may refer to the list of texts given at the beginning of the chapter notes to the first chapter.

We did not fully address the practical issue of memory allocation and reclamation for headers and vertices. In an explicit representation with arrays, we used a special list FREE to link together all unused locations. When pointer variables are used in practice in a language like C, one typically issues an instruction to allocate new memory when a new header or vertex is created and another instruction to reclaim the memory whenever a header or vertex is deleted. With some languages, instead of the user explicitly reclaiming memory, automatic "garbage collection" is used to recover memory that is no longer referenced by active pointers.

The POP(i) stack operation was essentially just an excuse to introduce the notion of amortized time analysis. See the book of Tarjan [1983] for more complicated examples.

3. Induction and Recursion

Introduction

Consider again the function $n!$. We have already seen how to compute it with a loop that multiplies the numbers 1 through n. Another way to compute $n!$ is with a *recursive* program that calls itself:

> **function** FACT(n)
>   **if** $n \leq 1$
>   **then return** 1
>   **else return** $n * FACT(n-1)$
>   **end**

Because $n! = n*(n-1)!$, it can be computed by solving a smaller version of the same problem and then multiplying by n. In this very simple example, the recursive program corresponds directly to a loop. If the recursion is "unwound", to compute $n!$, we compute $(n-1)!$, to compute $(n-1)!$ we compute $(n-2)!$, and so on until we get down to 1; then we work backwards computing $2*1 = 2$, then $3*2 = 6$, then $4*6 = 24$, and so on. This form of recursion, often called *tail* recursion, calls itself on the "tail" of the problem and then does some simple computation (in this case, a multiplication) to incorporate the "head" of the problem. In fact, it is such a simple form of recursion that a "smart" compiler can detect it and translate it to a loop that uses $O(1)$ space.

One may wonder why we want to express a computation this way. With FACT, it is arguable that we may not, because it was already a very simple and understandable program. However, in general, recursion will allow us to express large problems in terms of smaller problems of the same type, and can be a tool for discovering better algorithms as well as a powerful way to express complex algorithms.

At first glance, one might worry that a recursive program cannot work because recursive calls "trample" over the current values of the local variables. For example, when the recursive call FACT($n-1$) is made, we might worry that the memory location that stores the argument n (which is an argument to FACT and hence local to FACT) is now modified to have the value $n-1$, and after FACT($n-1$) returns, we will not be multiplying the result returned by n. The key assumption we make to insure that a recursive is well defined is that:

> *The values of all local variables (including the values of the arguments passed) at the point that a function call returns (including a recursive call), are identical to what they were just before the call.*

Of course, if we choose to, global variables or data that is pointed to by local variables could be modified (but the value of the pointer itself, which is a local variable, is not modified). The point is, that calling a function recursively is no different than calling any other function; we expect that the function call will not corrupt any local variables that have been defined inside the calling function (and are not supposed to be available to code that is outside that function).

Typically, the compiler maintains the integrity of local variables by making new copies of all arguments and local variables when a call is made. The copies are stored in a stack and popped off when the function returns and they need to be restored to the values they had just before the call. In some sense, one can think of the call FACT($n-1$) as invoking a completely different piece of code; it has the same flow of control, but uses a new set of memory locations to store variables. We will examine the details of how the compiler does this; not because we are concerned with eliminating recursion or do not "like" it, but rather to be sure that we understand the process and that we know how to analyze the time and space used. In practice, although there is some "hidden" space that is used by the compiler to implement recursion, except for programs like FACT that have exceptionally simple non-recursive implementations, this hidden space will be dominated by the space used by data structures employed by the algorithm in question.

To compute the running time $T(n)$ of a recursive program on an input of size n, it is typical to express $T(n)$ as a *recurrence relation*; that is $T(n)$ is expressed as a function of terms like $T(n-k)$ (the time for a problem that is k smaller) and $T(n/k)$ (the time for a problem that is a factor of k smaller). For example, for the recursive version of factorial, we shall see that the number of multiplications performed, $T(n)$, is 0 if $n \leq 1$, and in general can be expressed as $T(n)=T(n-1)+1$. Many standard equations such as this and their solutions can be found in mathematics and computer science reference books. We will derive a few (including this one) as we need them.

If the solution to a recurrence relation is not known, one can first experiment by expanding terms to "guess" at the solution and then prove the guess correct by *induction*. Induction is the mathematical equivalent of recursive programming and an important tool to analyze the time and space of recursive programs (and for proving correctness of recursive programs). If we know the assertion to be true for a particular integer (the "base" value), the idea is to prove the assertion true for all integers larger than the base value by showing that we can always work "backwards" down to the base value. That is, we prove the assertion true for some value k, and then show that for any integer $i>k$, if the assertion is true for all integers in the range k through $i-1$, then it must be true for i.

Example: Binary Search

Binary search was introduced in the first chapter. Recall that the problem is, given a portion of an array $A[a]$ ... $A[b]$ that is already arranged in increasing order, and a value x, find an i, $a \leq i \leq b$, such that $x=A[i]$, or determine that no such i exists. The iterative algorithm presented in the first chapter cut the problem in half by computing the midpoint m between a and b, checking whether $x \leq A[m]$, and repeating this process on the appropriate half of the array.

Iterative binary search algorithm:

    while $a<b$ **do begin**
        $m := \lfloor (a+b)/2 \rfloor$
        if $x \leq A[m]$ **then** $b := m$ **else** $a := m+1$
    end
    if $x=A[a]$ **then** {x is at position a} **else** {x is not in A}

We saw that this iterative algorithm is logarithmic time and constant space in addition to that used for $A[a]...A[b]$. It is so simple that it makes sense in practice to implement it with a loop. However, it is in some sense naturally a recursive algorithm that works by applying the same method to a problem of half the size. Since we are assuming that $0 \leq a \leq b$, we use the convention that returning a value of -1 means that x is not in A.

Recursive binary search algorithm:

    function RBS(a,b)

        if $a<b$ **then begin**
            $m := \lfloor (a+b)/2 \rfloor$
            if $x \leq A[m]$ **then return** RBS(a,m) **else return** RBS($m+1,b$)
        end

        else if $x=A[a]$ **then return** a

        else return -1

    end

Time: Let n denote the number of elements (i.e., the initial value of $b-a+1$); and as with the analysis in the first chapter, we can assume that n is a power of 2. Each time RBS is entered, it spends $O(1)$ time in addition to the recursive call. Since each call divides $a-b+1$ in half, RBS calls itself $\lceil \log_2(n) \rceil$ times, and runs in $O(\log(n))$ time.

Space: Like the factorial example presented in the introduction to this chapter, this is an example of tail recursion. As we shall see later in this chapter, the space may be $O(1)$, or may be as much as $O(\log(n))$, depending on how well the compiler implements "hidden" bookkeeping to implement the recursion.

Example: Traversing Linked Lists

Idea: In our examples for linked lists, lists were typically traversed with a loop that started at the first vertex and followed then *next* pointers. An equivalent approach is to visit the first vertex and then use a recursive call to visit the remainder of the list (i.e., use tail recursion). Below are a few of the examples presented for linked lists rewritten as recursive programs. We use the notation of the basic list operations.

Search a list: Because pointer-based implementations of lists were defined to have a list header, we first define a function that searches from a specified vertex forward and then call it on the first vertex of the list.

> **function** SEARCHTAIL(d,v,L)
>> **if** $v=nil$ **then return** *nil*
>> **else if** d=DATA(v) **then return** v
>> **else return** SEARCHTAIL(d,NEXT(v),L)
>> **end**

> **function** SEARCH(d,L)
>> SEARCHTAIL(d,FIRST(L),L)
>> **end**

Sum the data in vertices v through w:

> **function** SUM(v,w,L)
>> **if** $v=w$
>>> **then return** DATA(w)
>>> **else return** DATA(v)+SUM(NEXT(v),w,L)
>> **end**

Reverse a singly linked list:

> **procedure** REVERSE(L)
>> **if** (SIZE(L)>1) **then begin**
>>> d := DELETE(FIRST(L),L)
>>> REVERSE(L)
>>> INSERT(d,LAST(L),L)
>>> **end**
>> **end**

Example: Fast Computation of Integer Exponents

Idea: To compute x^i, instead of simply multiplying x times itself i times, we reduce the problem to computing $x^{i/2}$ and then doing one or two additional multiplications.

Recursive Program: x^i can easily be expressed in terms of $x^{i/2}$:

$$x^i = \begin{cases} \left(x^{i/2}\right)^2 & \text{if } i \text{ is even} \\ x\left(x^{(i-1)/2}\right)^2 & \text{if } i \text{ is odd} \end{cases}$$

The equivalent recursive program is:

function POWER(x, i):
    if $i=0$ **then return** 1
    else if i is even **then return** POWER$(x, i/2)^2$
    else return $x*$POWER$(x, (i-1)/2)^2$
    end

Equivalent iterative program: Consider the binary representation of i. Each time a bit is appended to the right of i, the value of i is doubled if this bit is 0 or doubled and incremented if this bit is 1 (because appending a bit to the right shifts all the other bits one position to the left). Hence x^i is squared if this bit is 0 or squared and multiplied by x if it is 1. If wc "unwind" the recursion of the procedure above, starting with a value of 1, we proceed from left to right in the binary representation of i, and at each stage, we square the value and then multiply it by x if the current bit is 1:

function iterativePOWER(x, i):
    Let $s[1]...s[m]$ be the bits of i written in binary.
    $v := 1$
    for $k := 1$ **to** m **do begin**
        $v := v^2$
        if $s[k]=1$ **then** $v := x*v$
        end
    return v
    end

Correctness: Proof by induction can be used (see the sample exercises).

Time: Both the recursive and iterative versions perform at most 2 multiplications per bit in the binary representation of i. Hence the time is $O(\log(i))$. This analysis makes the standard unit-cost assumption that a multiplication can be done in $O(1)$ time, which may not be realistic as i gets large. In any case, $O(\log(i))$ multiplications is better than the $O(i)$ multiplications that the straightforward approach uses.

(*) Example: Converting a String to a Number

Problem: Numbers are often provided as character strings (e.g., when someone types them) and have to be converted to a value that is stored in memory.

Stack representation: To simplify our presentation and to be compatible with the next example, we assume that a real number is provided to us on a stack, one digit per frame where the leftmost digit is on top of the stack. We use POP as both a function that removes and returns the top item and as a procedure that simply deletes the top item.

Getting a real number: *GetInteger* gets the integer to the left of the decimal point and *GetFraction* gets the fraction to the right of the decimal point. Here we chose to express GetInteger iteratively and GetFraction recursively, but both can easily be written either way; see the sample exercises. *GetReal* gets a real number by calling *GetInteger* and, if there is a decimal point, skipping it and calling *GetFraction*. An empty expression or one consisting of just a decimal point is taken to be zero. An error is not reported if there is other data stored deeper in the stack, although *GetReal* could check this.

> **function** GetInteger
> $i := 0$
> **while** the stack is not empty and TOP is a digit **do** $i := i*10 +$ POP
> **return** i
> **end**

> **function** GetFraction
> **if** the stack is empty or TOP is not a digit **then return** 0
> **else return** (POP+GetFraction)/10
> **end**

> **function** Getreal
> **if** the stack is empty or TOP is not a digit or a decimal point **then** *ERROR*
> **else begin**
> $\quad$ $i :=$ GetInteger
> $\quad$ **if** the stack is empty or TOP is not a decimal point **then** $f := 0$
> $\quad$ **else begin**
> $\quad\quad$ POP
> $\quad\quad$ $f :=$ GetFraction
> $\quad\quad$ **end**
> $\quad$ **return** $i+f$
> $\quad$ **end**
> **end**

Note: As usual, we are assuming that conditions and expressions are evaluated from left to right so that, in particular, the return statement of GetFraction does POP *before* the recursive call (see the sample exercises).

(*) Example: Evaluating a Prefix Expression

Infix vs. prefix notation for arithmetic expressions: Recall that standard *infix* notation has the form *operand operator operand* (e.g., $a+b$). With *prefix* notation, it is *operator operand operand* (e.g., $+ab$). Like postfix, an advantage of prefix is that parentheses are not needed and evaluation is always unique; for example $\bullet+a\bullet bcd$ means $(a+(b\bullet c))\bullet d$.

Idea: In the previous chapter, we saw how to evaluate a postfix expression from left to right with an iterative function *postEVAL* that used an auxiliary stack. Here, we evaluate a prefix expression from left to right by employing a recursive function. Although no auxiliary stack is used, as we shall see later, the compiler uses one to implement the recursion.

Recursive approach: Given a stack containing a prefix expression (each item of the stack is a *value*, an *operator*, or a *parenthesis*), *preEVAL* pops a binary operator and then recursively pops its two arguments.

> **function** preEVAL
> > **if** the stack is empty **then** *ERROR — missing value*
> > **else if** TOP is a value **then return** POP
> > **else if** TOP is a binary operator **then begin**
> > > $op :=$ POP
> > > $x :=$ preEVAL
> > > $y :=$ preEVAL
> > > **return** the result of applying op to x and y
> > > **end**
> > **else** *ERROR — illegal input*
> > **end**

> *Note*: The longest legal initial portion is evaluated. The calling code can check if the stack is not empty, and if so report "*ERROR — missing operator*".

Values that span more than one stack frame: *preEVAL* assumes that values are contained in a single stack frame. The first *else* statement can be modified to get values that span several stack frames. For example, if values are real numbers that are stored one character per frame, we can replace the first *else* statement with one that calls the GetReal procedure of the previous example:

> **else if** TOP is a digit or a decimal point **then return** GetReal

Unary operators: This algorithm can be generalized to handle operators that take only one operand (e.g., leading minus, a percent, or an exponent); see the sample exercises.

Correctness: A proof of correctness for this algorithm can also be used to argue that the evaluation of a prefix expression is always unique; see the sample exercises.

(*) Example: Converting Infix to Prefix or Postfix

Operators: We limit our attention to the binary operations $\bullet$, /, +, − and the usual conventions that $\bullet$ and / take precedence over + and −, and when an operator is missing it is assumed to be $\bullet$; for example, $a+bc$ means $a+(b\bullet c)$.

Recursive postfix conversion algorithm: An *expression* is a series of *terms* separated by + or −, a term is a series of *values* separated by $\bullet$ or /, and a value is a *basic value* (in a single stack frame) or an expression in parentheses. We assume that the infix expression is provided in an input stack S, where each frame has a *value*, a binary *operator*, or a *parenthesis* (like the previous example, this can be generalized to values that span more than one frame). We put an equivalent postfix expression in an output stack P.

    procedure PushExpression
        PushTerm
        while S is not empty and TOP(S) is not) **do begin**
            if TOP(S) is + or − **then** $op := $ POP(S) **else** *ERROR — illegal input*
            PushTerm
            PUSH(op,P)
            end
        end
    procedure PushTerm
        PushValue
        while S is not empty and TOP(S) is /, $\bullet$, or a basic value **do begin**
            if TOP(S) is / or $\bullet$ **then** $op := $ POP(S) **else** $op := \bullet$
            PushValue
            PUSH(op,P)
            end
        end
    procedure PushValue
        if TOP(S) is a basic value **then** PUSH(POP(S),P)
        else if POP(S) is (**then begin**
            PushExpression
            if POP(S) is not) **then** *ERROR — missing right parenthesis*
            end
        else *ERROR — missing value*
        end
    PushExpression
    if S is not empty **then** *ERROR — missing left parenthesis*

Correctness: See the exercises.

Conversion to prefix: A similar approach can be used (see the sample exercises).

Unary operators: Like prefix and postfix evaluation, this algorithm can be generalized to handle operators that take only one operand; see the sample exercises.

Proof by Induction

Idea: To prove a theorem true for all integers $\geq k$:

basis:

Prove it true for k.

inductive step:

Prove that for any $n > k$, true for all integers $k \leq i < n$ implies true for n.

Example: Summing Powers of 2

Idea: To compute $2^0 + 2^1 + 2^2 + \cdots + 2^n = 1 + 2 + 4 + \cdots + 2^n$, repeatedly double a variable p and add it on to a running total that is kept in the variable x:

input n
$x := p := 1$
for $i := 1$ **to** n **do begin**
    $p := 2*p$
    $x := x + p$
    end
output x

Induction can be used to show that for all inputs $n \geq 0$, the program computes $x = 2^{n+1} - 1$.

(*basis*)

For input $n=0$: $x = 1 = 2^{0+1} - 1$

(*inductive step*)

For input $n > 0$, assume the theorem true for all inputs in the range 0 through $n-1$ (so in particular, on input n-1 it computes $2^{(n-1)+1} - 1$). Since on input n the program executes the same instructions as for input $n-1$ and then executes one additional iteration of the loop (where it adds 2^n to the current value of x), it computes:

$$x = (2^{(n-1)+1} - 1) + 2^n = 2*2^n - 1 = 2^{n+1} - 1$$

Note: We have done this for the sake of example; proof by induction is more complex than necessary. This is a standard geometric sum; that is, if we let

$$S = 1 + 2 + 4 + \cdots + 2^n$$
$$2S = \quad\quad 2 + 4 + \cdots + 2^n + 2^{n+1}$$

then the terms in $2S$ are the same as for S except for the first term of S and the last term of $2S$, and hence $S = (2S - S) = 2^{n+1} - 1$ (see the appendix).

Example: Summing Odd Integers

Iterative version: The first n odd integers are of the form $2i-1$, $1 \leq i \leq n$; to sum them we use a simple loop that adds these values to a running total stored in the variable x:

input n
$x := 0$
for $i=1$ **to** n **do** $x := x+(2i-1)$
output x

Induction can be used to show that for all inputs $n \geq 0$, the program computes $x=n^2$.

(*basis*)

For input $n=0$, the *for* loop is not executed and $x = 0 = 0^2$ is computed.

(*inductive step*)

For inputs $n>1$, assume the theorem true for all inputs in the range 0 through $n-1$, so in particular, on input $(n-1)$ it computes $(n-1)^2$. Since on input n the program executes the same instructions as for input $n-1$ and then executes one additional iteration of the loop (where it adds $2n-1$ to the current value of x), it computes:

$$x = (n-1)^2 + (2n-1) = (n^2-2n+1) + (2n-1) = n^2$$

Recursive version: The loop of the iterative version can be expressed with simple tail recursion:

function SUMODD(n)
if $n=0$ **then return** 0 **else return** $SUMODD(n-1)+(2n-1)$
end

Induction can again be used to show that for all inputs $n \geq 0$, the program computes $x=n^2$.

(*basis*)

For input $n=0$, SUMODD returns 0.

(*inductive step*)

For inputs $n>1$, assume the theorem is true for all inputs in the range 0 through $n-1$, so in particular, on input $(n-1)$ it computes $(n-1)^2$. Since the sum of the first n odd numbers is the same as the sum of the first $n-1$ plus the addition term $(2n-1)$, the program correctly returns:

$$(n-1)^2 + (2n-1) = (n^2-2n+1) + (2n-1) = n^2$$

Note: As with summing powers of 2, proof by induction is more complex than necessary. This is just a standard arithmetic sum; that is, it is just the number of terms n, times the average value of each term $((2n-1)+1)/2 = n$, for a total of n^2 (see the appendix).

Example: Correctness of Binary Search

Consider again the recursive version of binary search presented earlier in this chapter:

function RBS(*a*,*b*)
    if *a*<*b* **then begin**
        m := $\lfloor (a+b)/2 \rfloor$
        if $x \leq A[m]$ **then return** RBS(*a*,*m*) **else return** RBS(*m*+1,*b*)
        end
    else if *x*=*A*[*a*] **then return** *a*
    else return −1
    end

Proof by induction of the correctness of the function *RBS*:

We show that *RBS* correctly finds the index of *x* in *A*[*a*] ... *A*[*b*] by using induction on the quantity (*b*−*a*).

    (*basis*)

        If (*b*−*a*)=0, and hence *a*=*b*, then the *if* statement is not executed and the *else if* and *else* statements correctly return *a* if *x*=*A*[*a*]=*A*[*b*] or −1 otherwise.

    (*inductive step*)

        For some *k*>0, assume that the algorithm works correctly if (*b*−*a*)<*k*, and consider the case (*b*−*a*)=*k*.

        Since *k*>0, it must be that *a*<*b* and hence the body of the first *if* statement is executed, which begins with computing the midpoint:

        m := $\lfloor (a+b)/2 \rfloor$

        Since *a*<*b* and the computation of *m* rounds (*a*+*b*)/2 downward, it must be that *a* ≤ *m* < *b*, and hence both halves defined by *m* are smaller than the original; that is, (*m*−*a*) < (*b*−*a*) and (*b*−(*m*+1)) < (*b*−*a*).

        Since *A* is sorted in increasing order, if *x* ≤ *A*[*m*], it cannot be that *x*=*A*[*i*] for some *m* < *i* ≤ *b*, unless *x*=*A*[*m*] (i.e., it could be that *A*[*m*] = *A*[*m*+1] = ... = *A*[*i*]), and hence it is ok to restrict the search to *A*[*a*] ... *A*[*m*]. Similarly, if *x*>*A*[*m*] it cannot be that *x*=*A*[*i*] for some *a* ≤ *i* ≤ *m*, and hence it is ok to restrict the search to *A*[*m*+1] ... *A*[*b*].

        Thus, the recursive calls are made correctly to problems that are at least one element smaller (that is when the recursive calls are made, the quantity (*b*−*a*) will be less than *k*). Hence, correctness follow by induction.

Note: The proof of correctness that was used in the first chapter for the iterative version is essentially a slightly less formal version of the proof presented here.

Example: Towers of Hanoi Puzzle

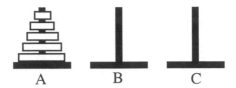

A   B   C

Problem: You are given three posts labeled A, B, and C.

On Post A there are n rings of different sizes, in the order of the largest ring on the bottom to the smallest one on top.

Posts B and C are empty.

The object is to move the n rings from Post A to Post B by successively moving a ring from one post to another post that is empty or has a larger diameter ring on top.

Inductive proof that the puzzle has a solution:

(*basis*) One ring can always be moved.

(*inductive step*) Assume that we can move $n-1$ rings from any given post to any other post. Since any of the rings 1 through $n-1$ can be placed on top of ring n, all n rings can be moved as follows:

    1. Move $n-1$ rings from Post A to Post C.

    2. Move ring n from Post A to Post B.

    3. Move $n-1$ rings from Post C to Post B.

Example: The solution for 3 disks takes 7 steps:

```
 1                                                    1
 2        2                    1         1       2    2
 3 _ _    3 1     3 1 2    3 _ 2     _ 3 2   1 3 2  1 3 _    _ 3
A B C    A B C   A B C    A B C    A B C   A B C  A B C    A B C
```

Recursive algorithm: The proof by induction is constructive because it tells you how to solve a problem with n rings in terms of solutions to problems with $n-1$ rings; it corresponds directly to a recursive program. The procedure TOWER(n,x,y,z) moves n rings from Post x to Post y (using Post z as a "scratch" post):

```
procedure TOWER(n,x,y,z)
    if n>0 then begin
        TOWER(n−1,x,z,y)
        write "Move ring n from x to y."
        TOWER(n−1,z,y,x)
        end
    end
```

Elimination of Recursion

To understand asymptotic time and space complexity of recursive programs, it is useful to understand how a compiler can translate a recursive program to machine language that corresponds to the RAM model; that is, how recursion is "removed" from a program.

Idea:

> All local variables and formal parameters for the procedure are stored on top of a stack. When a recursive call is made, new copies are pushed.

How procedures are removed:

> Step 1: Place a label at the start of each procedure.
>
> Step 2: Place a label directly after each procedure call.
>
> Step 3: Replace each call $P(x_1...x_n)$ that is followed by a label L by code to push the current state and *goto* the label for P:
>
> > PUSH L
> >
> > PUSH the parameters $x_1...x_n$
> >
> > PUSH space for local variables declared in P
> >
> > **goto** to the label associated with P
>
> Step 4: End a procedure (and replace each return statement) by code that discards from the stack the current state and goes to that return label.

Functions: A number of conventions can be used for functions that return a value; for example, the return value can be placed in a special variable associated with that function.

Time: The time required to implement the calls and returns of a given procedure is proportional to the number of formal parameters and local variables, which is typically a constant (but may not be in all cases for languages that allow a variable number).

Space: Each call must allocate new space on the stack for the parameters and local variables. The total space used will be the product of this and the maximum depth of the recursion. This space is "hidden" space that is being used by code added by the compiler, and is in addition to the space that is explicitly used by the program's data structures.

Example: Eliminating Recursion from n!

Recall the recursive version of factorial:

> **function** FACT(n)
>> **if** $n \leq 1$
>>> **then return** 1
>>> **else return** $n*$FACT($n-1$)
>> **end**

Idea: FACT has one parameter and no local variables, so the return address and current value of n are the only items that have to be pushed onto the stack when FACT is called. That is, $stack[top]$ is (as usual) the next available empty stack frame, $stack[top-1]$ contains n, and $stack[top-2]$ contains the return address.

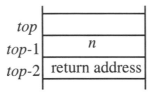

At compile time, we assume that space is created for a special variable *FactRetrunValue* that is used to store the most recent value returned by FACT. Recall that we use POP as both a procedure that simply removes the element on top of the stack and as a function that removes the top element and returns it.

The function FACT is converted to:

> *L1*: **if** $stack[top-1] \leq 1$ **then begin**
>> *FactReturnValue* := 1
>> POP   — discard n
>> **goto** POP   — remove a label and go to it
>> **end**
> **else begin**
>> PUSH(*L2*)   — push the return point (so n is now in $stack[top-2]$)
>> PUSH($stack[top-2]-1$)   — push $n-1$
>> **goto** *L1*
>> *L2*: *FactReturnValue* := *FactReturnValue* $*$ $stack[top-1]$
>> POP   — discard n
>> **goto** POP   — remove a label and goto it
>> **end**

The call x := FACT(n) is converted to:

> PUSH(*L3*); PUSH(n)
> **goto** *L1*
> *L3*: x := *FactReturnValue*

Example: Complexity of Recursive n!

Express the time with a recurrence relation: For $n=1$, *FACT* takes an amount of time that is bounded by some constant a. For $n>1$, FACT takes whatever time is consumed by the call to $FACT(n-1)$, plus an additional amount of time to perform the multiplication by n that is bounded by some constant b. Hence, if $T(n)$ denotes the running time of $FACT(n)$, and we let c be the larger of a and b, then:

$$T(n) \leq \begin{cases} c \text{ if } n \leq 1 \\ T(n-1)+c \text{ otherwise} \end{cases}$$

Experiment with expanding terms: Informally, we can see that for all $n \geq 1$:

$$\begin{aligned} T(n) &\leq T(n-1)+c \\ &\leq (T(n-2)+c)+c \\ &\leq ((T(n-3)+c)+c)+c \\ &= c+c+\cdots+c \ (n \text{ times}) \\ &= cn \\ &= O(n) \end{aligned}$$

Theorem: For all $n \geq 1$, $T(n) \leq cn$.

Proof:

(*basis*)

    For $n=1$, $T(n) \leq c = cn$.

(*inductive step*)

    For $n>1$, assume that $T(i) \leq ci$ for $1 \leq i < n$.

    Then $T(n) \leq T(n-1)+c \leq c(n-1)+c = cn$.

Corollary: $T(n)$ is $O(n)$.

Space: A "smart" compiler would see that this is just tail recursion and translate the program to a simple loop that uses $O(1)$ space. However, if we simply apply mechanical recursion elimination, we end up with a stack that grows to n frames, each using $O(1)$ space, for a total of $O(n)$ space.

Example: Eliminating Recursion from Towers of Hanoi

Recall the recursive version of Towers of Hanoi:
>   **procedure** *TOWER*(*n*,*x*,*y*,*z*)
>   **if** *n*>0 **then begin**
>   TOWER(*n*−1,*x*,*z*,*y*)
>   **write** "Move ring *n* from *x* to *y*."
>   TOWER(*n*−1,*z*,*y*,*x*)
>   **end**

Idea: Tower has four parameters and no local variables, so the return address will always be in *stack*[*top*−5] and the current values of *n*, *x*, *y*, and *z* will always be in locations *stack*[*top*−1] down to *stack*[*top*−4]. However, they move down when new values are being pushed on during the process of simulating a recursive call.

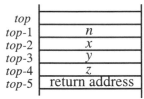

The procedure *TOWER* is converted to:
>   *L1*: **if** *stack*[*top*−1]>0 **then begin**
>   PUSH(*L2*) — push the return point
>   PUSH(*stack*[*top*−4]) — push *y*, which is now one position lower
>   PUSH(*stack*[*top*−6]) — push *z*, which is now two positions lower
>   PUSH(*stack*[*top*−5]) — push *x*, which is now three positions lower
>   PUSH(*stack*[*top*−5]−1) — push *n*−1, where *n* is now four positions lower
>   **goto** *L1*
>   *L2*: **write** "Move ring *stack*[*top*−1] from *x* to *y*."
>   PUSH(*L3*) — push the return point
>   PUSH(*stack*[*top*−3]) — push *x*, which is now one position lower
>   PUSH(*stack*[*top*−5]) — push *y*, which is now two positions lower
>   PUSH(*stack*[*top*−7]) — push *z*, which is now three positions lower
>   PUSH(*stack*[*top*−5]−1) — push *n*−1, where *n* is now four positions lower
>   **goto** *L1*
>   **end**
>   *L3*: POP; POP; POP; POP   — discard *n*, *x*, *y*, and *z*
>   **goto** POP   — remove and goto the label on top of the stack

The call *TOWER*(*n*,*A*,*B*,*C*) is converted to:
>   PUSH(*L4*); PUSH(*C*); PUSH(*B*); PUSH(*A*); PUSH(*n*)
>   **goto** *L1*
>   *L4*:

Example: Complexity of TOWER

Recurrence relation for the number of moves: For $n=1$, the two calls for $n-1$ do nothing and exactly one move is made. For $n>1$, twice whatever the number of moves required for $n-1$ are made plus the move made by the *write* statement. Hence, the number of moves made by TOWER on input n is:

$$T(n) = \begin{cases} 1 \text{ if } n = 1 \\ 2T(n-1)+1 \text{ otherwise} \end{cases}$$

Theorem: For $n \geq 1$, $TOWER(n,x,y,z)$ makes $2^n - 1$ moves.

Proof:

For $n=1$:

$$T(1) = 1 = 2^1 - 1$$

Now assume that TOWER works correctly for all values in the range 0 to $n-1$. Then:

$$\begin{aligned} T(n) &= 2T(n-1)+1 \\ &= 2(2^{n-1}-1)+1 \\ &= 2^n - 1 \end{aligned}$$

Time: The time for TOWER(n,x,y,z) is $\Theta(2^n)$ because it is proportional to the number of moves made. That is, every call to TOWER with $n \geq 1$ executes the write statement exactly once, and there are exactly two calls with $n=0$ for every call for $n=1$.

Space: The configuration of the rings is not explicitly stored (it is implicit in the reordering of x,y,z in the recursive calls). The total space used is dominated by the space for the recursion stack. Let c denote the space used by a copy of the local variables. Since the space used by the first call is released when it returns, the recurrence relation for the space used by TOWER(n,x,y,z) is:

$$S(n) = \begin{cases} c \text{ if } n = 1 \\ S(n-1)+c \text{ otherwise} \end{cases}$$

As we have already seen in the factorial example, the solution to this recurrence relation is $O(n)$.

Example: Non-Recursive Towers of Hanoi Algorithm

Idea: If we unwind the recursion, it is not hard to see that the recursive Towers of Hanoi algorithm alternates between moving the smallest ring and one of the other rings, and that the smallest rings moves in a regular clockwise or counterclockwise fashion.

Lemma 1: In any minimal length solution to the Towers of Hanoi puzzle, the first and every other move is with the smallest ring.

Proof: Two consecutive moves with the smallest ring could be combined into a single move. Two consecutive moves of a larger ring must be from a peg back to itself (since the third peg must have the smallest peg on top), and hence could be eliminated.

Lemma 2: In any minimal length solution to the Towers of Hanoi puzzle, for odd n, the small ring always moves in a clockwise direction (A to B to C to A ...) and for even n in a counterclockwise direction (A to C to B to A ...).

Proof: The move in the TOWER procedure goes from Peg x to Peg y. For $n=1$, $x,y=A,B$ and the move is clockwise. If we inductively assume the lemma true for $n-1$ rings, then it follows for n by observing that both recursive calls reorder the pegs so that the relationship between the source and destination pegs (clockwise or counterclockwise) reverses.

Algorithm to move rings clockwise one post:

> **if** n is odd **then** d := *clockwise* **else** d := *counterclockwise*
>
> **repeat**
>> Move the smallest ring one post in direction d.
>>
>> Make the only legal move that does not involve the smallest ring.
>>
>> **until** all rings are on the same post

Time: $\Theta(2^n)$ — Exactly the same sequence of moves as the recursive version.

Space: $O(n)$ — Unlike the recursive version, the configuration of the rings must be explicitly represented; it suffices to use 3 stacks, each holding a maximum of n rings.

Observation: Both the mechanical translation of the recursive program to a non-recursive one and the "hand coded" version employ the stack data structure and use $O(n)$ space, although the constants in practice for both time and space are likely to be smaller for the hand-coded version.

Sample Exercises

70. Consider the recursive and iterative versions of *POWER(x,i)*, to compute x^i, that were presented.

A. Explain why both versions use $O(\log(i))$ time, under standard unit-cost assumption can integers can be multiplied in $O(1)$ time.

B. Explain why both versions use $O(\log(i))$ space, under standard unit-cost assumption that integers consume $O(1)$ space.

C. Use induction to prove the correctness of the recursive version.

D. Use induction to prove the correctness of the iterative version.

E. The iterative version could be implemented directly if, for example, the programming language allowed you to access individual bits or if *i* were read from the input stream as a character string; it is also quite easy to write a subroutine which constructs a bit string from an integer in time proportional to the length of the string (which is proportional to log of the integer). Alternately, show how the procedure can be rewritten to use arithmetic operations by first computing a value *p* that is the largest power of 2 that is less than or equal to *i* and then successively dividing *p* by 2 to sequence through the bits of *j*. Analyze the time and space used and compare it to the other two versions of POWER.

Solution:

A. The recursive version makes calls by at least halving the value of *i*. Hence, if $T(i)$ denotes the time for input *i*, then for some constant *c* (that reflects all of the time but the recursive call), $T(0) \le c$ and in general:

$$T(i) \le T(i/2) + c$$
$$= T(i/4) + c + c$$
$$= c + c + \dots + c \quad (O(\log(i)) \; terms)$$
$$= O(\log(i))$$

The iterative version is a simply loop that is executed once for each of the $O(\log(i))$ bits in the binary representation of *i*.

B. In the recursive version, since the recursion can go as deep as the number of times *i* can be divided by 2 to get a value less than 1, the space for the recursion stack is $O(\log(i))$, and dominates the space used by the algorithm. For the iterative version, the space is dominated by $O(\log(i))$ space for the array *s*.

C. We use induction on *i*. A nice aspect of the recursive version is that the proof and the program are almost the same. For *i*=0, POWER correctly returns 1. For *i*>0, if we assume that *POWER* works correctly for values less than *i*, then the two *else* conditions precisely express the answer in terms of smaller values of *i*. This is because adding another bit to the right of a binary number doubles it (since all the other bits are shifted one position to the left) and then adds one if the bit is 1, and doubling i has the effect of squaring the value of x^i.

D. We use induction on m, the number of bits in i. If $m=1$, the *for* loop is executed once and sets v to 0 or 1 appropriately. For $m>1$, if we assume that POWER works correctly on values of i with less than m bits, then for m bits, the computation is identical as that for $m-1$ bits (i.e., the integer that has the same bits as the first $m-1$ bits of i) and then on the mth iteration, v is squared and depending on whether the rightmost bit is 0 or 1, it is then multiplied by x. This is precisely what happens to x^i when i is doubled (its bits are shifted left) and then 0 or 1 is added to i.

E. The first *while* loop calculates p and the second *while* loop performs essentially the same computation as the iterative version presented in the chapter:

> **procedure** POWER(x,i):
> $\quad p := 1; ;$ **while** $(2p<i)$ **do** $p := 2p$
> $\quad v := 1$
> $\quad$ **while** $p \geq 1$ **do begin**
> $\qquad v := v^2$
> $\qquad$ **if** $\lfloor i/p \rfloor$ is odd **then** $v := x*v$
> $\qquad p := \lfloor p/2 \rfloor$
> $\qquad$ **end**
> $\quad$ **end**

Only $O(1)$ space is used. As we saw in Part B, the other two versions of POWER use $O(\log(i))$ space. If n denotes the length of the input, $\log(i)$ can be $\Omega(n)$ in the case that the bits used for i are at least proportional to the bits used for x.

71. For the recursive and iterative versions of *POWER(x,i)*, consider the size of the numbers computed by POWER. About how many bits does it take to represent the quantity x^i that is computed by POWER when x is n bits and i is m bits? What value results when x is 10 bits and i is 2 bits? What about when both x and i are 10 bits? What about when x is 10 bits and i is 250 bits? What can you say about how reasonable the unit-cost assumption is for this problem.

Solution: If x is n bits, then $x \leq 2^n-1$ and similarly $i \leq 2^m-1$. Hence, since the number of bits need to represent a value k is $\lfloor \log_2(x^k) \rfloor + 1$, the number of bits to represent x^i is:

$$\leq \left\lfloor \log_2\left((2^n-1)^{(2^m-1)}\right) \right\rfloor + 1$$
$$= \left\lfloor (2^m-1)\log_2(2^n-1) \right\rfloor + 1$$
$$\leq (2^m-1)\log_2(2^n)$$
$$= 2^m n - n$$

This is about what we would expect when we consider how POWER works. Since each iteration about doubles the number of bits in v (squaring a number of k bits yields a number of $2k-1$ or $2k$ bits), the final value has about $2^m n$ bits. for example:

If x has 10 bits and i has 2 bits, then $2^2 10-10 = 30$.

If x and i each have 10 bits, then $2^{10} 10-10 = 10{,}230$.

In terms of the *size of the input*, a small number of input bits for i can yield a very large value (e.g., $i=250$ uses only eight bits but yields a value of over 2^{250}).

72. Suppose that very large integers (possibly millions of digits each) are represented in binary as singly linked lists, one bit per vertex, with the highest order bit at the beginning of the list and the lowest order bit at the end.

   A. Describe in English and give pseudo-code to add 1 to a number in this representation.

   B. Describe how to generalize your solution to add two numbers in this representation.

Solution:

   A. One approach is to reverse the list, traverse the list to perform the addition, and then reverse the list to its original state. A simple recursive approach (which effectively does this reversing process for you) is to recursively add 1 to the tail of the list, where the recursive procedure returns the carry bit, and then add the carry bit to the first element; if the incrementing process returns with carry bit 1, a new first element is created (with value 1):

```
function INCREMENT(list)
    if (list is empty) then return 1
    else begin
        carry := INCREMENT(NEXT(list))
        if (carry=1) then begin
            if (FIRST(list)=0) then carry := 0
            complement the bit stored in FIRST(list)
        end
        return carry
    end
end
```

carry := INCREMENT(list)

   if (*carry*=1) **then** insert a new vertex at the head of *list* with value 1

   B. If the two lists have the same length, then the solution is almost identical to the one above; we recursively add the tails and then the two current bits and the carry bit can be added to produce a new carry bit. When the two lists are not the same length, we can first compute their lengths if they are not already known, then set a variable *position* to the longer length, and decrement *position* each time a recursive call is made; whenever *position* is larger than the length of the smaller list, we simply define its bit to be 0.

73. Consider the example that was presented for converting a string to a number where the function *GetReal* made use of the function *GetInteger* to get the integer to the left of the decimal point (which worked iteratively) and the function *GetFraction* to get the fraction to the right of the decimal point (which worked recursively).

    A. Write a recursive version of *GetInteger*.

    B. Explain why *GetFraction* relies upon the conventions we have been using for pseudo-code that assume left to right evaluation of conditions and expressions.

    C. Write an iterative version of *GetFraction*.

    D. Write a combined version of *GetReal*, *GetInteger*, and *GetFraction* that reads in all the digits as a single big integer and then divides by the power of 10 corresponding to the position of the decimal point.

    E. Modify your solution to part D to assume that the expression is in the string $S[1]...S[n]$ starting at a position i, rather than at the top of a stack; the value of i should be modified to be the next position in S after the real number (or $n+1$ if the number goes to the end of S).

    F. Describe how your solution to Part F can be modified to consider as an error a value of i out of range, or an empty expression, or an expression consisting only of single decimal point.

Solution:

    A. GetInteger makes use of GetPowerAndInteger which returns two values, the integer i and the power of ten p that corresponds to the next digit to the left:

> **function** GetPowerAndInteger
>   **if** the stack is empty or TOP is not a digit **then return** $0, 1$
>   **else begin**
> d := POP
> i,p := GetPowerAndInteger
>   **return** $d*p+i$, $p*10$
>   **end**
>   **end**

> **function** GetInteger
> i,p := GetPowerAndInteger
>   **return** i
> **end**

    B. The compiler must evaluate expressions from left to right; otherwise if the recursive call is made before the POP operation, the code goes into an infinite loop that endlessly calls itself and never does any POP operations. This code is also assuming that the compiler evaluates conditions from left to right since if the stack is empty, we do not want to call the TOP function (although TOP could be implemented to gracefully return a false condition when the stack was empty). Some languages (such as C) specify evaluation from left to right of things like the arguments to a function and the parts of conditions and expressions (which makes things easier for programmers) but others (such as Pascal) do not (which gives the compiler designer flexibility in writing the compiler). Whether expression evaluation is from left to right or right to left may depend on the compiler being used, and some simple ways of recursively implementing expression evaluation can lead to right to left evaluation. A "safe" version of the pseudo-code that translates more directly to any compiler is:

> **function** GetFraction
>> **if** the stack is empty **then return** 0
>>
>> **else if** TOP is not a digit **then return** 0
>>
>> **else begin**
>>> $d :=$ POP
>>>
>>> **return** $(d +$ GetFraction$)/10$
>>>
>>> **end**
>>
>> **end**

C. GetFraction keeps track of the power of ten for which the next digit to the right must be divided by:

> **function** GetFraction
>> $p := 1$
>>
>> $f := 0$
>>
>> **while** the stack is not empty and TOP is a digit **do begin**
>>> $p := p * 10$
>>>
>>> $f := f + POP/p$
>>>
>>> **end**
>>
>> **return** f
>>
>> **end**

D. We initialize divisor to 0 and when the decimal point is encountered we set divisor to 1. Each time a digit is read, divisor is divided by 10 (which does nothing until a decimal point has been seen and divisor is set to 1). The condition for the *while* loop has been carefully stated so that it continues past a decimal point but stops if a second decimal point is encountered.

> $r := 0$
>
> $divisor := 0$
>
> $power := 1$
>
> **while** stack not empty and (TOP is a digit or ($divisor$=0 and TOP='.')) **do begin**

>   **if** *TOP="."* **then** *divisor* := 1
>   **else begin**
>   *r* := *r*power* + *TOP*
>   *power* := *power**10
>   *divisor* := *divisor*/10
>   **end**
>   POP
>   **end**
> **if** *divisor*>0 **then** *r* := *r*/*divisor*

E. We use essentially the same code except that a variable i (initialized to 1) is used to sequence through S ($S[i]$ replaces TOP and POP and $i := i+1$ replaces POP). The condition $1 \leq i \leq n$ of the *while* loop will cause 0 to be returned when a value of i is given that is out of range.

>   *r* := 0
>   *divisor* := 0
>   *power* := 1
>   **while** $1 \leq i \leq n$ and ($S[i]$ is a digit or (*divisor*=0 and $S[i]$='.')) **do begin**
>   **if** $S[i]$="." **then** *divisor* := 1
>   **else begin**
>   *r* := *r*power* + *S[i]*
>   *power* := *power**10
>   *divisor* := *divisor*/10
>   **end**
>   *i* := *i*+1
>   **end**
>   **if** *divisor*>0 **then** *r* := *r*/*divisor*

F. At the beginning, we can add the test

>   **if** i<1 or i>n or ($s[i]$="." and (i=n or $S[i+1]$ is not a digit))) **then** ERROR

where ERROR denotes code that does whatever is appropriate to report an error and then exits.

74. Recall the algorithm presented to recursively evaluate a prefix expression containing only binary operators.

A. Prove that a prefix expression (containing only binary operators) can always be evaluated correctly by working from right to left. That is, by repeatedly applying the rightmost operator to the two operands to its right.

B. Prove that the algorithm correctly evaluates a legal expression.

110   SAMPLE EXERCISES

C. Prove that the function always reports an error for an illegal expression, or returns with a non-empty stack (which can be checked and reported as an error by the code calling the function).

D. Generalize the algorithm to allow *unary operators* (operators that take only one operand, like a leading minus, a percent, or an exponent). You may assume that the same symbol is not used for both a unary and binary operator. So for example, an expression such as $a-(-b)$ would have to be represented in prefix by either subtracting from 0 (i.e., $-a-0b$), or by using a different symbol for a unary minus sign (e.g., if _ denotes unary minus, then we would have $-a\_b$).

Solution:

A. By definition, a binary prefix operator takes the two operands to its right. If an operator x is the rightmost operator in the prefix expression, then there are no other operators to the right of x that could cause two operands to be combined into a single one. Hence, the only possibility for x in a legal evaluation of the expression is that there are at least two values to the right of x and that x is applied to the two values to its right. Applying x to these two values has the effect of replacing x and these two values by a single value, resulting is a shorter legal prefix expression. This reasoning can be repeated until we are left with a single operand equal to the value of the expression, or until we discover that the expression is illegal.

B. Since recursive calls are made to get the arguments before the operator is applied to them, if we "unwind" the recursion, a legal expression is processed by working from right to left. Another way to prove this is by induction on the length of the expression:

   (*basis*)

   A legal expression of length 1 must be a value, and is correctly returned by the first *else if* statement.

   (inductive step)

   Assume that the algorithm works correctly for an expression of length less than n for some $n>1$, and consider a legal expression of length n. It must consist of an operator followed by two operands, which may each be values or any legal prefix expression of length less than n. The second *else if* statement pops this operator and then makes two recursive calls, which by the inductive hypothesis must be evaluated correctly.

C. If a stack frame is encountered that does not contain a binary operator or an operand, then the two *else if* statements will fail, and the final *else* statement will report an error. Now suppose that all stack frames contain legal binary operators or operands. Since by Part B we know that the algorithm evaluates the expression from right to left, it will always be able to successfully evaluate the expression, except when an operator is encountered but there are not two items below it on the stack to which it can be applied; this is detected by the first *if* statement. If evaluation is successful, but the stack is not empty (so the expression did not have enough operators), then this can be detected after the function returns by a final check by the code that called it.

C. We can insert the following *else if* statement between the two current *else if* statements:

> **else if** TOP is a unary operator **then begin**
> $op :=$ POP
> $x :=$ preEVAL
> **return** the result of applying op to x
> **end**

Note that this preserves the right to left evaluation (and Parts A, B, and C can be generalized).

75. We have already seen how to convert an infix expression with binary operators to a postfix expression with the three procedures *PushExpression*, *PushTerm*, and *PushValue*. This algorithm assumes • and / take precedence over + and −, and a missing operator to defaults to •; for example, $ab+c$ means $(a•b)+c$.

A. Generalize the algorithm to allow *unary operators* (operators that take only one operand) by letting $\$$ denote a unary operator that precedes its argument and $\%$ denote a unary argument that follows its argument, where $\%$ takes precedence over $\$$, and unary operators take precedence over binary operators. For example $\$(ab\%)\%$ is $ab\%•\%\$$ in postfix notation and $\$\%•a\%b$ in prefix notation.

B. Describe how the infix to postfix conversion algorithm for binary operators (so the generalization of Part A is not allowed) can be used to convert infix to prefix notation.

C. Generalize Part B to allow unary operators as defined in Part A.

Solution:

A. The procedures *PushExpression* and *PushTerm* are unchanged; we can generalize the *PushValue* procedure by adding a line at the beginning that discards $\$$'s and saves a *count* of how many there were, and code at the end that pushes all $\%$'s encountered and then pushes $\$$'s a number of times given by the value save in *count*:

> **procedure** PushValue
> $count := 0;$ **while** TOP(S) is $\$$ **do begin** POP(S); $count := count+1$ **end**
> **if** TOP(S) is a basic value **then** PUSH(POP(S),P)
> **else if** POP(S) is (**then begin**
> PushExpression
> **if** POP(S) is not) **then** *ERROR — missing right parenthesis*
> **end**
> **else** *ERROR — missing value*

while TOP(S) is % **do** PUSH(POP(S),P)

for $i := 1$ **to** *count* **do** PUSH($\$$,P)

end

B. To convert an infix expression in a stack S to a prefix expression in a stack P, we can:

    1. Reverse S. For example, $a-(b+c)/d+e$ becomes $e+d/(c+b)-a$.

    2. Convert S to a postfix expression in P.

    3. Reverse P.

Since the definitions of prefix and postfix are symmetric, reading a postfix expression from right to left gives a valid prefix expression; this is effectively done by steps 2 and 3. However, evaluation of this prefix expression may not be equivalent to evaluation of the original postfix expression because the order of operands has been reversed (and operators like $-$ and $/$ are not commutative because in general $a-b \neq b-a$ and $a/b \neq b/a$). Step 1 corrects this by first reversing the expression.

Note: We can combine these three steps into a single pass by modifying the infix to prefix algorithm to process the expression from right to left (treat $S[1]...S[n]$ as an array that is scanned from n down to 1) and adopt the convention that the top of P is the left end of the expression.

C. Parts A and B do not quite work now, because we do not want to reverse the order of % and $\$$. However, we can modify the version of *PushValue* presented in the solution to Part A to expect the positions of % and $\$$ to be reversed, by counting %'s at the start instead of $\$$'s, and then coding the *for* and *while* loops at the end so that %'s are still pushed before $\$$'s:

procedure PushValue

    count := 0; **while** TOP(S) is % **do begin** POP(S); *count* := *count*+1 **end**

    $\vdots$

    for $i := 1$ **to** *count* **do** PUSH(%,P)

    while TOP(S) is $\$$ **do** PUSH(POP(S),P)

    end

76. Prove by induction the following arithmetic sums defined for $n \geq 0$:

A. $\displaystyle\sum_{i=0}^{n} i = \tfrac{1}{2}n^2 + \tfrac{1}{2}n$

B. $\displaystyle\sum_{i=0}^{n} i^2 = \tfrac{1}{3}n^3 + \tfrac{1}{2}n^2 + \tfrac{1}{6}n$

C. $\displaystyle\sum_{i=0}^{n} i^3 = \tfrac{1}{4}n^4 + \tfrac{1}{2}n^3 + \tfrac{1}{4}n^2$

Solution:

A. For $n=0$, we see: $0 = \frac{1}{2}0^2 + \frac{1}{2}0$

For $n>0$, if we assume the theorem true for 0 through $n-1$, then:

$$\sum_{i=0}^{n} i = n + \sum_{i=0}^{n-1} i$$

$$= n + \left(\frac{1}{2}(n-1)^2 + \frac{1}{2}(n-1)\right) \quad \text{— inductive hypothesis}$$

$$= \frac{1}{2}n^2 + \frac{1}{2}n$$

B. For $n=0$, we see: $0 = \frac{1}{3}0^2 + \frac{1}{2}0^2 + \frac{1}{6}0$

For $n>0$, if we assume the theorem true for 0 through $n-1$, then:

$$\sum_{i=0}^{n} i^2 = n^2 + \sum_{i=0}^{n-1} i^2$$

$$= n^2 + \left(\frac{1}{3}(n-1)^3 + \frac{1}{2}(n-1)^2 + \frac{1}{6}(n-1)\right) \quad \text{— inductive hypothesis}$$

$$= n^2 + \frac{1}{3}(n^3 - 3n^2 + 3n - 1) + \frac{1}{2}(n^2 - 2n + 1) + \frac{1}{6}(n-1)$$

$$= \frac{1}{3}n^3 + \frac{1}{2}n^2 + \frac{1}{6}n$$

C. For $n=0$: $0 = \frac{1}{4}0^4 + \frac{1}{2}0^3 + \frac{1}{4}0^2$

For $n>0$, if we assume the theorem true for 0 through $n-1$, then:

$$\sum_{i=0}^{n} i^3 = n^3 + \sum_{i=0}^{n-1} i^3$$

$$= n^3 + \left(\frac{1}{4}(n-1)^4 + \frac{1}{2}(n-1)^3 + \frac{1}{4}(n-1)^2\right) \quad \text{— inductive hypothesis}$$

$$= n^3 + \frac{1}{4}(n^4 - 4n^3 + 6n^2 - 4n + 1) + \frac{1}{2}(n^3 - 3n^2 + 3n - 1) + \frac{1}{4}(n^2 - 2n + 1)$$

$$= \frac{1}{4}n^4 + \frac{1}{2}n^3 + \frac{1}{4}n^2$$

77. Prove that for $k \geq 1$, $n = 2^k$, and any real number x:

$$\sum_{i=0}^{n-1} x^i = \prod_{i=0}^{k-1} \left(1 + x^{\left(2^i\right)}\right)$$

Also, work out the value for $k=3$ and $x=2$.

Solution: Use induction on k.

basis: For $k=1$:

$$\sum_{i=0}^{n-1} x^i = \sum_{i=0}^{1} x^i = x^0 + x^1 = (1+x) = \prod_{i=0}^{0}(1+x) = \prod_{i=0}^{k-1}\left(1 + x^{\left(2^i\right)}\right)$$

inductive step: Assume that the theorem is true for $k-1$, divide the sum into the even and odd terms, simplify to get $(1+x)$ times a single sum going from $i=0$ to $n/2-1$, and then use the inductive hypothesis:

$$\sum_{i=0}^{n-1} x^i = \sum_{i=0}^{(n/2)-1} x^{2i} + \sum_{i=0}^{(n/2)-1} x^{2i+1}$$

$$= (1+x) \sum_{i=0}^{(n/2)-1} x^{2i}$$

$$= (1+x) \sum_{i=0}^{(n/2)-1} \left(x^2\right)^i$$

$$= (1+x) \prod_{i=0}^{k-2} \left(1 + \left(x^2\right)^{\left(2^i\right)}\right) \qquad \text{— by the inductive hypothesis}$$

$$= (1+x) \prod_{i=1}^{k-1} \left(1 + x^{\left(2^i\right)}\right)$$

$$= \prod_{i=0}^{k-1} \left(1 + x^{\left(2^i\right)}\right)$$

For $k=3$, we have

$$1 + x + x^2 + x^3 + x^4 + x^5 + x^6 + x^7 = (1+x)(1+x^2)(1+x^4)$$

and then if $x=2$, we are just adding up powers of 2:

$$1 + 2 + 4 + 8 + 16 + 32 + 64 + 128 = (1+2)(1+4)(1+16) = (3)(5)(17) = 255$$

78. For the Towers of Hanoi problem, we saw that TOWER made 2^n-1 moves and used $O(n)$ space. For this problem, simply assume that the input and output is encoded in some reasonable way (e.g., numbers are encoded in binary and characters use one byte each).

    A. How much time and space is used as a function of the size of the input?

    B. How much time is used as a function of the size of the output?

    C. How much space is used as a function of the size of the output?

Solution:

    A. Since the number of bits needed to represent n is $O(\log(n))$ and the characters to specify the posts use $O(1)$ space, the input has length $O(\log(n))$. Hence, as a function of the length of the input, the time is double-exponential and the space is exponential.

    B. Since the time is proportional to the length of the output, as a function of the length of the output, the time is linear.

    C. Since each line of output uses at most $O(\log(n))$ bits to specify the number of disks and $O(1)$ space for the rest of the line (the text and the names of the pegs), each line of output uses $O(\log(n))$ space and hence the total space used by the output is $O(2^n\log(n))$; since this function is asymptotically (much) larger than the $O(n)$ space used by the recursion stack, as a function of the length of the output, the space used is linear.

Exercises

79. Similar to factorial, we saw how the iterative program to sum odd integers could be expressed recursively with simple tail recursion. Give a recursive program to sum powers of two that is based on the iterative program given as an example.

Hint: It is ok to have a function return a pair of values.

80. Given two positive integers x and y, the *greatest common divisor of x and y*, denoted $gcd(x,y)$ is the largest integer that divides both x and y. Let $(x\ MOD\ y)$ denote the remainder when x is divided by y; that is, $(x\ MOD\ y) = x - \lfloor x/y \rfloor * y)$.

  A. Prove that for two integers $x, y > 1$, $gcd(x,y) = gcd((x\ MOD\ y), y)$.

  B. Based on Part A, give a simple iterative algorithm to find the greatest common divisor of two positive integers (this algorithm is known as *Euclid's Algorithm*).

  C. Define $gcd(0,x) = 0$, and give an equivalent recursive algorithm for Part B.

  D. What can you say about the worst case asymptotic time and space used by your algorithms for Parts B and C as a function of the number of characters in the input. Assume the input is a pair of integers separated by a comma; so for example, 1423,5286 has length 9.

  E. Generalize your algorithm of Parts B and C to compute two integers i and j such that:

$$ix + jy = gcd(x,y)$$

81. The recursive version of binary search worked by dividing the problem in half.

  A. Give pseudo-code for *recursive ternary search* that divides the problem into three parts. That is, it makes at most 2 comparisons and then works on a problem of size $n/3$.

  B. Generalize your solution to Part A to a *recursive k-ary search* algorithm that divides the problem into k parts. That is, it makes at most $k-1$ comparisons and then works on a problem of size n/k.

  C. For any fixed value of k, explain why your solution to Part B is $O(\log(n))$ time.

82. Recall from the chapter on lists, the example that presented *postEVAL*, a function to evaluate a postfix expression that uses a *while* loop to sequence through the expression and makes use of a stack S to push operands until they are needed by the appropriate operator.

  A. It is interesting to compare *postEVAL* with the recursive function *preEVAL* that was presented in this chapter to evaluate a prefix expression. The fact that *postEVAL* assumes its input is in an array and *preEVAL* assumes its input in a stack is really just a convenience of presentation for the pseudo-code. The real difference is that *postEVAL* uses a stack for its computation whereas *preEVAL* uses recursion (where the compiler is maintaining a "hidden" stack).

- Compare what is being saved on the stack used by *postEVAL* to what is being saved on the hidden stack used by *preEVAL*.

- Discuss why iteration is natural for evaluation of postfix and recursion is natural for evaluation of prefix.

- Describe how to implement postfix evaluation with recursion and how to implement prefix evaluation with iteration (directly, not by mechanical elimination of recursion).

B. Like the examples presented in this chapter, for simplicity *postEVAL* assumed that values occupy a single stack frame. Make use of the *GetReal* function that was presented in this chapter to modify the code for *postEVAL* so that values are real numbers stored one character at a time. Since *postEVAL* assumed that the input was in the array $P[1]...P[n]$, you will need to modify notation to make use of *GetReal*.

83. In the sample exercises we saw how to generalize the recursive algorithm to evaluate a prefix expression to allow unary operators, as long as the same symbol was not used for both a unary and binary operator. Suppose that we have only two operators, minus and plus, where we allow the minus symbol to be both unary and binary.

A. Explain why the prefix expression

$$+ + + + + a\, b\, c\, d - e\, f$$

can be evaluated in only one way:

$$((((a + b) + c) + d) + (- e)) + f$$

B. Explain why the prefix expression

$$- - a\, b$$

can be evaluated in two different ways that yield different answers:

$$- (a - b) = - a + b \quad \text{or} \quad (- a) - b = - a - b$$

Also, explain why the symmetric postfix expression

$$a\, b - -$$

can be evaluated in two different ways that yield different answers:

$$- (a - b) = - a + b \quad \text{or} \quad a - (- b) = a + b$$

C. Explain why the prefix expression

$$+ + - a\, b - c\, d$$

can be evaluated in two different ways that yield *different* answers:

$$((- a) + b) + (c - d) = - a + b + c - d$$

or

$$((a-b)+(-c))+d=a-b-c+d$$

Also, explain why the symmetric postfix expression

$$a\,b-c\,d-++$$

can be evaluated in two different ways that yield the *same* answer:

$$(a-b)+(c+(-d))=a-b+c-d$$

or

$$a+((-b)+(c-d))=a-b+c-d$$

Explain why symmetric formulas can behave differently with prefix and postfix. *Hint:* The binary minus operator is not commutative; that is, $a-b \neq b-a$.

D. Give an algorithm that determines if such a prefix expression has exactly one legal evaluation.

E. Give an algorithm that produces a legal evaluation if one or more legal evaluations is possible (it may arbitrarily choose which of the legal evaluations to produce).

84. The sample exercises showed how to use induction to prove that the algorithm presented in this chapter to evaluate a prefix expression is correct.

A. In a similar spirit, use induction to prove that the algorithm to convert infix to postfix is correct.

B. Generalize your proof for part A to expressions with unary operators as presented in the sample exercises.

85. To evaluate an arithmetic expression in normal infix notation, we could employ the two examples presented (first convert to prefix or postfix notation and then evaluate that expression). Give pseudo-code that evaluates an infix expression directly by sequencing through the expression with a *while* loop and making recursive calls when parentheses are encountered. Assume that the only operators are multiplication, division, addition, and subtraction, but allow the minus symbol to denote both the binary and unary operation.

86. We have seen that both the recursive and iterative versions of the Towers of Hanoi Problem use $O(n)$ space. Compare the constants that would be expected in practice, charging one unit of space to store an integer.

87. For the recursive version of factorial, we saw that $FACT(n)$ ran in $O(n)$ time and used $O(n)$ space for a stack that grew to a depth of n. Assuming that the input and output is encoded in binary:

A. How much time and space is used as a function of the size of the input?

B. How much time is used as a function of the size of the output?

C. How much space is used as a function of the size of the output?

88. What is wrong with the following "proof" that all horses are the same size.

Show by induction that all horses in any set of $i \geq 1$ horses are the same size.

(*basis*)

For $i=1$, in any set containing only one horse, all have the same size.

(*inductive step*)

For $i>1$, assume that for any set of $i-1$ horses, all have the same size:

Consider a large coral containing a set S of i horses.

Remove a horse from S to obtain a set X of $i-1$ horses.

Put this horse back and remove a different one to obtain another set Y of $i-1$ horses.

By the inductive hypothesis, all horses in X are the same size.

By the inductive hypothesis, all horses in Y are the same size.

Since horses of S that are in both X and Y cannot have two different sizes, it must be that the size of horses in X is the same as the size of horses in Y.

Hence, all horses in S must have the same size.

89. The *Fibonacci numbers* can be defined recursively for the non-negative integers as:

$$F(n) = \begin{cases} 0 & \text{if } n = 0 \\ 1 & \text{if } n = 1 \\ F(n-1) + F(n-2) & \text{if } n \geq 2 \end{cases}$$

that is, $F(0)$, $F(1)$, $F(2)$, ... is the sequence: 0, 1, 1, 2, 3, 5, 8, 13, 21, ...

A. Suppose that we program this recursively in the straightforward way as:

function FIB(n)
    if $n \leq 0$ **then return** 0
    else if $n=1$ **then return** 1
    else return $FIB(n-1)+FIB(n-2)$
    end

Prove that this algorithm uses exponential time as a function of n; that is, show that the time is $\Omega(2^{cn})$ for some constant $c>0$.

Hint: Since any constant $c>0$ will do, argue why FIB($n-2$) can take no more time than FIB($n-1$), and then analyze the time used by a recursive program that uses two calls to problems of size 2 smaller in a way that is similar to the analysis of TOWER that had two problems of size 1 smaller.

B. For the function FIB of Part A, if we let N denote the size of the input when written in binary (i.e., the number of bits in the binary representation of n), explain why the algorithm of Part A consumes double-exponential time as a function of N; that is for some constant $c > 0$, the time is:

$$\Omega\left(2^{2^{cN}}\right)$$

C. Assuming that storage of one integer is $O(1)$ space, and assuming that recursion is implemented with a stack in the usual way, explain why the space used by the algorithm of Part A is $O(n)$.

Hint: Like TOWER, the space used by one recursive call is reused after it returns and the second recursive call is made, so it is the maximum depth of recursion that determines the space used by the stack to implement the recursion.

D. Give an algorithm to compute $F(n)$ that uses $O(n)$ time and space by working from 2 to n to build a table of the values of F.

E. We can express the recursive definition of F as a multiplication with a 2 by 2 matrix:

$$\begin{pmatrix} 0 & 1 \\ 1 & 1 \end{pmatrix} \begin{pmatrix} F(n) \\ F(n+1) \end{pmatrix} = \begin{pmatrix} F(n+1) \\ F(n+2) \end{pmatrix}$$

By simply applying the definition of matrix multiplication given in the first chapter, we see that this equation is simply another way of expressing the two equations:

$$F(n+1) = F(n+1)$$

$$F(n) + F(n+1) = F(n+2)$$

However, by expressing the recursive relationship as a matrix multiplication, we effectively in one equation "save" the value of $F(n+1)$ while at the same time compute $F(n+2)$; that is, the two values that are needed to compute $F(n+3)$ are being stored in a single equation. If we denote this 2 by 2 matrix by x and define

$$x^0 = \begin{pmatrix} 1 & 0 \\ 0 & 1 \end{pmatrix}$$

then we see that for $i \geq 0$:

$$\begin{pmatrix} 0 & 1 \\ 1 & 1 \end{pmatrix}^i \begin{pmatrix} 0 \\ 1 \end{pmatrix} = \begin{pmatrix} F(n) \\ F(n+1) \end{pmatrix}$$

Hence, $F(n)$ can be computed by computing a 2 by 2 matrix to the nth power, doing a single multiplication by a 2 by 1 matrix, and then reading the top component. Using essentially the same method as used by the fast computation of integer exponents presented in this chapter, give an $O(\log(n))$ time recursive algorithm to compute $F(n)$. This approach relies on the fact that a sequence of matrix multiplications is *associative* that is, $A(BC)=(AB)C$. Include with your answer a proof that multiplication of 2 by 2 matrices is associative. Also, give an example to show that 2 by 2 matrix multiplication is not *commutative* (i.e., it is not necessarily true that $AB=BA$) and explain why this is not a problem for this approach.

Note: In general, multiplication of matrices of any dimension is associative. You can prove that here, or prove just the case of 2 by 2 matrices.

F. Assuming that storage of one integer is $O(1)$ space, and assuming that recursion is implemented with a stack in the usual way, explain why the space used by the algorithm of Part E is $O(\log(n))$.

G. Again, motivated by the presentation of POWER, give an iterative version of your algorithm for Part E, and, assuming that storage of one integer is $O(1)$ space, explain why it uses only $O(1)$ space.

H. The number

$$\alpha = \frac{1+\sqrt{5}}{2} = 1.61803...$$

is often called the "golden ratio" because if one starts with a 1 by α rectangle and removes a 1 by 1 rectangle from it, the remaining α–1 by 1 rectangle has the same aspect ratio:

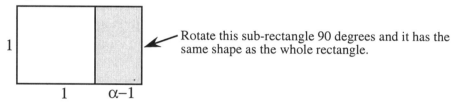

Rotate this sub-rectangle 90 degrees and it has the same shape as the whole rectangle.

That is:

$$\frac{1}{\alpha} = \frac{\alpha-1}{1}$$

The constant α is a solution to the equation $x^2-x-1=0$. The other solution (there are only two since this is a quadratic equation) is:

$$\beta = \frac{1-\sqrt{5}}{2} = 1-\alpha = -0.61803...$$

Give a proof by induction that:

$$F(n) = \frac{1}{\sqrt{5}}\left(\frac{1+\sqrt{5}}{2}\right)^n - \frac{1}{\sqrt{5}}\left(\frac{1-\sqrt{5}}{2}\right)^n$$

Hint: Verify $F(0)$ and $F(1)$, then write

$$F(n) = \frac{1}{\sqrt{5}}\left(\alpha^n - \beta^n\right)$$

and use the identities $\alpha^2 = \alpha+1$ and $\beta^2 = \beta+1$ (which follow from the fact that α and β are solutions to $x^2-x-1=0$) to show the inductive step.

Note: The golden ratio is often called φ in the literature.

I. Although the expression of Part H contains irrational numbers (i.e., the square root of 5 cannot be expressed exactly with a finite number of digits), everything must cancel out when the expression is computed to get an integer answer. We can rewrite the expression for F as:

$$F(n) = \frac{1}{2^n\sqrt{5}}\left[\left(1+\sqrt{5}\right)^n - \left(1+\left(-\sqrt{5}\right)\right)^n\right]$$

Using the expression developed for the binomial sum, show that for $n>0$:

$$F(n) = \frac{1}{2^n\sqrt{5}}\left[\sum_{m=0}^{n}\binom{n}{m}\sqrt{5}^m - \sum_{m=0}^{n}\binom{n}{m}\left(-\sqrt{5}\right)^m\right]$$

$$= \frac{1}{2^{n-1}}\sum_{m=0}^{\lfloor(n-1)/2\rfloor}\binom{n}{2m+1}5^m$$

Also, assuming that multiplication of two integers is $O(1)$ time, give an algorithm to compute this expression in $O(n)$ time.

J. To compute $F(n)$, we can compute the square root of 5 to some number of digits, compute the expression of Part H in $O(\log(n))$ time by using the fast integer exponent algorithm, and round off to the nearest integer to get $F(n)$. But we first need to know how many digits as a function of n must be used to insure a correct answer. Let:

    $b(n) =$ Number of bits to the right of the decimal point in the binary representation of α and β used in the computation of $F(n)$.

    $L(n) =$ Smallest integer such that if $b(n) \geq L(n)$ then the computation of $F(n)$ is correct.

One approach to computing a sufficient value for $L(n)$ is to use the series expansion of $ln(1+x)$ for $0<x<1$ (see the chapter notes), where ln denotes the natural logarithm:

$$\text{for } 0 < x < 1, \quad \ln(1+x) = x - \frac{1}{2}x^2 + \frac{1}{3}x^3 - \frac{1}{4}x^4 + \frac{1}{5}x^5 + \text{L}$$

Explain each of the following steps:

- From this expansion it follows that: $x - \frac{1}{2}x^2 < \ln(1+x) < x$

- Write $\alpha = M + d$ where M is α written to $b(n)$ bits of precision and $0<d<1$ is a real number. If we can show that $(M+d)^n - M^n < 1/4$, then since $-\beta$ is less than α, the total error of $F(n) = \alpha^n - \beta^n$ will be less than $1/2$, and hence the computation of $F(n)$ rounded to the nearest integer must be correct.

- Let e denote the natural logarithm base. Hence, since $M = e^{ln(M)}$ it suffices that d satisfy:

$$e^{n\ln\left(M\left(1+\frac{d}{M}\right)\right)} - M^n < \frac{1}{4} \quad \text{which simplifies to} \quad e^{n\ln\left(1+\frac{d}{M}\right)} < 1 + \frac{1}{4M^n}$$

- Hence, by upper bounding the logarithm it suffices that d satisfy:

$$e^{n\left(\frac{d}{M}\right)} < 1 + \frac{1}{4M^n}$$

- Hence, by taking logarithms of both sides and lower bounding the logarithm on the right, it suffices that d satisfy:

$$\frac{nd}{M} < \frac{1}{4M^n} - \frac{1}{32M^{2n}} \quad \text{which simplifies to} \quad d < \frac{8M-1}{32nM^{2n-1}}$$

- The right hand side of this expression is a real number less than 1, and we want $b(n)$ to be larger than the number of 0's to the right of the decimal point in the binary representation of the right hand side. Hence, it suffices for $b(n)$ to be larger than $\log_2$ of the inverse of the right side:

$$b(n) > \log_2\left(32nM^{2n-1}\right) - \log_2(8M-1)$$

- Since dropping the second term on the right only makes the right side bigger, and since $\log_2(n) < n$, $M < \alpha < 1.62$, and $\log_2(1.62) < 0.7$, it suffices that $b(n) \geq 1.4n + 5$ and hence for $n > 8$, $L(n) < 2n$.

K. A classic method for computing the square root r of a real number $x > 0$ to within a specified accuracy d is to start with a guess for r, and then repeatedly average this guess with x/r:

Guess a real number $0 < r < x$.

while $\left|x - r^2\right| > d$ **do** $r := \frac{1}{2}\left(r + \frac{x}{r}\right)$

Assuming that numbers are represented in binary, explain why this method converges, and once the first bit of r is correct, why each iteration increases the accuracy of r by at least one bit, and hence the algorithm uses $O(n)$ time in the worst case to achieve n bits of accuracy.

Note: In fact, although the algorithm starts off slower or faster depending on the quality of the initial guess, once the first digit is correct, it is possible to show that each iteration approximately doubles the number of correct digits; see the chapter notes.

Chapter Notes

Recursive programs are presented in virtually all texts on data structures and algorithms; the reader may refer to the list of texts given at the beginning of the chapter notes to the first chapter. For more detail about how a compiler implements recursive procedures, the reader may refer a standard text on the subject, such as Aho, Sethi, and Ullman [1986].

For further reading on proof by induction, in addition to texts on data structures and algorithms, the reader may refer to the list of texts on discrete mathematics given in the chapter notes to the first chapter.

See texts on discrete mathematics such as Liu [1985] for a presentation of the Fibonacci numbers and how to solve recurrence relations of this type. The web page of Knott [2001] is devoted to the Fibonacci numbers and the related mathematics.

For the iterative square root algorithm that guesses a value r for the square root of x and then repeatedly does

$$r := \tfrac{1}{2}\left(r + \tfrac{x}{r}\right)$$

it can be shown that if r currently has c correct digits, then after the next iteration r will have at least $2c-1$ correct digits; it is an example of Newton's algorithm (see a text on numerical analysis such as Dahlquist and Bjorck [1974]). For a proof of the series expansion of $ln(1+x)$ for $0<x<1$, see an algebra or calculus textbook such as Thomas [1968].

4. Trees

Introduction

A *tree* corresponds to a hierarchy. At the top there is the root, the root can have children, the children can have children, and so on. For example, on a typical personal computer, there is the hard drive (the root), inside it are directories and files, inside directories there are other directories and files, etc. Directories correspond to vertices that have children and files correspond to "leaf" vertices that do not have any children (empty directories contain no files and also correspond to leaves).

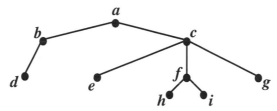

We shall begin by assuming trees to be *unordered*; that is, the order in which siblings are attached to a vertex is arbitrary (just an artifact of the data structure) does not change the data being represented by the tree. We shall then consider *binary* trees, where each non-leaf vertex can have a left child, a right child, or both a left and right child (so a vertex with a single left child is different than one with a single right child). Binary trees can be used to efficiently store and access a set of data items for which an order is defined (e.g., a set of names with the standard English alphabetical ordering). Although we won't need to here, we could generalize the notion of a binary tree to define arbitrary *ordered* trees that, for example, can have a first and third child but no second child. In fact, the *trie* data structure that is presented in the chapter on strings is essentially such a generalization.

Although a linked list is a special case of tree, a key "theme" will be to store items in a tree in a way that causes it to remain "balanced" so that root to leaf paths in a tree of n vertices tend to be on the order of $\log(n)$, not n. This allows operations that take time proportional to the length of a root to leaf path to work quickly. Here we will present techniques to keep trees "balanced" that work in the average case or in an "amortized" sense; in a later chapter we will study methods to insure that trees stay balanced at all times.

Tree Terms

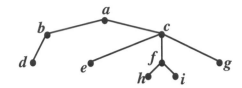

Tree, vertex,

root, parent: A tree can be defined recursively as a set of vertices consisting of a root r for which no other vertex is designated as its parent, together with a (possibly empty) set S of disjoint trees, where r is the parent of each of the roots of the trees in S. In the figure, vertices are labeled a through i, a is the root, a is the parent of b and c, b is the parent of d, c is the parent of e, f, and g, and f is the parent of h and i.

child: v is the parent of w if and only if w is a child of v.

internal vertex: A vertex that is not the root and has at least one child; in the figure, b, c, and f are internal vertices.

leaf: A vertex with no children; in the figure, d, e, h, i, and g are leaves.

sibling: If two vertices are children of the same vertex, then they are siblings. In the figure, b and c are siblings, e, f, and g are siblings, and h and i are siblings.

ancestor: A vertex v is an ancestor of a vertex w if $v=w$ or v can be reached from w by following the parent relationship; if $w \neq v$, then w is a *proper* ancestor of v.

descendant: w is a descendant of v if and only if v is an ancestor of w; if $w \neq v$, then w is a *proper* descendant of v.

subtree: The children of a vertex v are the roots of the subtrees of v. In the figure, removing the root a leaves two subtrees (one subtree rooted at b and one subtree rooted at c).

DEPTH(v): The number of edges on the path from the root to vertex v (the depth of the root is 0).

HEIGHT(v): The number of edges on a longest path from vertex v to a leaf (leaves have height 0).

MIN-HEIGHT(v): The number of edges on a shortest path from vertex v to a leaf (leaves have min-height 0).

LEVEL(v): LEVEL(v) = HEIGHT(*root*) – DEPTH(v)

Representing Trees

LMCHILD-RSIB Representation: Assign an arbitrary ordering to the children of each vertex of a tree (so one child is the "leftmost" child). A tree T is a just a pointer to its root vertex. Each vertex v contains four fields:

DATA(v,T):   The data stored at v.

PARENT(v,T):   Pointer to the parent of v (*nil* if v is the root or $v=nil$).

LMCHILD(v,T):   Pointer to leftmost child of v (*nil* if v is a leaf or $v=nil$).

RSIB(v,T):   Pointer to the sibling of v that is directly to its right (*nil* if v is the root, v is a rightmost sibling, or $v=nil$).

Simplified notation: When T is understood, we may omit it as an argument.

Example: Suppose we draw the four fields of a vertex from left to right:

data	parent pointer	left child pointer	right sibling pointer

The figure below shows a tree and its corresponding LMCHILD-RSIB representation:

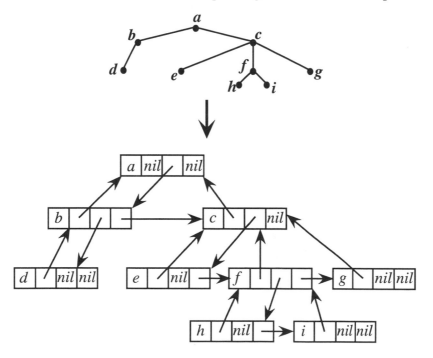

Pre-Order Traversal

Idea: To visit all vertices in the subtree rooted at v, pre-order traversal uses the rule: "Visit v and then (recursively) visit the subtrees of v."

>   **procedure** PRE(v):
>   {visit v}
>   **for** each child w of v do PRE(w)
>   **end**
>   PRE(*root*)

Post-order traversal: To visit all vertices in the subtree rooted at v, post-order traversal uses the rule: "(Recursively) visit the subtrees of v and then visit v"

Unique ordering: Because the *for* loop does not specify in what order children are visited, many different traversals of the tree may satisfy the preorder and post-order definitions. In practice, a particular representation of the tree in memory will give rise to a natural "standard" order. For example, if we are using the LMCHILD-RSIB representation, a unique traversal of the tree results from replacing the *for* loop by a *while* loop that follows the RSIB links from first to last. Below is the entire procedure rewritten in this way, and renamed *first-to-lastPRE*:

>   **procedure** first-to-lastPRE(v):
>   {visit v}
> $w := \text{LMCHILD}(v)$
>   **while** (w is not *nil*) **do begin**
>   first-to-lastPRE(w)
> $w := \text{RSIB}(w)$
>   **end**
>   **end**

Example: Starting at the root, assuming children are visited in alphabetical order:
>   pre-order: $a, b, d, c, e, f, h, i, g$
>   post order: $d, b, e, h, i, f, g, c, a$

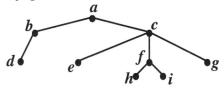

Example: Height of a Vertex v

>   **function** HEIGHT(v):
> $h := 0$
>   **for** each child w of v **do** $h := \text{MAXIMUM}(h,\text{HEIGHT}(w)+1)$
>   **return** h
>   **end**

Level-Order Traversal

Idea: Visit the children of a vertex v before going deeper into the subtrees, so that vertices are visited from highest to lowest level.

> **procedure** LEV(v)
>> Initialize a queue to contain the root.
>> **while** queue is not empty **do begin**
>>> v := DEQUEUE
>>> {visit v}
>>> **for** each child w of v **do** ENQUEUE(w)
>>> **end**
>> **end**
>> LEV(*root*)

Unique ordering: Like pre-order, the *for* loop does not specify the order in which children are visited. Again, if we are using the LMCHILD-RSIB representation, a natural unique traversal of the tree results from replacing the *for* loop by a *while* loop that follows RSIB links from left to right:

> w := LMCHILD(v)
> **while** (w is not *nil*) **do begin**
>> ENQUEUE(w)
>> w := RSIB(w)
>> **end**

Example: Assuming that children are visited in alphabetical order, level order starting at the root is $a, b, c, d, e, f, g, h, i$.

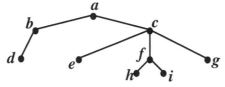

Example: MIN-HEIGHT of a Vertex v

> **function** MIN-HEIGHT(v)
>> $h := 0$
>> Initialize a queue to contain the pair v,h.
>> **while** v is not a leaf **do begin**
>>> v,h := DEQUEUE
>>> **for** each child w of v do ENQUEUE($w,h+1$)
>>> **end**
>> **return** h
>> **end**

Binary Search Trees

Idea:

Each vertex has at most two children.

This is an example of an ordered tree structure; it is possible to have a right child but no left child.

For each vertex, all items in its left sub-tree are smaller and all items in its right subtree are larger.

Example: If items are the names of animals and standard English alphabetical ordering is used, a legal binary search tree could be:

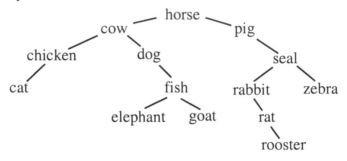

Representation of BST's: Each vertex v has four fields:

          DATA: The data stored at v.

   PARENT: A pointer to the parent of v.

    LCHILD: A pointer to the left child of v (*nil* if no left child).

    RCHILD: A pointer to the right child of v (*nil* if no right child).

LMCHILD and RSIB operations: For compatibility with algorithms already presented (e.g., pre-order traversal), we observe that both are $O(1)$ time:

    function LMCHILD(v)

       if (LCHILD(v)$\neq$*nil*)

           then LCHILD(v)

           else RCHILD(v)

      end

    function RSIB(v)

       if (v=*root*) **or** (v=RCHILD(PARENT(v)))

           then *nil*

           else RCHILD(PARENT(v))

      end

Basic Binary Search Tree Operations

MEMBER(d, T): Return a pointer to the vertex of T containing d:
  Move down from the root, going left when d is smaller and right when d is greater, until a vertex v with data d is found or you get "stuck" at a leaf (and return *nil*).

INSERT(d, T): Insert a new vertex in T with data d:
  Proceed like a MEMBER operation except when you get stuck at a leaf v, create a child of v with data d (or create a new root if T was empty).

MIN(v, T): Return a pointer to the smallest vertex in the subtree of T rooted at v:
  Move down from v following left child pointers until you find a vertex v with no left child; return v.

MAX(v, T): Symmetric to MIN.

DELETEMIN(v, T): Delete and return smallest data in the subtree of T rooted at v:
  First compute $w = MIN(v, T)$, and let p be the parent of w and r the right child of w. As depicted on the following page, if w is a left child (and it might be that $w = v$) then delete w and make r the left child of p (the left child of p becomes nil if r is *nil*). There are also two "degenerate" cases when $w = v$ and w is a right child or the root.

  $w := MIN(v, T)$

  $p := PARENT(w)$

  $r := RCHILD(w)$

  if $r \neq nil$ **then** make the parent of r be p

  if w is a left child **then** make the left child of p be r
  else if w is a right child **then** make the right child of p be r
  else make r the new root (and make T be the empty tree if $r = nil$)

  reclaim the memory for w and **return** the data that was stored in w

DELETEMAX(v, T): Symmetric to DELETEMIN.

DELETE(v, T): Delete a vertex v of T by getting new data from an appropriate leaf (the largest in the left subtree or the smallest in the right subtree) and then deleting that leaf:
  if v is a leaf **then** remove v
  else if v has a left child **then** change data of v to DELETEMAX(LCHILD(v),T)
  else change data of v to DELETEMIN(RCHILD(v),T)

Simplified Notation: We omit the argument T when it is understood. Also, for MIN, MAX, DELETEMIN, and DELETEMAX, we omit the argument v when it is understood that the search starts from the root (so DELETEMIN(T) returns the smallest item in T).

Details of the DELETEMIN Operation

$w := \text{MIN}(v, T)$
$p := \text{PARENT}(w)$
$r := \text{RCHILD}(w)$

if $r \neq nil$ **then** Make the parent of r be p.

if w is a left child **then** Make the left child of p be r.

Note: It might be that $v=p$ or $v=w$, in which case we have the simpler figure on the right; if $r=nil$, then the left child of p becomes *nil*:

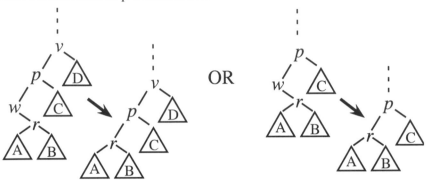

OR

else if w is a right child **then** Make the right child of p be r.

Note: Now it must be that $v=w$; if $r=nil$, then the right child of p becomes *nil*:

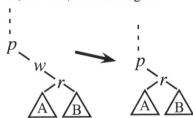

else Make r the new root (and make T be the empty tree if $r=nil$).

Note: Now it must be that $v=w$ is the root, and r is a right child of the root; if $r=nil$, then the tree becomes empty:

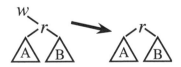

Reclaim the memory for w and **return** the data that was stored in w.

Example: Some Sample Binary Search Tree Manipulations

Starting with an empty tree, inserting *h*, *c*, *d*, *i*, *b*, *f*, *a*, *l*, *m*, *e*, *n*, *g*, *j*, *k* produces:

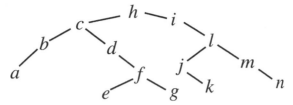

DELETEMIN, DELETEMAX produces:

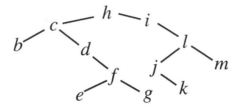

DELETE(*h*) produces:

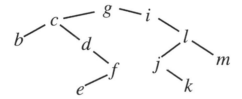

DELETE(*i*) produces:

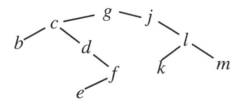

Note: We could optimize the delete procedure (see the homework problems) to take advantage of a vertex with only one child, so for example, DELETE(*i*) could have done:

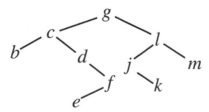

In-Order Traversal of a Binary Search Tree

Idea: Visit the left subtree, visit the root, visit the right subtree.

> **procedure** IN(*v*)
>   **if** *v≠nil* **then begin**
>   IN(LCHILD(*v*))
>   {visit *v*}
>   IN(RCHILD(*v*))
>   **end**
>   **end**
> IN(*root*)

Sorting: In-order traversal visits the data in sorted order.

Example: In-order for the following tree visits the vertices in alphabetical order *a* through *n*.

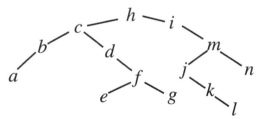

In-order traversal of unordered trees:

Pre-order and post-order traversal work equally well for unordered trees as for ordered trees like binary trees, because a vertex is visited before or after *all* of its children.

We have defined in-order traversal specifically for binary trees (where it makes sense to have a right child but no left child) and hence the traversal down from a vertex *v* with only one child *w* will be different for the case when *w* is a left child (where *w* comes before *v*) and when *w* is a right child (where *v* comes before *w*).

We can generalize in-order traversal to unordered trees by adopting the convention that we first traverse the leftmost subtree, then visit the root, then traverse the remaining subtrees.

Example: Evaluating An Arithmetic Expression

Representing arithmetic expressions with binary trees: Consider an arithmetic expression with the usual convention that parentheses group terms but in their absence, multiplication and division take precedence over addition and subtraction (equal precedence goes from left to right). Such an expression can be represented by a tree that reflects the order of operations. A vertex is labeled by an operation that is applied it to its children. For the case of a leading minus sign, the minus vertex has only a right child.

Example:

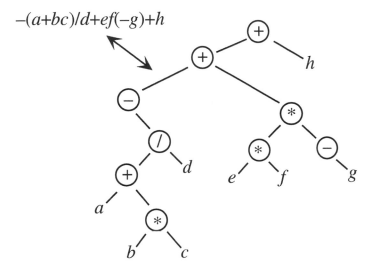

$-(a+bc)/d+ef(-g)+h$

In-order traversal of this tree gives back the expression; pre-order traversal changes the expression to *prefix notation* (the operator comes before the operands) and post-order traversal gives *postfix notation* (the operator comes after the operands).

Evaluating an expression tree with in-order traversal:
    function EVAL(v)
        if $v=nil$ **then return** 0
        else if v is a leaf **then return** the value stored at v
        else begin
            OP := the operator stored at v
            return OP(EVAL(LCHILD(v)),EVAL(RCHILD(v)))
            end
        end
    EVAL(*root*)

Efficient construction of the expression tree: A recursive procedure similar to the example given in the chapter on induction and recursion for converting infix to prefix notation can be used to construct the expression tree in linear time (see the exercises).

(*) Joining and Splitting Binary Search Trees

JOIN(T_1,d,T_2): A single tree with root d is formed from T_1, d, and T_2.

     ** Always assumes that items in T_1 are $<d$ and items in T_2 are $>d$.

SPLIT(d,T): Delete d and split T into items $<d$ and items $>d$.

     ** For ease of presentation, we assume that d is in T.

     T_1 and T_2 are initialized to the left and right subtrees of v.

     As described on the next page, delete the vertex v that contains d and move up the path to the root to "unzip" the tree into two trees; T_1 the tree of all vertices $<d$ and T_2 the tree of all vertices $>d$.

     Each time we follow a left child edge up to a vertex w, w has data that is larger than everything seen thus far (and w should be merged into T_2).

     Each time we follow a right child edge up to a vertex w, w has data that is smaller than everything seen thus far (and so w should be merged into T_1).

Example of the SPLIT operation: Vertices 1 through 5 are shown, connected to them are subtrees of arbitrary size labeled A through F, and a SPLIT on vertex 3 is depicted. T_1 is initialized to C and T_2 to D. We then move up from 3 to the root, placing 2 and B in T_1 (by making the root of C a right child of 2), placing 4 and E in T_2 (by making the root of D a left child of 4), placing 5 and F in T_2 (by making 4 a left child of 5), and finally placing 1 and A in T_1 (by making 2 a right child of 1).

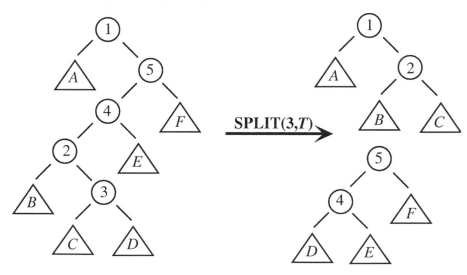

Detailed description of SPLIT

Notation: Analogous to the notation used for low-level list operations, for low-level tree operations we use *q.left*, *q.right*, and *q.parent* to denote the pointer fields associated with vertex *q*:

procedure SPLIT(*d*, *T*):

    Binary search down to the vertex *v* that contains *d*.

    x := *v.left*
    y := *v.right*

    while *v.parent*≠*nil* **do**

        if (*v.parent*).*left*=*v* **then begin** {*v* is a left child}
            v := *v.parent*
            v.left := *y*
            y.parent := *v*
            y := *v*
        end

        else begin {*v* is a right child}
            v := *v.parent*
            v.right := *x*
            x.parent := *v*
            x := *v*
        end

    x.parent := *nil*
    y.parent := *nil*

    return the two trees rooted at *x* and *y*

    end

Correctness: Can be verified by showing that each iteration of the *while* loop preserves the fact that *x* and *y* are roots of trees that contain elements less than and greater than *d* respectively; see the exercises.

Time: Proportional to the length of the path from *v* to the root.

Space: Since pointer fields are simply changed to form two trees from *T*, only $O(1)$ space is used in addition to the space used to store *T*.

Indexing a Binary Search Tree

Given that an in-order traversal of a binary search tree T produces the elements in sorted order, it is natural to ask for an element's index in this sorted list. Define:

INDEX(d, v, T): Return the index of d in the sorted list of data items stored the subtree of T rooted at v (where 1 is the index of the first element), or 0 if d is not in this subtree.

Idea: For each vertex w, COUNT(w) stores the number of vertices in the subtree rooted at w (including w). At each vertex v, if $d<$DATA(v) then the index of d in the subtree rooted at v is the same as the index of d in the subtree rooted at LCHILD(v), if $d=$DATA(v) then it is COUNT(LCHILD(v))+1, otherwise compute the index of d in the right subtree of v and add that to COUNT(LCHILD(v))+1.

Recursive algorithm: Define $COUNT(nil)=0$.
```
    function INDEX(d,v)
        if v=nil then return 0
        else if d<DATA(v) then return INDEX(d,LCHILD(v))
        else if d=DATA(v) then return COUNT(LCHILD(v))+1
        else begin
            i := INDEX(d,RCHILD(v))
            if i=0 then return 0 else return  COUNT(LCHILD(v))+1+i
            end
        end
```

Non-recursive algorithm: Move down from v, keeping a total of counts added thus far in the variable i; return the index of d when it is found, or return 0 if the *while* loop falls out the bottom of the tree.
```
    function INDEX(d,v)
        i := 0
        while v≠nil do begin
            if d<DATA(v) then v := LCHILD(v)
            else begin
                i := i+COUNT(LCHILD(v))+1
                if d=DATA(v) then return i else v := RCHILD(v)
                end
            end
        return 0
        end
```

Maintaining the COUNT fields: The INSERT and DELETE operations can be augmented to increment or decrement the counts along the corresponding root-to-leaf path; operations like JOIN and SPLIT can also be adapted (see the exercises).

(*) Binary Search Tree Ordering Lemma

Idea: The shape of a binary tree for a given set of data items is not unique. In fact, depending on the order that n elements are inserted, any possible shape for a binary tree of n vertices can result.

Definition: Two binary trees T_1 and T_2 have the same *shape* if they both are empty or there is a one-to-one correspondence between the vertices of T_1 and T_2 such that:

1. The root of T_1 corresponds to the root of T_2.

2. If v is a vertex of T_1 and w a vertex of T_2, then either both LCHILD(v) and LCHILD(w) are *nil*, or LCHILD(v) corresponds to LCHILD(w).

3. Symmetric to Condition 2 for RCHILD.

Lemma: Let T be *any* binary tree of n vertices and let $S = \{s_1, ..., s_n\}$ be any set of n distinct data items (with some ordering $\leq$ defined for every pair of elements). Then there exists a sequence of n INSERT operations (where each one inserts a distinct element of S) that constructs a binary search tree that has the same shape as T.

Proof: We use induction on n.

(*hypothesis*)

> The elements of S may be inserted into an initially empty tree X so that after all elements have been inserted, X has the same shape as T.

(*basis*)

> If $n=0$, then X and T are both empty and have the same shape by definition.

(*inductive step*)

> Let T be any binary tree of $n>0$ vertices.
>
> Let i be the number of vertices in the left subtree of T.
>
> Since the root is one vertex, it must be that $i < n$ (i.e., the left subtree of T has $<n$ vertices), and the right subtree of T has $n-i-1 < n$ vertices.
>
> Hence, if the $i+1^{th}$ largest element of S is inserted first, then by induction, the i elements of S that must go to the left of the root of X can be inserted in an order that makes the left subtree of X have the same shape as the left subtree of T.
>
> Similarly, the $n-i-1$ elements of S that must go to the right of the root of X can be inserted in an order that makes the right subtree of X have the same shape as the right subtree of T.

(*) Average Time to Build a Binary Search Tree

Theorem: The expected time to build a binary tree from a set of n distinct data items is $O(n\log(n))$, under the assumption that the order in which data items are inserted is random (i.e., all possible orderings are equally likely).

Proof: We count the number of comparisons to build the tree (the total time spent will be within a constant factor of this).

Let $T(n)$ denote the number of comparisons to build a tree from i randomly ordered distinct INSERT instructions.

For $n \leq 1$, no comparisons are used to build a tree for zero elements (an empty tree) or one element (a tree with just one vertex, which is the root).

For $n \geq 2$, suppose that the first element to be inserted is the ith largest of the elements. Then the $i-1$ smaller elements will be compared with the root (a total of $i-1$ comparisons) and go into the left subtree of the root (where $T(i-1)$ comparisons are used to build it) and the $n-i$ greater elements will be compared with the root (a total of $n-i$ comparisons) and go into the right subtree (where $T(n-i)$ comparisons are used to build it).

Hence, there are a total of $(i-1)+(n-i) = n-1$ comparisons with the root plus $T(i-1)$ comparisons to build the left subtree plus $T(n-i)$ comparisons to build the right subtree.

Since the first element is equally likely to be the ith largest for any $1 \leq i \leq n$, we can add up the time for all possible values of i and then divide by n:

$$T(0) = T(1) = 0$$

For $n \geq 2$:

$$T(n) = (n-1) + \tfrac{1}{n}\sum_{i=1}^{n}\left(T(i-1) + T(n-i)\right)$$

$$= (n-1) + \tfrac{1}{n}\sum_{i=1}^{n}T(i-1) + \tfrac{1}{n}\sum_{i=1}^{n}T(n-i)$$

$$= (n-1) + \tfrac{1}{n}\sum_{i=1}^{n}T(i-1) + \tfrac{1}{n}\sum_{i=1}^{n}T(i-1)$$

$$= (n-1) + \tfrac{2}{n}\sum_{i=1}^{n}T(i-1)$$

$$= (n-1) + \tfrac{2}{n}\sum_{i=0}^{n-1}T(i)$$

Induction may be now used to show $T(n)$ is $O(n\log(n))$.

Proof by induction that for $n \geq 2$ and some constant $k \geq 1$, $T(n) \leq kn\log_2(n)$:

For the basis, the terms in the summation are all 0 and $T(2) = n{-}1 = 1 < kn\log_2(n)$.

For the inductive step, assume that for $2 \leq i < n$, $T(i) \leq kn\log_2(n)$. Then:

$$T(n) = \frac{2}{n}\sum_{i=0}^{n-1} T(i) \; + \; n \; - \; 1 \quad \text{—by the definition of } T(n)$$

$$< \frac{2k}{n}\sum_{i=2}^{n-1}\left(i\log_2(i)\right) \; + \; n \quad \text{—by the inductive hypothesis}$$

For $k=4$, we can continue without calculus:

$$\leq \frac{8}{n}\left(\sum_{i=2}^{\lceil n/2 \rceil} i\lceil \log_2(i)\rceil \; + \; \sum_{i=\lceil n/2 \rceil+1}^{n-1} i\lceil \log_2(i)\rceil\right) + n$$

$$\leq \frac{8}{n}\left(\sum_{i=2}^{\lceil n/2 \rceil} i\left(\lceil \log_2(n)\rceil - 1\right) \; + \; \sum_{i=\lceil n/2 \rceil+1}^{n-1} i\lceil \log_2(n)\rceil\right) + n$$

$$< \frac{8}{n}\left(\lceil \log_2(n)\rceil\sum_{i=0}^{n-1} i \; - \; \sum_{i=0}^{\lceil n/2 \rceil} i\right) + n$$

$$\leq \frac{8}{n}\left(\lceil \log_2(n)\rceil\frac{n(n-1)}{2} - \left(\frac{n(n+2)}{8}\right)\right) + n \quad \text{— since } \frac{n}{2} \leq \lceil n/2 \rceil$$

$$< 4n\lceil \log_2(n)\rceil$$

For a better approximation, we can bound the sum with an integral (see the beginning of the appendix) using $k = 2ln(2) < 1.4$, where ln denotes the natural logarithm (that is, log to the base e, where $e = 2.7182...$ — see the first chapter):

$$< \frac{4\ln(2)}{n}\int_{i=2}^{n}\left(i\log_2(i)\right)di \; + \; n$$

$$= \frac{4\ln(2)}{n}\left[\frac{i^2\log_2(i)}{2} - \frac{i^2}{4\ln(2)}\right]_{i=2}^{i=n} + \; n$$

$$= \frac{4\ln(2)}{n}\left(\left(\frac{n^2\log_2(n)}{2} - \frac{n^2}{4\ln(2)}\right) - \left(\frac{2^2\log_2(2)}{2} - \frac{2^2}{4\ln(2)}\right)\right) + \; n$$

$$= 2\ln(2)n\log_2(n) - \frac{1}{n}\left(8\ln(2) - 4\right)$$

$$< 2\ln(2)n\log_2(n)$$

$$< 1.4n\log_2(n)$$

The Rotation Operation for Binary Search Trees

Idea: Change a few pointers at a particular place in the tree so that one subtree becomes less deep in exchange for another one becoming deeper. A sequence of rotations along a root to leaf path can help to make a binary search tree more "balanced". *RR* stands for "rotate right" and *RL* stands for "rotate left".

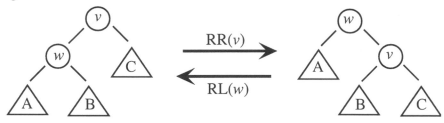

Notation: *q.left*, *q.right*, and *q.parent* denote the pointer fields associated with a vertex *q*.

RR(*v*) is just a few assignment statements:

> *u* := *v.parent*
> *w* := *v.left*
> *x* := *w.right*
> *v.left* := *x*
> *x.parent* := *v*
> *w.right* := *v*
> *v.parent* := *w*
> **if** *u* ≠ *nil* **then begin**
>   **if** *u.left* := *v* **then** *u.left* := *w*
>   **else** *u.right* := *w*
>   **end**
> *w.parent* := *u*

RL(*w*) is symmetric to RR(*v*); exchange "*v*" with "*w*" and "*left*" with "*right*":

> *u* := *w.parent*
> *v* := *w.right*
> *x* := *v.left*
> *w.right* := *x*
> *x.parent* := *w*
> *v.left* := *w*
> *w.parent* := *v*
> **if** *u* ≠ *nil* **then begin**
>   **if** *u.right* := *w* **then** *u.right* := *v*
>   **else** *u.left* := *v*
>   **end**
> *v.parent* := *u*

Example: A Sequence of Rotations

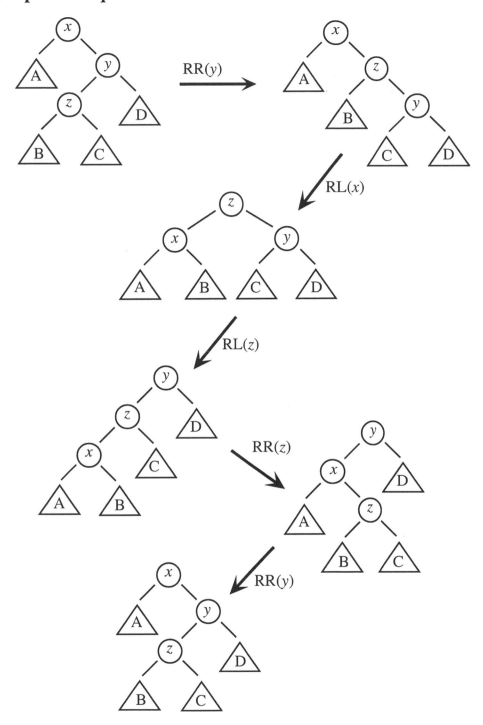

Self-Adjusting Binary Search Trees

Idea: Each time an operation such as INSERT walks down to a vertex x, do a sequence of rotations to "push" x up to the root. SPLAY-STEP(x) uses two rotations to reduce the depth of x by two (except it uses 1 rotation when x has depth 1). If x has depth d, SPLAY(x) performs a sequence of $\lceil d/2 \rceil$ SPLAY-STEP's to make x become the root:

> SPLAY(x): **while** x is not the root **do** SPLAY-STEP(x)

The SPLAY-STEP procedure: Apply one of steps 1L, 2L, or 3L (or three symmetric cases 1R, 2R, and 3R where the roles of right and left are reversed):

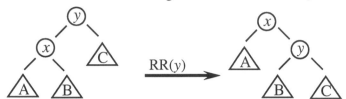

Case 1L: x is a left child and PARENT(x) is the root.

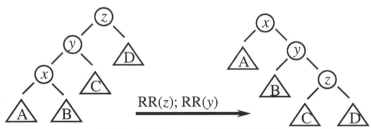

Case 2L: x is left child and PARENT(x) is a left child.

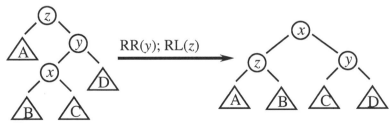

Case 3L: x is a left child and PARENT(x) is a right child.

Advantages of self-adjusting BST's:

Splaying can be applied to any binary search tree because SPLAY-STEP needs no additional information over the normal PARENT, LCHILD, and RCHILD fields.

When a SPLAY is performed after every basic binary search tree operation, the tree tends to stay balanced. Although an individual operation can be time consuming, a sequence of $O(n)$ operations is always $O(n\log(n))$.

Example, some sample SPLAY-STEPS:

Below is a binary search tree and the four resulting trees for each of the four SPLAY-STEPS of SPLAY(4) (numbers denote vertices and letters denote subtrees):

(*) Amortized Complexity of Self-Adjusting Binary Search Trees

Idea: In a similar but more complex spirit to the amortized analysis of the POP(i) operation presented in the chapter on lists, starting with an empty tree, we consider n operations. Our analysis holds for any operations that traverse a path to a vertex v, do $O(1)$ additional work, and then SPLAY(v); e.g., INSERT, MEMBER, MIN, DELETE, DELETEMIN. We assign $O(\log(n))$ "credits" to each operation, where a credit pays for $O(1)$ computing time, and show that the $O(n\log(n))$ credits suffice to perform the n operations. It is ok if an individual operation uses more than the credits assigned to it as long as it can "borrow" the extra credits from operations that do not use all their credits.

Rank of a vertex x:

$$weight(x) \quad = \text{number of descendants of } x \text{ (including } x \text{ itself)}$$
$$rank(x) \quad = \lfloor \log_2(weight(x)) \rfloor$$

Credit lemma: $3(rank(root)-rank(x))+1$ credits, in addition to whatever credits are left over from previous SPLAY operations, suffice to perform a SPLAY operation.

Proof: Define the *credit invariant* to associate with every non-leaf vertex x, $rank(x)$ credits. Let *newrank* denote the rank function after one splay step. Since SPLAY(x) always moves x up at least one level, it must always be that for the parent p of x, $newrank(x) \geq rank(p)$. Hence, summing the values of $newrank(x)-rank(x)$ along a path from a vertex v up to the root gives an amount $\leq rank(root)-rank(v)$. Thus it suffices to show that one splay step on a vertex x takes $3[newrank(x)-rank(x)]$ credits plus one additional credit if this is the last splay step, since summing these terms along a path from x to the root gives the desired $3(rank(root)-rank(x))+1$ credits. We consider Cases 1L, 2L; the reasoning for Case 3L is similar to 2L, and Cases 1R, 2R, and 3R are symmetric:

Case 1L:

This is the last step of the splay; the extra credit pays for the step itself. Since the rank of the root stays the same, the number of credits that must be added to the tree to maintain the credit invariant is:

$$newrank(y)-rank(x) \leq newrank(x)-rank(x) < 3(newrank(x)-rank(x))$$

Case 2L:

Again, since the rank of the root stays the same, the number of credits that must be added to maintain the invariant is:

$$(newrank(y)+newrank(z)) - (rank(y)+rank(x)) \leq 2(newrank(x)-rank(x))$$

If $rank(x)<newrank(x)$, then since $3(newrank(x)-rank(x))$ credits are allowed, we have an extra credit to pay for the step itself.

Otherwise, it must be that $newrank(x) = rank(x) = rank(y) = rank(z)$ and more than half the vertices are in subtrees A and B.

Hence $newrank(z)<rank(z)$, and by maintaining the credit invariant, a credit is freed up to pay for the step itself.

Theorem: The time for n operations is $O(n\log(n))$.

Proof: The time for all n operations is proportional to the time for all splaying, which by the credit lemma is $\leq 3\lfloor \log_2(n) \rfloor +1$ (since all vertices have rank $\leq \lfloor \log_2(n) \rfloor$).

Example: Tree Sort

Tree sort algorithm: Given a sequence of n input items to be sorted:

    Initialize an empty binary search tree T.

    for each input item x **do** INSERT(x,T)

    while T is not empty **do** output DELETEMIN(T)

Standard binary search tree:

Expected time $O(n\log(n))$: If input items are randomly ordered, we have already seen that the expected time to insert the items, which is proportional to the sum of the depths of all the vertices, is $O(n\log(n))$. Since each deletion has the effect of traversing a path to a vertex and removing that vertex, the total time for deletions is proportional to the sum of the depths of all vertices, and hence is also $O(n\log(n))$.

Worst case time $\Omega(n^2)$: If input arrives in order of largest to smallest, the *for* loop constructs a chain of $n-1$ left children where the largest is the root and the smallest is the only leaf (essentially a linked list), and both the *for* and *while* loops are $O(n^2)$.

Self-adjusting binary search tree is worst case $O(n\log(n))$ time:

Starting with an empty self-adjusting binary search tree, a sequence of $O(n)$ INSERT and DELETEMIN operations is always $O(n\log(n))$.

Example: Using a self-adjusting binary search tree, consider again the case when the input arrives in reverse sorted order.

The first item inserted becomes the root. After that each successive item is made the left child of the root and then a SPLAY makes it the root; so in only $O(n)$ time the *for* loop constructs a root with a chain of $n-1$ right children (again, essentially a linked list).

This is now a very easy $O(n)$ time case for the *while* loop because each delete simply removes the root and the SPLAY does nothing

However, even if the problem was to sort from largest to smallest, the self-adjusting tree does well. It starts by using $O(n)$ time to delete a leaf of depth $n-1$, but as time goes on, the tree gets more and more balanced and DELETEMIN's get less and less expensive, where the total time sums to only $O(n\log(n))$. For example, the first three deletions for $n=16$ go as follows:

Joining and Splitting Self-Adjusting Binary Search Trees

JOIN(T_1,d,T_2):

A single tree with root d is formed from T_1, d, and T_2.

** Always assumes that all items in T_1 are less than d and all items in T_2 are greater than d.

SPLIT(d,T):

Delete d and split T into items $<d$ and items $>d$ by searching for d (which SPLAY's d to make it the root) and then detaching the left and right subtrees of d (discard d).

** For ease of presentation, we assume that d is in T.

(*) **Complexity:** Referring to the amortized analysis of self-adjusting binary search trees presented earlier:

JOIN preserves the credit invariant since the $3\lfloor \log_2(n) \rfloor + 1$ credits allocated per operation are more than enough to pay for this operation (1 credit) and place $rank(d) = \lfloor \log_2(n) \rfloor$ credits on d (the credit invariant for all vertices in T_1 and T_2 is unaffected by this operation).

SPLIT preserves the credit invariant since if it is true for T after the SPLAY, then it must be true in the two subtrees of d.

Hence a sequence of INSERT, DELETE, MEMBER, JOIN, and SPLIT operations is always $O(n\log(n))$ time.

Sample Exercises

90. Consider the two definitions of a tree below, one the recursive one used in the tree terms at the beginning of this chapter and one that is non-recursive:

> *recursive*: A *tree* can be defined recursively as a collection of *vertices* consisting of a *root r* for which no other vertex is designated as its *parent*, together with a (possibly empty) set S of disjoint trees, where r is the parent of each of the roots of the trees in S.

> *non-recursive*: A *tree* is a collection of *vertices*. Exactly one vertex is the *root*. The root has no *parent*; for all other vertices v, there is exactly one vertex $w \neq v$ that is the parent of v. Furthermore, the parent relationship has no *cycles*; that is, it is not possible for the following algorithm to set w equal to v:
>
> $w := \text{PARENT}(v)$
>
> **while** $w \neq v$ and w is not the root **do** $w := \text{PARENT}(w)$

Prove that these two definitions are equivalent.

Solution: We proceed in two parts, one to show that a set of vertices satisfying the recursive definition must also satisfy the non-recursive definition, and the other part showing that a set of vertices satisfying the non-recursive definition must also satisfy the recursive definition.

> *recursive definition -> non-recursive definition*: Suppose that T is a set of vertices that satisfies the recursive definition. We proceed by induction on the number of vertices in T. If T has just one vertex, then it satisfies the non-recursive definition (since a single vertex is a tree). Now suppose that T is a tree (by the recursive definition) of more than one vertex, and assume that all sets of fewer vertices that satisfy the recursive definition also satisfy the non-recursive definition. By the recursive definition, there must be exactly one vertex r of T that is the root of T and has no parent, and the other vertices of T are partitioned into a set S of disjoint trees whose root vertices have parent r in T. Since all of the trees in S have at least 1 fewer vertices than T (since no tree in S includes r), by the inductive hypothesis, they all satisfy the non-recursive definition. Hence, since r is the parent of all roots of trees in S, all vertices in T except r have exactly one parent, since all vertices in the trees of S that are not roots have one parent by the non-recursive definition. In addition, since r has no parent, there are no cycles in T, since there are no cycles in the trees of S by the non-recursive definition. Hence T satisfies the non-recursive definition.

> *non-recursive definition -> recursive definition*: Suppose that T is a set of vertices that satisfies the non-recursive definition. We again proceed by induction on the number of vertices in T. If T has just one vertex, then it satisfies the recursive definition (since a single vertex is a tree). Now suppose that T is a tree (by the non-recursive definition) of more than one vertex, and assume that all sets of fewer vertices that satisfy the non-recursive definition also satisfy the recursive definition.

By the non-recursive definition, there must be exactly one vertex r of T that is the root of T and has no parent. For each vertex x of T such that r is the parent of x, we can form the set T_x as follows:

Initialize T_x to contain x.

while there is a vertex v in T but not in T_x whose parent is in T_x **do**

Add v to T_x.

Since each vertex except r has exactly one parent and cycles are not allowed, the sets T_x constructed above together with r form a partition of the vertices of T. Since each such tree T_x satisfies the non-recursive definition of a tree and has fewer vertices than T, by the inductive hypothesis, T_x also satisfies the recursive definition of a tree. Hence T satisfies the recursive definition of a tree.

91. Consider the tree:

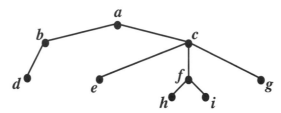

A. Starting at the root, and with the convention that children are visited in *reverse* alphabetical order (i.e., from right to left in the figure), list the order that the vertices are visited in preorder.

B. Same as Part A but for post-order.

C. Same as Part A but for level-order.

D. List the order that vertices are visited by an in-order traversal, using the following conventions:

- If a vertex v has only one child w that is attached by an edge going to the left, then w is visited before v.

- If a vertex v has only one child w that is attached by an edge going to the right, then w is visited after v.

- If a vertex v has two or more children, then leftmost child (as drawn in the figure) is visited first, then v is visited, then the remaining children are visited from left to right.

Solution:

A. *a, c, g, f, i, h, e, b, d*

B. *g, i, h, f, e, c, d, b, a*

C. *a, c, b, g, f, e, d, i, h*

D. *d, b, a, e, c, h, f, i, g*

92. Examples were given of how to compute height with pre-order traversal and min-height with level order traversal.

    A. Describe in English and give pseudo-code to compute the min-height with a pre-order traversal.

    B. Refine your pseudo-code of Part A to work specifically for the LMCHILD-RSIB representation.

Solution:

    A. Almost the same as computing the height; traverse the tree and compute 1 plus the minimum height of a child:

        procedure MIN-HEIGHT(v):

            $h := 0$

            for each child w of v **do** $h := $ MINIMUM(h,MIN-HEIGHT(w)+1)

            return h

        end

    B. We replace the *for* loop with a *while* loop that follows the RSIB links:

        procedure MIN-HEIGHT(v):

            $h := 0$

            $w := $ LMCHILD(v)

            while (w is not *nil*) **do begin**

                $h := $ MINIMUM(h,MIN-HEIGHT(w)+1)

                $w := $ RSIB(w)

            end

            return h

        end

93. For an arbitrary tree (not necessarily binary), describe in English and give pseudo-code for how to label each vertex with its depth. Analyze the time and space used.

Solution: Modify a pre-order traversal to take as an argument the current depth, so that initially, we call LABEL($root$,0):

    procedure LABEL(v,d):

        Label v with depth d.

        for each child w of v **do** LABEL(w,d+1)

    end

This pre-order traversal visits each vertex once and hence uses time linear in the number of vertices in the tree. It uses only $O(1)$ space in addition to the space used to store the tree and the space used by the recursion (which is proportional to the height of the tree and hence at most $O(n)$ for a tree of n vertices).

94. Explain why the following program visits a tree in pre-order:

> Initialize a *stack* to contain the *root*.
>
> **while** *stack* is not empty **do begin**
>> POP a vertex *v*
>>
>> {visit *v*}
>>
>> **if** *v* has a right sibling **then** PUSH(right sibling of *v*)
>>
>> **if** *v* is not a leaf **then** PUSH(leftmost child of *v*)
>
> **end**

Solution: The stack is used to "remember" the vertices that we have encountered thus far but have yet to be visited. After visiting a vertex *v*, its left child is pushed (which will cause all descendants of *v* to be visited); however, first the right sibling of *v* is pushed so that after the descendants of *v* have been visited, the right sibling of *v* will be next.

95. The following program visits the vertices of a tree:

> Initialize a *stack* to contain the *root*.
>
> **while** *stack* is not empty **do begin**
>> POP a vertex *v*
>>
>> {visit *v*}
>>
>> **for** each child *w* of *v* **do** PUSH(*w*)
>
> **end**

- A. Explain why this program visits a tree in pre-order.

- B. If *stack* is replaced by *queue*, POP is replaced by DEQUEUE, and PUSH is replaced by ENQUEUE, explain why this program performs level-order traversal.

Solution:

- A. If we view POP as entering the procedure and PUSH as making a recursive call, this code is virtually identical to the recursive procedure for pre-order traversal. If we assume that we have a standard LMCHILD-RSIB representation where children are visited from left to right by following RSIB links, the only difference between this version and the standard one is that the order of the siblings is reversed when they are pushed. So although both versions visit vertices in an order that satisfies the definition of a pre-order traversal, they are in general different orderings.

- B. The code becomes identical to that presented for level-order traversal.

96. For an arbitrary tree (not necessarily binary), describe in English and give pseudo-code for how to compute its *diameter*, the length of the longest path (which does not necessarily go through the root). Analyze the time and space used.

Solution: We already know how to label each vertex with its height (it is also possible to incorporate the computation of heights directly into the solution). The following function

DIAMETER takes a pointer to a vertex and returns the diameter of the corresponding subtree:

> **procedure** DIAMETER(p)
>
>> **if** p is a leaf **then return** 0
>>
>> **else if** p has only one child q **then return**
>>   MAXIMUM{HEIGHT(q)+1,DIAMETER(q)}
>>
>> **else begin**
>> i = largest height of a child of p
>> j = second largest height of a child of p
>> k = largest diameter of a child of p
>>   **return** MAXIMUM{$i+j+2,k$}
>>   **end**
>
> **end**

Note: By defining HEIGHT(nil) = -1, the first *else* statement is not needed.

The values i, j, and k can be computed by a single pass through the children of p, with a recursive call to DIAMETER on each child; hence, DIAMETER(p) amounts to nothing more than a traversal of the subtree rooted at p that works in time proportional to the number of vertices in that subtree. $O(1)$ space is used in addition to that used by the tree.

97. For simplicity, the operation SPLIT(T,d) for binary search trees assumed that d was in T. Modify the SPLIT algorithm to work for when d is not in T.

Solution: At the start of the SPLIT procedure, when searching down to look for the vertex containing d, if we get to a leaf and find that d is not in T, add v as a new child of this leaf. It will be ignored and deleted by the SPLIT algorithm, but this will save us the trouble of having a special starting condition for the *while* loop of the SPLIT procedure.

98. Prove that a *complete binary tree of height k*, where all non-leaf vertices have two children and all leaves have the same depth, has 2^k-1 non-leaf vertices and 2^k leaves.

Solution: We use induction on k:

For $k = 0$, a tree of height 0 has $0 = 2^0-1$ non-leaf vertices and $1 = 2^0$ leaf.

For $k > 0$ assume these assertions are true for complete trees of height less than k, and consider a complete tree of height k that is composed of a root and two subtrees of height $k-1$. By induction, the two subtrees each have $2^{k-1}-1$ non-leaf vertices and 2^{k-1} leaves.

Hence the tree as a whole has $2(2^{k-1}-1)+1 = 2^k-1$ non-leaf vertices and $2(2^{k-1}) = 2^k$ leaves.

Exercises

99. Give pseudo-code for inserting an item d as a child of a vertex v in an unordered tree stored in the LMCHILD-RSIB representation ($v=nil$ to make a new root).

100. Suppose that to represent arbitrary ordered trees, the LMCHILD-RSIB representation is augmented by adding an index field to each vertex that labels siblings with a distinct integer ≥ 1. So, for example, if a vertex has two children with indices 17 and 46, then it is considered to be missing children 1 through 16 and 18 through 45, and not to have any children numbered higher than 46. Give pseudo-code to insert (appropriately generalized to insert with any index) and delete a vertex under the assumption that sibling lists must be in increasing order of index; discuss how the complexity of these operations compares with the unordered case.

101. Consider the following approaches for representing a tree with an array:

parent array: Each position of the array stores a data item and the index of its parent.

child array: The array is organized in blocks containing a data item followed by indices to all of its children.

Analyze the complexity of basic tree operations for these two representations.

102. Define the *child number* of a tree to be the maximum number of children that any vertex has. Let T be an unordered tree stored with the LMCHILD-RSIB representation. Describe in English and give pseudo-code (using the PARENT, LMCHILD, and RSIB functions) to compute the child number of T.

103. Give a proof by induction that the following algorithm computes the height of the subtree rooted at v in a binary search tree:

function BHEIGHT(v):
    if $v=nil$ **then return** -1
    else return $1+$MAXIMUM{BHEIGHT(LCHILD(v)),BHEIGHT(RCHILD(v))}
    end

104. For $k \geq 0$, define the *k-height* of a tree to be the number of edges on a longest possible path from the root to a vertex that has at least k children such that every vertex on the path has at least k children (-1 if there is no such path). For example, the 0-height of a tree is simply the height of the tree. Let T be an unordered tree stored with the LMCHILD-RSIB representation.

A. Describe in English and give pseudo-code (using the PARENT, LMCHILD, and RSIB functions) for the function K-HEIGHT(k,T) to compute the k-height of T.

B. Suppose that the definition of k-height did not include the restriction that every vertex on the path has at least k children. Give pseudo-code for this alternate definition.

105. For $k \geq 0$, define the *k-min-height* of a tree to be the number of edges on a shortest possible path from the root to a vertex that has at most k children such that every vertex on the path has at most k children (-1 if there is no such path). For example, the 0-*min-height* of a tree is the shortest distance from the root to a leaf, that is, the MIN-HEIGHT of the root. Let T be an unordered tree stored with the LMCHILD-RSIB representation.

A. Describe in English and give pseudo-code (using the PARENT, LMCHILD, and RSIB functions) for the function K-MIN-HEIGHT(k,T) to compute the k-min-height of T.

B. Suppose that the definition of k-min-height did not include the restriction that every vertex on the path has at most k children. Give pseudo-code for this alternate definition.

106. Starting with an empty binary search tree, insert the following strings (in this order):

horse, cow, pig, seal, rat, dog, goat, elephant, fish, rooster, zebra, roach, cat, hen, llama, aardvark, hog, donkey, rhino, hippo, tiger, lamb, lion, leopard, lynx, kitty, ant, ape, animal

A. Draw the resulting tree.

B. Delete llama and draw the tree.

C. Starting from Step B, delete aardvark and draw the tree.

D. Starting from Step C, delete horse and draw the tree.

107. In this chapter, we took the opposite approach to linked lists, and did not include headers as part of the tree data structure. Discuss the advantages and disadvantages of the use of headers with trees, and how their use would affect the implementation of the basic operations.

108. As DELETE(v,T) for binary search trees was presented, v is replaced by either the largest element in its left subtree or the smallest element in its right subtree. However, if v has only one child, we can simply delete v and move its only child up to take its place. Give pseudo-code for the DELETE operation that incorporates this improvement; your code may make use of the MIN, MAX, DELETEMIN, and DELETEMAX operations.

109. Consider the problem of constructing a binary tree representing an arithmetic expression over $*$, $/$, $+$, and $-$ where $*$ and $/$ take precedence over $+$ and $-$. Using an approach similar to the example given in the chapter on induction and recursion for converting infix to prefix notation, present pseudo-code that constructs a tree for which in-order traversal evaluates the expression. Your code should allow a minus sign to work as a unary operator (e.g., $-2+3$).

110. For the INDEX operation, it was noted that the binary search tree INSERT (DELETE) operation can be augmented to increment (decrement) each of the counts along the corresponding root-to-leaf path.

A. Give a detailed description of the modifications needed for the INSERT and DELETE operations to maintain the COUNT fields.

B. Give a detailed description of the modifications needed for the JOIN and SPLIT operations to maintain the COUNT fields.

111. Consider an application where a binary search tree T is first constructed from a set of data items $D=\{d_1...d_n\}$ to form a database and then only MEMBER instructions are performed to access the data. For all that follows, we assume that D is indexed such that:

$$d_1 < d_2 < d_3 ... < d_n$$

Suppose that we are given a set $P=\{p_1...p_n\}$ of fixed probabilities such that for each data item d_i the probability that the next MEMBER instruction will be d_i is p_i; that is, the p_i's do not change over time and the sum of all n p_i's is 1. In addition, suppose that we are given a set $Q=\{q_0,q_1,...q_n,q_{n+1}\}$ such that q_0 is the probability that the next MEMBER instruction asks for an item $<d_1$, q_i, $1 \leq i < n$, is the probability that the next MEMBER instruction asks for an element between d_i and d_{i+1}, and q_n is the probability that the next MEMBER instruction asks for an element $>d_n$. An *optimal binary search tree* is defined as follows:

> T is *optimal for D with respect to P and Q* if the sum over each element of its depth in T times its probability is $\leq$ that for any binary search tree representing these elements.

Define:

T_{ij} = minimum cost tree for $\{d_i, d_{i+1}, ..., d_{i+j-1}\}$

An approach for computing an optimal tree is to build a triangular table (the first column has all n entries, the second column has only the first $n-1$ entries defined, ..., the last column has only the first entry defined) of all values of T_{ij} by working from smaller to larger values of j and filling in each column by using entries from a previous column (it is easy to initialize the first column since entries are just the costs for trees with a single root vertex). In addition to the cost of $T_{i,j}$ entry (i,j) stores a vertex *weight* that is the sum of the probabilities from P and Q corresponding to its subtree:

$$w_{ij} = q_{i-1} + (p_i+q_i) + (p_{i+1}+q_{i+1}) + ... + (p_{i+j-1}+q_{i+j-1})$$

The root of $T_{i,j}$ is also stored with entry (i,j) so that the optimal tree can be computed from the table after it is constructed.

A. Describe how to compute an entry (i,j) in $O(n)$ time by computing the cost for each of $d_i...d_{i+j-1}$ being the root (and then using entries from earlier columns to compute the optimal left and right subtrees of each potential root). Then present pseudo-code that fills in each of the $O(n^2)$ entries in a total of $O(n^3)$ time.

B. Instead of trying all possible roots $d_i...d_{i+j-1}$ to compute an entry (i,j), show that the search can be limited to roots that are between the root for $T_{i-1,j-1}$ and $T_{i,j-1}$ Then argue why this reduces the total time to $O(n^2)$.

C. Give pseudo-code to construct the optimal binary search tree for D with respect to P and Q from the table constructed in Part A or B that runs in $O(n)$ time.

112. Give a proof of correctness for the algorithm presented for the SPLIT operation for binary search trees.

Hint: Show that each iteration of the *while* loop preserves the fact that x and y are roots of trees that contain elements less than and greater than d respectively.

113. Fill in the details of the amortized analysis for self-adjusting binary search trees; specifically:

A. Annotate the argument given for Case 1L.

B. Annotate the argument given for Case 2L.

C. Give a detailed argument for Case 3L.

D. Give more details for the main theorem.

114. Give pseudo-code for splitting a self-adjusting binary search tree that works for both the cases that d is in T and d is not in T.

115. Consider tree sort using a regular binary search tree:

A. When the input arrives in order of largest to smallest, prove that both the *for* and *while* loops are $\Omega(n^2)$.

B. When the input arrives in order of smallest to largest, explain why the *while* loop becomes $O(n)$ but the *for* loop remains $\Omega(n^2)$.

116. Complete the example of performing tree sort on 16 elements with a self-adjusting binary search tree.

Chapter Notes

Basic tree data structures are presented in virtually all texts on data structures and algorithms; for further reading, the reader may refer to the list of texts given at the beginning of the chapter notes to the first chapter. Similar to linked lists, practical tree implementations must address the issue of reclaiming memory when vertices are deleted.

Self-adjusting binary search trees, which are often called *splay trees*, were introduced by Sleator and Tarjan [1983]; the presentation here is based on the presentation in the book of Tarjan [1983]. For additional reading, see also the books of Denenberg and Lewis [1991] and Mehlhorn [1986] (Volume 1).

Optimal binary search trees are constructed from a stationary set of probabilities, and are presented in many texts on data structures and algorithms; the hints given in the exercise on this problem are from the presentation in Aho, Hopcroft, and Ullman [1974]. See also Gilbert and Moore [1959], Knuth [1971]. Hu and Tucker [1971] consider the case where data is stored only in the leaves. In contrast, *dynamic binary search trees* keep the tree optimal for the probability distribution defined by the MEMBER instructions already processed; see the book of Mehlhorn [1984] (Volume 1) for further reading on this and related subjects.

Bentley [1975] considers multidimensional binary search trees for associative searching.

5. Algorithm Design

Introduction

There are hosts of algorithm design techniques that have been studied:

    divide and conquer

    dynamic programming

    randomization

    greedy algorithms and matroids

    transformation to a graph algorithm (network flow, etc.)

    optimization algorithms (hill climbing, simulated annealing, etc.)

    approximation algorithms

    linear programming

    heuristic search

    genetic algorithms

    "software tools"

    etc.

Here we focus primarily on the first two, *divide and conquer* and *dynamic programming*. Both solve a large problem by combining solutions to smaller ones of the same type.

Divide and conquer differs from dynamic programming in the way in which sub-problems are computed. It is typically used when there is a large space of possible sub-problems, but only a relatively small number are needed and of those that are needed, each is typically used only once in the solution of the whole problem. Divide and conquer algorithms are typically recursive programs that work "top down", making a recursive call to solve a sub-problem whenever the solution to a sub-problem is needed.

In contrast, dynamic programming is typically used when the space of all possible sub-problems is relatively small and it may be that a particular sub-problem is used many times. Dynamic programming algorithms typically work iteratively "bottom up" to store the solution to all sub-problems in a table where solutions to larger sub-problems make use of already computed solutions to smaller problems.

We finish this chapter with *randomized algorithms*, *greedy algorithms*, and *simulated annealing*. They are approaches that can stand by themselves or used in conjunction with divide and conquer or dynamic programming. Randomized algorithms make random choices (e.g., coin flipping) to achieve good expected time for a correct solution or good worst case time for a solution that is highly likely to be correct.

Greedy algorithms attempt to find the best solution to an entire problem by always choosing the most desirable solution to sub-problems. For example, to travel from one city to another by car, at each intersection we could choose the road that goes closest in the direction of the destination city; although there may be times that by consulting a map, we would discover that it is best to travel in a slightly worse direction for a short while in order to get to a major highway that goes quicker to the destination. We shall see examples like this, where greedy may be fast but not always the best, as well as examples where it can be proved to yield an optimal solution.

Simulated annealing is often used when no efficient solution is known but the problem can be formulated as finding a solution in a large space of solutions by starting with a relatively poor solution and successively making small changes to work towards a better solution. The danger, like with the greedy approach, is that one might get stuck at a local minimum (a poor solution that cannot be improved with small changes). Simulated annealing attempts to avoid this by initially exploring changes that may lead to worse solutions and then, over time, settling down to more conservative choices.

Divide and Conquer

Idea:

1. Divide a large problem into a number of smaller problems of the same type.
2. Solve the smaller problems (recursively).
3. Combine solutions to the smaller problems into a solution to the original problem.

Examples we have already seen:

Binary Search: A "degenerate" case where the original problem is "divided" into only one smaller problem (finding an item in a list of $n/2$ items).

Tree Traversal Algorithms: Pre-order, in-order, and post-order all traverse a tree by recursively traversing smaller sub-trees.

Balanced sub-problems: To get an efficient algorithm, it is typically important that the smaller problems be of approximately the same size. For example, the following algorithm for sorting a list is just tail recursion that essentially performs an insertion sort in $O(n^2)$ time (so we have gained nothing over brute-force sorting).

1. Divide the list into its first element and the remaining $n-1$ elements.
2. Sort the $n-1$ elements recursively.
3. Insert the first element into this sorted list.

Base case size: Although divide and conquer typically works for small problems, it may make sense to solve sub-problems of small sizes by a simpler (but asymptotically less efficient) method. For example, when we presented a recursive version of binary search in the chapter on induction and recursion called *RSB*, although the process did not stop until the endpoints became the same (and we were searching a list of size 1), we could just as well, for any $k \geq 0$, stopped when the list is size $1+k$ or smaller:

> **function** kRBS(a,b)
>   **if** $(a+k)<b$ **then begin**
> $m := \lfloor (a+b)/2 \rfloor$
>   **if** $x \leq A[m]$ **then return** kRBS(a,m) **else return** kRBS($m+1,b$)
>   **end**
>   **else begin**
>   **for** $i := a$ **to** b **do if** $x=A[i]$ **then return** i
>   **return** -1
>   **end**
>   **end**

We usually do not bother with this level of detail in our pseudo-code descriptions of algorithms, but it may be important in practice to perform additional analysis or experiments to optimize the base case size for a particular implementation, especially when the recursion is more complex than the simple tail recursion of this example.

Example: Merge Sort

Idea: To sort a list, divide it in half, sort the two halves independently, and then merge the two sorted lists into a single sorted list. To simplify our presentation, we begin by assuming that linked lists are used.

Sort a list:

    procedure MSORT(*list*):
        if *list* is empty or has only one element **then return** *list*
        else begin
            Divide *list* into two equal size lists, *list1* and *list2*.
            return MERGE(MSORT(*list1*),MSORT(*list2*))
            end
        end

Merge two sorted lists in linear time and space:

    function MERGE(*list1*,*list2*)
        list3 := the empty list
        while both *list1* and *list2* are not empty **do begin**
            Compare the first elements of *list1* and *list2* and whichever is smaller
            (ties can broken arbitrarily), delete it and append it to *list3*.
            end
        append *list1* and *list2* to the end of *list3*
        return *list3*
        end

Time to sort a list of n elements: A linked list of n elements can be divided in $O(n)$ time by going to the middle (if the length is not known, first traverse it to count the number of elements) or by extracting every other element; if n is odd, the first half can have one more element than the second half. If $T(n)$ denotes the time to sort n elements, then if $n \leq 1$, $T(n) \leq a$ for some constant a. For $n>1$, two recursive calls on problems of half the size take time $2T(\lceil n/2 \rceil)$, and for some constant b, bn time is used for the dividing and merging. If N is the smallest power of 2 that is $\geq n$, and c the maximum of a and b then:

$$T(n) \leq 2T\left(\frac{N}{2}\right) + cN$$

$$= 4T\left(\frac{N}{4}\right) + 2c\frac{N}{2} + cN$$

$$= 8T\left(\frac{N}{8}\right) + 4c\frac{N}{4} + 2c\frac{N}{2} + cN$$

$$= cN + cN + \cdots + cN \quad \text{---} \quad \log_2(N)+1 \text{ terms}$$

$$= cN(\log_2(N)+1)$$

$$= O(n\log(n))$$

Array Implementation of Merge Sort

Idea: If lists are represented as arrays, MSORT can take a pair of indices i,j indicating the range of elements to be sorted and do recursive calls on the two halves of this range (we assume the array name is A; a pointer to A could be passed as a third argument).

Sort: To sort $A[i]...A[j]$, *arrayMSORT* computes the midpoint m between i and j by doing $m := \lfloor (i+j)/2 \rfloor$. If $A[i]...A[j]$ is an odd number of elements (i.e., $j-i+1$ is odd), one more element is in the left half $A[i]...A[m]$ than the right half $A[m+1]...A[j]$; otherwise, the the same number of elements are in the two halves (i.e., $(j-i+1)/2$ elements each). Recursive calls are made on each half, and then the two halves are merged together. To sort $A[1]...A[n]$, we simply call *arrayMSORT*$[1,n]$.

```
procedure arrayMSORT(i,j)
    if i<j then begin
        m := ⌊(i+j)/2⌋
        arrayMSORT(i,m)
        arrayMSORT(i+1,j)
        arrayMERGE(i,j,m)
        end
    end
```

Merge: To merge $A[i]...A[m]$ with $A[m+1]...A[j]$, MERGE places items in sorted order into $B[i]...B[j]$ and then copies this sequence back into $A[i]...A[j]$. Essentially the same algorithm as for linked lists is used; the variables x and y are the current positions in the two lists $A[i]...A[m]$ and $A[m+1]...A[j]$, and the variable z goes from i to j to fill up B.

```
procedure arrayMERGE(i,j,m)
    x := i
    y := m+1
    z := i
    while z ≤ j do begin
        if y>j then begin B[z] := A[x]; x := x+1 end
        else if x>m then begin B[z] := A[y]; y := y+1 end
        else if A[x]<A[y] then begin B[z] := A[x]; x := x+1 end
        else begin B[z] := A[y]; y := y+1 end
        z := z+1
        end
    for k := i to j do A[k] := B[k]
    end
```

Complexity: Since only the arrays A and B are used, the space is linear. The time analysis is identical to the linked list version.

Example: Quick Sort

Idea: To sort $A[i]...A[j]$, partition A into the two portions that are $<$ and $>$ a *pivot element*, and recursively sort the two halves. We choose $A[i]$ as the pivot element (we shall see later that a random element may be better). Work from the outsides towards the middle, exchanging elements when the one on the left is $\geq A[i]$ and the one on the right is $<A[i]$ until the variables x and y "cross" (i.e., $x=y+1$). Then move $A[i]$ between the two halves (so the recursive calls will not include $A[i]$, which is already correctly positioned in the sorted order). Quick sort is *in-place*, using only $O(1)$ space in addition to A.

A:

	items $<A[i]$	items yet to be processed	items $\geq A[i]$
i		x	y	j

Sort $A[i]...A[j]$:

> **procedure** arrayQSORT(i,j)
>   **if** $i<j$ **then begin**
> a,b := PARTITION(i,j)
>   QSORT(i,a)
>   QSORT(b,j)
>   **end**
>   **end**

Partition in-place $A[i]...A[j]$ with respect to $A[i]$:

> **function** PARTITION(i,j)
> x := $i+1$
> y := j
>   **while** $x \leq y$ **do**
>   **while** $x \leq y$ and $A[x]<A[i]$ **do** x := $x+1$
>   **while** $x \leq y$ and $A[y] \geq A[i]$ **do** y := $y-1$
>   **if** $x<y$ **then begin** exchange $A[x]$ and $A[y]$; x := $x+1$; y := $y-1$ **end**
>   **end**
>   Exchange $A[i]$ and $A[y]$.
>   **return** $y-1$ and $y+1$ (that is, $A[i]...A[y-1] < A[y] \leq A[y+1]...A[j]$)
>   **end**

Theorem: The expected time of arrayQSORT is $O(n\log(n))$, under the assumption that the input items are distinct and randomly ordered.

Proof: Time is proportional to the number of comparisons in A. Partitioning compares $n-1$ items to $A[1]$ and, if $A[1]$ is the ith largest element, $T(i-1)$ comparisons are used to sort elements $< A[1]$ and $T(n-i)$ comparisons are used to sort elements $>A[1]$. Since $A[1]$ is equally likely to be anywhere in the sorted order, add up for all n values of i and divide by n to get the *same* recurrence as for building a binary search tree:

$$T(n) = (n-1)+\frac{1}{n}\sum_{i=1}^{n}\left(T(i-1)+T(n-i)\right) = O\left(n\log(n)\right)$$

CHAPTER 5

Example: Finding the kth Largest Element

Idea: The kth largest element in a list L of $n \geq 1$ elements, is the kth element when L is arranged in non-decreasing order; for example, the 4th largest of $(3,1,8,7,2,3,4)$ is 3. Instead of sorting L, find an element m which is close to the median of L, partition L into the elements $<m$, $=m$, and $>m$, and then recursively look in only one of these three lists.

function KLARGEST(k,L)

    if $|L|<20$ **then** sort L and return the kth element

    else begin

        1.   Divide L into $\lfloor n/5 \rfloor$ "mini-lists" of 5 elements each, and at most 4 leftovers.

        2. $M :=$ a list of the medians of the mini-lists

        3. $m :=$ KLARGEST$(\lceil |M|/2 \rceil, M)$

        4.   Form the lists A, B, and C of the elements in L that are $<m$, $=m$, and $>m$.

        5.   **if** $k \leq |A|$ **then return** KLARGEST(k,A)

            else if $k \leq (|A|+|B|)$ **then return** m

            else return KLARGEST$((k-|A|-|B|),C)$

    end

end

Lists A and C must have $<3/4n$ elements: Divide and conquer algorithms typically work best when sub-problems have equal size. Although m may not be the median of L, neither A nor C gets too large a fraction of L. Since at least $\lceil |M|/2 \rceil$ of the mini-lists contain a median that is $\leq m$ and at least two other elements that are $\leq m$, at least $\lceil 3|M|/2 \rceil$ $\geq \lfloor 3n/10 \rfloor$ of the elements of L are $\leq m$, and so $|C| \leq \lceil 7n/10 \rceil$. Similarly, since at least

$$|M| - \lceil |M|/2 \rceil + 1 \; = \; \lfloor |M|/2 \rfloor + 1 \; \geq \; \lceil |M|/2 \rceil$$

of the mini-lists contain a median that is $\geq m$ and at least two others that are $\geq m$, it follows that $|A| \leq \lceil 7n/10 \rceil$. Hence, both $|A|$ and $|C|$ are $\leq (3/4)n$, since $n \geq 20$ implies:

$$\lceil 7n/10 \rceil \; \leq \; (7n/10)+1 \; \leq \; (7n/10) + 0.05n \; = \; 0.75n \; = \; (3/4)n$$

Time and space used by KLARGEST: The space is $O(n)$. There is a constant c such that $T(n) \leq cn$ if $n<20$, and for $n>20$, since there is one recursive call in Step 3 on a list of size $\leq (n/5)$, one in Step 5 on a list of size $\leq (3/4)n$, and everything else is $O(n)$:

$$T(n) \leq T(\tfrac{1}{5}n) + T(\tfrac{3}{4}n) + cn$$

A proof by induction now shows that $T(n) \leq 20cn$. For $n<20$, $cn<20cn$, and for $n \geq 20$, by applying the inductive hypothesis, $T(n) \leq 4cn+15cn+cn = 20cn$.

Practical considerations: Intuitively, time is linear because $(1/5)+(3/4)<1$. However, because the inequality is close, the constant is poor. We shall see a practical randomized approach later.

Example: Polynomial Multiplication

Idea: Although we shall see faster ways to multiply polynomials when we study the discrete Fourier transform, a simple strategy is to divide the polynomial in half and recursively work with the two pieces.

Notation:

$$P(x) = P[n]x^n + P[n-1]x^{n-1} + \cdots + P[1]x + P[0]$$

$$P_{left}(x) = P[n]x^{\lceil n/2 \rceil} + P[n-1]x^{\lceil n/2 \rceil - 1} + \cdots + P[\lfloor n/2 \rfloor + 1]x$$

$$P_{right}(x) = P[\lfloor n/2 \rfloor]x^{\lfloor n/2 \rfloor} + P[\lfloor n/2 \rfloor - 1]x^{\lfloor n/2 \rfloor - 1} + \cdots + P[1]x + P[0]$$

Recursive decomposition: Since $P(x) = x^{\lfloor n/2 \rfloor}P_{left} + P_{right}$ it follows that for two n^{th} degree polynomials P and Q:

$$
\begin{aligned}
PQ &= (x^{\lfloor n/2 \rfloor}P_{left} + P_{right})(x^{\lfloor n/2 \rfloor}Q_{left} + Q_{right}) \\
&= x^n P_{left}Q_{left} + x^{\lfloor n/2 \rfloor}(P_{left}Q_{right} + P_{right}Q_{left}) + P_{right}Q_{right}
\end{aligned}
$$

Assuming that recursive calls are used to multiply polynomials of degree $n/2$ and that two polynomials of degree n are added or scaled by a power of x in $O(n)$ time, this straightforward decomposition leads to a divide and conquer algorithm with a running time given by $T(1) = O(1)$ and in general $T(n) = 4T(n/2) + O(n) = O(n^2)$, which is no better than the standard method. Note that all powers of x can be computed in a total of $O(n)$ time with a simple *for* loop.

Equivalent computation with only 3 multiplicatons:

$$
\begin{aligned}
A &= P_{left}Q_{left} \\
B &= P_{right}Q_{right} \\
C &= (P_{left} + P_{right})(Q_{left} + Q_{right}) \\
PQ &= x^n A + x^{\lfloor n/2 \rfloor}(C - A - B) + B
\end{aligned}
$$

Assume that n is a power of two. For some constant c, $T(1)=c$ and for $n>1$:

$$T(n) \le 3T\left(\tfrac{n}{2}\right) + cn$$

$$= 3\left(3T\left(\tfrac{n}{4}\right) + \tfrac{c}{2}n\right) + cn = 3\left(3\left(3T\left(\tfrac{n}{8}\right) + \tfrac{c}{4}n\right) + \tfrac{c}{2}n\right) + cn$$

$$= cn\left(1 + \tfrac{3}{2} + \left(\tfrac{3}{2}\right)^2 + \left(\tfrac{3}{2}\right)^3 + \cdots \left(\tfrac{3}{2}\right)^{\log_2(n)}\right)$$

$$= cn\left(\text{a geometric series in } 3/2 \text{ with largest term } \log_2(n)\right) \quad \text{— see the appendix}$$

$$< 3cn\left(\tfrac{3}{2}\right)^{\log_2(n)} = 3cn\left(n^{\log_2(3/2)}\right) = 3cn\left(n^{\log_2(3)-1}\right) = 3cn^{\log_2(3)} = O\left(n^{1.59}\right)$$

When n is not a power of two, the actual running time will be no worse than that for the next power of 2, which is $<2n$, and does not change the asymptotic complexity.

CHAPTER 5

Example: Strassen's Algorithm for Matrix Multiplication

Idea: Recall from the first chapter the definition of multiplying a m by w matrix X with a w by n matrix Y to form a m by n matrix Z. Although there are asymptotically faster ways to multiply matrices, *Strassen's Algorithm* was the first to beat the mwn running time (n^3 for square matrices) of the standard algorithm by decomposing the matrices into sub-matrices of half the size.

Recursive decomposition: In a spirit similar to the previous example for polynomial multiplication, divide X and Y into four sub-matrices with half the vertical and horizontal dimensions. It can be shown (see the exercises) that X and Y can be multiplied just as if they were 2 by 2 matrices, where recursive calls and matrix addition (pair-wise addition of corresponding elements) are used to multiply and add sub-matrices. This decomposition by itself does not improve the running time, since after unwinding the recursion, the same mwn multiplications are performed as with the standard algorithm.

$$\begin{array}{c}\\ \lceil m/2 \rceil \\ \lfloor m/2 \rfloor \end{array}\overset{\lceil w/2 \rceil \quad \lfloor w/2 \rfloor}{\begin{bmatrix} X_{1,1} & X_{1,2} \\ \hline X_{2,1} & X_{2,2} \end{bmatrix}} * \begin{array}{c} \lceil w/2 \rceil \\ \lfloor w/2 \rfloor \end{array}\overset{\lceil n/2 \rceil \quad \lfloor n/2 \rfloor}{\begin{bmatrix} Y_{1,1} & Y_{1,2} \\ \hline Y_{2,1} & Y_{2,2} \end{bmatrix}} = \begin{array}{c} \lceil m/2 \rceil \\ \lfloor m/2 \rfloor \end{array}\overset{\lceil n/2 \rceil \quad \lfloor n/2 \rfloor}{\begin{bmatrix} Z_{1,1} & Z_{1,2} \\ \hline Z_{2,1} & Z_{2,2} \end{bmatrix}}$$

For example:

$$\begin{pmatrix} 1 & 1 & 1 & 1 \\ 2 & 2 & 2 & 2 \\ 3 & 3 & 3 & 3 \end{pmatrix} * \begin{pmatrix} 1 & 5 \\ 2 & 6 \\ 3 & 7 \\ 4 & 8 \end{pmatrix} = \begin{pmatrix} \begin{pmatrix} 1 & 1 \\ 2 & 2 \\ (3 & 3) \end{pmatrix} & \begin{pmatrix} 1 & 1 \\ 2 & 2 \\ (3 & 3) \end{pmatrix} \end{pmatrix} * \begin{pmatrix} \begin{pmatrix} 1 \\ 2 \\ 3 \\ 4 \end{pmatrix} & \begin{pmatrix} 5 \\ 6 \\ 7 \\ 8 \end{pmatrix} \end{pmatrix} = \begin{pmatrix} 10 & 26 \\ 20 & 52 \\ 30 & 78 \end{pmatrix}$$

Using 7 multiplications instead of 8:

$$\begin{aligned} A &= (X_{11} + X_{22}) * (Y_{11} + Y_{22}) \\ B &= (X_{12} - X_{22}) * (Y_{21} + Y_{22}) & Z_{1,1} &= A + B - D + F \\ C &= (X_{11} - X_{21}) * (Y_{11} + Y_{12}) & Z_{1,2} &= D + E \\ D &= (X_{11} + X_{12}) * Y_{22} & Z_{2,1} &= F + G \\ E &= \quad\quad X_{11} * (Y_{12} - Y_{22}) & Z_{2,2} &= A - C + E - G \\ F &= \quad\quad X_{22} * (Y_{21} - Y_{11}) \\ G &= (X_{21} + X_{22}) * Y_{11} \end{aligned}$$

Assuming that two $n/2$ by $n/2$ matrices are multiplied recursively with this construction, for square matrices the recurrence is similar to the previous example (and the analysis can be generalized to non-square matrices — see the exercises):

$$T(n) = 7T\left(\tfrac{n}{2}\right) + cn^2 = O\left(n^{\log_2(7)}\right) = O\left(n^{2.81}\right)$$

Note: Because $n^{2.81}$ is close to n^3, Strassen's algorithm may only be practical for relatively large matrices, and "tuning" the base case size may be important.

Divide and Conquer Recurrence Relations

Idea: Perhaps the most common type of divide and conquer algorithm is where there are a number of problems of a fraction of the original size, and the time, exclusive of the recursive calls, for dividing and combining is a "common" function like n, n^2, or $n\log(n)$.

Theorem: For $a,e,h \geq 0$, $c,f > 0$, $b \geq 1$, $g > 1$ real numbers and $d \geq 2$, $n \geq 1$ integers, if $T(n)$ is a non-decreasing function ($i<j$ implies $T(i) \leq T(j)$) that satisfies

$$T(n) \leq \begin{cases} a & \text{if } n \leq b \\ cT\left(\left\lceil \frac{n}{d} \right\rceil\right) + en^f \left(\log_g(n)\right)^h & \text{if } n > b \end{cases}$$

then:

$$T(n) = \begin{cases} O\left(n^f \log(n)^h\right) & c < d^f \\ O\left(n^f \log(n)^{h+1}\right) & c = d^f \\ O\left(n^{\log_c(d)}\right) & c > d^f \end{cases}$$

Proof: We first consider the case that $n \geq 1$ is an exact integer power of d:

$$T(n) \leq cT\left(\frac{n}{d}\right) + en^f \log_g(n)^h$$

$$= c\left(cT\left(\frac{n}{d^2}\right) + e\left(\frac{n}{d}\right)^f \log_g\left(\frac{n}{d}\right)^h \right) + en^f \left(\log_g(n)\right)^h$$

$$= c\left(c\left(cT\left(\frac{n}{d^3}\right) + e\left(\frac{n}{d^2}\right)^f \log_g\left(\frac{n}{d^2}\right)^h \right) + e\left(\frac{n}{d}\right)^f \log_g\left(\frac{n}{d}\right)^h \right) + en^f \log_g(n)^h$$

$$= an^{\log_d(c)} + e \sum_{i=0}^{\log_d(n)-1} \left(\frac{n}{d^i}\right)^f \log_g\left(\frac{n}{d^i}\right)^h c^i$$

If $c \leq d^f$, then $n^{\log_d(c)} \leq n^f$, and we can factor out n^f and $\log_g(n)^h \geq \log_g\left(n/d^i\right)^h$ to get

$$T(n) = O\left(n^f \log_g(n)^h \sum_{i=0}^{\log_d(n)-1} \left(\frac{c}{d^f}\right)^i \right)$$

from which the theorem follows; i.e., for $c < d^f$ the resulting geometric sum is $O(1)$ and for $c = d^f$, each term is 1, and so the sum contributes an additional factor of $O(\log(n))$.

For $c > d^f$, since for some real number $\varepsilon > 0$, $n^f \log_g(n)^h = O\left(n^{\log_d(c)-\varepsilon}\right)$, we have:

$$T(n) = O\left(n^{\log_d(c)} + \sum_{i=0}^{\log_d(n)-1} \left(\frac{n}{d^i}\right)^{\log_d(c)-\varepsilon} c^i \right) = O\left(n^{\log_d(c)} + n^{\log_d(c)-\varepsilon} \sum_{i=0}^{\log_d(n)-1} \left(d^\varepsilon\right)^i \right)$$

from which the theorem follows since this geometric sum is $O(n^\varepsilon)$.

For n not an integer power of d, let m be the smallest integer $>n$ that is an integer power of d; then $m<dn$, and since $T(n)$ is non-decreasing and $T(n)$ is always $O(n^k)$ for some k, the analysis for m is within a constant factor of that for n, since $(dn)^k = (d^k)n^k = O(n^k)$.

Dynamic Programming

Idea: Store the answers to all sub-problems in a table; compute this table by working from smaller sub-problems to larger ones. Examples we have already seen include:
- Pascal's triangle of binomial coefficients.
- Optimal binary search tree construction.

Example: The Knapsack Problem

You walked miles to a remote beach and collected many shells. Although their weights vary greatly, per pound you like them all equally well. Unfortunately, due to back problems, your doctor has set a strict limit on how much weight you can carry. You want to carry as much as you can without exceeding this limit. Here, we consider only the following simple question:

> Given a set of positive integer weights $W[1]$, $W[2]$, ... $W[m]$ and an integer $n \geq 1$, is there a subset of the weights that adds up to exactly n?

Define a truth array $T[i,j]$, $1 \leq i \leq m$ and $0 \leq j \leq n$, to be $T[i,j]=true$ if and only if there is a subset of $W[1]...W[i]$ that sums to exactly j (the empty subset is defined to sum to 0).

For the first row, $T[1,j]=true$ if and only if $j=0$ or $j=W[1]$. For higher numbered rows, $T[i,j]=true$ if and only if at least one of the following two conditions holds:

1. $T[i-1,j]=true$ (do not include the ith item)
2. $W[i] \leq j$ and $T[i-1,j-W[i]]=true$ (include the ith item and have $j-W[i]$ weight left).

Hence, higher numbered rows can be computed from lower numbered ones:

for $0 \leq j \leq n$ **do if** $j=0$ or $j=W[1]$ **then** $T[1,j] := true$ **else** $T[1,j] := false$
for $i := 2$ **to** m **do**
    for $0 \leq j \leq n$ **do**
        if $T[i-1,j]=true$ or $T[i-1,j-W[i]]=true$ **then** $T[i,j] := true$ **else** $T[i,j] := false$
print $T[m,n]$

The algorithm uses $O(mn)$ time to fill in a table that uses $O(mn)$ space.

The following example with $n=13$ shows the true entries and leaves the false ones blank:

	0	1	2	3	4	5	6	7	8	9	10	11	12	13
W[1]=2	T		T											
W[2]=7	T		T					T		T				
W[3]=5	T		T			T		T		T			T	
W[4]=4	T		T		T	T	T	T		T		T	T	T
W[5]=12	T		T		T	T	T	T		T		T	T	T

Note: Table height is proportional to the number of weights m, but table width is proportional to n, not the number of digits to represent n. Thus, depending on the relationship of n to m, the width of the table could be very large as compared to the length of a string representing the input. No efficient way to compute the optimal solution for arbitrary classes of inputs is known (see the chapter notes).

Example: The Paragraphing Problem

Problem: Given words $w_1...w_m$ and a line length L, break the sequence into lines subject to a penalty function $\alpha(i,j) \geq 0$ that measures how "bad" words $w_i...w_j$ look on a line that is both right and left justified (or in the case of the last line, when $j=m$, only left justified). Minimize the sum of the penalty function over all lines of the paragraph.

Idea: Compute from right to left $cost[i]$, the cost of paragraphing $w_i L w_m$. Only the values of $cost[j]$, $j>i$, are needed to compute $cost[i]$. To compute $cost[i]$, we try all ways of breaking off the first line of $w_i L w_m$ and adding the cost of that line to the cost of paragraphing the remaining words (already computed). An array $break$ can store the best break points found.

Computing the *cost* and *break* arrays:
    Input and store the words $w_1...w_m$ in a list of strings
    $cost[m+1] := 0$
    for $i := m$ **downto** 1 **do begin**
        $cost[i] := \infty$
        $break[i] := m$
        $j := i$
        while $j \leq m$ and $w_i...w_j$ can fit on a line **do begin**
            if $(\alpha(i,j)+cost[j+1])<cost[i]$ **then begin**
                $cost[i] := \alpha(i,j)+cost[j+1]$
                $break[i] := j$
            end
            $j := j+1$
        end
    end

Formatting the paragraph:
    $i := 1$
    while $i \leq m$ **do begin**
        Output a line consisting of w_i through $w_{break[i]}$
        $i := break[i]+1$
    end

Complexity: Let n be the sum of the lengths of all input words. Reading the input is $O(n)$. Under the assumption that $\alpha(i,j)$ can be computed in $O(1)$ time, computing the *cost* and *break* arrays is at worst $O(mL) \leq O(nL)$, which is $O(n)$ assuming that L is constant with respect to n. Formatting the paragraph takes an additional $O(n)$ time. In addition to the space to store the input, $O(m)$ space is used to store the *cost* and *break* arrays.

In fact, it is not necessary for this analysis that $\alpha(i,j)$ can be computed in $O(1)$ time, only that it can be computed in $O(1)$ time from values already computed.

Example: Optimal Ordering of Matrix Multiplications

Problem: We have already seen how to multiply two matrices. Since a sequence of matrix multiplications is *associative,* that is $A(BC)=(AB)C$ (see the exercises), it can be computed by performing the multiplications in any order, and the order chosen can greatly affect the total number of arithmetic operations performed (for simplicity, we just count the number of multiplications performed).

Example: Suppose we wish to multiply matrices of the following dimensions:

$(A = 1x5) (B = 5 \ x \ 25) (C = 25 \ x \ 75) (D = 75 \ x \ 1) (E = 1 \ x \ 150)$

With the standard algorithm that uses *mwn* integer multiplications to multiply a m x w matrix by a w x n matrix to produce a m x n matrix:

$A(B(C(DE)))$ uses 312,000 integer multiplications

$((AB)(CD))E$ uses $2,175$ integer multiplications

Optimal matrix ordering algorithm: For a sequence of matrices $M_1 \cdots M_n$, let $D[1]...D[n+1]$ be the dimensions, that is M_i has dimensions $D[i]$ by $D[i+1]$, and let $M[i,j]$ be the minimum number of integer multiplications needed to compute the matrix product $M_i...M_{i+j-1}$ (we wish to compute $M[1,n]$). Then $M[i,1]=0$, and for $2 \le j \le n$, $M[i,j]$ is the minimum over all k ways to choose the final matrix multiplication to be performed, which can be expressed as the time to multiply $M_i \cdots M_{i+k-1}$ (which is $M[i,k]$) that produces a matrix of dimension $D[i]$ by $D[i+k]$ plus the time to multiply $M_{i+k} \cdots M_{i+j-1}$ (which is $M[i+k,j-k]$) that produces a matrix of dimension $D[i+k]$ by $D[i+j]$, plus time $D[i]D[i+k]D[i+j]$ for the final multiplication.

> **for** $1 \le i \le n$ **do** $M[i,1] := 0$
> **for** $j := 2$ **to** n **do**
>   **for** $i := 1$ **to** $n-j+1$ **do**
> $M[i,j] = \underset{1 \le k < j}{\text{minimum}} \{ M[i,k] + M[i+k, j-k] + D[i]D[i+k]D[i+j] \}$

Complexity: $O(n^3)$ time, $O(n^2)$ space.

Example: Consider the matrices A, B, C, D, and E of the previous example. Dynamic programming assigns $M_1=A$, $M_2=B$, $M_3=C$, $M_4=D$, $M_5=E$ and finds the optimal solution of 2,155 integer multiplications for $(A(B(CD)))E$ as follows:

		1	2	3	4	5
	1 (M_1=A)	0	125	2,000	2,505	2,155
	2 (M_2=B)	0	9,375	2,000	3,250	
i	3 (M_3=C)	0	1,875	2,750		
	4 (M_4=D)	0	11,250			
	5 (M_5=E)	0				

j

$D[1] = 1$
$D[2] = 5$
$D[3] = 25$
$D[4] = 75$
$D[5] = 1$
$D[6] = 150$

Example: Context-Free Language Recognition

Definition: A *context-free grammar* in *Chomsky normal form* is a set of rules of the form $A \rightarrow BC$ or $A \rightarrow d$, where A, B, and C denote *variables* and d denotes a *character*. The strings defined by a variable A, called a *context free language*, are all character strings that can be produced by starting with A and applying a sequence of rules.

> *Note*: Although rules can in general have arbitrary right sides, Chomsky normal form limits right sides to either a pair of variables or a single character; any context-free grammar can be put in this form (see the chapter notes).

Example: S generates all strings consisting of zero or more a's followed by a single b:

$$S \rightarrow AS$$
$$S \rightarrow b$$
$$A \rightarrow a$$

Notation:

w = input string of length $n>0$
$w(i,j)$ = substring of w at position i of length j
$V[i,j]$ = set of all variables that can produce $w(i,j)$

CYK algorithm: For $j>1$ a variable A in $V[i,j]$ can only produce $w(i,j)$ by a rule of the form $A \rightarrow BC$ where B produces part of $w(i,j)$ and C produces the remainder of $w(i,j)$; that is, for some k, B is in $V[i,k]$ and C is in $V[i+k,j-k]$. Hence $V[i,j]$ can be computed from sets with smaller second index. The goal is to compute $V[1,n]$, all variables from which w can be constructed. The CYK algorithm has essentially the same structure as the optimal matrix ordering algorithm.

for $1 \le i \le n$ **do** $V[i,1] := \{A : A \rightarrow w(i,1)\}$
for $j := 2$ **to** n **do**
    for $i := 1$ **to** $n-j+1$ **do**
        $V[i,j] = \bigcup_{1 \le k < j} \{A : A \rightarrow BC,\ B \text{ in } V[i,k],\ C \text{ in } V[i+k, j-k]\}$

Complexity: $O(|G|n^3)$ time, $O(|G|n^2)$ space.

Example: On the left is a grammar, on the right is the table for $V[i,j]$ for input *abaab*:

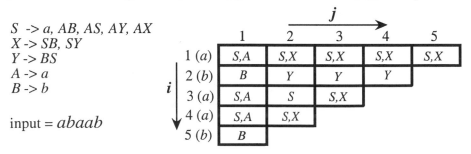

```
S  -> a, AB, AS, AY, AX
X  -> SB, SY
Y  -> BS
A  -> a
B  -> b

input = abaab
```

	1	2	3	4	5
1 (a)	S,A	S,X	S,X	S,X	S,X
2 (b)	B	Y	Y	Y	
3 (a)	S,A	S	S,X		
4 (a)	S,A	S,X			
5 (b)	B				

Dynamic Programming Sums

Idea: Often when filling in entries in a table in a dynamic programming algorithm, indices of inner loops go up to the index of the next outer loop; for example:

> **for** $k := 1$ **to** n **do**
>> **for** $j := 1$ **to** k **do**
>>> **for** $i := 1$ **to** j **do**
>>>> Do something that takes $O(1)$ time.

Theorem:

$$S_1(n) = \sum_{i=1}^{n} 1 = n$$

$$S_2(n) = \sum_{j=1}^{n}\sum_{i=1}^{j} 1 = \tfrac{1}{2}n^2 + \tfrac{1}{2}n$$

$$\text{for } k > 2, \quad S_k(n) = \sum_{i_k=1}^{n} \cdots \sum_{i_2=1}^{i_3} \sum_{i_1=1}^{i_2} 1 = \tfrac{1}{k!}n^k + \Theta\!\left(n^{k-1}\right)$$

Proof: By induction.

(*basis*) For $k=2$, the second of the two sums is i, and hence $S_2(n)$ is a simple arithmetic sum, which sums to $\tfrac{1}{2}n^2 + \tfrac{1}{2}n$.

(*inductive step*) For $k>2$, we have:

$$
\begin{aligned}
S_k(n) &= \sum_{i=1}^{n} S_{k-1}(i) \\
&= \sum_{i=1}^{n}\left(\tfrac{i^{k-1}}{(k-1)!} + \Theta\!\left(i^{k-2}\right)\right) \quad \text{--- by induction} \\
&= \tfrac{1}{(k-1)!}\sum_{i=1}^{n} i^{k-1} + \sum_{i=1}^{n}\Theta\!\left(i^{k-2}\right) \\
&= \tfrac{1}{k!}n^k + \Theta\!\left(n^{k-1}\right) \quad \text{--- bounds for an arithmetic sum (see the appendix)}
\end{aligned}
$$

Since we are using Θ, this is really two proofs. The Θ bounds from the inductive hypothesis and the arithmetic sum give pairs of constants for O and Ω; see the exercises.

Another proof: Another proof is to show that $S_k(n)$, $k \geq 1$, is the binomial coefficient

$$S_k(n) = \binom{n+k-1}{k}$$

and to show this implies the bound; the exercises work through this proof and also give the exact polynomials for $S_3(n)$ and $S_4(n)$.

Randomized Algorithms

Idea: Make random choices to achieve some desirable outcome with high likelihood; two common applications are:

Good expected performance: The algorithm employs random choices in such a way that it has good worst case time and produces the correct result with high probability.

Good expected time: The algorithm employs random choices in such a way that it always produces the correct result and runs in good expected time (where the analysis does not rely upon probabilistic assumptions about the input).

A nice aspect of this sort of expected time is that if your luck is bad and things are taking too long, you can simply stop and restart the algorithm. For example, if there is a 50% probability that the algorithm will finish in a given time t, then:

With probability $1 - .5 * .5 = .75\%$, in 2 tries, at least one will finish in time t,

With probability $1 - .5 * .5 * .5 = 87.5\%$, in 3 tries, at least one will finish in time t.

With probability $1 - (1/2)^n$, in n tries, at least one will finish in time t.

Example: Statistical Sampling

Suppose that to advertise a product, you would like to purchase a large list of email addresses, but you have heard that many vendors, who charge proportional to the length of the list, pad their lists with some percentage of invalid addresses. You would like to check a list before committing to sending out a large volume of email, and return it to the vendor if it contains too many invalid addresses.

You could write a program that takes as input a list of length n, in $O(n)$ time checks the validity of each address, and reports the number of invalid address.

Another approach is to choose $f(n)$ addresses at random, where $f(n) << n$, and check only these. For example, if $n = 1,000,000$ and $f(n) = \sqrt{n}$, then the algorithm checks only 1,000 addresses. If half of the addresses were invalid, since each address checked is equally likely to be valid or invalid, there is only a 1 in 2^{1000} chance that you will not find an invalid address. In general, if p is the percentage of addresses that are invalid, it is possible to show that with high likelihood, you will find a number of invalid addresses that is relatively close to $pf(n)$.

So here we have an example of guaranteed worst case time of $O(f(n))$ and good expected performance in the sense that if there is a fraction of the addresses that are invalid, you are likely to detect them.

Example: Randomized Quick Sort

The PARTITION algorithm presented earlier for "standard" quick sort uses the first element as the pivot element, which may in practice not be a "typical" element (e.g., consider the case that the data is already sorted). Also, although it works for duplicates, performance can degrade because all equal elements will end up on the left side and may cause the recursion to be unbalanced (e.g., consider the case when all elements are the same).

Idea: Use a slightly more complex randomized "3-partitioning" procedure. Instead of choosing the pivot element to be $A[i]$, choose a random index $i \leq r \leq j$ and exchange $A[r]$ with $A[i]$. A third index variable z allows for <, =, and > regions:

$$A: \boxed{\quad | \text{items} <A[i] \quad | \quad \text{items yet to be processed} \quad | \quad \text{items} =A[i] \quad | \text{items} >A[i] \quad |}$$

$i \qquad \qquad x \qquad\qquad\qquad\qquad y \qquad\qquad z \qquad\qquad j$

$A[r]$ is first exchanged with $A[i]$

3PARTITION is similar to PARTITION except that the second inner *while* loop makes two comparisons: if $A[y]=A[i]$, then y is decremented as usual, but if $A[y]>A[i]$, then the = region is effectively shifted left to make a new position in the > region. Again, we assume the array name is A; a pointer to A could be passed as an argument.

function 3PARTITION(i,j)
    r := a random integer in the range $i \leq r \leq j$
    Exchange $A[i]$ and $A[r]$.
    $x := i+1$
    $y := z := j$
    while $x \leq y$ **do**
        while $x \leq y$ and $A[x]<A[i]$ **do** $x := x+1$
        while $x \leq y$ and $A[y] \geq A[i]$ **do**
            if $A[y]=A[i]$ **then** $y := y-1$
            else if $A[y]>A[i]$ **then begin** exchange $A[y]$ and $A[z]$; $y := y-1$; $z := z-1$ **end**
        if $x<y$ **then begin** exchange $A[x]$ and $A[y]$; $x := x+1$; $y := y-1$ **end**
        end
    Exchange $A[i]$ and $A[y]$.
    return $y-1$ and $z+1$ (that is, $A[i]...A[y-1] < A[y]...A[z] < A[z+1]...A[j]$)
    end

Complexity: Since the time for 3PARTITION differs by a constant factor from PARTITION, the expected time for quick sort remains $O(n\log(n))$; however, now this bound is independent of the input distribution, even when duplicates are allowed.

An alternate approach: Use a random pivot element and modify PARTITION so that the equal elements are distributed equally between the left and right. We no longer have the benefit of smaller sub-problems when there are many equal elements, but on the other hand, we do not have the overhead of 3-partitioning when there are not many equal elements. See the exercises.

Example: Randomized k^{th} Largest Element

Although the algorithm presented earlier for finding the kth largest element is a clever example of divide and conquer, it may not be practical because the linear running time has a relatively large constant and the algorithm does not work in place.

Idea: In a spirit similar to quick sort, rather than go through all the work of constructing the mini-lists to get a good pivot element m, simply choose a random element of the list. We use the procedure 3*PARTITION*(i,j) of randomized quick sort; based on a random value in the range i through j it rearranges A and returns a pair of integers a,b such that $A[i]...A[a] < A[a+1] = ... = A[b-1] < A[b]...A[j]$.

Randomized in-place version of KLARGEST: Steps 1 through 4 are collapsed to a single call to *3PARTITION* that uses $O(1)$ space in addition to that used by A:

> **function** randKLARGEST(k,i,j)
>> **if** $k<1$ or $k>(j-i+1)$ **then** ERROR — k is out of range
>>
>> **else if** $i=j$ **then return** $A[i]$
>>
>> **else begin**
>>> 1. a,b := 3PARTITION(i,j)
>>> 2. **if** $k \le (a-i+1)$ **then return** KLARGEST(k,i,a)
>>>   **else if** $k \le (b-i+1)$ **then return** $A[a+1]$
>>>   **else return** KLARGEST$((k-b+i),b,j)$
>>
>> **end**
>
> **end**
>
> randKLARGEST$(k,1,n)$

Expected time: The random element chosen by *3PARTITION* is equally likely to be any $1 \le i \le n$ and, depending on k, a recursive call is made on $i-1$ or $n-i$ elements. Since *3PARTITION* is $O(n)$, for some constant c, $T(1) \le 4c$, $T(2) \le 8c$, and for odd $n>1$:

$$T(n) \le \underset{1 \le k \le n}{MAX}\left\{ \frac{1}{n}\left(\sum_{i=k+1}^{n} T(i-1) + \sum_{i=1}^{k-1} T(n-i) \right) \right\} + cn$$

$$= \underset{1 \le k \le n}{MAX}\left\{ \frac{1}{n}\left(\sum_{i=k}^{n-1} T(i) + \sum_{i=n-k+1}^{n-1} T(i) \right) \right\} + cn$$

$$\le \frac{2}{n} \sum_{i=(n+1)/2}^{n-1} T(i) + cn \quad \text{— the maximum occurs at } k = \frac{n+1}{2} \text{ (see the exercises)}$$

Induction may be now used to show $T(n) \le 4cn$.

Proof by induction for odd n

For the basis, when $n=1$, $T(1) \le 4c = 4cn$. For the inductive step, for an odd $n>1$ assume that for odd values $\le n$ that

$$T(n) \le \frac{2}{n} \sum_{i=(n+1)/2}^{n-1} T(i) + cn$$

and consider $n+2$:

$$T(n+2) \le \frac{2}{n+2} \sum_{i=(n+3)/2}^{n+1} T(i) + c(n+2) \quad \text{— by definition}$$

$$\le \frac{8c}{n+2} \sum_{i=(n+3)/2}^{n+1} i + c(n+2) \quad \text{— by induction}$$

$$= \frac{8c}{n+2}\left(\frac{(n+1)+\frac{n+3}{2}}{2}\right)\left((n+1)-\frac{n+3}{2}+1\right) + c(n+2)$$

$$= \frac{8c}{n+2}\left(\frac{3n+5}{4}\right)\left(\frac{n+1}{2}\right) + c(n+2)$$

$$< c(3n+6) + c(n+2)$$

$$= 4c(n+2)$$

Proof by induction for even n:

By using the odd analysis for the next larger value of n, we get at most $4c(n+1)$ time. For a more careful analysis observe that not only does k take its maximum at $n/2$ (see the exercises), but the combined sum adds in an extra copy of the term $T(n/2)$ and hence we have the recurrence:

$$T(n) \le \frac{2}{n} \sum_{i=(n/2)}^{n-1} T(i) - \frac{1}{n}T\left(\frac{n}{2}\right) + cn$$

For the basis, when $n=2$, $T(2) \le 8c = 4cn$. For the inductive step, for an even $n>2$ assume that it is true for even values $\le n$ and consider $n+2$:

$$T(n+2) \le \frac{2}{n+2} \sum_{i=(n+2)/2}^{n+1} T(i) - \frac{1}{n+2}T\left(\frac{n+2}{2}\right) + c(n+2) \quad \text{— by definition}$$

$$\le \frac{8c}{n+2} \sum_{i=(n+2)/2}^{n+1} i - \frac{4c}{n+2}\left(\frac{n+2}{2}\right) + c(n+2) \quad \text{— by induction}$$

$$= \frac{8c}{n+2}\left(\frac{(n+1)+\frac{n+2}{2}}{2}\right)\left((n+1)-\frac{n+2}{2}+1\right) - 2c + c(n+2)$$

$$= \frac{8c}{n+2}\left(\frac{3n+4}{4}\right)\left(\frac{n+2}{2}\right) + cn$$

$$< c(3n+6) + c(n+2)$$

$$= 4c(n+2)$$

Greedy Algorithms

Idea: When trying to find an optimal solution to a problem, be "greedy" and construct partial solutions that work towards the goal as quickly as possible. For some problems, greedy algorithms lead to optimal solutions. For other problems, although greedy approaches may often give good solutions, more complex strategies can give better solutions in the worst case.

Example: Bin Packing

You walked miles to a remote beach and collected many shells. Unfortunately, due to excesses of the past, you have back problems and your doctor has forbidden you to carry anything. A hermit offers to carry back your shells in buckets ("bins"); he charges a fixed cost per bin and will only load each bin up to a maximum weight W. Your problem is to pack your shells into as few bins as possible. Formally:

> Given a maximum weight W and a set of items $S = \{x_1, ..., x_n\}$, where associated with each item x_i is an integer weight $w(x_i)$, $0 < w(x_i) \leq W$, partition S into disjoint subsets $S_1, ..., S_k$ such that the sum of the weights of the elements in each S_i is $\leq W$ and k is as small as possible.

One approach for a greedy algorithm to solve this problem is to consider the items one at a time and use a "first-fit" strategy; that is, place the next item in the first bin in which it fits (or start a new bin if it will not fit into any of the existing ones). It is not hard to construct examples where the first-fit strategy does not yield a minimum number of bins (see the exercises). On the other hand, at most one of the bins filled by the first-fit strategy can be less than half full (i.e., it cannot be that first-fit finishes with two bins that are less than half full since the items in the higher numbered one of these two bins could have been placed in the lower numbered one), and hence the number of bins used by first-fit is at most double that of an optimal solution (since in the best possible case, the minimum solution is a set of bins that all have weight W). In fact, if M denotes the minimum number of bins that can be used to solve the problem, it can be shown (see the chapter notes) that the number of bins used by the first-fit strategy is

$$\leq \frac{17}{10} M + 2$$

and that there is an infinite class of problems for which first-fit does about this poorly.

Note: Like the knapsack problem, no efficient way to compute the optimal solution for arbitrary classes of inputs is known (it is another example of a *NP-hard* problem — see the chapter notes). The greedy first-fit strategy is a fast and simple approach to get a reasonable solution (in fact, probably much better than 17/10 of the minimal solution on typical problems).

Example: Huffman Trees

Idea: Given a set of n *weights*, find a binary tree with n leaves that best stores these weights in the sense that bigger weights should be closer to the root.

Definition: A tree is a *Huffman* tree if:

1. Each non-leaf vertex has exactly two children.
2. Each leaf has a weight (real number) in the range 0 to 1; the sum of all leaf weights is 1.
3. The sum over all the leaves of the leaf's depth times its weight is minimum among all binary trees with the same set of leaf weights.

Example: If the number of weights n is a power of two and all weights are identical (i.e., each weight is $1/2^n$), then the Huffman tree is a complete binary tree of height $\log_2(n)$. On the other extreme, if the first weight is 1/2 (equal to the sum of all remaining weights), the second is 1/4 (equal to the sum of all remaining weights), and so on, then the Huffman tree has height $n-1$. In general, the shape is somewhere in-between. The figure below shows Huffman trees for three different sets of 8 weights:

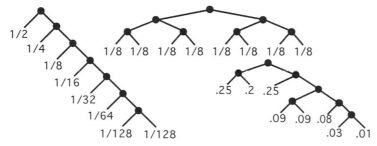

Huffman tree construction algorithm: A "bottom-up" greedy algorithm can be used:

Initialize a FOREST of single-vertex binary trees, one for each weight.

while the FOREST has more than one tree **do begin**

Let X and Y be tries in the FOREST of lowest weight (ties broken arbitrarily).

Create a new root r and make the roots of X and Y the children of r.

Define the weight of r to be the sum of the weights of the roots of X and Y.

end

Correctness: Consider the smallest weight leaf x (ties broken arbitrarily). We can assume that there is no other leaf deeper than x since otherwise a tree of the same or smaller weight can be constructed by exchanging x with that leaf. If x's sibling y was a vertex (possibly not a leaf) of weight greater than some other leaf z, the weight of the tree could be reduced by exchanging z with the subtree rooted at y. Thus, there is always a Huffman tree that has the two vertices of smallest weight as sibling leaves, and correctness of the Huffman algorithm follows by induction.

Example: Shortest Common Superstrings

A *superstring* for a set of strings S is a string for which all strings of S are a substring. For example, *aaabbabaabc* is a superstring for {*abc, baab, aabba, bab, abaa, aaa*}. For simplicity, we always assume that no string u in S is a substring of another string v in S (clearly, a superstring for $S-\{u\}$ is a superstring for S).

Overlap lemma: For strings a and b, $OL(a,b)$ denotes the longest string that is both a suffix of a and a prefix of b. As shown in the figure below, if a, b, and c are strings that can be overlapped as shown below, then $|OL(a,c)| = |OL(a,b)| + |OL(b,c)| - |b|$. The proof observes that $|OL(a,b)|+|OL(b,c)|$ counts $|c|$ twice (see the exercises).

Greedy Superstring Construction Algorithm:
> **while** there is more than one string in S **do begin**
>> Let u and v be two strings in S such that $OL(u,v)$ is as large as possible.
>> Overlap u with v to form a single string of length $|u|+|v|-|OL(u,v)|$.
> **end**

Theorem: The number of characters *saved* by the greedy algorithm (i.e., the sum of all string lengths minus the length of the superstring) is at least half that saved by an optimal algorithm that constructs a shortest possible superstring.

Proof: Let m be the number of characters saved on the first iteration of greedy by choosing strings u and v. Now suppose that from this point on, we make the same choices as the optimal algorithm, but skip a step if it is not possible. If there is a single string w such that the optimal algorithm overlaps ... u w v ..., then the only steps that cannot be done are overlapping u with w and w with v, and in the end, we gained m on the first iteration but lost the savings for these two overlaps later on, and since m is maximal, we gained at least half of what was lost. Otherwise, we will not be able to overlap u with some string p for some savings x and some string q with v for some savings y. Also, if u comes after v in the optimal ordering, we have to omit the step that overlaps v with u for some savings z. Thus, we have gained m but lost $x+y+z$. Since m is maximal, if $x+y \leq m$, we have gained at least half of what was lost. Otherwise, since $|OL(u,v)|=m$, by the overlap lemma, in a final step we can overlap the string ending with q with the string starting with p for a savings of at least $x+y-m$ ($b=OL(u,v)$ in the overlap lemma), and again we have gained at least half of what was lost. We carry this reasoning to subsequent iterations to see that the savings from greedy is at least 1/2 of optimal.

Note: No efficient way to compute the optimal solution for arbitrary classes of inputs is known (see the chapter notes).

CHAPTER 5

Simulated Annealing

Idea: With problems that have no known efficient solution, we sometimes must resort to searching a large space of potential solutions, by starting with a relatively poor one and making successive attempts to improve it. *Simulated annealing* can help such an approach avoid getting "stuck" at a *local minimum*; that is, help avoid solutions that are poor but cannot be improved by making small changes. It allows large changes at the beginning and then as time goes on, lowers the probability that large changes will be allowed. The probability function that determines when large changes are allowed is motivated by the cooling process used for annealing of materials. Another analogy is filling a jar with beans; by shaking the jar violently at first and then less and less, the beans may "settle down" in a way that more beans fit than if they were simply poured in.

Key components of a simulated annealing algorithm:
- A formulation of the problem as a space of *solutions S*.
- An *initial solution* that is typically a random member of *S*.
- An *objective function F* that maps solutions in *S* to non-negative values.
- A *perturbation function P* that defines for *s* in *S*, a set of solutions *P(s)*.

Example: Suppose that a package delivery service wishes to contract two trucking companies to make its deliveries, and hopes to have as few customers as possible have to deal with more than one company. Given a list of computer records of the form "a package has been sent between Customers *A* and *B* at some point in the past", you wish to divide the customers into two equal size sets that minimizes the number of records that have customers in different sets. Initially, the customers can be divided arbitrarily, the objective function is the number of records that have a customer in each half, and the perturbation function could be the one which, for each solution, gives the set of all solutions that can be obtained by exchanging two customers that are in opposite halves.

A typical simulated annealing algorithm:

Input an initial solution s, an objective function F, and a perturbation function P.

Initialize t ("temperature"), $r < 1$ ("cooling ratio"), and f ("freezing point").

while $t > f$ **do begin**

$\sigma :=$ a random member of $P(s)$

$\Delta := F(\sigma) - F(s)$

$q :=$ a random real number between 0 and 1

if $\Delta < 0$ or $q < \dfrac{1}{2^{\Delta/t}}$ **then** s $:= \sigma$

$t := r * t$

end

Output s.

Choosing the parameters t, r, and f: Tuning these parameters is problem dependent. Typically, t should be high enough to cause poor changes to be chosen with high probability, f should be low enough so that poor changes have low probability, and $r < 1$ should be made as close to 1 as available computation time allows.

Exercises

117. A sorting method is *stable* if elements that are equal remain in the same relative order after the sort is complete. Determine which of these sorting methods are stable: bubble sort, merge sort, quick sort, randomized quick sort.

118. Although merge sort was presented as a recursive process, it has a simple pattern.

  A. Assuming that the size of the list is a power of 2, write a non-recursive version of arrayMSORT that works from the "bottom up" by merging the halves of sub-lists of size two, then sub-lists of size 4, and so on until there is a single sorted list.

  B. Describe how to modify your solution to Part A for arbitrary size lists.

119. Like merge sort, the recursive operation of quick sort has a simple pattern.

  A. Consider a list $A[1]...A[n]$ where n is a power of 2, all items in A are distinct, and we are given a function MEDIAN(i,j) that returns a real number that is greater than exactly half the numbers in A (so if $j-i+1$ is even, then MEDIAN(i,j) is not equal to any of $A[i]...A[j]$). Present pseudo-code for a non-recursive version of quick sort that works from the "top down" by partitioning the whole list, partitioning two lists of half the size, partitioning 4 lists of 1/4 the size, and so on.

  B. Modify your solution to Part A to work for normal quick sort where n is not necessarily a power of two and the function MEDIAN is not available, by using a queue or a stack to keep track of the blocks into which A is currently partitioned. Explain why this is just a "streamlined" version of what would be produced by mechanical recursion elimination.

120. Consider the following simpler version of *PARTITION* for quick sort:

function simplePARTITION(i,j)

> $x := i+1$
>
> $y := j$
>
> **while** $x \leq y$ **do**
>> **if** $A[x] \leq A[i]$ **then** $x := x+1$
>> **else if** $A[y] > A[i]$ **then** $y := y-1$
>> **else** exchange $A[x]$ and $A[y]$
>> **end**
>
> Exchange $A[i]$ and $A[y]$.
>
> **return** $y-1$ and $y+1$
>
> **end**

  A. Prove that it correctly partitions.

  B. Prove that *simplePARTITION* can in the worst case use at most twice as many comparisons as *PARTITION*.

121. Consider the procedure PARTITION(i,j) presented for quick sort.

A. Prove that it correctly rearranges A and returns a pair of values a,b such that:

1. $b = a+2$
2. $A[i]...A[a]$ are less than the original value of $A[i]$.
3. $A[a+1]$ equals the original value of $A[i]$.
4. $A[b]...A[j]$ are greater than or equal to the original value of $A[i]$.

B. Prove that it performs at most $j-i$ comparisons of elements of A.

C. Prove that it performs at most $\lfloor(j-i)/2\rfloor+1$ exchanges of elements of A.

122. Consider the procedure *3PARTITION(i,j)* presented for randomized quick sort.

A. Prove that it correctly rearranges A and returns a pair of values a,b such that:

1. $b \leq a+2$
2. $A[i]...A[a]$ are less than the original value of $A[r]$.
3. $A[a+1]...A[b-1]$ equal to the original value of $A[r]$.
4. $A[b]...A[j]$ greater than or equal to the original value of $A[r]$.

B. Prove that even in the worst case, *3PARTITION* cannot perform more than twice as many comparisons or exchanges of elements of A as *PARTITION*.

C. Give a precise expression for the worst case number of comparisons of elements of A and the worst case number of exchanges of elements of A that are performed by *3PARTITION*.

123. Give pseudo-code for the following version of 2-partitioning:

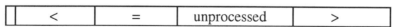

124. Give pseudo-code for the following version of 3-partitioning:

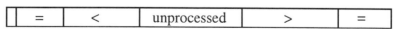

125. Consider the following version of 3-partitioning:

$=$	$<$	unprocessed	$>$	$=$

Give pseudo-code that partitions in this manner and then rearranges the array so that it can return a valid pair of values expected by QSORT (that is, a single < block followed by a single = block followed by a single > block).

126. Although 3-partitioning is an effective way to deal with duplicates, they can be handled with 2-partitioning.

A. Modify the *PARTITION* procedure used by quick sort so that it leaves the same number of equal elements on both sides. That is, if there are no duplicates, it works the same, if there are an even number of duplicates then it returns with $b = a+1$ and an equal number of duplicates on each side, and if there are an odd number of duplicates, it returns with $b = a+2$ and an equal number of duplicates on each side.

B. Further modify *PARTITION* to use the duplicates to make the two sides as equal as possible. That is, place more duplicates on one side than on the other if it helps to make $a-i$ closer to $j-b$.

C. Explain why with either Part A or Part B, the expected-time analysis for the case of distinct elements still holds.

127. We have presented quick sort as working in-place on an array. However, if we were using linked lists, we could implement the algorithm directly on them.

A. Give pseudo-code for quick sort using 2-partitioning that works on a singly linked list.

B. Give pseudo-code for randomized quick sort using 3-partitioning that works on a singly linked list.

C. Suppose that you were given a pointer to a singly linked list and wanted to sort that list. Discuss how in practice, your solutions to A and B compare with copying the list to array, doing the sort in-place, and then copying back to an array. That is, under various practical assumptions, is the "overhead" of copying back and forth to an array worth the speed of the array version of the algorithm?

D. In the same spirit as Part C, make comparisons between the list and array versions of merge sort and between merge sort and quick sort.

128. Explain how to modify the functions KLARGEST and randKLARGEST to return a pointer to the kth largest element, rather than the element itself.

129. For constants a, b, and c, suppose that in some practical implementation KLARGEST ran in time an, randKLARGEST ran in (expected) time bn, quick sort ran in time $cn\log_2(n)$, and randomized quick sort ran in (expected) time $dn\log_2(n)$.

A. For $a=25$, $b=4$, $c=2$, and $d=4$ compute for what value of n KLARGEST becomes faster than quick sort. Also, for this value of n, compute for what value of $m>n$ randKLARGEST can handle an array of m elements in the same amount.

B. Consider the following program to find the maximum of an array:
   function arrayMAX(A)
       $m := A[1]$

```
for i := 2 to n do if m<A[i] then m := A[i]
return m
end
```

Using *arrayMAX* as a benchmark, and assuming that the programming style, computer used, etc. is more or less equivalent in all cases, if *arrayMAX* runs in time qn, give estimates for a, b, c, and d assuming that the input is an array of n elements.

C. In the function KLARGEST, the constant 20 arises in two ways. The number 20 is the constant that comes out of the analysis. It is also the threshold the algorithm uses to decide when to apply divide and conquer (the *if* statement tests if the list has fewer than 20 elements, and if so, it simply finds the k largest element by sorting the list). This threshold is in some sense arbitrary, it just happened to work for the analysis to make it 20, but any value larger than 20 will also work. Discuss whether it might be better in practice to make the threshold larger than 20.

130. In the recurrence used in the analysis for randKLARGEST, explain why the expression is maximized when $k=n/2$ for even n and $k=(n+1)/2$ for odd n.

Hint: Explain why $T(i)$ must be a non-decreasing function of i and then consider what value of k maximizes the smallest term on both sides.

131. The *median* of a set of $n \geq 1$ distinct integers, where n is an odd, is an integer m in the set such that there are $\lfloor n/2 \rfloor$ elements $<m$ and $\lfloor n/2 \rfloor$ elements $> m$.

A. Describe how to construct a function MEDIAN that takes as input a linked list of n distinct integers and, using $O(n)$ time and space, returns a pointer to the median element.

B. Generalize your answer to Part A to allow n to be even and elements to be equal, and explain how MEDIAN can be used to make quick sort run in worst case time $O(n\log(n))$.

132. With the unit-cost RAM model, multiplication of two integers is $O(1)$ time no matter how large they are. However, assume now that you are designing computer hardware or designing a program to multiply very large integers of n bits each (e.g., thousands or millions of bits), and you can only manipulate $O(1)$ bits in $O(1)$ time.

A. Assume that it takes $O(1)$ time to perform operations on individual bits of the input and output integers such as adding two bits or comparing two bits (but it is ok to have normal integer variables for flow of control statements like *for i := 1 to n*). Generalize the divide and conquer multiplication algorithm for polynomials that was presented in this chapter to multiply two n-bit integers.

B. Generalize your solution to Part *A* for when, for a parameter m, you can multiply m bits on $O(1)$ time. Analyze your running time as a function of both m and n.

133. A n by n *Toeplitz matrix* is a n by n matrix T such that $T[i,j] = T[i-1,j-1]$, $2 \leq i,j \leq n$. Show how to generalize the divide and conquer multiplication algorithm for polynomials that was presented in this chapter to multiply a Toeplitz matrix times a column vector (we shall see asymptotically better ways to do it when we consider the fast Fourier transform algorithm, but this is still an interesting exercise).

Hint: A Toeplitz matrix can be represented by the $2n-1$ values that are along its first row and first column.

134. Recall Strassen's Algorithm for multiplying a m by w matrix X by a w by n matrix Y to produce a m by n matrix Z.

   A. Prove that matrix addition is both *commutative* (i.e., $A+B=B+A$) and *associative* (i.e., $A+(B+C)=(A+B)+C$).

   B. Prove that matrix multiplication is associative but not in general commutative.

   C. Prove that multiplication based directly on the recursive decomposition works and that exactly the same mwn multiplications of elements are performed by "unwinding the recursion".

   D. For the case of square matrices, show that for some constant $k \geq 0$, multiplication based directly on the recursive decomposition gives the recurrence:

   $$T(n) = 8T(n/2) + kn^2$$

   Show that this recurrence is $O(n^3)$ in two ways. First, in a fashion similar to the analysis presented in the example for multiplying two polynomials, expand it and express it as a geometric sum. Second, explain how it meets the criteria for the simple type of divide and conquer recurrence relation that was presented in the chapter.

   E. Prove that Strassen's computation of A through G followed by the computation of $Z_{1,1}$, $Z_{1,2}$, $Z_{2,1}$, $Z_{2,2}$ gives the correct result.

   F. Generalize the time analysis to show that the algorithm is:

   $$O(MIN(m,w,n)^{2.81} + MAX(m,w,n)^2)$$

   Hint: Think of the computation as having two phases, in which the second phase starts when the recursion gets down to the point where at least one of the dimensions of sub-matrices is 1.

   G. Give a more careful analysis that Part F that presents tighter bounds in terms of m, w, and n.

135. Strassen's method uses 7 multiplications and 18 additions of sub-matrices; show how to do it with 7 multiplications and less than 18 additions.

136. Consider the proof for generic divide and conquer recurrence relations.

    A. Explain why the analysis for the cases $c < d^f$ and $c > d^f$ is tight to within a constant factor by showing that a single term dominates the expression.

    B. For the case $c = df$, show that the recurrence expands to exactly

$$T(n) = n^f \left(a + e \sum_{i=0}^{\log_d(n)-1} \log_g \left(\frac{n}{d^i} \right)^h \right)$$

    and then explain why the analysis was tight to within a constant factor; that is, show that the right-hand side in the expression above is $\Omega(n^f \log(n)^{h+1})$.

    Hint: One approach is to separate the sum into two sums by using the identity $\log(a/b) = \log(a) - \log(b)$; another is to simply throw out the last half of the terms and show that the remaining $O(\log(n))$ terms are each $O(\log(n)^h)$.

137. Consider the dynamic programming algorithm presented for the knapsack problem.

    A. Modify it to output a set of weights that sums to n.

    Hint: Store with entries how they were constructed and work back at the end.

    B. Describe how to modify the algorithm so that it takes time proportional to the number of entries that are true, and present pseudo-code.

    Hint: Represent a row by a linked list and show how to generate the list for a row from the previous row.

    C. As can be seen from the example presented, it is possible that a true in the last column could be found before the last row was reached, and the computation time might be shortened by stopping when this was the case. Discuss other possible heuristics for shortening the computation. Can sorting the weights in increasing or decreasing order help? If so, assuming that it takes time $O(m\log(m))$ to sort the weights, under what conditions does the sorting not affect the worst case asymptotic time? Would it make sense to use bucket sort, and if so how does that affect the worst case asymptotic time?

    D. Assume that the input is represented as a long string with one character for each bit of the binary representations of the m weights and for n. Show that for $k \geq 0$, as a function of the number of characters in the input, this algorithm is $\Omega(n^k)$.

138. The *partition problem* is the special case of the knapsack problem where the sum of the weights is exactly $2n$; that is, partition the weights into two sets that sum to the same amount. Show how an algorithm to solve it can be used to solve the knapsack problem in the same amount of asymptotic time.

Hint: If S denotes the sum of all of the weights $W[1]...W[m]$ in the original knapsack problem, add two new weights, $W[m+1]=n+1$ and $W[n+2]=S-n+1$ and show that the weights $W[1]...W[m+2]$ can be partitioned if and only if there is a subset of the weights $W[1]...W[m]$ that sums to n.

139. In the paragraphing problem, we intentionally left the definition of $\alpha(i,j)$ vague; it may be that available space is just spread evenly between words, or it may be something more complicated.

    A. Fill in the details of the complexity analysis of why the time is $O(mL)$ under the assumption that $\alpha(i,j)$ can be computed in $O(1)$ time.

    B. If $\alpha(i,j)$ simply spreads available space evenly, explain how $\alpha(i,j)$ can be computed incrementally in $O(1)$ time; that is, $\alpha(i,i)$ can be computed in $O(1)$ time and $\alpha(i,j)$ can be computed from $\alpha(i+1,j)$ in $O(1)$ time.

    C. Fill in the details of the complexity analysis of why the time is $O(mL)$ under the assumption that $\alpha(i,j)$ can be computed incrementally in $O(1)$ time.

    D. Under the assumption that computing $\alpha(i,j)$ takes time proportional to the sum of the lengths of $w_i...w_j$, show that the time is at most $O(mL^2)$.

    E. Suppose that α is an increasing function as you move in either direction from optimum; that is, for any i, there is a j (where j could be m) such that:

$$\alpha(i,i) \geq \alpha(i,i+1) ... \geq \alpha(i,j) \leq \alpha(i,j+1) ... \leq \alpha(i,m)$$

Discuss how this property of α might be helpful to the complexity of the paragraphing problem under various assumptions about the complexity of computing α and how it measures quality.

140. The Optimal Matrix Ordering Algorithm relied on the fact that matrix multiplication is associative. Give a formal proof of this fact; that is, prove that for any three matrices A, B, and C of dimensions p by q, q by r, and r by s respectively, $p,q,r,s \geq 1$, that:

$$A(BC) = (AB)C$$

141. Consider the Optimal Matrix Ordering Algorithm.

    A. Modify the pseudo-code to store along with $M[i,j]$ the value of k for which the minimum was found, and add additional pseudo-code to use this information to output the ordering that produced the minimum value.

    B. For the example presented, show how your pseudo-code computes the ordering $(A(B(CD)))E$ that produces the minimum number of integer multiplications.

142. Consider the CYK Algorithm.

    A. Modify the pseudo-code to store, along with each variable stored in $V[i,j]$, a value of k for which a right-hand side was found, and add additional pseudo-code to use this information to output a sequence of rules that produces w from S.

    B. For the example presented, show how your pseudo-code computes the sequence of rules:

S->AY->ABS->$ABAS$->$ABAAB$->$aBAAB$->$abAAB$->$abaAB$->$AbaaB$->$abaab$

C. It is possible for there to be more than one way to generate w from S. Modify the CYK algorithm to keep a list for each variable stored in $V[i,j]$ all values of k for which a right-hand side was found, and then use this information to list all possible ways that S can produce w.

D. Consider the time complexity of Part C. Show that the asymptotic complexity of building the table is not changed and that the time to output all derivations of w from S is linear in the size of that output. On the other hand, prove that for an input string of n a's, the size of the output for the following grammar is $\Omega(2^n)$:

$$S \rightarrow a, XX, SS$$

$$X \rightarrow a, XX, SS$$

E. Explain why the space used by Part C is $O(|G|n^3)$ and give an example of a grammar that uses this much space.

143. The computation of Pascal's triangle in the first chapter is an example of dynamic programming. It computed each row from the previous row, using the identity:

$$\binom{n}{m} = \binom{n-1}{m-1} + \binom{n-1}{m}$$

We can also base a recursive divide and conquer algorithm on this identity as follows:

> **function** CHOOSE(i,j)
> > **if** $j=0$ or $i=j$
> > > **then return** 1
> > > **else return** CHOOSE($i-1,j-1$)+CHOOSE($i-1,j$)
> > **end**

A. Show that CHOOSE($n,n/2$) is $\Omega\left(2^{n/2}\right) = \Omega\left(2^{\sqrt{n}}\right)$.

Hint: Show that the recursive calls go down at least $n/2$ levels, where each level has twice as many calls as the preceding one.

B. Clearly CHOOSE is a far more expensive way to compute Pascal's triangle, even though it is a seemingly similar approach. Give an intuitive explanation of why dynamic programming works better for this problem.

144. Discuss how dynamic programming could be implemented with recursive calls if previously computed sub-problems were "remembered" for you. In particular, describe how a compiler can generate code to efficiently keep track of recursive calls that have been made (and what was computed); what assumptions are required?

145. Recall the analysis of typical dynamic programming sums that showed:

$$S_1(n) = \sum_{i=1}^{n} 1 = n$$

$$S_2(n) = \sum_{j=1}^{n}\sum_{i=1}^{j} 1 = \tfrac{1}{2}n^2 + \tfrac{1}{2}n$$

$$\text{for } k > 2, \quad S_k(n) = \sum_{i_k=1}^{n} \cdots \sum_{i_2=1}^{i_3}\sum_{i_1=1}^{i_2} 1 = \tfrac{1}{k!}n^k + \Theta\!\left(n^{k-1}\right)$$

A. Show that:

$$S_3(n) = \sum_{k=1}^{n}\sum_{j=1}^{k}\sum_{i=1}^{j} 1 = \tfrac{1}{6}n^3 + \tfrac{1}{2}n^2 + \tfrac{1}{3}n$$

Hint: Explain why

$$S_3(n) = \sum_{i=1}^{n} S_2(n)$$

and then plug in the expression for $S_2(n)$ and use the formulas from the appendix for the two resulting arithmetic sums.

B. In a similar fashion to Part A, show:

$$S_4(n) = \sum_{l=1}^{n}\sum_{k=1}^{l}\sum_{j=1}^{k}\sum_{i=1}^{j} 1 = \tfrac{1}{24}n^4 + \tfrac{1}{4}n^3 + \tfrac{11}{24}n^2 + \tfrac{1}{4}n$$

C. Give a proof by induction that $S_k(n)$ is a polynomial of degree k that has no constant term; that is, for some constants $a_1, a_2, \ldots a_k$:

$$S_k(n) = a_k n^k + a_{k-1}n^{k-1} + \cdots + a_2 n^2 + a_1 n$$

D. The expression presented for $S_k(n)$ gave an exact expression of $1/k!$ for the constant for the n^k term and then simply said that the rest of the expression for $S_k(n)$ was $\Theta(n^{k-1})$. For many arguments, simply showing that $S_k(n)$ is $\Theta(n^k)$ suffices. Clearly, the sum is bounded above by n^k since it is only made larger by making all the upper limits n, and hence it is $O(n^k)$. A crude approach to show it is $\Omega(n^k)$ is to simply throw out terms; for example for $k=3$:

$$\sum_{k=1}^{n}\sum_{j=1}^{k}\sum_{i=1}^{j} 1 \;\geq\; \sum_{k=\frac{n}{2}}^{n}\sum_{j=\frac{n}{4}}^{k}\sum_{i=\frac{n}{8}}^{j} 1 \;\geq\; \tfrac{n}{2}\tfrac{n}{4}\tfrac{n}{8} \;=\; \tfrac{n^3}{64} \;=\; \Omega\!\left(n^3\right)$$

Explain this equation and generalize it to any k.

E. The proof presented in this chapter for $S_k(n)$ noted that the proof is really two proofs. One to show that

$$S_k(n) = \tfrac{1}{k!}n^k + O\!\left(n^{k-1}\right)$$

and one to show that:

$$S_k(n) = \frac{1}{k!}n^k + \Omega\left(n^{k-1}\right)$$

Fill in the details.

F. In the chapter, it was noted that another proof that

$$S_k(n) = \frac{1}{k!}n^k + \Theta\left(n^{k-1}\right)$$

is based upon showing that:

$$\text{for } k \geq 1,\ S_k(n) = \binom{n+k-1}{k}$$

Write a detailed proof by proceeding as follows:

Step 1: Explain why $S_k(n)$ is the number of k-tuples that satisfy:

$$1 \leq i_1 \leq i_2 \leq i_3 \leq \cdots \leq i_k \leq n$$

Step 2: Explain why $S_k(n)$ is the number of k-tuples that satisfy:

$$1 < i_1 < \left(i_2+1\right) < \left(i_3+2\right) < \cdots \leq \left(i_k+k-1\right) \leq n$$

Step 3: Explain why $S_k(n)$ is the number of ways to choose k items from $n+k-1$ items when repetitions are not allowed.

Hint: Since repetitions are not allowed, you can always list a chosen k-tuple in a unique sorted order. Think of the leftmost summation as trying all possibilities for the largest item in a tuple.

Step 4: Explain why Step 3 implies:

$$\text{for } k \geq 1,\ S_k(n) = \binom{n+k-1}{k} = \frac{(n+k-1)!}{(n-1)!k!} = \frac{1}{k!}\prod_{i=0}^{k-1}(n+i)$$

Step 4: Explain why Step 4 implies:

$$S_k(n) = \frac{1}{k!}n^k + \Theta\left(n^{k-1}\right)$$

If you expand the product you get a "messy" polynomial of degree k; explain why the coefficient of n^k must be 1.

G. For the general case of $k>2$, the presentation in this chapter used a proof by induction that employed the analysis of an arithmetic sum that is used in the appendix, which gives the correct coefficients for both the first and second terms. We can also prove this by only using a basic upper and lower bounding of a sum of powers to get a correct first coefficient:

$$\sum_{i=1}^{n} i^{k-1} > \int_{x=0}^{n} x^{k-1} = \tfrac{1}{k} n^k$$

$$\sum_{i=1}^{n} i^{k-1} = \sum_{i=1}^{n-1} i^{k-1} + n^{k-1} < \int_{x=0}^{n} x^{k-1} + n^{k-1} = \tfrac{1}{k} n^k + n^{k-1}$$

Given these two bounds, we can use two proofs by induction to get the desired Θ bound, one for Ω and one for O. To show that

$$S_k(n) \geq \tfrac{1}{k!} n^k + \Omega\!\left(n^{k-1}\right)$$

it suffices to show that:

$$S_k(n) \geq \tfrac{1}{k!} n^k + \tfrac{1}{k!} n^k$$

(*basis*) For $k=2$:

$$S_3(n) = \tfrac{1}{2} n^2 + \tfrac{1}{2} n = \tfrac{1}{2!} n^2 + \tfrac{1}{2!} n^{(2-1)}$$

(*inductive step*) For $k>2$:

$$S_k(n) = \sum_{i=1}^{n} S_{k-1}(i)$$

$$= \sum_{i=1}^{n} \left(\tfrac{1}{(k-1)!} i^{k-1} + \tfrac{1}{(k-1)!} i^{k-2} \right)$$

$$= \tfrac{1}{k!} n^k + \frac{1}{(k-1)(k-1)!} n^{k-1}$$

$$> \tfrac{1}{k!} n^k + \tfrac{1}{k!} n^{k-1}$$

Give a similar proof by induction to show that:

$$S_k(n) \leq \tfrac{1}{k!} n^k + O\!\left(n^{k-1}\right)$$

Hint: Use the inductive hypothesis:

$$S_k(n) \leq \tfrac{1}{k!} n^k + \tfrac{1}{k} n^{k-1}$$

146. We have considered here *binary* Huffman trees. Generalize the construction to the case where all non-leaf vertices have k children, for any $k \geq 2$.

Hint: Since each step of the construction algorithm reduces the number of tries in the forest by $k-1$, if $k-1$ does not evenly divide $n-1$, then a first step can combine the appropriate number of smallest weight trees into a single one.

147. Recall the greedy algorithm for the shortest common superstring problem and for any $n \geq 1$, consider the following set of strings:

$$S_n = \{ab^n, b^{n+1}, b^n a\}$$

A. Give a superstring for S_n that saves $2n$ characters (i.e., the sum of the lengths of the strings in S_n minus the length of the superstring is $2n$).

B. Explain why the greedy algorithm can save only n characters.

C. Now consider the string:

$$T_n = \{aab^n a, b^n ab^n, ab^n aa\}$$

Explain why an optimal algorithm saves $2n+2$ characters and the greedy algorithm, independent of how ties are broken, always saves only $n+2$ characters.

148. Give a precise statement of the overlap lemma for the superstring problem (i.e., the conditions that the strings a, b, and c must satisfy) along with a detailed proof..

Hint: It may help to think of the lemma as $|b| = |OL(a,b)| + |OL(b,c)| - |OL(a,c)|$; that is, $|b|$ is the sum of $|OL(a,b)|$ and $|OL(b,c)|$ less a correction for what was counted twice.

149. Recall the greedy algorithm for the shortest common superstring problem. Consider an even simpler greedy approach that works sequentially through S. That is, if $S = \{x_1, ..., x_n\}$, then the simple greedy algorithm is:

> $t :=$ the empty string
>
> **for** $i := 1$ **to** n **do begin**
>
> > Combine t and x_i into a single string by overlapping the right end of t with the left end of x_i as much as possible.
>
> **end**

A. Give an infinite set of strings S_1, S_2, ... for which simple greedy saves $O(n)$ characters on S_n but an optimal algorithm saves $O(n^2)$ characters on S_n (i.e., the sum of the lengths of the strings in S_n minus the length of the superstring is $O(n)$ for greedy but $O(n^2)$ for an optimal algorithm).

B. Show that this factor of $O(n)$ difference in savings can still happen if the simple greedy algorithm is "improved" to take at each step the smaller of overlapping t with x_i and overlapping x_i with t.

150. Give an algorithm to find a minimal length superstring for a set of strings S, where all strings in S are exactly two characters.

151. Suppose that a Huffman trie was used to define codes for a set of k-character strings, $k \geq 1$. Describe how a superstring could be used to store the leaves of the trie, and under what circumstances space would be saved.

152. Without techniques like simulated annealing (even with them), iterative improvement algorithms (local search algorithms, hill-climbing algorithms, etc.) can get "stuck" at a local minimum. Consider the following approach to solving the knapsack problem:

First construct an initial solution by randomly including elements until it is not possible to include other elements. Then successively try to find three elements a, b, and c such that a is currently in the solution and b and c are not, $a < b+c$, and replacing a by b and c results in a legal solution.

Give an example of a solution resulting from this approach that is not optimal but for which no further improvement is possible.

Chapter Notes

The divide and conquer approach is presented in most texts on algorithms, such as those listed at the start of the notes to the first chapter.

Merge sort and quick sort are also presented in most texts on algorithms. The book of Knuth [1969] is a classic source for detailed analysis of many sorting methods, including merge sort and quick sort (both 2 and 3 partitioning).

Parberry [1998] presents a precise count of the number of comparisons made by merge sort by showing that the number of comparisons used in the worst case is given by:

$$T\left(\left\lceil \tfrac{n}{2} \right\rceil\right) + T\left(\left\lfloor \tfrac{n}{2} \right\rfloor\right) + n - 1 = n\left\lceil \log_2(n) \right\rceil - 2^{\left\lceil \log_2(n) \right\rceil} + 1$$

Quick sort was presented by Hoare [1962], with improvements for choosing the pivot element suggested by Singleton [1969] and Frazer and McKellar [1970]. For finding the kth largest element, the expected linear time algorithm was presented in Hoare [1961], the worst case linear time algorithm was presented in Blum, Floyd, Pratt, Rivest, and Tarjan [1972], and Floyd and Rivest [1975] give improvements for the average time approach by partitioning around an element that is obtained by recursive sampling. The presentation of quick sort and finding the kth largest element given here is along the lines of that given in Aho, Hopcroft and Ullman [1974].

The $O(n^{1.59})$ polynomial multiplication algorithm can also be used to multiply integers (with $O(n^{1.59})$ bit operations; it was presented in Karatsuba and Ofman [1962], and generalizations were presented in Winograd [1973]. The presentation here is along the lines of that for integer multiplication in Aho, Hopcroft, and Ullman [1974]. This algorithm can be viewed as a first step towards an $O(n\log(n))$ algorithm using the discrete Fourier transform, and the reader may wish to refer to that chapter for further reading.

Strassen's algorithm for matrix multiplication was presented in Strassen [1969]. The exercises to Chapter 6 of the book of Aho, Hopcroft, and Ullman [1974] observe that Strassen's algorithm can be implemented with only 15 matrix additions (Exercise 6.5), and consider alternate derivations of using 7 multiplications of matrices of half the size (Exercise 6.21). Winograd [1970] shows that matrix multiplication is no harder than matrix inversion (i.e., an algorithm for inversion can be used for multiplication); it is only necessary to observe that if A and B are n by n matrices and I is the n by n identity matrix (1's on the main diagonal and 0's everywhere else), then the matrix product AB can be computed by inverting a $3n$ by $3n$ matrix as follows:

$$\begin{bmatrix} I & A & 0 \\ 0 & I & B \\ 0 & 0 & I \end{bmatrix}^{-1} = \begin{bmatrix} I & -A & AB \\ 0 & I & -B \\ 0 & 0 & I \end{bmatrix}$$

In the book of Aho, Hopcroft and Ullman [1974], the reverse direction is presented (an algorithm for multiplication can be used for inversion in the same asymptotic time); it also shown how to use matrix multiplication to perform a "LUP decomposition" of a n by n matrix, where the matrix is expressed as the product of:

L: A n by n unit lower triangular matrix; that is, all elements above the main diagonal (all entries (i,j) such that $i<j$) are 0, and all elements on the main diagonal (all entries (i,j) such that $i=j$) are 1.

U: A n by n upper triangular matrix; that is, all elements below the main diagonal (all entries (i,j) such that $i>j$) are 0.

P: A n by n permutation matrix; that is, a matrix of 0's and 1's such that each row and each column has exactly one 1 (multiplying a n by n matrix by a n by n permutation matrix simply permutes its rows).

Given that matrix multiplication can be used to compute LUP decomposition, it follows that the determinant of a matrix can be computed in the same time as matrix multiplication (the determinant of a triangular matrix is just the product of its main diagonal, and the determinant of a permutation matrix is 1 or -1 depending on whether it represents an odd or even permutation, which can be checked by simply performing the permutation by successively exchanging rows). LUP decomposition can also be used to solve sets of linear equations, because once the LUP decomposition has been computed, in $O(n^2)$ time, two back-substitution steps (as presented in Chapter 1) and a permutation complete the solution; so any matrix multiplication algorithm that is asymptotically faster than $O(n^3)$ gives an asymptotically faster method to solve a set of linear equations than was presented in Chapter 1.

Matrix multiplication underlies the complexity of many computation problems, and there has been a succession of asymptotically faster and faster methods for matrix multiplication since the introduction of Strassen's algorithm. Coppersmith and Winograd [1987] present an $O(n^{2.376})$ algorithm to multiply two n by n matrices.

Also widely studied is the special case of Boolean matrix multiplication, where all entries are 0 or 1, and the logical OR and logical *AND* operations are used instead of addition and multiplication. Fischer and Meyer [1971] observe that general matrix multiplication can be used for Boolean matrix multiplication (and all computations can be done modulo $n+1$); this is also noted in the books of Aho, Hopcroft, and Ullman [1974] and Cormen, Rivest, Lieserson, and Stein [2001].

The book of Aho, Hopcroft, and Ullman [1974] shows that Boolean matrix multiplication can be used to perform transitive closure in the same asymptotic time; it also presents the interesting "Four Russians Algorithm" to multiply two Boolean matrices in $O(n^3/\log(n))$ steps, due to Arlazarov, Dinic, Kronrod, and Faradzev [1970]. It is also true that transitive closure can be used to perform Boolean matrix multiplication in the same asymptotic time. If A and B are n by n matrices and I is the n by n identity matrix, then the matrix product AB can be computed by performing transitive closure on a $3n$ by $3n$

matrix, where only the first three of the matrices in the sum defined by the closure are non-zero:

$$
\begin{bmatrix} I & A & 0 \\ 0 & I & B \\ 0 & 0 & I \end{bmatrix}^{*} = \begin{bmatrix} 0 & A & 0 \\ 0 & 0 & B \\ 0 & 0 & 0 \end{bmatrix}^{0} + \begin{bmatrix} 0 & A & 0 \\ 0 & 0 & B \\ 0 & 0 & 0 \end{bmatrix}^{1} + \begin{bmatrix} 0 & A & 0 \\ 0 & 0 & B \\ 0 & 0 & 0 \end{bmatrix}^{3} + \begin{bmatrix} 0 & A & 0 \\ 0 & 0 & B \\ 0 & 0 & 0 \end{bmatrix}^{3} + \cdots
$$

$$
= \begin{bmatrix} I & 0 & 0 \\ 0 & I & 0 \\ 0 & 0 & I \end{bmatrix} + \begin{bmatrix} 0 & A & 0 \\ 0 & 0 & B \\ 0 & 0 & 0 \end{bmatrix} + \begin{bmatrix} 0 & 0 & AB \\ 0 & 0 & 0 \\ 0 & 0 & 0 \end{bmatrix} + \begin{bmatrix} 0 & 0 & 0 \\ 0 & 0 & 0 \\ 0 & 0 & 0 \end{bmatrix} + \cdots
$$

$$
= \begin{bmatrix} I & A & AB \\ 0 & I & B \\ 0 & 0 & I \end{bmatrix}
$$

For further reading about computations involving matrices, see the books of Pan [2001], Bini and Pan [1994], and Golub and Van Loan [1983].

Generic analysis of recurrence relations is given in most books on algorithms, such as those listed at the beginning of the chapter notes to the first chapter. The book of Cormen, Leiserson, Rivest, and Stein [2001] includes analysis with the integer round-up and down operators.

Dynamic programming is presented in most texts on algorithms, such as those listed at the start of the chapter notes to the first chapter. The book of Bellman [1957] is devoted to dynamic programming. Giegerich [2000] considers ways to control ambiguity in dynamic programming and gives further references on the subject of dynamic programming as it applies to problems in computational biology.

The knapsack problem is an example of a *NP-hard* problem, where no algorithm of polynomial asymptotic time is known for solving the problem in general; see the section on this subject in the chapter on graph algorithms, as well as the notes to that chapter. The knapsack problem is shown NP-hard in Karp [1972]. The dynamic programming algorithm for knapsack is presented in the book of Garey and Johnson [1979]. Lin and Storer [1991] present parallel processor efficient implementations of the dynamic programming algorithm presented here.

The paragraphing problem and matrix multiplication ordering examples are commonly used in algorithms texts, such as many of those listed at the beginning of the chapter notes to the first chapter. Motta, Storer, and Carpentieri [2000] adapt the paragraphing algorithm to improve scene cut quality in a sequence of compressed video frames, whereas, Carpentieri and Storer [1994] show that other types of frame alignment are very hard.

Context-free languages have played a central role in the development of programming languages and compilers, because they can be used to formally specify the syntax of a programming language, although restrictions are typically placed on the type of grammars that are used so that faster compilers can be written (see for example, the text of Aho, Sethi, and Ullman [1986]). In general, context-free grammars allow the right-hand sides of productions to be arbitrary sequences of variables and characters (right-hand sides may also be the empty string). Usually, one of the variables is designated as the *start* variable, and the language generated by the grammar is defined as the language generated by the start variable. It can be shown that any context free grammar that generates a non-empty set of strings can be converted to a *Chomsky normal form* (CNF) grammar that generates the same set of strings. The CYK algorithm was introduced by J. Cocke and is presented in Kasami [1965] and Younger [1967]; the book of Hopcroft and Ullman [1979] refers to this algorithm as CYK and contains a presentation of the algorithm, as well as basic concepts relating to context-free languages and an algorithm to convert to Chomsky Normal Form. Earley [1970] presents an algorithm that is still $O(n^3)$ worst case for arbitrary grammars, but is $O(n^2)$ worst case for unambiguous grammars (ones where no string in the language has two different ways to produce it from the start symbol); see also Kasami and Torii [1969]. Valiant [1975] shows that general context-free language is no harder asymptotically that matrix multiplication. Graham, Harrison, and Ruzzo [1976] give an $O(n^3/\log(n))$ time algorithm for on-line general context-free language recognition.

For further reading on randomized algorithms and probabilistic analysis of algorithms, see the books of Rajasekaran, Pardalos, Reif, and Rolim [2001], Motwani and Raghavan [1995] and Hofri [1987].

Greedy algorithms are presented in most texts on algorithms. *Matroids* provide a general framework when greedy algorithms are optimal; see for example the book of Papadimitriou and Steiglitz [1982].

The book of Garey and Johnson [1979] presents the bin packing problem in detail. Like the knapsack problem, it is an example of a NP-hard problem. The 17/10 bound for first-fit bin packing is presented in Johnson, Demers, Ullman, Garey, and Graham [1974]. Johnson [1973] shows that under a number of standard distributions of input sizes that first-fit bin packing averages only about 7% worse than optimal.

For further reading and references on Huffman trees, see the notes to the chapter on strings.

The shortest common superstring problem is shown NP-hard in Maier and Storer [1978] and Gallant, Maier, and Storer [1980]; they also give an efficient algorithm for the case that all strings have length ≤ 2. Turner [1989] showed that the savings by greedy is at least 1/2 that of optimal, although the presentation here is based on a modification of the simpler factor of 4 proof presented in the book of Storer [1988], which also presents a number of variations of the greedy approach (including the simple greedy approach presented in the exercises and the $O(n)$ gap between the worst-case performance of this algorithm and an optimal one). More efficient implementation of the greedy algorithm can be achieved using pattern matching techniques (see the exercises for the chapter on strings). A related question is approximating the length (as opposed to the savings) of a shortest superstring; this question and many related issues are addressed in Chapter 8 of the book edited by Apostolico and Galil [1997] (the chapter is by M. Li and T. Jiang). Gusfield, Landau, and Schieber [1991] employ suffix tries to find maximal overlaps between strings at the end of one to the beginning of another, which has applications to the superstring approximation algorithms.

We have only presented the most simple version of simulated annealing here. In practice, complex "cooling schedules" may be employed, as well as carrying along a population of potential solutions. The book of Laarhaven and Aarts [1987] is devoted to simulated annealing. The package delivery problem is an example of the *graph partitioning problem*, which is also known to be NP-hard (see the chapter on graph algorithms). Johnson [1985] presents experiments with simulated annealing used to solve graph partitioning. Storer, Nicas, and Becker [1985] use simulated annealing for routing in VLSI layout.

6. Hashing

Introduction

We have already seen how binary search trees can support operations such as DELETEMIN in addition to the three basic operations of INSERT, DELETE, and MEMBER in $O(n\log(n))$ expected time for n operations. In many applications, the three basic operations (and in some cases just INSERT and MEMBER) are all we need. When this is the case, hashing provides a way to reduce the expected time to $O(1)$ expected time per operation.

The ideas behind hashing come naturally if we approach the problem in the straightforward fashion and then work around the memory problem we encounter. Let n be the number of data items to be stored and m the number of possible data items. Ideally, we would like to employ an array as follows:

1. Initialize all locations of a (large) array $A[0...m-1]$ to contain a *nil* value.
2. Associate a unique integer $0 \leq h(d) < m$ with each possible input item d.
3. Place d in position $A[h(d)]$.

There are two potential problems with this approach. The first is that it could be time consuming to initialize A. This is not really a problem for a number of reasons. Special purpose hardware may be available, and there are ways to perform "on-the-fly" initialization which in $O(1)$ time per access keeps track of the positions of A that have been used without actually initializing A (an additional $O(m)$ space is needed — see the chapter on sets over a small universe). In any case, this first problem will become moot after we address the second problem, where m will become $O(n)$ and the time to initialize A can be amortized against the time to insert the elements into A.

The second problem is that m could be very large. For example, consider storing a set of n names, each represented by a sequence of 25 characters. Even if n is relatively small (say a few thousand or million), the number of possible sequences of 25 characters (that determines the value of m), even if they are limited to the 26 lower case characters of the English alphabet, is $26^{25} > 2^{100}$. The key idea of hashing is to overcome this problem by making h a function that is not one-to-one so that h "crowds" the set of possible input items into a range 0 to $m-1$, where m is chosen to be relatively small. In practice m is typically the same or only a small constant times n (e.g., $m = 1.5n$). Of course this means that many possible input items get mapped to any particular value in the range 0 to $m-1$. The hope is that the actual input items that are stored will get spread fairly evenly, and in fact, we shall see that when the input distribution is random, they are likely to. Nevertheless, there will be items that get sent to the same location, and dealing with such "collisions" is part of the hashing process.

Basic Hashing Algorithm

Application: Maintain a set of items and support the three operations of INSERT, MEMBER, and DELETE.

Idea: Use a *hash function h* to map a data item *d* into a hash table array $A[0]...A[m-1]$. Because more than one item may go to the same position, each position is called a *bucket*, and contains a pointer to a linked list of all items that have hashed to that position.

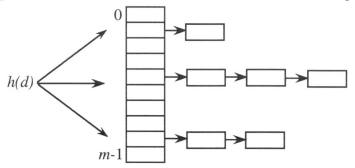

Notation:

$A[0] ... A[m-1] =$ The *hash table*; each location of the array (called a "bucket") contains a pointer to a linked list of all items stored there.

$h =$ A function that maps a data item *d* to an integer $0 \leq h(d) < m$.

Basic operations:

MEMBER(*d*): Search the bucket at $A[h(d)]$.

INSERT(*d*): Do MEMBER(*d*) and if *d* is not present, add *d* to the bucket at $A[h(d)]$.

DELETE(*d*): Do MEMBER(*d*) and if *d* is present, remove *d* from the bucket at $A[h(d)]$.

The MOD hash function: *a* MOD *b* denotes the remainder when *a* is divided by *b*:

$$(a \text{ MOD } b) = a - \left\lfloor \frac{a}{b} \right\rfloor b$$

It is always an integer in the range 0 to $b-1$ when *a* and *b* are positive integers (see the exercises for more general uses of the MOD function). When *d* is an integer, the standard MOD hash function computes:

$h(d) = d \text{ MOD } m$

To avoid patterns when *d* and *m* have factors in common, *m* should be prime (i.e., no integers evenly divide *m* except 1 and *m*). Alternately, for a large prime $q > m$, compute:

$h(d) = (d \text{ MOD } q) \text{ MOD } m$

Hash Functions for Data Items with Many Bits

Idea: In principle, since any data item can be represented as a sequence of bits, it can be viewed as a non-negative integer, and the standard MOD hash function can be used. However, it is convenient to work with integers that fit into a single machine word. To convert a data item into an index into the hash table:

1. Choose a large prime q that can be represented and manipulated with the word size on the machine you are using.
2. Use arithmetic operations to combine pieces of the data item into a single large number, doing a MOD q as necessary to prevent arithmetic overflow.
3. MOD the result by m to get a hash table index between 0 and $m-1$.

Modeling data items as strings: Although the division of a data item into pieces of manageable sizes can be tailored to the application in question, a simple approach is, for some $k \geq 1$, to partition the bits of each data item into k blocks of $b \geq 1$ bits (if a data item has fewer than kb bits, pad to the left with 0's). For example, $b=8$ corresponds to partitioning into bytes. Thus, all data items can be viewed as strings of k characters over the alphabet 0 to $b-1$. We assume that both b and k are constants $\leq m$.

Example, the weighted hash function: Choose a small prime p and a large prime q such that summing two values of size pq does not cause an integer overflow on your machine; for example, for p a prime of 10 bits ($p<1024$) and a maximum positive integer of 31 bits, 20 bits would do for q. Choose a sequence of weights $W[0]...W[k-1]$, each in the range 0 to $p-1$ (e.g., randomly pick distinct values). To hash a given string s, let $s[0]...s[k-1]$ denote its characters, add up the characters of s times the corresponding weight, and MOD the result by m to get a valid index into the hash table:

> **function** $h(s)$
> $\quad z := 0$
> $\quad$ **for** $i := 0$ **to** $k-1$ **do** $z := (z+s[i]W[i])$ MOD q
> $\quad z := z$ MOD m
> $\quad$ **return** z
> **end**

Example, the polynomial hash function: Instead of using individual weights, replace the *for* loop of the weighted hash function by a loop that uses Horner's rule to compute the polynomial $h(s) = s[k-1]p^{k-1} + \cdots + s[1]p + s[0]$. For example, choosing $p \approx b$ corresponds to shifting bits to the left by about the size of a character.

> **for** $i := k-1$ **downto** 0 **do** $z := (pz+s[i])$ MOD q

Note: For data items represented by long strings, computation of the hash function can dominate the total running time. As presented in the chapter on strings, for some applications a *trie* data structure can be a practical alternative.

Complexity of Hashing

For simplicity, we limit our attention to the complexity of a sequence of n INSERT operations followed by n MEMBER operations, one for each of the items inserted into a hash table of size $m \geq n$.

Assumptions:

- The elements to be inserted are chosen at random from the set of all possible elements.

- The hash function is an "ideal" one where for any i in the range of $0 \leq i < m$, if x is randomly chosen, then the probability that $h(x)=i$ is $1/m$. Equivalently, if x and y are randomly chosen, then the probability that $h(x)=h(y)$ is $1/m$.

Space: $O(m+n)$

The array uses $O(m)$ space and the total space used by linked list vertices is $O(n)$.

Worst case time: $O(n^2)$

In the worst case, all elements could hash to the same position.

The n MEMBER instructions will take (n^2) time because for each value of $1 \leq i \leq n$, one of the member instructions will examine i linked list buckets, and hence the total number of comparisons is given by the arithmetic sum (see the Appendix):

$$1+2+3+\cdots+n = \tfrac{1}{2}n^2 + \tfrac{1}{2}n = \Omega\left(n^2\right)$$

Expected time: $O(n)$

Since a new item can be inserted at the front of a bucket, each INSERT is $O(1)$ in the worst case.

Since each INSERT is equally likely to hash to any of the m buckets and $n \leq m$ elements are inserted, then on inserting the ith element, the expected length of the list on which it is placed is $(i-1)/m < 1$. Hence the expected bucket size is $O(1)$, and the expected time for each MEMBER operation is $O(1)$.

The Constant e

Idea: Recall from the first chapter that the constant e is the natural logarithm base; that is, it is the unique real number such that the function e^x is its own derivative. Although we typically use base 2 logarithms in algorithm analysis (due to the use of binary numbers, divide and conquer, etc.), the constant e arises often in the kind of expected analysis used for various aspects of hashing, where we are sometimes interested in the precise values of constants.

Basic facts about e (see the exercises, where hints outline the derivations):

A. For all real numbers x: $e^x = \sum_{i=0}^{\infty}\left(\dfrac{x^i}{i!}\right) = 1 + x + \dfrac{x^2}{2!} + \dfrac{x^3}{3!} + \dfrac{x^4}{4!} + \dfrac{x^5}{5!} + \cdots$

B. $e = 2.71821828459\ldots$ and $1/e = 0.36787944\ldots$

C. For any integer $n>1$, let $e_{<n} = \left(\dfrac{n+1}{n}\right)^n$ and $e_{>n} = \left(\dfrac{n}{n-1}\right)^n$.

   Then $e = \lim\limits_{n\to\infty} e_{<n} = \lim\limits_{n\to\infty} e_{>n}$ and for all $n \geq 6$:

   $$2.5 < e_{<n} < e_{<(n+1)} < e < e_{>(n+1)} < e_{>n} < 3$$

   $$\frac{1}{3} < \frac{1}{e_{>n}} < \frac{1}{e_{>(n+1)}} < \frac{1}{e} < \frac{1}{e_{<(n+1)}} < \frac{1}{e_{<n}} < 0.4$$

D. For $x \geq 0$ a real number: $e^x = \lim\limits_{n\to\infty}\left(\dfrac{n+x}{n}\right)^n = \lim\limits_{n\to\infty}\left(\dfrac{n}{n-x}\right)^n$

E. For all real numbers $0<x<1$, $e^x < 1 + x + \dfrac{x^2}{2(1-x)}$

   and for $0 < x \leq 1/2$: $e^x < 1 + x + x^2 \leq 1 + \dfrac{3}{2}x$

F. For all real numbers $0<x<1$: $\dfrac{1}{e^x} > 1 - x$

G. To switch between logarithms in base 2 and base e (designated as ln):
   $\log_2(e) = 1.442695\ldots$
   $\ln(2) = 0.693147\ldots$
   $\ln(x) = \log_2(x)/\log_2(e) < \log_2(x)/1.443$
   $\log_2(x) = \ln(x)/\ln(2) < \ln(x)/0.7$

Expected Number of Empty Buckets

Theorem: For $c \geq 1$, if $n > 1$ items are inserted into a hash table of size cn, the expected number of empty buckets is $\Theta(n)$.

Proof: First consider the case $c = 1$.

The expected number of empty buckets is $\Omega(n)$: Let k be such that after item $(n/2) + k$ is inserted, exactly $n/2$ buckets are non-empty (if k does not exist, then more than $n/2$ buckets must be empty). For each of the remaining $(n/2)-k$ items, there is at most probability $1/2$ that it is placed in an empty bucket. Hence, the expected number of empty buckets is at least:

$$\frac{n}{2} - \frac{1}{2}\left(\frac{n}{2} - k\right) = \frac{n}{4} + k \geq \frac{n}{4}$$

The expected number of non-empty buckets is $\Omega(n)$: Since at most $i-1$ buckets are full after $i-1$ items have been inserted, the probability that the ith item is placed in an empty bucket is $\geq (n-(i-1))/n$. Hence, using an arithmetic sum (see the appendix), the expected number of non-empty buckets is at least:

$$1 + \frac{n-1}{n} + \frac{n-2}{n} + \cdots + \frac{1}{n} = \frac{1}{n}(1 + 2 + \cdots + n) = \frac{n+1}{2} > \frac{n}{2}$$

Since both empty and non-empty buckets are $\Omega(n)$, the theorem follows for $c=1$.

Although the above approach can be generalized to $c > 1$ (see the exercises), for a more precise analysis we derive an expression for $E(i)$, the expected number of empty buckets after i items have been inserted into a hash table of size $m = cn$ (and so $m - E(i)$ is the expected number of non-empty buckets). The expected number of empty buckets after i items have been inserted is $E(i-1)$. $E(i)$ is equal to $E(i-1)$ minus the expected number of additional buckets to be made non-empty by the insertion of the ith item, which is $E(i-1)/m$ (since the ith item is equally likely to be inserted into any of the m buckets), and hence we have the simple recurrence relation

$$E(i) = \begin{cases} m & \text{if } i = 0 \\ E(i-1) - \frac{E(i-1)}{m} = \left(\frac{m-1}{m}\right)E(i-1) & \text{if } i > 0 \end{cases}$$

which has the solution:

$$E(i) = m\left(\frac{m-1}{m}\right)^i$$

For $c=1$, $E(n) = n/e_{>n}$, and hence for $n \geq 6$, $n/3 < E(n) < n/e$. Thus for large n:

$$E(n) \approx \frac{n}{e}$$

Similarly, for $c > 1$ and large n:

$$E(n) = m\left(\frac{m-1}{m}\right)^n = cn\left(\left(\frac{m-1}{m}\right)^m\right)^{1/c} \approx \frac{cn}{e^{1/c}} = \frac{m}{e^{n/m}}$$

Chernoff Bound

Idea:

The classic *Chernoff bound* provides a limit on the deviation from the average value of the number of heads in a sequence of independent coin tossings. For example, if a coin is tossed 100 times, on the average we expect heads to come up 50 times, but each time we do 100 coin tosses, the precise number of heads will typically vary between somewhat less than 50 and somewhat more.

The Chernoff bound is useful for deriving precise bounds about how hashing behaves with high probability (where the probability gets higher as the number of items hashed increases).

Statement of the Chernoff bound: Let:

N = number of coin tossings

p = the probability of coming up heads ($1-p$ for tails)

X = number of heads

u = mean value of $X = pN$

e = the natural logarithm base = $2.7182\ldots$

Then for any real number $\alpha > 0$, the probability that $X \geq (1+\alpha)u$ is at most:

$$\left(\frac{e^{\alpha}}{(1+\alpha)^{(1+\alpha)}} \right)^{u}$$

See the chapter notes for references to derivations of this and related bounds.

General approach: Model the hashing of an item to one of the m buckets in a hash table as tossing a m-sided coin, and use the Chernoff bound to get a better understanding of the (high probability) behavior of hashing.

Notation for using the Chernoff bound to analyze hashing:

bn = the number of items that are hashed into the table, $b>0$

cn = the size of the table (so $m=cn$), $c \geq 1$

p = the probability that an item goes to bucket i, $p = 1/cn$

X = the number of elements that hash to the ith bucket

u = the mean value of $X = p(bn) = b/c$

Size of the Largest Bucket

Idea:

Hashing is not perfect, there will usually be buckets with more than $O(1)$ elements in them, even when $b<c$.

However, with high likelihood, the size of the largest bucket is not too large.

For $a \geq 1$ we use the Chernoff bound to derive a general expression for the probability that the ith bucket contains more than $a\log_2(n)$ elements.

Different questions can be answered by plugging in different values for a, b, and c, including values that are functions of n.

Choose α: For any $a \geq 1$, let $\alpha = (ac\log_2(n)/b)-1$.

Apply the Chernoff bound: We are using the Chernoff bound with $N=bn$, $p=1/cn$, and $u=b/c$. Assuming $\log_2(n)>b/ac$ (so $\alpha>0$), the probability that $X \geq a\log_2(n)$ is the same as the probability that $X \geq (1+\alpha)u$, which is:

$$\left(\frac{e^{\alpha}}{(1+\alpha)^{(1+\alpha)}} \right)^{\mu}$$

$$< \left(\frac{e^{a\frac{c}{b}\log_2(n)}}{\left(a\frac{c}{b}\log_2(n)\right)^{\left(a\frac{c}{b}\log_2(n)\right)}} \right)^{(b/c)}$$

$$= \frac{1}{\left(\frac{1}{e}a\frac{c}{b}\log_2(n)\right)^{a\log_2(n)}}$$

$$= \frac{1}{2^{a\log_2(n)\log_2(\frac{1}{e}a\frac{c}{b}\log_2(n))}}$$

$$= \frac{1}{n^{a(\log_2\log_2(n)+\log_2(a)-\log_2(b)+\log_2(c)-\log_2(e)}}$$

Example, $\log_2(n)$ is a reasonable practical upper bound:

If n items are hashed into a table of size $n \geq 2{,}048$, the probability of a particular bucket receiving more than $\log_2(n)$ items is given by setting $a = b = c = 1$:

$$< \frac{1}{n^{a\left(\log_2\log_2(n)+\log_2(a)-\log_2(b)+\log_2(c)-\log_2(e)\right)}}$$

$$< \frac{1}{n^{1\left(\log_2\log_2(2{,}048)+\log_2(1)-\log_2(1)+\log_2(1)-1.443\right)}}$$

$$= \frac{1}{n^{\log_2(11)-1.443}}$$

$$< \frac{1}{n^2}$$

Hence the probability of *any* bucket receiving more than $\log_2(n)$ items is less than $1/n$.

Example, $O(\log(n)/\log\log(n))$ is the asymptotic bound:

If n items are hashed to a table of size $2n$, the probability of a particular bucket receiving more than $2\log_2(n)/\log_2\log_2(n))$ items is given by setting $a=2/\log_2\log_2(n)$, $b=1$, and $c=2$:

$$< \frac{1}{n^{a\left(\log_2\log_2(n)+\log_2(a)-\log_2(b)+\log_2(c)-\log_2(e)\right)}}$$

$$= \frac{1}{n^{\left(\frac{2}{\log_2\log_2(n)}\right)\left(\log_2\log_2(n)+\log_2\left(\frac{2}{\log_2\log_2(n)}\right)-\log_2(1)+\log_2(2)-\log_2(e)\right)}}$$

$$= \frac{1}{n^{\left(\frac{2}{\log_2\log_2(n)}\right)\left(\log_2\log_2(n)-\log_2\log_2\log_2(n)+2-\log_2(e)\right)}}$$

$$< \frac{1}{n^{\left(\frac{2}{\log_2\log_2(n)}\right)\left(\log_2\log_2(n)-\log_2\log_2\log_2(n)+0.6\right)}}$$

$$< \frac{1}{n^{\left(2-\left(\frac{\log_2\log_2\log_2(n)}{\log_2\log_2(n)}\right)\right)}}$$

$$\to \frac{1}{n^2} \quad \text{as } n \to \infty$$

Hence the probability of *any* bucket receiving more than $2\log_2(n)/\log_2\log_2(n))$ items approaches $1/n$ as $n \to \infty$.

Overfilling a Hash Table

Idea:

Even when we "crowd" a hash table of size n with $n\log_2(n)$ elements, we can use Chernoff bounds to show that with high probability, the size of the largest bucket is only $O(\log_2(n))$.

That is, the distribution of bucket sizes simply "flattens", and the size of the largest bucket becomes the same order as the average size of a bucket.

Example: If $n\log_2(n)$ items are hashed into a table of size n, the probability of a particular bucket receiving more than $4\log_2(n)$ items is given by setting $a=4$, $b=\log_2(n)$, and $c=1$:

$$< \frac{1}{n^{a\left(\log_2\log_2(n)+\log_2(a)-\log_2(b)+\log_2(c)-\log_2(e)\right)}}$$

$$= \frac{1}{n^{4\left(\log_2\log_2(n)+\log_2(4)-\log_2\log_2(n)+\log_2(1)-\log_2(e)\right)}}$$

$$= \frac{1}{n^{4\left(2-\log_2(e)\right)}}$$

$$< \frac{1}{n^{4(0.6)}}$$

$$< \frac{1}{n^2}$$

Hence the probability of *any* bucket receiving more than $4\log_2(n)$ items is less than $1/n$.

Resizing a Hash Table

Problem: Suppose that you are receiving data items one at a time and inserting them into a hash table, but you do not know in advance the number of data items n.

Idea: Rehash into a larger table when the current table "fills" and amortize that cost against the work used to fill up the current table.

Resizing algorithm: Start with a table of size of $m=1$ (in practice, m can be initialized to a larger value, such as an estimate of the number of elements that are likely to be inserted).

> $n=0$
>
>   **for** each new element that has to be added to the table **do begin**
>> $n := n+1$
>>
>>   **if** $n>m$ **then begin**
>>> $m := 2m$
>>>   Allocate a new hash table of size m.
>>>   Rehash all the existing elements into the new table.
>>>   Reclaim the memory for the old table.
>>
>>   **end**
>>
>>   Hash the new element into the table.
>
>   **end**

Time: After all n elements have been inserted, the final hash table has size $n \leq m < 2n$. Hence, if N denotes the largest power of 2 that is less than n, then the total number hash operations is n (the number of hashes to put items in the table for the first time) plus the number of rehashes, which is given by a geometric sum (see the appendix):

$$n + (1 + 2 + 4 + \llcorner + N) = n + 2N - 1 \leq n+2(n-1)-1 \leq 3n - 3$$

Hence the expected time is $O(n)$.

Space: Since the space for smaller hash tables is reclaimed, the number of buckets used is bounded by the number of buckets in the final table, which is $2N<2n$. Space for the n linked list vertices used to store the n items in the buckets is $O(n)$, and hence the total space used is $O(n)$.

Universal Hashing

Idea: We may want to avoid the possibility that there is a bad match between the hash function and the input (or perhaps an adversary that has selected data that hashes to a single bucket). Universal hashing randomizes the choice of the hash function.

Universal hashing: A set of hash functions H is *universal* if for each pair of distinct data items d_1 and d_2, the number of hash functions h in H for which $h(d_1)=h(d_2)$ is $|H|/m$, where m is the size of the hash table. Given a universal set of hash functions H, at the time the hash table is initialized, *universal hashing* chooses a hash function h from H at random.

Benefits of universal hashing when $n \leq m$: Consider *any* data item d (even one chosen by an adversary who knows H but not our hash function). For any other data item x, the probability that x is hashed to the same bucket as d is only $1/m$ (since by definition of universal hashing, only this fraction of the hash functions in H map two items to the same bucket). Hence, by summing over the other $n-1$ data items (assuming that $n \leq m$), it follows that the expected number of items that hash to the same bucket as d is $(n-1)/m<1$.

Example, the weighted universal hash function: We have already considered the weighted hash function for a string $s[0]s[1]...s[k-1]$ that is computed by

$$h(d) = (s[0]W[0] + ... + s[k-1]W[k-1] \text{ MOD } q) \text{ MOD } m$$

for a large prime q and weights $W[0]...W[k-1]$ in the range 0 to $p-1$. If $m=p=q$ is a prime number and the weights are randomly chosen, then the weighted hash function satisfies the definition of a universal hash function (assuming that $k \leq m$), with $|H|=m^k$. To see this, consider any pair of data items $x[0]...x[k-1]$ and $y[0]...y[k-1]$ such that for some $0 \leq i<k$, $x[i] \neq y[i]$ (they may differ at additional positions as well). If $h(x)=h(y)$, then:

$$\leq W[i]\big(x[i]-y[i]\big)=\sum_{j\neq i} W[j]\big(y[j]-x[j]\big) \text{ MOD } m$$

We can divide by $x[i]-y[i]$ (since it cannot be zero) to get a unique solution for $W[i]$ in the integers MOD m (see the exercises) for each of the m^{k-1} choices for the other weights. Thus, for only $|H|/m$ functions h in H is it the case that $h(x)=h(y)$.

Practical considerations for the weighted universal hash function: Using a prime $q>m$ for use with a practical hash table of size m (not necessarily prime) works well in practice. There are also variations of the weighted hash function that satisfy our definition of universal hashing for m not prime (see the chapter notes).

CHAPTER 6

Twin Hashing

Idea: We have already seen how universal hash functions can be used to prevent a dependency between the input and the hash function chosen. Another issue, at least theoretically, is the size of the largest bucket. If we are unlucky and after many items have been placed into the table there are many member instructions to a single bucket, it would be nice to know that even in the worst case when this was the largest bucket, it cannot be too large. We have already seen that asymptotically the expected size of the largest bucket is $O(\log(n)/\log\log(n))$. *Twin hashing* reduces by an exponential factor the expected size of the largest bucket to only $O(\log\log(n))$.

Twin hashing algorithm: Select two different hash functions h_1 and h_2.

> MEMBER(d):   Search the buckets at $A[h_1(d)]$ and $A[h_2(d)]$.
>
> INSERT(d):   Do MEMBER(d) and if d is not present, add d to the smaller of the buckets at $A[h_1(d)]$ and $A[h_2(d)]$ (if they are the same size, choose $A[h_1(d)]$).
>
> DELETE(d):   Do MEMBER(d) and if d is present, remove d from its bucket.

Complexity: The time analysis is the same as for a single hash function; the overhead of using two hash functions and searching two buckets only increases time by a constant factor. No additional space is needed; although we could store the size of each bucket (which would only increase space by a constant factor), it is not necessary for a worst case analysis since the two buckets must be searched in any case for a MEMBER operation (and their size can be counted).

Size of the largest bucket: The expected size of the largest bucket is only $O(\log\log(n))$ under the assumption that h_1 and h_2 are ideal hash functions that distribute items uniformily and randomly (in fact, performance is improved slightly if h_1 and h_2 place elements uniformily and randomly into the left and right halves of the table respectively, so that $0 \le h_1(d) < \lceil m/2 \rceil \le h_2(d) < m$ for a table that goes from 0 to m-1); see the chapter notes.

Practical considerations: Twin hashing basically doubles the expected time for hashing. The asymptotic "insurance" it provides may not be worth this cost in many typical applications. In any case, however, it is a nice theoretical tool.

Bloom Filters

Bloom filters provide a way to use less space in practice for a sequence of insert and yes-no member operations, if we are willing to accept a small probability of a *false positive* (the answer is "yes" when it should be no):

ynMEMBER(d):   Return "yes" if d is present or "no" otherwise.

Example: A spell checker can be implemented by inserting all words into a dictionary and then using ynMEMBER queries to check the spelling of each word in a document; we may be willing to accept a small probability that a spelling error is not detected if the size of the dictionary is much smaller. Similarly, we may want to keep track web pages that are cached locally; if we think we have a page locally and then when we go to retrieve it we find out that it is not there (and so we have to go out to the web for the original), the time wasted is not a big issue if this happens rarely.

Idea: To store n items, use $k \geq 1$ hash functions to an array A[0]...A[m-1] of bits (initially all 0's). INSERT(d) computes each of the k hash functions and sets the corresponding positions of A to 1. ynMEMBER(d) checks each of the corresponding k hash positions; if any are 0, the answer is "no", otherwise it is "yes". The figure below depicts $k=5$ and $m=20$, and shows how the first item might be placed into an empty table. Subsequent insertions will make more bits 1 (and may again set some or all of these five bits to 1).

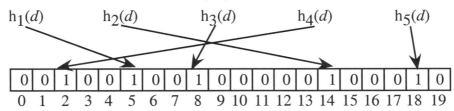

Choosing k: After all n items are hashed to A, the probability that a given bit is still 0 is

$$p = \left(1 - \frac{1}{m}\right)^{kn} = \left(\left(\frac{m-1}{m}\right)^m\right)^{kn/m} = \left(\frac{1}{e_{>m}}\right)^{kn/m} \approx \frac{1}{e^{kn/m}}$$

or equivalently $k \approx \ln(1/p)m/n$. The probability of a false match is given by $f = (1-p)^k$; notice that k appears twice in f — once as the exponent of the entire expression (where larger k is better) and once in the exponent of the denominator of p (where smaller k is better). Calculus can be used (see the chapter notes) to show that f is minimized when:

$p = 1/2$ and $k \approx .7m/n$ (k must be rounded to the nearest integer)

Example: Suppose n is 100,000 items and m is 1 million bits. Choose $k = 7$ and there is less than a 1% probability of a false match. Assuming that each item has a substantial size (e.g., a character string), a million bits is much more compact than a list of the n data items (or even a list of their hash values, if enough bits are used per hash value to insure a low probability of a false match — see the exercises).

Exercises

153. Using the basic linked list operations presented in the chapter on lists, give pseudo-code for the hashing MEMBER, INSERT, and DELETE operations.

154. A *field* is specified by five items:

1. A set S.

2. An element of S called 0.

3. An element of S called 1.

4. A binary operator + defined on S that is commutative (i.e., for all a and b in S, $a+b = b+a$), associative (i.e., for all a, b, and c in S, $a+(b+c) = (a+b)+c$), for which 0 is an identity element (i.e., for all a in S, $a+0 = 0$), and for which every element has an additive inverse with respect to 0 (i.e., for all a in S there exists a unique element b such that $a+b = 0$).

5. A binary operator $*$ defined on S that is commutative, associative, for which 1 is an identity element, for which every non-zero element has an multiplicative inverse with respect to 1 (i.e., for all $a{\neq}0$ in S there exists a unique element b such that $a*b = 1$), and which distributes over + (i.e., for all a, b, and c in S, $a*(b+c) = (a*b)+(a*c)$ and $(a+b)*c = (a*c)+(b*c)$.

Prove that the integers module p, for any prime number $p \geq 2$, are a field.

155. We have seen that practical hash functions often employ the MOD function. Prove that for any three positive integers x, y, and z:

A. $(x + y)\ \mathrm{MOD}\ z = ((x\ \mathrm{MOD}\ z) + (y\ \mathrm{MOD}\ z))\ \mathrm{MOD}\ z$

B. $(x - y)\ \mathrm{MOD}\ z = ((x\ \mathrm{MOD}\ z) - (y\ \mathrm{MOD}\ z))\ \mathrm{MOD}\ z$

C. $(x * y)\ \mathrm{MOD}\ z = ((x\ \mathrm{MOD}\ z) * (y\ \mathrm{MOD}\ z))\ \mathrm{MOD}\ z$

D. Suppose the definition of MOD is generalized to all real numbers (positive, 0, or negative) by defining $(x\ \mathrm{MOD}\ y) = x$ if $y=0$ and otherwise:

$$(x\ MOD\ y) = x - \left\lfloor \tfrac{x}{y} \right\rfloor y$$

Are Parts A through C still true?

156. Although it may have limited practical use, the classic method for testing "by hand" without long division whether a number is a multiple of 3 or 9 is a nice example for getting experience with the MOD function. Given an integer n, let $T(n)$ denote the sum of its digits when written in standard decimal notation. For example, $T(123) = 6$.

A. Prove that for any positive integer n:

$$(n\ \mathrm{MOD}\ 9) = (T(n)\ \mathrm{MOD}\ 9)$$

B. Explain why Part A implies that to check whether a positive integer is divisible by 9, one can repeatedly add up the digits (i.e., add up the digits of n, then the digits of $T(n)$, then the digits of $T(T(n))$, etc.) until you are down to a single digit (if it is 9, then the answer is yes, otherwise it is no).

Hint: First prove that $(10^i \text{ MOD } 9) = 1$, then represent an integer n as the sum of its base 10 digits times the corresponding power of 10, then simplify with modular arithmetic.

C. Explain why the algorithm of Step B also works to test if a number is a multiple of 3, except at the end you check whether the single digit remaining is 3 or 9.

157. Prove the facts about e that were presented earlier in this chapter:

A. For all real numbers x: $e^x = \sum_{i=0}^{\infty} \left(\frac{x^i}{i!} \right) = 1 + x + \frac{x^2}{2!} + \frac{x^3}{3!} + \frac{x^4}{4!} + \frac{x^5}{5!} + \cdots$

Hint: Recall the standard *Taylor Series* for a continuously differentiable function

$$f(x) = f(0) + f'(0)x + \frac{f''(0)x^2}{2!} + \frac{f'''(0)x^3}{3!} + \frac{f''''(0)x^4}{4!} + \cdots$$

where f' denotes the first derivative of f, f'' the second derivative, etc.,

B. $e = 2.71821828459...$ and $1/e = 0.36787944...$

Hint: Compute the first 6 terms from Part A to show that $e > 2.71$. Then show that $e < 2.72$ by bounding the remaining terms by the 7th term times a geometric sum with $r=1/7$. Now describe how to generalize this process.

C. For any integer $n>1$, let $e_{<n} = \left(\frac{n+1}{n} \right)^n$ and $e_{>n} = \left(\frac{n}{n-1} \right)^n$.

Then $e = \lim_{n \to \infty} e_{<n} = \lim_{n \to \infty} e_{>n}$ and for all $n \geq 6$:

$$2.5 < e_{<n} < e_{<(n+1)} < e < e_{>(n+1)} < e_{>n} < 3$$

$$\frac{1}{3} < \frac{1}{e_{>n}} < \frac{1}{e_{>(n+1)}} < \frac{1}{e} < \frac{1}{e_{<(n+1)}} < \frac{1}{e_{<n}} < 0.4$$

Hint: Compute the binomial expansion of $(1+1/n)^n$ and show that as n approaches infinity, it gets arbitrarily close to the sum of Part A for $x=1$. Then show that the facts about $e_{<n}$ and $e_{>n}$ follow. Finally, explain why the condition that n be an integer is not necessary.

D. For $x \geq 0$ a real number: $e^x = \lim_{n \to \infty} \left(\frac{n+x}{n} \right)^n = \lim_{n \to \infty} \left(\frac{n}{n-x} \right)^n$

Hint: Given $e_{>n}$ from Part C, for an integer $x \geq 1$, use the fact that

$$\left(\frac{n}{n-x}\right)^n = \left(\left(\frac{(n/x)x}{(n/x)x-x}\right)^{n/x}\right)^x = \left(e_{>(n/x)}\right)^x$$

to explain why these limits hold for $x \geq 1$. Finally, explain why the above limits still hold for any real number $x \geq 0$.

E. For all real numbers $0 < x < 1$, $e^x < 1 + x + \dfrac{x^2}{2(1-x)}$

and for $0 < x \leq 1/2$: $e^x < 1 + x + x^2 \leq 1 + \dfrac{3}{2}x$

Hint: Use the Taylor series from Part A and for all terms past the third, replace the denominator by 2, and replace these terms with the formula for a geometric sum. For $0 < x < 1/2$, the worst case occurs at $x = 1/2$.

F. For all real numbers $0 < x < 1$: $\dfrac{1}{e^x} > 1 - x$

Hint: Using Part A, show that for $0 < x < 1$, $1/e^x > 1/(1+x)$. Then prove that for $0 < x < 1$, $1/(1+x) = 1 - x + x^2 - x^3 + x^4 - x^5 + x^6 - \ldots$ (multiply both sides by $1+x$ and show that all terms but the 1 on the right cancel out). Then explain why for all real numbers $0 < x < 1$, $1/e^x > 1-x$ (after $1-x$, group each positive term with the following negative term to get a sequence of positive terms).

G. To switch between logarithms in base 2 and base e (designated as $\ln$):
$\log_2(e) = 1.442695\ldots$
$\ln(2) = 0.693147\ldots$
$\ln(x) = \log_2(x)/\log_2(e) < \log_2(x)/1.443$
$\log_2(x) = \ln(x)/\ln(2) < \ln(x)/0.7$

Hint: Using the facts about logarithms and exponentials from the first chapter.

158. Recall that the analysis of the expected number of empty buckets first gave a proof that employed only simple mathematics for the case $c=1$. Generalize this proof to $c \geq 1$.

A. Show that the expected number of empty buckets is $\Omega(n)$ by again letting k be such that after item $(n/2)+k$ is inserted, exactly $n/2$ buckets are non-empty (if k does not exist, then more than $cn-(n/2)$ buckets must be empty), and explaining why for each of the remaining $(n/2)-k$ items, there is at most probability $(cn-(n/2))/cn$ that it is placed in an empty bucket, and hence the expected number of empty buckets is at least:

$$\left(cn - \tfrac{n}{2}\right) - \left(\frac{cn - \tfrac{n}{2}}{cn}\right)\left(\tfrac{n}{2} - k\right) = \left(\frac{(2c-1)^2}{c}\right)\tfrac{n}{4} + \left(1 - \tfrac{1}{2c}\right)k \geq \tfrac{n}{4}$$

B. Show that the expected number of non-empty buckets is $\Omega(n)$ by again observing that at most $i-1$ buckets are full after $i-1$ items have been inserted, and then explaining why the probability that the ith item is placed in an empty bucket is $\geq (cn-(i-1))/(cn)$, and hence, the expected number of non-empty buckets is at least:

$$\frac{cn}{cn} + \frac{cn-1}{cn} + \cdots + \frac{cn-n+1}{cn} = n - \frac{1+2+3+\cdots(n-1)}{cn}$$
$$= n - \frac{(n-1)(n/2)}{cn} \quad \text{— arithmetic sum (see appendix)}$$
$$> \left(1 - \frac{1}{2c}\right)n$$
$$\geq \frac{n}{2}$$

159. For appropriate values of b and c we have seen that $\log_2(n)$ is about as large as a bucket gets with high probability. Show that the distribution of bucket sizes from largest to smallest goes down geometrically (i.e., something like less than half with >1 element, less than 1/4 with >2 elements, less than 1/8 with >3 elements, etc.).

160. Generalize the hash table re-sizing algorithm to a factor of $k>1$, and incorporate a parameter I for the initial size of the table.

161. It is natural to ask how many items must be hashed to make the probability of there being at least one bucket with two items > 1/2. This is an example of the classic "birthday paradox" that states that there is an even chance that two in a group of 23 people have the same birthday (assuming birthdays are distributed uniformly at random). Derive the birthday paradox and generalize your answer to a hash table of size m.

Hint: Explain why two people have a different birthday with probability (1–1/365), the probability that a third person has a different birthday than either of these two is (1–2/365), and so on, and then find the least i such that the product of i terms is < 1/2.

162. Suppose a Bloom filter uses $m = 10n$ bits. Compare this to storing a list of b bit hash values (i.e. at total of bn bits), where b is large enough to give the same probability of a false match. Generalize your answer to arbitrary n and m.

Hint: The probability of a false match is $(1-1/2^b)^n$.

163. Suppose that deletion is added to a bloom filter by storing counts in A instead of bits (INSERT increments counts and DELETE decrements them). Give an expression for how many bits are needed for counts, with high probability.

164. In this chapter we have presented *open* hashing, where all items that hash to the same position are stored in a bucket that is represented by a linked list. With *closed* hashing, whenever an item is hashed to a position that is already occupied, search for some other place in the table to put it; so all buckets have ≤ 1 element, and items (or pointers to them) are stored directly in the table.

  A. Present pseudo-code for the basic MEMBER, INSERT, and DELETE operations with *linear probing*, where you simply scan from that position forward (wrapping around if you get to the end of the table) until an empty position is found.

  B. Generalize Part A to *quadratic probing* where you scan forward by quadratic amounts (in the hope of avoiding bunching up of entries).

Chapter Notes

Hashing is a standard topic for most books on data structures and algorithms; for further reading the reader may refer to the list of texts given at the beginning of the notes to the first chapter. In particular, the books of Knuth [1969] (Volume 3) and Gonnet and Baeza-Yates [1991] present detailed analysis of the constants involved with many variations. See also Morris [1968].

For an introduction to modular arithmetic, see a text on discrete mathematics such as Graham, Knuth, and Patashnik [1989] or a text on number theory such as Rosen [2000].

Here, all items that hashed to the same location were stored in a bucket that was represented by a data structure (linked lists) that was in addition to the hash table itself. This type of hashing is usually called *open* hashing in the literature. Another commonly used hashing technique is *closed* hashing, where whenever you try to hash an element to a position that is already occupied, you search for some other place in the table to put it. For example, *linear probing* simply scans from that position forward (wrapping around if it gets to the end of the table) until an empty bucket is found, and *quadratic probing* scans forward by non-linear amounts (in the hope of avoiding bunching up of entries). Closed hashing can have advantages in some applications where memory is critical and you are "fine tuning" constants. Many texts that present open hashing present closed hashing as well; in addition to the texts of Knuth [1969] (Volume 3) and Gonnet and Baeza-Yates [1991] that were mentioned earlier, the reader can also refer to, for example, the books of Horowitz and Sahni [1976] and Cormen, Leiserson, Rivest, and Stein [2001].

For presentation of the Chernoff bound, see the paper of Hagerup and Rub [1989] which presents in a short easy to read form the proof of the bound as it was stated here, and gives references to the original work of Chernoff and improvements made by others; the book of Hofri [1987] gives a more general presentation of inequalities for sums of bounded random variables. See also the book of Habib, McDiarmid, Ramirez-Alfonsin, and Reed [1998], where a proof of a different statement of the Chernoff bound is presented in the chapter by McDiarmid.

In this chapter we only addressed upper bounds on the expected size of the largest bucket. It can be shown under a number of assumptions that the expected size of the largest bucket is also $\Omega(\log(n)/\log\log(n))$; see for example, Gonnet [1981], Mehlhorn and Vishkin [1984], and Siegel [1989].

The question of whether one can do better than $O(1)$ expected time for hashing has been extensively studied. In addition to addressing lower bounds on the problem, Dietzfelbinger, Karlin, Mehlhorn, Heide, Ronhert, and Tarjan [1994] present a dynamic perfect hashing technique that use linear space, constant time for membership, and constant expected amortized time for insertion and deletion; their work is based on that of

Fredman, Komlos, and Szemeredi [1984] which gives linear space and constant time membership for the static problem (they take non-linear time to initially insert the items).

Carter and Wegman [1979] present universal hash functions. The presentation here is along the lines of the presentation in first edition of the book of Cormen, Leiserson, and Rivest [2001]; in their second edition, they specifically address the case that m is not prime.

Azar, Broder, Karlin, and Upfal [2000] show that when n items are hashed into n buckets using twin hashing, the expected size of the largest bucket is

$$\frac{\ln\ln(n)}{\ln(2)} + O(1) < 1.45 * \ln\ln(n) + O(1)$$

where ln denotes the natural logarithm; when twin hashing is generalized to $k \geq 2$ hash functions, the bound becomes $\ln\ln(n)/\ln(k)+O(1)$. They also extend this bound to the dynamic case where deletions can be done uniformly and randomly. Berenbrink, Czumaj, Steger, and Vocking [2000] show that the performance holds even in the heavily loaded case (and they also generalize to $k \geq 2$ hash functions); they show that when n items are hashed into a table of size m, where n can be much larger than m, with twin hashing the expected size of the largest bucket is $(n/m)+O(\ln\ln(m))$ (i.e., only $O(\ln\ln(m))$ above the average bucket size). Vöcking [1999] shows that twin hashing is improved if the table is partitioned into two halves to reduce ties; that is, if h_1 distributes items uniformly and randomly in the range $0 \leq h_1(d) < \lceil m/2 \rceil$ and h_2 distributes items uniformly and randomly in the range $\lceil m/2 \rceil \leq h_2(d) < m$, then the expected size of the largest bucket is improved to

$$\frac{\ln\ln(n)}{2\ln(\alpha)} + O(1) = \frac{\ln\ln(n)}{2\ln(1.61803...)} + O(1) < 1.05 * \ln\ln(n) + O(1)$$

where α denotes the golden ratio (the golden ratio is often refereed to as φ — see the exercises of the chapter on induction and recursion for the use of the golden ratio with the Fibonacci numbers); when $k \geq 2$ buckets are used, the bound becomes $\ln\ln(n)/(k\ln(\alpha_k))+O(1)$, where α_k is the exponent of growth for a generalized Fibonacci sequence. Mitzenmacher, Richa, and Sitaraman [2001] give a comprehensive survey of twin hashing and many related ideas, generalizations, and applications, and includes an extensive list of references; they use the name "the power of two random choices", which comes from the Ph.D. thesis of Mitzenmacher [1996].

Bloom filters are presented in Bloom [1970]. Fan, Cao, Almedia, and Broder [1998] use Bloom filters for Web caches, which is also considered by Mitzenmacher [2001c]. Mitzenmacher [2001b,c], upon which the presentation here is based, presents compressed Bloom filters and includes a review of Bloom filters (that contains an analysis of the optimal value of k) and additional references.

7. Heaps

Introduction

We have seen how hashing provides improved performance for applications that only need MEMBER, INSERT, and DELETE and do not need operations such as DELETEMIN that are also supported by binary search trees. Heaps provide improved performance when we only need the operations of INSERT and DELETEMIN.

A typical application that only needs the operations of INSERT and DELETEMIN is a *priority queue* (a queue where the next item to be removed is the one of highest priority). As we have already seen with tree sort, one of the simplest uses of a priority queue is sorting, where all the INSERT operations are done before all of the DELETEMIN operations to output the items in sorted order:

> **while** input items remain **do**
> INSERT(the next item of the unsorted list)
>
> **while** the priority queue is not empty **do**
> output DELETEMIN

Observe that if INSERT and DELETEMIN are implemented to be $O(\log(n))$ time per operation, then this sorting algorithm is $O(n\log(n))$.

The key ideas behind heaps are:

- Use a tree structure that represents a partial ordering where a parent is always smaller than its children.

- Because we are interested in only the INSERT and DELETEMIN operations, and because this partial ordering is more "flexible" than a total ordering (e.g., a binary search tree), pointers are not necessary for practical implementations. Instead, we can pack items into an array in such a way that arithmetic operations can be used to move around the tree and exchange operations can be used to implement INSERT and DELETEMIN operations.

- The array implementation is the basis for *heap sort*, a worst case $O(n\log(n))$ algorithm that works in place on an array containing the input items.

Complete *k*-ary Trees

Definition: In a *complete k-ary tree* of n vertices, all non-leaf vertices have exactly $k \geq 2$ children and all leaves have the same depth.

Lemma: A complete *k*-ary tree of height h has k^h leaves.
Proof: Each level has k times as many vertices as the previous level. Hence, the number of vertices at depth i is k^i, and hence the number of leaves (vertices at depth h) is k^h.

Lemma: A complete *k*-ary tree of height h has $(k^{h+1}-1)/(k-1)$ vertices. Furthermore, if a complete *k*-ary tree of height k has x vertices, the one of height $h+1$ has $kx+1$ vertices.
Proof: Sum the vertices at each level to get a geometric sum (see the appendix):

$$1 + k + k^2 + \cdots + k^h \;=\; \sum_{i=0}^{k} k^i \;=\; \frac{k^{h+1}-1}{k-1}$$

Furthermore, multiplying this expression by k and adding 1 gives the expression for $h+1$.

Lemma: The number of leaves in a complete *k*-ary tree of n vertices is $((k-1)n+1)/k$.
Proof: If $L(n)$ denotes the number of leaves in a complete tree of n vertices, then $L(1)=1$ and for $n>1$ a tree of n vertices is constructed from a root and k equal size trees of $(n-1)/k$ vertices each, and so we can use induction where $L(1) = 1 = ((k-1)1+1)/k$, and for $n>1$:

$$L(n) \;=\; kL\!\left(\frac{n-1}{k}\right) \;=\; k\left(\frac{(k-1)\left(\dfrac{n-1}{k}\right)+1}{k}\right) \;=\; \frac{(k-1)n+1}{k}$$

Lemma: The height of a complete *k*-ary tree of $n \geq 1$ vertices is $\lfloor \log_k(n) \rfloor$.
Proof: Since there are k^h leaves in a tree of height h, the height as a function of n is the logarithm to the base k of $L(n)$:

$$h = \log_k(L(n))$$
$$= \log_k\!\left(\frac{(k-1)n+1}{k}\right)$$
$$= \log_k((k-1)n+1)-1$$
$$= \left(\log_k(k-1) + \log_k(n)\right)-1$$
$$= \left(1 + \lfloor \log_k(n) \rfloor\right)-1$$
$$= \lfloor \log_k(n) \rfloor$$

The second to last step follows because $\log_k(k-1)<1$ and since the entire quantity $\log_k(k-1) + \log_k(n)$ is an integer, $\log_k(n)$ is not an integer, so rounding $\log_k(k-1)$ up to 1 and $\log_k(n)$ down to $\lfloor \log_k(n) \rfloor$ must give the same quantity.

Full k-ary Trees

Definition: For $k \geq 2$, in a *full k-ary tree* of n vertices, all non-leaf vertices have exactly k children and all levels are full except possibly for some rightmost portion of the bottom level (if a leaf at the bottom level is missing, then so are all of the leaves to its right). For $1 \leq m \leq k$, we define the new operation $m\text{CHILD}(v)$ that returns the mth child of v. The *rightmost bottom leaf* (*RB leaf*) is the rightmost leaf of the bottom level.

Observation: Unlike complete k-ary trees, for every $n \geq 1$ there is a full k-ary tree with n vertices. When traversed in level order, full trees are like complete trees except that the traversal stops sooner.

The height of a full k-ary tree of n vertices is $\leq \lceil \log_k(n) \rceil$: Let h denote the height of a full k-ary tree, and let $m \geq n$ be the number of vertices in a complete k-ary tree of height h. We have already shown that $h = \lfloor \log_2(m) \rfloor$. Hence if $n = m$ then $h \leq \lceil \log_2(n) \rceil$. Otherwise, the largest complete tree contained in the full tree has $\leq (n-1)$ vertices, and so from the previous page we know that $m \leq k(n-1)+1 < kn$. If n is a power of k then

$$h = \lfloor \log_k(m) \rfloor < \lfloor \log_k(kn) \rfloor \quad \text{and} \quad \lfloor \log_k(kn) \rfloor = \lfloor \log_k(n) \rfloor + 1 = \lceil \log_k(n) \rceil + 1$$

from which it follows that $h \leq \lceil \log_k(n) \rceil$. If n is not a power of k, then:

$$h = \lfloor \log_k(m) \rfloor \leq \lfloor \log_k(kn) \rfloor = \lfloor \log_k(n) \rfloor + 1 = \lceil \log_k(n) \rceil$$

Idea: Pack the tree in level order into an array starting at 0:

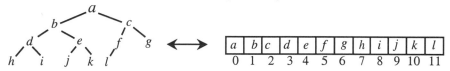

The mth child of a vertex with index i has index $ik+m$ (see the exercises). Because a rightmost portion of the last level may be missing, we check that we have not computed an index that is past the end of the tree (and so the corresponding child does not exist).

Basic $O(1)$ time operations:

> PARENT(i): **if** $i > 0$ **then** $\lfloor (i-1)/k \rfloor$ **else** *nil*
>
> LCHILD(i): **if** $((ik)+1) \leq n$ **then** $(ik)+1$ **else** *nil*
>
> mCHILD(i): **if** $((ik)+m) \leq n$ **then** $(ik)+m$ **else** *nil*
>
> RSIB(i): **if** $i < n$ **and** i not a multiple of k **then** $i+1$ **else** *nil*

Heap Implementation with Full Trees

Idea: The heap is a full binary tree in the array $H[0]...H[n-1]$, where an item is always $\le$ its children (so the item at the root is the smallest). To simplify our presentation, the variable *nextRB*, which is initially 0, denotes the position directly after that of the RB leaf ($nextRB = RB+1$):

INSERT: Place the new item after the RB leaf and "percolate up":

> **procedure** PERCUP(i):
>> **while** $i>0$ and $H[i]<H[PARENT(i)]$ **do begin**
>>> exchange $H[i]$ with $H[PARENT(i)]$
>>> $i := PARENT(i)$
>>> **end**
>> **end**

> **procedure** INSERT(d):
>> **if** $nextRB \ge n$ **then** *ERROR — heap is full*
>> **else begin**
>>> $H[nextRB] := d$
>>> PERCUP($nextRB$)
>>> $nextRB := nextRB+1$
>>> **end**
>> **end**

DELETEMIN: Exchange the root with the RB leaf and "percolate down":

> **procedure** PERCDOWN(i)
>> **while** i is not a leaf **and** $H[i]$ is greater than one of its children **do begin**
>>> Determine the child j of i such that $H[j]$ is the smallest.
>>> Exchange $H[i]$ with $H[j]$.
>>> $i := j$
>>> **end**
>> **end**

> **function** DELETEMIN:
>> **if** $nextRB \le 0$ **then** *ERROR — heap is empty*
>> **else begin**
>>> $nextRB := nextRB-1$
>>> exchange $H[0]$ and $H[nextRB]$
>>> PERCDOWN(0)
>>> **return** $H[nextRB]$
>>> **end**
>> **end**

Note: The deleted element is left in $H[nextRB]$, the position after the new *RB* leaf.

Building a Heap in Linear Time

Idea:

A sequence of insert operations builds the heap from the top down, and hence most of the inserts have to traverse a path $O(\log(n))$ deep (over half the vertices are in the last two levels of a full tree, or just the last level of a complete tree).

Instead, if all elements are to be inserted at the same time, a full binary tree of the elements in an arbitrary order can be formed, and the tree can be rearranged from the bottom up to be a heap.

Visit the vertices in reverse level order (working from $H[n-1]$ down to $H[0]$), percolating each vertex down to make that subtree into a heap (since the two subtrees must already be heaps).

Linear time heap construction algorithm:

    procedure BUILDHEAP:
        Store n items into $H[0]$ ... $H[n-1]$ in any order.
        for $i := (n-1)$ **downto** 0 **do** PERCDOWN(i)
        end

Time: The time to visit all vertices is proportional to n plus the number of exchanges performed by PERCDOWNs. For $n=1$ no exchanges are made, so assume $n>1$ and let N be the number of vertices on level 1. Then, since the portion of the tree from level 1 up is a complete k-ary tree with N leaves, in the entire tree, the level of the root is $\log_k(N)+1$ and, for $i \geq 1$, there are N/k^{i-1} vertices on level i. Since PERCDOWN from a vertex at level $i \geq 1$ makes at most i exchanges, the total number of exchanges for all PERCDOWN operations is:

$$\leq \sum_{i=1}^{\log_k(N)+1} \left(i \frac{N}{k^{i-1}} \right)$$

$$= kN \sum_{i=1}^{\log_k(N)+1} \left(\frac{i}{k^i} \right)$$

$$< kN \frac{1/k}{\left(1-\frac{1}{k}\right)^2} \quad \text{— weighted geometric sum (see the appendix)}$$

$$= \frac{N}{\left(1-\frac{1}{k}\right)^2}$$

Hence, since k is constant and for any $k \geq 2$, $N \leq n$, the number of exchanges is $O(n)$. Note that as k gets large, the numerator approaches n, the denominator approaches 1, and so the expression approaches n; for $k=2$, $N \leq (n+1)/2$, and because the derivation above contains an inequality, the total number of exchanges is $\leq 2n+1$.

Heap Sort

We consider in detail heap sort using a full binary tree. The entire algorithm works in-place, using only $O(1)$ space in addition to the space used by the heap array. The time is $O(n\log(n))$ to sort n items.

1. Define PERCDOWN using "low-level" operations:

```
procedure PERCDOWN(i):
    while 2i+1<nextRB do begin
        j := 2i+1
        if j+1<nextRB and H[j+1]<H[j] then j := j+1
        if H[i]<H[j] then return
        else begin
            exchange H[i] and H[j]
            i := j
        end
    end
end
```

2. Read in the n input items into the array H in $O(n)$ time.

```
n := 0
while input items remain do begin
    H[n] := next input item
    n := n+1
end
nextRB := n
```

3. Build the heap in $O(n)$ time.

```
for i := n−1 downto 0 do PERCDOWN(i)
```

4. DELETEMIN each element in a total of $O(n\log(n))$ time.

```
while nextRB>0 do begin
    nextRB := nextRB−1
    exchange H[0] and H[nextRB]
    PERCDOWN(0)
end
```

5. H could have been output as part of Step 4 (by outputting the value of $H[0]$ at the start of each iteration). Alternately, H is now sorted from right to left, so it can be output in $O(n)$ time with a simple loop.

```
for i := n−1 downto 0 do output H[i]
```

Implementing Heaps with Pointers

Problem: Although array implementations of heaps are efficient, in practice a standard pointer-based tree facility (each vertex has *parent*, *lchild*, *rchild*, and *data* fields) may already be available "for free" and we might want to implement a heap using the operations provided. Also, it is a useful exercise to understand how the array implementation can be simulated with pointers.

Idea: A pointer to the RB leaf is maintained, and the key issue is how to find the next or previous leaf in level order. Half the time, the RB leaf is a left child, and the next leaf in level order will be its right sibling, or the RB leaf is a right child and the previous leaf in level order is its left sibling. However, in general, to find the next or previous leaf in level order, we follow parent links up to the lowest common ancestor (which could be the root in the worst case), and walk back down via left-child links.

Adding a new rightmost leaf (removal is similar):

> p = RB leaf
>
> **while** $p \neq root$ and p is a right child **do** p=PARENT(p)
>
> **if** $p \neq root$ **then begin**
> $\qquad p := \text{PARENT}(p)$
> $\qquad$ **if** RCHILD(p)$\neq nil$ **then** $p := \text{RCHILD}(p)$
> $\qquad$ **end**
>
> **while** p has two children **do** $p := \text{LCHILD}(p)$
>
> **if** LCHILD(p) := *nil*
> $\qquad$ **then** make the new RB leaf the left child of p
> $\qquad$ **else** make the new RB leaf the right child of p

Time: Since we do at most a walk up to the root and back down to a leaf, a distance of $O(\log(n))$, the $O(\log(n))$ time of the INSERT operation is not changed by more than a constant factor.

Lower Bounds on Heap Operations and Sorting

Idea: One might wonder if a sequence of n heap operations can be implemented in less than $O(n\log(n))$ time. Since the operations can easily be used to sort, if sorting cannot be done in less than $O(n\log(n))$, then neither can the heap operations. Although asymptotically sorting can be done faster by employing arithmetic operations in clever ways (see the chapter notes), for practical algorithms that are based on comparing elements, we can show an $\Omega(n\log(n))$ lower bound.

Decision trees: Suppose that besides input and output, a sorting program P consists entirely of comparisons of elements and flow of control (all of the sorting programs we have considered thus far can be expressed this way). For any particular value of n, we can "unwind" all loops and recursive calls to create a decision tree that expresses all possible ways the program can work for n input elements. For example, for $n=6$, the actions of P might be represented by the following tree:

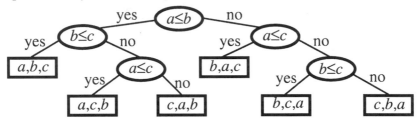

Lemma: The height of a binary tree of L leaves where every non-leaf vertex has exactly two children is at least $\log_2(L)$.

Proof: By induction on L. Clearly the lemma is true for $L=1$. For $L>1$, at least one of the two subtrees of the root has $\geq L/2$ leaves and hence by induction has height at least $\log_2(L/2) = \log_2(L)-1$. Since the root increases the height by 1, the lemma follows.

Theorem: Any program to sort, which for any particular value of n can be represented by a decision tree, runs in $\Omega(n\log(n))$ time.

Proof: Since there are $n!$ possible permutations of the input and depending on the input values any of these permutations could be the correct sort, the decision tree must have $n!$ leaves. Hence, by the lemma above, its height must be at least:

$$\log_2(n!) \geq \log_2\left(n(n-1)\cdots\left\lceil\frac{n}{2}\right\rceil\right) \geq \left(\frac{n}{2}\right)^{\left(\frac{n}{2}\right)} = \tfrac{1}{2}n\log_2(n)-\tfrac{1}{2}n = \Omega(n\log(n))$$

Or, an integral can be employed to get a better approximation (see the appendix):

$$\log_2(n!) = \left(\log_2(1) + \cdots \log_2(n)\right) > \int_1^n \log_2(x)dx = n\log_2(n) - \frac{n-1}{\ln(2)} > n\log_2(n) - 1.44n$$

CHAPTER 7

Generalizing the Sorting Lower Bound to the Expected Case

Problem: Given that $\Omega(n\log(n))$ is a lower bound on the worst case time for comparison based sorting, it is natural to ask whether algorithms could be designed that had, for example, expected time $O(n)$, even if the worst case time was greater.

Idea: In a spirit similar to our analysis of the expected performance of binary search trees and quick sort, we will formulate a proof by induction that sums over all possibilities for the left and right subtrees of the root.

Expected depth: The *expected depth* of a decision tree as the sum over all leaves of the probability of reaching that leaf times its depth.

Theorem: Under the assumption that all permutations of n input elements are equally likely to appear, any decision tree that sorts n elements must have an expected depth of at least $\log_2(n!) = \Omega(n\log(n))$.

Proof: We can assume that all leaves of a decision tree T to sort n elements are reachable (otherwise, we can delete all unreachable leaves that have probability 0 without changing the expected depth), and hence T has exactly $n!$ leaves. Let:

$D(T)$ = the sum of the depths of all leaves of a decision tree T with $n!$ leaves

$M(n)$ = the minimum value of $D(T)$ over all decision trees with $n!$ leaves

We now show by induction on n that $M(n) \geq n!\log_2(n!)$.

(*basis*)

For $n=1$, since $\log_2(1!) = \log_2(1) = 0$, it follows that $M(n) \geq 0 = n!\log_2(n!)$.

(*inductive step*)

For $n>1$ and $1 \leq i < n$, since in a decision tree with $n!$ leaves, there are i leaves in the left subtree of the root and $n!-i$ leaves are in the right subtree of the root, and the root adds 1 to all paths going down to its subtrees,

$$D(T) = \left(i + D\left(T_i\right)\right) + \left((n!-i) + D\left(T_{n!-i}\right)\right) = D\left(T_i\right) + D\left(T_{n!-i}\right) + n!$$

and hence:

$$M(n) = \underset{1 \leq i \leq n!}{MIN}\left\{M(i) + M(n!-i) + n!\right\}$$

$$\geq M(n) = \underset{1 \leq i \leq n!}{MIN}\left\{i\log_2(i) + (n!-i)\log_2(n!-i)\right\} + n! \quad \text{—— inductive hypothesis}$$

$$= n!\log_2\left(\frac{n!}{2}\right) + n! \quad \text{—— since the minimum occurs at } \frac{n!}{2}$$

$$= n!(\log_2(n!) - 1) + n!$$

$$= n!\log_2(n!)$$

Given that $M(n)$ is the *minimum* value of $D(T)$ over all decision trees with $n!$ leaves, we can divide by the number of leaves, $n!$, to see that the expected depth of a decision tree that sorts n elements is at least $\log_2(n!)$, which from the preceding page is $\Omega(n\log(n))$.

Exercises

165. To show that the number of vertices in a complete k-ary tree of height h is $(k^{h+1}-1)/(k-1)$, we observed that summing up the number of vertices at each level gave a simple geometric sum. Give a "direct" proof by induction.

Hint: Let $T(h)$ denote the number of vertices in a complete tree of height k, and explain why $T(0)=1$ and:

$$T(h) \; = \; kT(h-1)+1 \; = \; k\frac{\left(k^{(h-1)+1} - 1\right)}{k-1} + 1 \; = \; \frac{k^{h+1} - 1}{k-1}$$

166. We showed that if a complete k-ary tree of height h has x vertices, then one of height $h+1$ has $kx+1$ vertices, by simply multiplying the formula for height h by k, adding 1, and checking that the result simplifies to the formula for height $h+1$. Give an alternate proof that makes a correspondence between a vertex of a complete k-ary tree of height h and k vertices in one of height $h+1$.

167. Prove that in full tree, the mth child of a vertex with index i has index $ik+m$.

Hint: Since indices start at 0, the index of the first vertex at a given depth is the number of vertices in the complete k-ary tree consisting of all vertices of lower depth. If there are x vertices in this tree, we know that there are $kx+1$ vertices in the complete k-ary tree that includes this depth, which is the index of the first vertex at the next greater depth.

168. Starting with an empty binary heap using a full-tree implementation:

    A. Insert the integers 17, 3 ,4, 7, 2, 9, 11, 5, 19, 14, 20, 18 in that order, and after each step, show the heap array.

    B. Do DELETEMINs until the heap is empty, and show the contents of the array after each step.

169. For heaps represented with pointers, give a proof of correctness for the procedure presented to add a new vertex; be sure to consider the following cases:

    A. The RB leaf is a left child.

    B. The RB leaf is a right child; the bottom level of the tree is not full.

    C. The RB leaf is a right child; the bottom level of the tree is full.

170. In Step 5 of the low level presentation of heap sort it was noted that the sorted list could have been output in Step 4 by outputting the value of $H[0]$ at the start of each iteration. Suppose we did it this way and that furthermore, we do not need the sorted data for any other purpose; that is, our only task is to output the sorted list. In this case, another approach is to simply "fill down" by replacing $H[0]$ with the smallest of its children and then repeating this process on the vacated child until a leaf is reached.

Vacated leaves can be filled in with dummy values so that leaves can be detected as the tree gets smaller. Give pseudo-code for this approach and discuss how it might compare in practice to using PERCDOWN.

171. For heaps represented with pointers, pseudo-code was presented to add a new RB leaf. Present corresponding pseudo-code to remove the RB leaf and include a proof of correctness of your algorithm; be sure to consider the following cases:

A. The RB leaf is a right child.

B. The RB leaf is a left child; it is not the only leaf on the bottom level.

C. The RB leaf is a left child; it is the only leaf on the bottom level.

172. For heaps represented with pointers, present pseudo-code for building a heap in linear time.

173. For heaps represented with pointers, if we compare time required to that of an array implementation, both do $O(1)$ work per step of the PERCUP and PERCDOWN operations; the only real difference is the work to keep track of the RB leaf. In the array implementation, we simply increment or decrement a variable *nextRB*, whereas in the pointer implementation, we had to traverse a path in the tree that could, in the worst case, go all the way up to the root and back down. Fortunately, not only is a walk up to the root only $O(\log(n))$ time, but we often do not have to go even that far. In fact, for half the possible positions of the RB leaf (when it is a left child), the next RB leaf to be inserted will just be its right sibling and for half the possible positions of the RB leaf (when it is a right child), the next RB leaf after it is deleted will be its left sibling.

A. For heaps represented with pointers, give a worst case example of a sequence of heap operations where the time to keep track of the leftmost leaf is $\Omega(n\log(n))$.

B. For heaps represented with pointers, consider heap sort, where the heap has been built in linear time and now a sequence of n DELETEMIN operations removes the elements in sorted order. Prove that the total cost of keeping track of the RB leaf for these DELETEMIN operations is $O(n)$.

Hint: For the bottom level, 1/2 the time we traverse a subtree of height 1, 1/4 of the time a subtree of height 2, and so on; after computing the cost of a level, sum over the levels.

174. For the lower bound on sorting, we needed to bound the function $\log_2(n!)$, and saw that a more accurate bound could be achieved by expressing this as a sum and then bounding it below by an integral. Verify the details as follows:

A. Verify that for any constant c, the derivative of $n ln(n) - n + c$ is $ln(n)$, where ln denotes the natural logarithm function (see the facts about e in the chapter on hashing).

B. Using Part A, explain why:

$$\int_1^n \ln(x)dx = n\ln(n) - n + 1$$

C. Using Part B, explain why:

$$\int_1^n \log_2(x)dx = n\log_2(n) - \frac{n-1}{\ln(2)} > n\log_2(n) - 1.44n$$

D. Using Part C, for $n \geq 2$, show *Sterling's Approximation*:

$$n! \geq e\left(\frac{n}{e}\right)^n$$

175. Suppose that for k distinct elements, a sequence S is formed by including m_i copies of the ith element, $1 \leq i \leq k$.

A. Generalize the lower bound for worst case comparison based sorting to show that the number of comparisons required is:

$$\log_2\left(\frac{n!}{m_1!m_2!\cdots m_k!}\right) + n$$

B. Show that this bound also applies to the expected case.

C. Show how to modify merge sort to run in time proportional to this bound. Be sure to include a detailed analysis of the running time.
   Hint: When duplicates meet, change them into a value-count pair.

D. Show how to modify heap sort to run in time proportional to this bound.
   Hint: Observe that when it is time to take an element out, all of its duplicates will be contiguous at the top of the heap, and they can be removed together. Be sure to include a detailed analysis of the running time.

E. Explain why the expected time of quick sort is proportional to this bound..

176. In the proof for the lower bound on the expected time for comparison based sorting, we made use of the fact that for $x>1$, the expression

$$\underset{1\leq i\leq x}{MIN}\{i\log_2(i) + (x-i)\log_2(x-i)\}$$

takes its minimum at $i=(x/2)$. Explain why by employing the principle from calculus that a continuous function with a single minimum value achieves its minimum at the point when the derivative is 0 (i.e., a tangent line is horizontal).

177. Show that any worst case or expected time comparison based algorithm to find the kth largest element in a list of n elements uses at least $n-1$ comparisons by showing that in any decision tree for the problem, every reachable leaf has depth at least $n-1$.

178. Consider the problem of finding the largest and second largest element of a list of n elements:

A. Present an algorithm that uses only $n + \lceil \log_2(n) \rceil - 2$ comparisons.

   Hint: Use a tree to find the largest element, and then you only have to consider the elements to which the largest was compared to find the second largest element.

B. Prove that any algorithm to find the largest and second largest element must use $n + \lceil \log_2(n) \rceil - 2$ comparisons.

Chapter Notes

For further reading on the basic heap data structure as well as the heap sort algorithm, the reader may refer to virtually any text on data structures and algorithms, such as those listed at the beginning of the notes to the first chapter.

Many "fancy" heap data structures have been developed to solve specific problems more efficiently ("binomial" heaps, "Fibonacci heaps", "leftist heaps", etc.); see for example the books of Tarjan [1983] and Cormen, Leiserson, Rivest, and Stein [2001]. Chazelle [2001] presents "soft heaps" that support additional operations and give improved approximate performance.

The presentation on the lower bound for sorting by comparisons is along the lines of that in Aho, Hopcroft, and Ullman [1974], as is the exercise for when there are duplicates; the hints with that exercise are due to Maier [1990].

There has been considerable work on sorting in less than $O(n\log(n))$ time (by not limiting the algorithm to a comparison model). Fredman and Willard [1990] achieve $O(n\log(n)/\log\log(n))$ and $O(n)$ space by adapting a sort with a balanced tree so that vertices have a number of children that grows slowly with n (roughly $\log(n)/\log\log(n)$), thus yielding a tree of slightly lower height ($\log(n)/\log\log(n)$); they use arithmetic operations to quickly find the correct child of a vertex when moving down the tree. Andersson, Hagerup, Nilsson, and Raman [1995] consider improved bounds.

8. Balanced Trees

Introduction

Here we consider trees that can support basic operations (MEMBER, INSERT, DELETE, MIN, MAX, JOIN, SPLIT, etc.) in $O(\log(n))$ worst case time per operation. A naive approach would be to use something like a full k-ary tree as was used for heaps. Although for just the operations INSERT and DELETEMIN this approach worked to implement priority queues, for a wider range of operations, we need a type of tree with a reasonable degree of flexibility; that is, it must be possible to represent the data with trees of many shapes. The idea is to limit the set of possible tree shapes to ones that have height $O(\log(n))$, but with enough flexibility to make "adjustment" of the tree relatively easy after an operation disturbs its balance.

We have already seen how binary search trees can support a sequence of n operations in $O(n\log(n))$ expected time, and how self-adjusting binary search trees could improve this bound to $O(n\log(n))$ in the worst case by walking back up to the root to "adjust" the tree after each operation (by "juggling" vertices via left and right rotations). All of our approaches here will be in the same spirit, except that an extra bit will be stored and maintained with each vertex that will allow the "juggling" by rotations to have the more precise effect of $O(\log(n))$ *worst case time per operation.*

We begin with 2-3 *trees*, where each non-leaf vertex can have two or three children and all leaves have exactly the same depth. Here, flexibility comes from the ability to change the number of children of vertices on a root-to-leaf path (and once in a while increase or decrease the height of the entire tree by adding a new root when the existing one needs to have four children or deleting the existing root when it has only one child). We shall then see that 2-3 trees are equivalent to a class of binary trees, called *red-black trees*, where the number of vertices on a root to leaf path can differ by at most a factor of two. We shall also consider *AVL trees*, which work in a similar way, but with improved worst case height.

2-3 Trees

Definition: A 2-3 *Tree* satisfies the following conditions:
- A. All non-leaf vertices have 2 or 3 children.
- B. All leaves have the same depth.
- C. Data stored at vertices are organized as follows:

    Vertex with two children: Contains a single data item that is greater than all data in the left subtree and less than all data in the right subtree.

    Vertex with three children: Contains two data items; the smaller of the two is greater than all data in the left subtree and less than all data in the middle subtree, and the larger of the two is greater than all data in the middle subtree and less than all data in the right subtree.

    Leaves: Contain one or two data items.

Example:

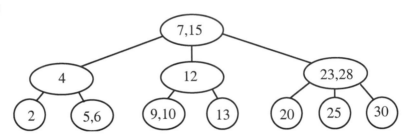

The height of a 2-3 Tree with n data items is between $\lfloor \log_3(n) \rfloor$ and $\lfloor \log_2(n) \rfloor$.

By adding items, any 2-3 tree of height $h \geq 0$ and n data items can be converted to a maximal 2-3 tree of height h with $m \geq n$ data items, where all non-leaf vertices have three children and all vertices contain two data items. It is straightforward to prove by induction (see the exercises) that it is a complete 3-ary tree with $(m+1)/3$ leaves and height $\log_3(m+1)-1$. Hence, since $m+1$ is a power of 3:

$$h = \log_3(m+1)-1 = \lfloor \log_3(m) \rfloor \geq \lfloor \log_3(n) \rfloor$$

Similarly, by deleting items, any 2-3 tree of height $h \geq 0$ and n data items can be converted to a minimal 2-3 tree of height h, with $m \leq n$ data items where all non-leaf vertices have two children and all vertices contain one data item. It is straightforward to prove by induction (see the exercises) that it is a complete binary tree with $(m+1)/2$ leaves and height $\log_2(m+1)-1$. Hence, since $m+1$ is a power of 2:

$$h = \log_2(m+1)-1 = \lfloor \log_2(m) \rfloor \leq \lfloor \log_2(n) \rfloor$$

Generalized binary search: For the MEMBER operation, standard binary search can be generalized to compare with two items at vertices with 3 children (MIN and MAX follow a leftmost or rightmost path).

Inserting into a 2-3 Tree

Idea: As depicted by the figure on the following page:

First search down to find the leaf where the new item belongs and add the new item to that leaf.

If the leaf only had one item before and now has two, then we are done.

Otherwise, we now have three items to work with: split off a new leaf to contain one of the items (and attach this new leaf to the parent), leave one item in the existing leaf, and give the third item to the parent.

Move up to the parent and continue this process until a vertex that previously had only one item and two children is reached (and so it is ok to add a third child and additional item) or until the top of the tree is reached and a new root can be created.

procedure INSERT(d, T):

Search down to place the new item d in the appropriate leaf v.

while v has 3 data items $a \leq b \leq c$ **do begin**

> **if** v is the root
> > **then** $p :=$ a new root with v as its leftmost child
> > **else** $p :=$ PARENT(v)
>
> Remove a and the two leftmost subtrees of v (which are *nil* if v is a leaf) to form a new subtree and attach this new subtree to p as the left sibling of v.
>
> Move b to p.
>
> $v := p$
>
> **end**

end

Example, insert 24 and 26 into the previous example:

insert 24
and then
insert 26:

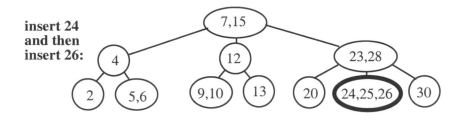

split:

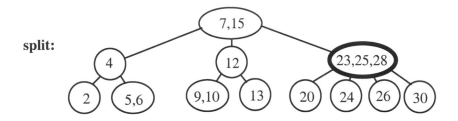

split:

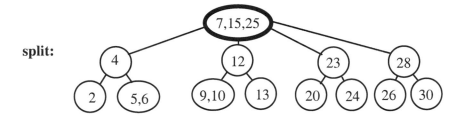

split and
make a
new root:

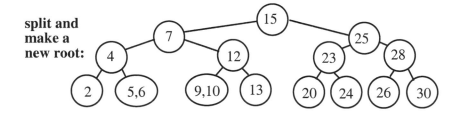

CHAPTER 8

Deleting from a 2-3 Tree

Idea: As depicted by the figures on the following two pages:

Search to the vertex v that contains the data item d to be deleted and remove d.

Similar to deletion in a binary search tree, if v is not a leaf, replace d with the largest item from the left subtree defined by d (or the smallest item from the right subtree), change v to be that leaf, and proceed as though d was removed from that leaf.

Move up the tree, giving data items to siblings, until we find a sibling with an extra item (and its associated subtree) that can be taken, or until the root is deleted.

procedure DELETE(d, T):

Search to the vertex v that contains the data item d to be deleted and remove d.

If v is not a leaf then remove an appropriate data item from a leaf to replace d (in the same fashion as deletion from a binary search tree) and change v to be that leaf.

while v contains no data items **do begin**

    if v=*root* **then** delete v and make its only child the new root

    else if v has an adjacent sibling with two items **then begin**
        p := PARENT(v)
        w := an adjacent sibling of v with two data items $a<b$
        c := the data item in p that separates v and w
        Move c to v.
        if w is a sibling to the left of v **then begin**
            if w is not a leaf **then** Move w's rightmost subtree to be v's left subtree.
            Move b to p.
            end
        else begin
            if w is not a leaf **then** Move w's leftmost subtree to be v's right subtree.
            Move a to p.
            end
        end

    else begin
        p := PARENT(v)
        w := an adjacent sibling of v with only one data item
        c := the data item in p that separates v and w
        Move c to w.
        if v is not a leaf **then** Add v's only subtree to w.
        Delete v.
        v := p
        end
    end
end

Example, delete leaf 24 from the previous example:

Remove leaf data:

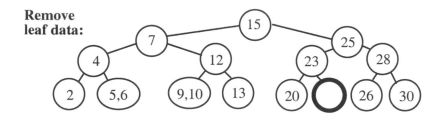

Give parent data to sibling:

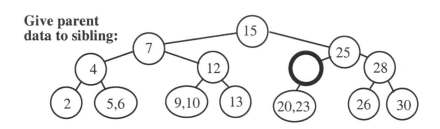

Give parent data and subtree to sibling:

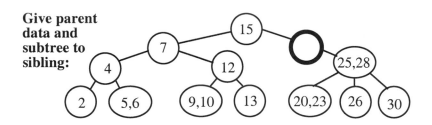

Give parent data and subtree to sibling (old root can now be deleted):

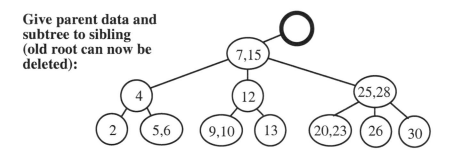

CHAPTER 8

Example, delete 13, then delete 12 from previous example:

**Remove
leaf data:**

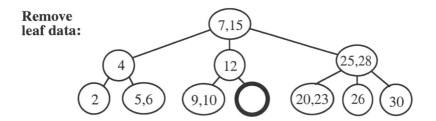

**Take data
from sibling:**

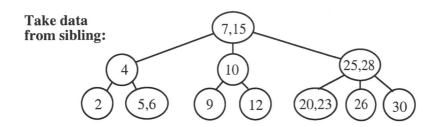

**Remove
leaf data:**

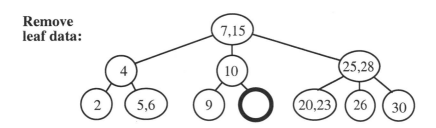

**Give parent
data sibling:**

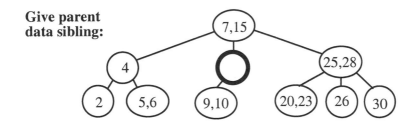

**Take data and
subtree from
sibling:**

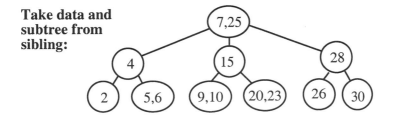

Joining 2-3 Trees

JOIN(T_1,d,T_2): A single 2-3 tree is formed from T_1, d, and T_2.

** Always assumes that all items in T_1 are less than d and all items in T_2 are greater than d.

Idea: Attach the root of the smaller tree to the vertex of height one greater on the side of the larger tree, and if a vertex with 4 children is created, proceed up as with the INSERT operation.

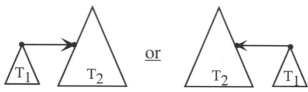

procedure JOIN(T_1,d,T_2):

    Case 1: $HEIGHT(T_1) \leq HEIGHT(T_2)$:

        if $HEIGHT(T_1)=HEIGHT(T_2)$ **then begin**
        Create new root d with left subtree T_1 and right subtree T_2.
        end

        else begin
        Attach T_1 as the left child of the vertex v of height $HEIGHT(T_1)+1$ on the leftmost path in T_2 and add d to v; if v now has 4 children, proceed like an INSERT.
        end

    Case 2: Symmetric to Case 1.

    end

Time: If tree heights are not known, they can be computed in $O(\log(n))$ time by traversing any path to a leaf. Moving down to the attachment and then back up for the INSERT takes time proportional to the difference of the tree heights, at most $O(\log(n))$ time.

Example, JOIN(T_1, 5, T_2):

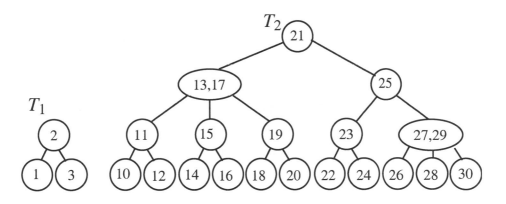

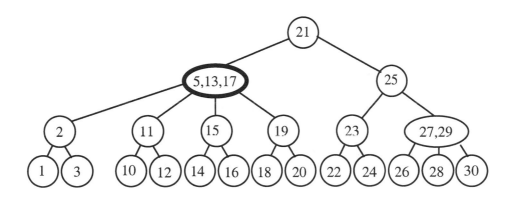

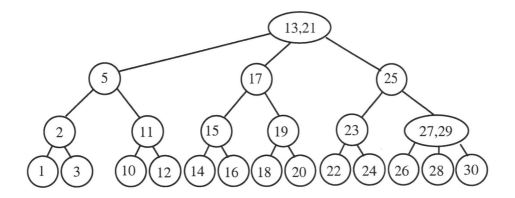

Splitting 2-3 Trees

SPLIT(*d*,*T*): Delete *d* and split *T* into items <*d* and items >*d*.
    ** For ease of presentation, we assume that *d* is in *T* (see the exercises).

Idea: Remove the vertices on the path from the root of *T* to *d* to create a left forest and a right forest; each forest can then be "zipped" together with JOINs that together only take $O(\log(n))$ time to form two new trees *A* and *B*:

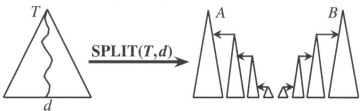

procedure SPLIT(*x*,*T*):

    Initialize *v* := *root* and two empty stacks *P* and *Q*.

    while *v* does not contain *d* **do**
        For each item in *v* that is <*d*, push it and the associated subtree onto *P*, and for each item in *v* that is >*d*, push it and the associated subtree onto *Q*.
        v := the root of the remaining subtree (where the search for *d* goes)
        end

    If *v* has a second item *x* in addition to *d*, push *x* and its associated subtree (or an empty tree if we are at a leaf) onto *P* if *x* is to the left of *d* or *Q* if *x* is to the right of *d*.

    A := left subtree of *d*
    while *P* not empty **do** pop tree *Z* and data *x*; *A* := JOIN(*Z*,*x*,*A*)

    B := right subtree of *d*
    while *Q* not empty **do** pop tree *Z* and data *x*; *B* := JOIN(*B*,*x*,*Z*)

    Discard *d*.

    return the two trees *A* and *B*

    end

    Note: Define HEIGHT(*nil*) = −1 so that JOINs work even if a tree is empty.

Complexity: The search down to construct *P* and *Q* is $O(\log(n))$; as trees are pushed, their heights can be stored with them (i.e., the height of *T*, which can be computed once in advance, less the current depth). Since *P* and *Q* initially have at most two trees of any given height, at most 3 trees of a given height are joined on the way up to form *A* and *B* (JOINs takes time proportional to the difference in heights) for a total of $O(\log(n))$ time.

Example, SPLIT(*T*,11):

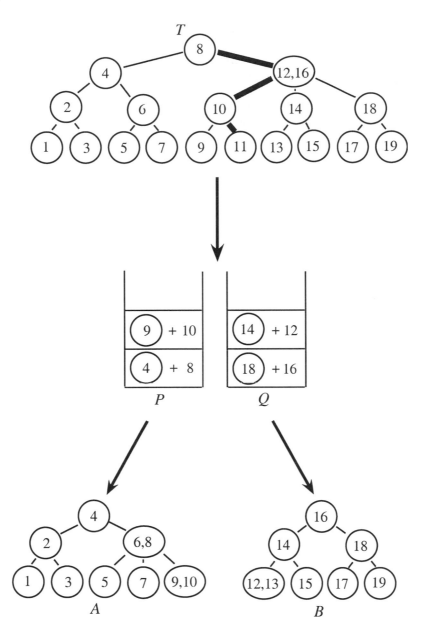

Red-Black Trees

Definition: A binary search tree is a *red-black tree* if:

A. Except for the root, every non-leaf vertex must have a sibling.

B. Every vertex is *red* or *black*; the root is always black.

C. All root to leaf paths have the same number of black vertices.

D. A red vertex may not have a red sibling and if it is not a leaf, it must have two black children.

E. Black leaves must have a sibling; red leaves may not have a sibling.

Example: All of the following are red-black trees:

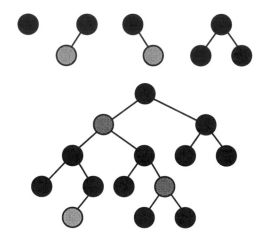

Red-black tree terminology:

- The *black height* of a vertex v is the number of black vertices on any path from v to a leaf (we shall see on the next page that the black height of v must always be well defined).

- A vertex is a *virtual leaf* if it is a leaf or has a single child.

- The *virtual min-height* of a vertex v is the length of a shortest path from v to a virtual leaf.

Properties of Red-Black Trees

1. **Black height is well defined, and siblings always have the same black height.**

   Proof: By Condition C, all root to leaf paths that pass through a vertex v have the same number of black vertices x. Hence if there are y black vertices on a path from the root to v, then all paths from a child of v to a leaf have $x-y$ black vertices. Thus, the black height of v is well defined and equal to $x-y$ if v is red or $x-y+1$ if v is black. Also, if v has two children, then these two siblings have the same black height $x-y$.

2. **Any subtree of a red-black tree with a black root is a red-black tree.**

   Proof: By Property 1, Condition C is satisfied. All other conditions are true by definition if they are true for the entire tree.

3. **All vertices on a shortest path from a black vertex to a virtual leaf are black.**

   (*So for every black vertex, there is at least one path of black vertices to a virtual leaf, and the black height of a black vertex is one more than its virtual min-height.*)

   Proof: We use induction on the black height h of a black vertex. If a black vertex has black height 1, then it must be a virtual leaf, and the statement is true. For $h>1$ assume that the statement is true for all black vertices of black height $<h$, and consider a black vertex v of black height h. Suppose that there was a red vertex on a minimal length path P from v to a virtual leaf q (and we will show this leads to a contradiction). Then there must be a black descendant w of v on P (it could be that $w=v$) that is not a virtual leaf and has a red child a that is on P and a black child b that is not on P. Since a is red, it must have a black child c of black height $<h$, and by induction, the vertices on the path from c to q are all black. Since the path from a to q can be no longer than the minimal length path from b to a virtual leaf, the path from c to q is at least one shorter. But this is a contradiction, since from Property 1 we know that if a and b are siblings, they must have the same black height.

4. **A non-leaf is red if and only if it has a black parent and its virtual min-height is greater than its sibling.**

   Proof: Let v be a vertex with a parent p and a sibling w. If v is red, p and w must be black; in addition, since by Property 3 all vertices on a minimal length path from w to a virtual leaf must be black, by Property 1, the virtual min-height of v is greater than that of w. Conversely, suppose p is black and the virtual min-height of v is greater than that of w; then v cannot be black and satisfy Property 3, since a path of black vertices from v to a virtual leaf would imply that the black height of v is greater than that of its sibling w.

5. **The coloring of vertices for a red-black tree is unique.**

   Proof: We use induction on the virtual min-height h of the tree. Coloring is clearly unique for $h=0$. For $h>1$ assume that coloring is unique for virtual min-height $<h$ and consider a red-black tree T of virtual min-height h. If the two children of the root have the same virtual min-height, they must both be black by Property 4, and by Property 2 the statement follows by induction. Similarly, if one child a has a greater virtual min-height than the other child b, a must be red by Property 4, and by Property 2 the statement follows by induction on b and the children of a.

Equivalence of Red-Black and 2-3 Trees

Theorem:

1. Every red-black tree of black height h and n black vertices can be converted to an equivalent 2-3 tree of height $h-1$ and n vertices.
2. Every 2-3 tree of height h and n vertices can be converted to an equivalent red-black tree of black height $h+1$ and n black vertices.

Proof of part 1: Merge each red vertex with its black parent to form a single vertex with two data items and 0 or 3 children (depending on whether or not it is a leaf).

Proof of Part 2: Color all vertices black. For each leaf with two items, remove the smaller one and put it in a left child colored red. For each non-leaf vertex v that contains two items (and has three subtrees), remove the smaller of the two items, create a new red vertex w with this item, remove the left and middle subtrees of v and make them the left and right subtrees of w, and make w the left child of v.

Note: This construction is not unique; left/right and smaller/larger can be exchanged.

Example: The figure below shows a 2-3 tree on the bottom and three equivalent red-black trees above it.

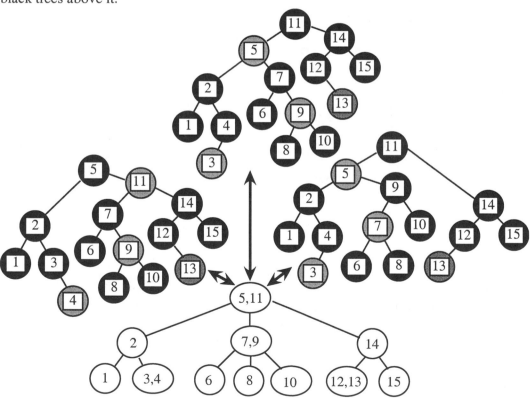

Example: Red-Black Tree Insertion Algorithm

Idea: Although we could convert to a 2-3 tree to perform operations on a red-black tree, a "direct" implementation is a simple traversal back to the root, performing a rotation or two at each step.

procedure INSERT(d,T):

Binary search and add a new red leaf v that contains the new data d.

while v has a red sibling or a red parent **do begin**

   if v has a red sibling **then begin**
      Color v and its sibling black.
      $v := \text{PARENT}(v)$
      Color v red.
   end

   else there are 4 cases for v and PARENT(v) being red:

      Case 1L: v is a right child and PARENT(v) is a left child:

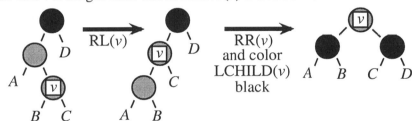

      Case 1R: symmetric to case 1L.

      Case 2L: v is a left child and PARENT(v) is a left child:
         $v := \text{PARENT}(v)$
         Do the second half of Case 1L (RR(v) and color LCHILD(v) black).

      Case 2R: symmetric to case 2L.

   end

if the root is red **then** color it black

end

Example: Inserting into a Red-Black Tree in Sorted Order

When the items 1, 2, 3, 4, 5, 6, and 7 are inserted into an initially empty red-black tree, we get the same result as if we had used a 2-3 tree; that is, a complete binary tree of 7 vertices that produces a sorted list when traversed in pre-order:

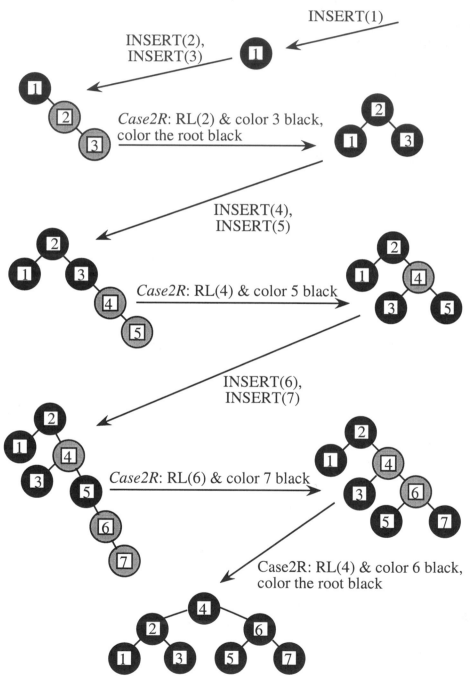

Height of a Red-Black Tree

Definition: A binary tree is *height balanced* if:
1. Except the root, all non-leaf vertices have a sibling.
2. For any vertex v, the number of vertices on any two paths from v down to a leaf differ by at most a factor of 2.

A height balanced binary tree of n vertices has height at most $2\lceil\log_2(n)\rceil - 3$.
Proof: If all vertices had a depth $\geq\lceil\log_2(n)\rceil - 1$, then either the tree is perfectly balanced (since a complete binary tree of 2^k-1 vertices has height $k-1$) or has one leaf that is one deeper than all of the others, which even in the degenerate case (a root with a single leaf), is height balanced. The other possibility is that there is at least one leaf of depth $\leq \lceil\log_2(n)\rceil - 2$, which has $\leq\lceil\log_2(n)\rceil - 1$ vertices on its root-to-leaf path. Hence, by the balance condition, there can be at most $2(\lceil\log_2(n)\rceil - 1)$ vertices on any root-to-leaf path, and hence the height of the tree is at most $2\lceil\log_2(n)\rceil - 3$.

Any red-black tree is height balanced.
Proof: Since red must alternate with black, the height is at most double the black height.

For infinitely many n there are red-black trees of n vertices, height $2\lceil\log_2(n)\rceil - 3$.
Proof: Examples for heights 3, 5, and 7 are trees of 6, 14, and 30 vertices:

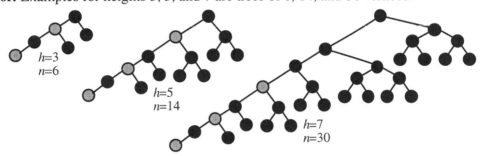

In general, a chain of $h+1$ vertices alternating black and red forms the leftmost path, with 2 copies of each of the complete black trees of heights from $i=0$ to $i=(h-3)/2$ (which have $2^{i+1}-1$ vertices each) hanging off of this path. Hence, the total number of vertices n is:

$$n = (h+1) + 2\sum_{i=0}^{(h-3)/2}\left(2^{i+1}-1\right) = 2 + 2\sum_{i=1}^{(h-1)/2} 2^i = 2^{(h+3)/2} - 2$$

Hence, for every odd integer $h \geq 3$, there are red-black trees of height h with n vertices such that $\log_2(n) = \log_2(2^{(h+3)/2}-2) < (h+3)/2$, which implies that $h > 2\log_2(n)-3$. Since h is an integer, $h\geq\lceil 2\log_2(n)\rceil-3$. Since a red-black tree is height balanced, $h=\lceil 2\log_2(n)\rceil-3$.

Note: Not every height balanced tree can have its vertices colored to be a red-black tree, unless we allow black vertices with two red children; see the exercises.

AVL Trees

Idea: Similar to a red-black tree, but a more precise balance condition that yields smaller worst case tree height.

Definition: A binary search tree is an *AVL tree* if every non-leaf vertex except the root has a sibling, and for every pair of siblings, their height differs by at most 1.

Terminology:

- A vertex v in a binary search tree is *light* if it has a sibling of greater height; vertices that are not light are said to be *heavy*.

- A vertex is *left-heavy* if it has a heavy left child and a light right child.

- A vertex is *right-heavy* if it has a light left child and a heavy right child.

Example: All of the following are AVL trees (heavy vertices are black, light vertices are shaded).

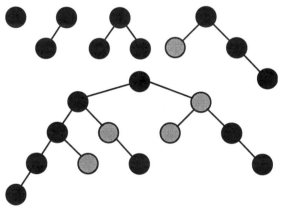

Note: In any AVL tree, from every vertex, there is at least one path of heavy vertices to a leaf (since every non-leaf has at least one heavy child).

The AVL Algorithm

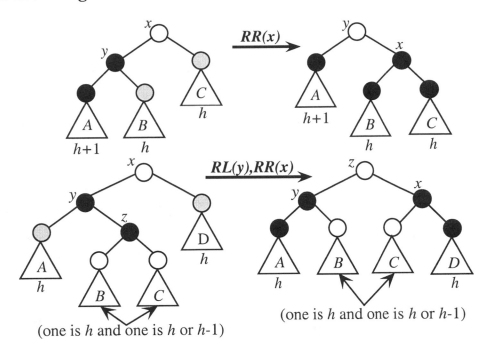

Idea: After an INSERT or DELETE, go up to the root from the deepest vertex x that was visited, and at each vertex, if it is left-heavy apply one of the two transformations shown above (right-heavy is symmetric). In these figures, relative heights are indicated for clarity, but we need only the *heavy-light* labels. We leave it as an exercise to implement JOIN and SPLIT (along the same lines as for 2-3 trees) by balancing in this fashion.

> **while** $x \neq nil$ **do begin**
>> *CASE* 1: x is left heavy:
>>> $y := \text{LCHILD}(x)$
>>> **if** y is left-heavy **then begin**
>>>> $\text{RR}(x)$
>>>> Label y the same as the label of x.
>>>> Label x heavy.
>>>> Label both children of x heavy.
>>>> **end**
>>> **else begin**
>>>> $z := \text{RCHILD}(y)$
>>>> $\text{RL}(y), \text{RR}(x)$
>>>> Label z the same as the label of x.
>>>> Label x heavy.
>>>> Label the left child of y and the right child of x heavy.
>>>> **end**
>>> **end**
>> *CASE* 2: x is right-heavy: Symmetric to CASE 1.
> **end**

Height of an AVL Tree

Theorem:

1. An AVL tree is height balanced.
2. There are infinitely many AVL trees with a height that is twice the min-height.

Proof of Part 1: We use induction on the height h of the AVL tree. All AVL trees of height $h \leq 1$ are height balanced. For $h>1$, assume that all AVL trees of height $<h$ are height balanced and consider an AVL tree T of height h. Since T has height h, one subtree of the root has height $h-1$ and the other has height $h-1$ or $h-2$, and hence both subtrees have at least $h-1$ vertices on a longest path to a leaf, and by induction, at least $(h-1)/2$ vertices are on a shortest path to a leaf. Thus, the worst case ratio of maximum over minimum number of vertices on a root-to-leaf path of T is $\leq (h+1)/((h-1)/2+1) = 2$.

Proof of Part 2: We define a sequence of AVL trees where T_i has height i and a minimum possible min-height. As shown below, T_0 is a single vertex and T_1 is a minimal size tree of height 1. For $i>1$, T_i is a root with sub-trees T_{i-2} and T_{i-1}.

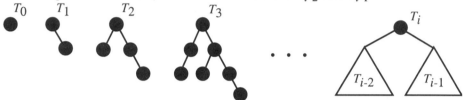

We leave as an exercise a formal proof by induction that T_i has min-height $i/2$ if i is even and min-height $(i+1)/2$ if i is odd. For example, T_4 and T_5 are:

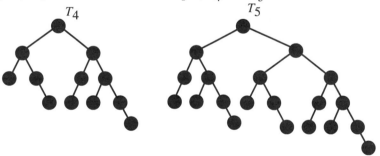

Idea: The above theorem says that although AVL trees are height balanced, in the worst case they can have root to leaf paths that differ by almost as much as possible. That is, if x is a leaf of depth h and y is a leaf of depth $2h$, then the number of vertices on the path from the root to x is $h+1$ and the number of vertices on the path from the root to y is $2h+1 = 2(h+1)-1$, or only one less than the maximum of $2(h+1)$ allowed for a height balanced tree. On the other hand, we shall see on the next page that the worst case height of an AVL tree of n vertices is less than 2/3 of the $2\lceil \log_2(n) \rceil - 3$ that is guaranteed for height balanced trees (and red-black trees).

A More Precise Analysis of the Height of an AVL Tree

Theorem: An AVL tree of n vertices has height $< 1.45\lceil \log_2(n) \rceil$.

Proof: If $T(h)$ denotes the minimum number of vertices possible in an AVL tree of height h, then $T(0)=1$, $T(1)=2$, and for $h>1$, $T(h)=T(h-1)+T(h-2)+1$. Let α be the *golden ratio*, the positive real number that satisfies $\alpha^2=(1+\alpha)$; see the chapter notes and the exercises to the chapter on induction and recursion:

$$\alpha = \frac{1+\sqrt{5}}{2} = 1.61803...$$

With induction we verify that $T(h) > 0.437\alpha^{h+3}-1$:

For $h=0$: $T(0) = 1 > 1.852-1 > 0.437\alpha^3-1$
For $h=1$: $T(1) = 2 > 2.996-1 > 0.437\alpha^4-1$
For $h>1$, assume the inequality true for smaller h and:

$$\begin{aligned}
T(h) &= T(h-1)+T(h-2)+1 \\
&> (0.437\alpha^{h+2}-1)+(0.437\alpha^{h+1}-1)+1 \\
&= 0.437\alpha^{h+1}(\alpha+1)-1 \\
&= 0.437\alpha^{h+1}(\alpha^2)-1 \\
&= 0.437\alpha^{h+3}-1
\end{aligned}$$

We now apply standard properties of logarithms (see the first chapter). For an AVL tree of n vertices with height h, given that $n > 0.437\alpha^{h+3}-1$, we can add one to both sides, take $\log_2$ of both sides (if $a>b$, then $\log(a) \geq \log(b)$), and because $\log(xy) = \log(x) + \log(y)$ and $\log(x^y) = y\log(x)$, we have:

$$\log_2(n+1) \geq \log_2(0.437)+(h+3)\log_2(\alpha)$$

Finally, we divide both sides by $\log_2(\alpha)$ and rearrange terms to see:

$$h \leq \frac{\log_2(n+1)}{\log_2(\alpha)} - \frac{\log_2(0.437)}{\log_2(\alpha)} - 3 < 1.4405\log_2(n+1)-1 < 1.45\lceil \log_2(n) \rceil$$

Note: Along similar lines as was used for the Fibonacci numbers in the exercises to the chapter on induction and recursion, a proof by induction can use the fact that $\alpha^2=(1+\alpha)$ to show that the minimum number of vertices in an AVL tree of height h satisfies:

$$T(h) = \left(1+\frac{2}{\sqrt{5}}\right)\left(\frac{1+\sqrt{5}}{2}\right)^h + \left(1-\frac{2}{\sqrt{5}}\right)\left(\frac{1-\sqrt{5}}{2}\right)^h - 1$$

Storing Data Only in the Leaves

Idea: With any of the search trees we have considered, it may be convenient to store data only in the leaves (e.g., when large portions of the data are in external storage) and view non-leaf vertices simply as "guides" to get to the data associated with the leaves (the leaves account for more than half the vertices anyway).

Example: When implemented this way, we can restrict leaves to have only one data item. We allow data items in internal vertices to be $\geq$ items in the corresponding subtree, rather than $>$ as for standard 2-3 trees. The first tree below is the result of inserting the items 1 through 15 into a 2-3 tree as we have defined it, and the second shows when non-leaf vertices store the smallest items contained in the leaves of the corresponding subtrees.

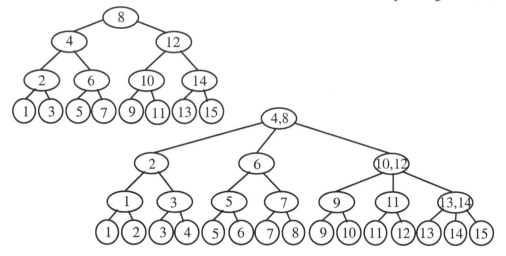

The basic 2-3 tree operations are essentially unchanged, except that the process may start off at a leaf in a slightly different way for INSERT and DELETE:

> INSERT(d, T): If the tree is currently empty, create a new one. Otherwise, search down to determine the leaf l where d belongs, let v be the parent of l (or let v be a new root if l is the root), add key d to v, and add a new child to v that has the key d. Essentially the same *while* loop to move up the tree can now be used.

> DELETE(l, T): If the tree contains only the single leaf l, we can make the tree empty. Otherwise, let v be the parent of l, and delete the leaf l and a key from v. The same *while* loop to move up the tree can now be executed.

The INDEX function: Most things that can be done in the standard representations can be done in a similar fashion when data is stored at the leaves. For example, in the chapter on trees, we saw how the INDEX function can be implemented for binary search trees by maintaining a count field with each vertex of the number of vertices in its subtree (and the same can be done for balanced trees). When data is stored in the leaves, we can let the counts be the number of leaves in the subtree.

Exercises

179. Starting with an empty 2-3 tree, insert the following strings (in this order) using the usual English alphabetical ordering:

> *horse, cow, pig, seal, rat, dog,*
>
> *goat, elephant, fish, rooster, zebra, roach, cat,*
>
> *hen, llama, aardvark, hog, donkey, rhino, hippo, tiger,*
>
> *lamb, lion, leopard, lynx, kitty, ant, ape, animal*

   A. Draw the resulting tree after *dog* has been inserted, after *cat* has been inserted, after *tiger* has been inserted, and after *animal* has been inserted.

   B. Delete *cow, fish, hen, lion, aardvark, ant,* and *ape*, and draw the tree after *hen* has been deleted, after *aardvark* has been deleted, and after *ape* has been deleted.

180. Consider the following tree:

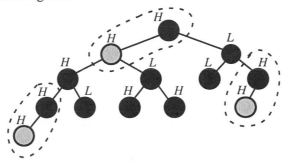

   A. Explain why this tree satisfies the definition of a red-black tree if solid vertices are taken to be *black* and shaded vertices are taken to be *red*.

   B. If the vertices are combined as indicated by the dashed lines, explain why a 2-3 tree results.

   C. Explain why this tree satisfies the definition of an AVL tree if the vertices are labeled *light* or *heavy* as indicated by the "L" and "H" labels.

181. Given pseudo-code for the MEMBER instruction for 2-3 trees.

182. In earlier chapters we have seen that a complete binary tree of height h (all leaves have depth h and all non-leaf vertices have two children) has 2^h-1 non-leaf vertices and 2^h leaf vertices for a total of $2^{h+1}-1$ vertices.

   A. Explain why this means that a complete binary tree containing n data items, one data item per vertex, has height exactly $\log_2(n+1)-1$.

B. Prove by induction that a complete 3-ary tree of height h (all leaves have depth h and all non-leaf vertices have three children) has $(3^h-1)/2$ non-leaf vertices and 3^h leaf vertices. Explain why this means that a complete 3-ary tree containing n data items, 2 data items per vertex, has height exactly $\log_3(n+1)-1$.

C. For $k \geq 2$, prove by induction that a complete k-ary tree of height h (all leaves have depth h and all non-leaf vertices have k children) has $(k^h-1)/(k-1)$ non-leaf vertices and k^h leaf vertices. Explain why this means that a complete k-ary tree containing n data items, $k-1$ data items per vertex, has height exactly $\log_k(n+1)-1$.

D. For $k \geq 2$, prove by induction that if a complete k-ary tree stores exactly $k-1$ data items at each vertex for a total of m data items, then it has $(m+1)/k$ leaves.

183. *B-trees* generalize the notion of a 2-3 tree trees where non-leaf vertices have more children.

A. Generalize the basic 2-3 algorithms for MEMBER, INSERT, and DELETE to K–J *trees*, where non-leaf vertices have between K and J vertices for fixed integers $K \geq 2$, and $J \geq 2K-1$.

B. Explain why the condition $J \geq 2K-1$ is needed.

C. Assuming $J = 2K-1$, give expressions for the minimum and maximum height of a K–J tree as a function of the number of vertices n.

E. Consider a model of a computer disk drive where it takes time x to move ("seek") to the start of a portion of the disk ("track") that contains a block of m data items, and once at the start of that block, it takes time y per data item to sequentially visit each item in the block (performing either a read or a write). Discuss the practical implications of storing n data items at just the leaves instead of at all vertices. In particular, consider the case where:

- x is much larger than y.

- Data items are records consisting of a short key together a large amount of associated data (e.g., a bank record where the key is the account number, and the associated data is the transaction history for the past year). So values at non-leaf vertices are just keys, whereas leaves store entire records.

In addition, discuss how changing the values of K and J might affect practical performance.

184. Write pseudo-code to delete a vertex from a red-black tree.

Hint: Think about how the corresponding algorithm for a 2-3 tree works.

185. Write pseudo-code to:

A. Perform a JOIN operation on a red-black tree.

B. Perform a SPLIT operation on a red-black tree.

Hint: Think about how the corresponding algorithm for a 2-3 tree works.

186. Prove that if it is possible to color the vertices of a binary search tree T to be a red black tree, then we can do so by working down from the root and coloring a vertex red only if it has a black parent and its virtual min-height is larger than its sibling (or it has no sibling):

procedure redblackCOLOR(v):

Color v black.

if v is a leaf **then** return

else if v has a single child x **then** Color x red.

else if both children x and y of v have the same virtual min-height **then begin**
redblackCOLOR(x)
redblackCOLOR(y)
end

else begin
Let x denote the child of v with smaller virtual min-height.
Let y denote the child of v with larger virtual min-height.
Let a and b denote the children of y.
Color y red.
redblackCOLOR(x)
redblackCOLOR(a)
redblackCOLOR(b)
end

end

Label each vertex with its virtual min-height.

redblackCOLOR(r)

Hint: Use the properties presented for red-black trees.

187. The definition of an AVL tree stipulated that every non-leaf vertex except the root has a sibling. If we define a missing subtree to have height -1, show that the following simpler definition is equivalent:

"A binary search tree is an *AVL tree* if for every pair of siblings, their height differs by at most 1."

188. Consider the relationship between the height and min-height of an AVL tree.

  A. Give an example of an AVL tree where the height of the left subtree of the root is greater than that of the right subtree of the root, but the min-height of the left subtree of the root is smaller than the min-height of the right subtree of the root.

  B. Generalize your answer to Part A to an infinite class of trees.

189. Recall the construction that was presented to show an infinite sequence of AVL trees T_1, T_2, ... such that T_i has height $2i$ if i is odd and height $2i-1$ if i is even.

  A. Prove that T_i has height $2i$ if i is odd and height $2i-1$ if i is even.

  B. Prove that T_i has the maximum height for an AVL tree with min-height i.

  C. Prove that T_i has the *minimum* number of vertices for an AVL tree that satisfies Parts A and B.

  D. Define an infinite sequence of AVL trees that has the *maximum* number of vertices for an AVL tree that satisfies Parts A and B.

190. Recall the sequence of AVL trees T_1, T_2, ... that were presented as "worst case" examples of AVL trees where T_i has height i and min-height $i/2$ if i is even or $(i+1)/2$ if i is odd:

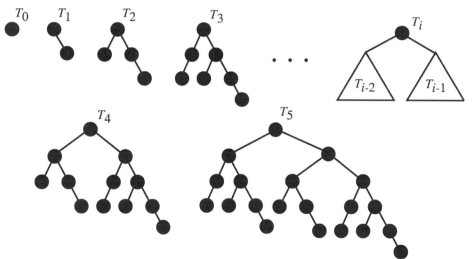

  A. Show how to correctly label the vertices of T_1, T_2, T_3, and T_4 as *light* or *heavy*.

  B. For any $i \geq 1$, give a recursive definition of how to correctly label the vertices of T_i as *light* or *heavy*.

  C. Show how to color the vertices of T_1, T_2, T_3, and T_4 *red* and *black* to form a legal red-black tree.

  D. For any $i \geq 1$, give a recursive definition of how to color the vertices of T_i *red* and *black* to form a legal red-black tree.

E. Show that the following tree is an AVL tree but cannot have its vertices colored to be a red-black tree:

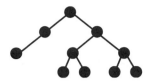

F. Show that there are infinitely many trees that are AVL trees but cannot have their vertices colored to be a red-black tree.

G. Explain why any AVL tree can have its vertices colored to be a relaxed red-black tree.

191. Prove by induction that the minimum number of vertices in an AVL tree of height h satisfies exactly:

$$T(h) = \left(1 + \frac{2}{\sqrt{5}}\right)\left(\frac{1+\sqrt{5}}{2}\right)^h + \left(1 - \frac{2}{\sqrt{5}}\right)\left(\frac{1-\sqrt{5}}{2}\right)^h - 1$$

Hint: Along the same lines as used in the exercise on Fibonacci numbers in the chapter on induction and recursion, make use of the positive real number

$$\alpha = \frac{1+\sqrt{5}}{2} = 1.61803...$$

that is often called the "golden ratio", and the negative real number:

$$\beta = \frac{1-\sqrt{5}}{2} = 1 - \alpha = -0.61803...$$

The real numbers α and β are the two solutions to $x^2 - x - 1 = 0$, and satisfy:

$$\alpha^2 = \alpha + 1 \quad \text{and} \quad \beta^2 = \beta + 1$$

Use these identities to simplify the mathematics for the inductive step.

192. Write pseudo-code to:

A. Perform a JOIN operation on an AVL tree.

B. Perform a SPLIT operation on an AVL tree.

193. A *relaxed* red-black tree is like a standard red-black tree except that:

- Condition D is relaxed to allow a red vertex to have a red sibling.

- Condition E is relaxed to allow a black leaf to have no sibling and to allow a red leaf to have a red sibling.

(But we still require that a non-leaf red vertex must have two black children and that a red leaf may not have a sibling that is a black leaf.)

Clearly a relaxed red-black tree is height balanced, since like a red-black tree, for every black vertex on a root to leaf path there is at most one red vertex.

A. Show that there are infinitely many values of n for which a height balanced tree of n vertices cannot have its vertices colored to form a legal red-black tree.

Hint: We have already seen that a complete binary tree must be made a red-black tree by coloring all of its vertices black (and we know this coloring is unique from the properties presented for red-black trees). Consider a binary search tree T that has a root where one subtree has virtual min-height h and the other subtree is a complete subtree of height $\geq h+2$. Explain why the root of this subtree may be red, but all of its other vertices must be black, and why this implies that it is impossible to color T to be a red-black tree.

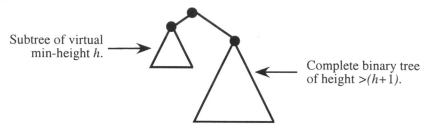

Subtree of virtual min-height h. →

← Complete binary tree of height $>(h+1)$.

B. Prove that the vertices of a height balanced tree can always be colored to form a relaxed red-black tree as follows:

procedure hbCOLOR(v,m):

Color v black.

if v is a leaf **then begin**
$\quad$ **if** $m>0$ **then** *ERROR — T is not height balanced*
end

else if v has a height 1 **then begin**
$\quad$ **if** $m=0$ **then** Color all children of v red.
$\quad$ **else if** $m=1$ **then** Color all children of v black.
$\quad$ **else** *ERROR — T is not height balanced*
end

else if both children x and y of v have virtual min-height $m-1$ **then begin**
$\quad$ hbCOLOR($x,m-1$)
$\quad$ hbCOLOR($y,m-1$)
end

else if the virtual min-heights of v's children are $m-1$ and $\geq m$ then begin
$\quad$ Let x denote the child of v with virtual min-height $m-1$.
$\quad$ Let y denote the child of v with virtual min-height $\geq m$.
$\quad$ Let a and b denote the children of y.

$\quad$ Color y red.
$\quad$ hbCOLOR($x,m-1$)

        hbCOLOR(*a*,*m*−1)
        hbCOLOR(*b*,*m*−1)
        end

    else if both children *x* and *y* of *v* have virtual min-height $\geq m$ **then begin**
        Color *x* and *y* red.
        for each child *z* of *x* and *y* **do** hbCOLOR(*z*,*m*−1).
        end

    end

Label each vertex with its min-height and let *M* be the min-height of the root.

hbCOLOR(*r*,*M*)

Hint: The parameter *m* in a call hbCOLOR(*v*,*m*) is the target min-height (the black height of the subtree rooted at *v* will be *m*+1). If *v* has min-height *m*, then *m* is reduced by one when the recursive call is on a child of min-height *m*−1 or a grandchild of min-height $\geq (m-1)$.

C. The coloring produced by the algorithm of Part C is not unique. Because in the worst case a height balanced binary search tree can have a shortest path from the root to a leaf *x* with *i* vertices and a longest path from the root to a leaf *y* with $2i$ vertices, the algorithm of Part C always colors in a "conservative" fashion that makes all vertices on the path from the root to *x* black, even if *x* has no sibling.

- Explain why the algorithm of Part C colors a complete binary tree of height *h* with black height *h*+1, but it is possible to color it as a relaxed red-black tree with black height (*h*+1)/2 when *h* is even and (*h*/2)+1 when *h* is odd.

- Prove that if there are several ways of coloring a binary search tree to be a relaxed red-black tree, then the algorithm of Part C colors in a way that gives the maximum black height.

D. Go through the properties presented for red-black trees and describe how they are affected if we are using relaxed red-black trees.

E. Give pseudo-code to insert an element into a relaxed red-black tree.

F. Give pseudo-code to delete an element from a relaxed red-black tree.

G. Define a 2-4 *tree* to be the generalization of a 2-3 tree where non-leaf vertices can have 2, 3, or 4 children (and vertices store 1, 2, or 3 values). Explain how:

- Any 2-4 tree of height *h* can be converted to a relaxed red-black tree of height $\leq 2h$.

- Any relaxed red-black tree of height *h* can be converted to a 2-4 tree of height $\leq h$, as long as we allow the exception that a vertex of height 1 can have only one child.

194. For 2-3 trees where data is stored only in the leaves (and values at non-leaf vertices are used to guide searches down from the root), give pseudo-code for the operations:

- A. MEMBER
- B. INSERT
- C. DELETE
- D. JOIN
- E. SPLIT

195. When balanced trees are implemented to store data only at the leaves (using values at non-leaf vertices as guides for searching down from the root) explain how both in-order and pre-order traversal of the tree can be used to list the data in sorted order.

Chapter Notes

Balanced trees are addressed in most texts on data structures and algorithms, such as those listed at the beginning of the chapter notes to the first chapter.

The book of Aho, Hopcroft, and Ullman [1974] presents 2-3 trees where all data is stored in the leaves. The presentation here of red-black trees, in a way that corresponds directly to 2-3 trees with data stored at all vertices, is along the lines of that in the book of Lewis and Denenberg [1991] (although they define black height in terms of edges so that the height of the corresponding 2-3 tree is the same, rather than one less as it is in our presentation). Red-black trees were presented by Bayer [1972]; Guibas and Sedgewick [1978] addressed theoretical issues and adopted the red-black naming convention. Red-black trees are often represented in a way that corresponds to 2-4 trees (where a black vertex may have two red children); see, for example, the books of Tarjan [1983] and Cormen, Leiserson, Rivest, and Stein [2001]. The basic notion of a binary tree being "balanced" if the shortest and longest distances from the root to a leaf differ by a factor of 2 (or some other amount) has been used by many (see, for example the discussion and references in the book of Tarjan [1983]). Here we have specifically framed the formulation in terms of the number of vertices on a root to leaf path, rather than the length of the path (which is one less), to be consistent with the equivalence between our definition of 2-3 trees and red-black trees, and to give a simple and "generous" definition for a tree that is balanced according to a factor of 2. This does, however, have the effect of making the definition of a "relaxed red-black tree" given in the exercises a bit ugly because it must allow a vertex to have a black leaf as its only child (which makes relaxed red-black trees slightly different than 2-4 trees).

AVL trees were presented by Adelson-Velskii and Landis [1962], and are presented in many texts on data structures and algorithms. Nievergelt and Reingold [1973] consider binary search trees of bounded balance. See also the books of Mehlhorn [1984] (Volume 1) and Tarjan [1983]. In the proof for the height of an AVL tree, we used an approximation for $T(h)$. As observed in the exercises, the minimum number of vertices in an AVL tree of height h satisfies

$$T(h) = \left(1 + \frac{2}{\sqrt{5}}\right)\left(\frac{1+\sqrt{5}}{2}\right)^h + \left(1 - \frac{2}{\sqrt{5}}\right)\left(\frac{1-\sqrt{5}}{2}\right)^h - 1 \approx 1.44\lceil \log_2(n) \rceil$$

where the real numbers

$$\alpha = \frac{1+\sqrt{5}}{2} = 1.61803... \quad \text{and} \quad \beta = \frac{1-\sqrt{5}}{2} = 1 - \alpha = -0.61803...$$

represent the classic "golden ratio" that is used in the solution to the expression for the *Fibonacci numbers* (see the exercises to the chapter on induction and recursion). See the book of Liu [1985] for a presentation of how to solve recurrence relations as well as the Fibonacci numbers in particular. The $1.44\lceil \log_2(n) \rceil$ upper bound is for the *worst case* height of an AVL tree; in practice, the constant for the expected height may be close to 1 for uniform distributions (see Section 6.4.2 of the book of Reingold, Nievergelt, and Deo [1977]).

The term "B-tree", proposed by Bayer and McCreight [1972], is often used to describe generalizations of 2-3 trees where non-leaf vertices have higher degrees (e.g., between k and $2k-1$ vertices for some fixed integer $k \geq 2$). Such generalizations can be useful for secondary storage applications where it may be desirable to trade off increased time to visit a vertex against the number of times you move from one vertex to another (and from one place to another in the secondary storage).

CHAPTER NOTES

9. Sets Over a Small Universe

Introduction

So far, when looking at implementations of generic operations like MEMBER, INSERT, and DELETE, we have been assuming that the set of all possible elements (the "universe") is much larger than the actual number of elements that will be stored. In fact, typically, the space of all possible items (e.g., names represented by strings of 25 characters) is larger than can possibly be stored.

There are, however, important applications where the number of possible items in the set is quite small. Such applications often involve the "internals" of an algorithm where the set of items being stored are related to indices of data already in memory. In this chapter, we let n denote the number of items to be stored and m the number of possible items. The basic assumption is that m is small enough so that it is practical to use $O(m)$ space; in fact, for many applications, it will be that m is $O(n)$.

We begin by considering the problem of initializing an array. We usually do not think much of this problem, because for typical applications, if the array has n locations, the running time of the algorithm is at least $\Omega(n)$, and so the $O(n)$ time to initialize the array does not affect the asymptotic complexity. In addition, initialization is typically a very simple loop or nested set of loops that has a very low constant. However, there can be times when it is convenient to declare a very large array, but use only a small portion of it. For example, in the next chapter, we may wish to represent a directed graph with only $O(n)$ vertices by an adjacency matrix with n^2 entries. Space is often "free" if we have it. However, we would like to avoid spending $O(n^2)$ time to initialize the array if we are then going to use an algorithm that is, for example, $O(n)$ and only uses $O(n)$ locations. The "on-the-fly" array initialization algorithm avoids explicitly initializing an array by using a "hand-shaking" protocol to determine whether an entry has already been used or needs to be initialized before being read. Because array initialization is so fast in practice, this algorithm is mostly of theoretical interest.

We next consider the problem of in-place permutation. That is, rearranging an array according to some permutation function with an algorithm that uses only $O(1)$ space in addition to the space used by the array itself. This can be useful for an application that has a very large array that barely fits in memory.

Bucket sorting is then presented. When the n elements to be sorted are small integers in the range 0 to n, bucket sorting allows us to sort in linear time. Such sorting problems often arise in the internal workings of an algorithm, and in such cases, bucket sorting is very fast and practical.

We then consider representing set with "bit vectors" that allow basic operations to be done in $O(1)$ time.

Finally, we consider the classic "UNION-FIND" problem. We are given a collection of disjoint sets of elements drawn from a small universe. In the basic version of the problem, there are initially n sets, where set i contains the single element i. We then need to process a sequence of UNION and FIND instructions, where UNION(i,j) unions the sets i and j, and FIND(i) determines to which set i currently belongs. The classic fast UNION-FIND algorithm is remarkable in that it is simple and practical, but an amortized analysis can be used to show that, although it is not linear, it is in some sense as about as close to linear as an algorithm can be. Formally, the running time for n operations is $O(nA^{-1}(n))$ where $A^{-1}(n)$ denotes the inverse of *Ackermann's function*. Although it is possible to define faster growing functions, $A(n)$ is a function that grows so fast that asymptotically $A^{-1}(n)$ is essentially a constant for practical purposes. Although the UNION-FIND problem seems artificial, it has been effectively used in the internal workings of a number of classic algorithms to get improved performance. In fact, the algorithm had been used in practice for years before its complexity was correctly analyzed.

On the Fly Array Initialization

Problem: We need to initialize a large array A, but may only use of a few of the locations (and do not want complexity to be dominated by initialization time).

Idea: Initialize an entry the first time it is used. Employ "hand shaking" with an auxiliary array B of the same dimensions as A and a stack S implemented as a one dimensional array. A position in A has been initialized if and only if the corresponding position in B points an active portion of S and that item of S points back.

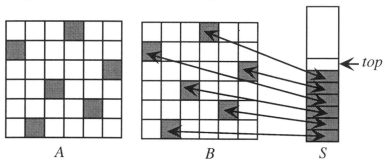

On-the-fly initialization algorithm: Initialize $top=0$ and precede each access to A by the following code, where x denotes a vector of the dimension of A; for example, for the figure above, x denotes a pair of values that designates a position in the two-dimensional array A.

    if $B[x]<0$ or $B[x] \geq top$ or $S[B[x]] \neq$ (a pointer to $B[x]$) **then begin**
        $B[x] := top$
        $S[top] :=$ a pointer to $B[x]$
        $top := top+1$
        Initialize $A[x]$.
    end

Correctness: We make no assumption about what values are initially present in B; it may be that just by chance an unused entry $B[x]$ points to an active portion of S (i.e., that $0 \leq B[x] < top$). First we check if $B[x]$ does not point into S, and if not, then $A[x]$ needs to be initialized. Second, if $B[x]$ points to a position i in S, then we check if $S[i]$ contains a pointer back to $B[x]$; if it does not, then again $A[x]$ needs to be initialized. On the other hand, if it does point back, then this location must have been previously initialized because that is the only way items are placed in S.

Time: $O(1)$ time per access to A.

Space: Since B has the same dimensions as A and S contains at most the number of elements in A, the space for A is increased by only a constant factor.

In-Place Permutation

Problem: Given an array of elements $A[1]$ through $A[n]$ and a one-to-one function f on the integers 1 to n, perform a sequence of exchanges so that the element in the ith location ends up in the $f(i)^{th}$ location. Use only $O(1)$ space in addition to the space used by A.

Idea: Any permutation is a collection of cycles (a goes to b goes to c ... goes back to a — see the exercises). Make a single pass through the array; each time an element is encountered that is part of a new cycle, "rotate" that cycle one position forward.

Method 1: $O(n)$ — "Cheat" slightly and "mark" the elements of A; this makes it easy to tell if we are at the start of a new cycle. When the first position i of a cycle is detected, for each successive position j of the cycle, exchange $A[i]$ and $A[j]$ (i.e., $A[i]$ is storing the previous position visited as j sequences through the cycle).

    for $i := 1$ **to** n **do** Label $A[i]$ "unmarked".
    for $i := 1$ **to** n **do**
        if $A[i]$ is not marked **then begin**
            Label $A[i]$ "marked".
            $j := f(i)$
            while $j > i$ **do begin**
                Label $A[j]$ "marked".
                Exchange $A[i]$ and $A[j]$.
                $j := f(j)$
            end
        end

Method 2: $O(n^2)$ — Like Method 1, but since we have not marked already processed positions, first use a *while* loop to check if the position is the leftmost in its cycle.

    for $i := 1$ **to** n **do begin**
        $j := f(i)$
        while $j > i$ **do** $j := f(j)$
        if $j = i$ **then begin**
            $j := f(i)$
            while $j > i$ **do begin**
                Exchange $A[i]$ and $A[j]$.
                $j := f(j)$
            end
        end
    end

Note: See the exercises for a different $O(n^2)$ method.

Bucket Sorting

Problem: Given n items in an input/output queue, *IOqueue*, where associated with each item is an integer "key" in the range 0 to $m-1$, rearrange them in increasing order of key (duplicate keys are allowed).

Idea: If m is not too large, in $O(n+m)$ time and space, declare an array of size m, do "perfect" hashing, and then visit each of the buckets in order to collect all of the n items.

Sorting integers (when items are their own keys): Declare an array B from 0 to $m-1$, initialize each entry to 0, and increment the ith position when i is read.

> 1. **for** $i := 0$ **to** $m-1$ **do** $B[i] := 0$
>
> 2. **for** $i := 1$ **to** n **do**
> $\quad j := \text{DEQUEUE}(IOqueue)$
> $\quad B[j] := B[j]+1$
>   **end**
>
> 3. **for** $i := 0$ **to** $m-1$ **do** enqueue to *IOqueue* $B[i]$ copies of i

Generic method (when items have an associated key): *IOqueue* contains pointers to the items rather than the items themselves (so items can always be moved in $O(1)$ time).

> 1. **for** $i := 0$ **to** $m-1$ **do** $Q[i] :=$ an empty queue
>
> 2. **while** *IOqueue* is not empty **do**
> $\quad I := \text{DEQUEUE}(IOqueue)$
> $\quad \text{ENQUEUE}(I, Q[k])$, where k is the key for I
>   **end**
>
> 3. **for** $i := 0$ **to** $m-1$ **do** enqueue to *IOqueue* all items in $Q[i]$

Use multiple bucket sorts to sort vectors: Given two dimension d vectors of integers $V=(v_1...v_d)$ and $W=(w_1...w_d)$, $V \le W$ means either $V=W$ or for some $1 \le i \le d$, $v_i < w_i$, and for $1 \le j < i$, $v_j = w_j$. For example, English character strings of length d can be viewed as vectors where $s \le t$ means the string s comes alphabetically before the string t. We assume that all vector components are in the range 0 to $m-1$. By working from $i=d$ down to $i=1$ and bucket sorting on the ith component, at the ith stage, vectors that have the same first $d-i$ components will be in the correct relative order. We assume that the ith vector, $1 \le i \le n$, is stored in $A[i,1]...A[i,d]$.

> ENQUEUE to *IOqueue* the integers 1, 2, ..., n.
>
> **for** $y := d$ **downto** 1 **do** Bucket sort *IOqueue* where the key for item x is $A[x,y]$.
>
> **while** *IOqueue* is not empty **do** Output DEQUEUE(*IOqueue*).

Complexity: Each bucket sort uses $O(n+m)$ time and space, for a total of $O(d(n+m))$ time and $O(n+m)$ space, in addition to the $O(dn)$ space to store the vectors. In the chapter on strings, we shall see how to improve this time to $O(dn+m)$.

Bit-Vector Representation of Sets

Problem: Given sets $S_1...S_n$ of elements in the range 0 to $m-1$, store them so that set initialization, insertion, deletion, and member operations can all be done in constant time.

Bit vector representation: Associate with each set S_i, an array $S_i[0]...S_i[m-1]$ where $S_i[j]$ is 1 if S_i contains j and 0 otherwise ($S_i[j]$ can also contain a pointer to any other data associated with j):

INITIALIZE(S_i):   By employing on-the-fly array initialization, we can assume that S_i is initially empty (that is, the array associated with S_i is initialized to 0).

INSERT(j,S_i):   Add j to the set S_i (or effectively do nothing if S_i already contains j) by setting $S_i[j]$ to 1.

DELETE(j,S_i):   Delete j from the set S_i (or effectively do nothing if S_i does not contain j) by setting $S_i[j]$ to 0.

MEMBER(j,S_i):   Determine if j is an element of S_i (and if so, return any data associated with j) by checking whether $S_i[j]$ is equal to 1.

Disjoint sets: If the sets to be stored will always be disjoint, a single array $S[0]...S[m-1]$ suffices, where $S[j]$ is 0 if j is not in any set, and otherwise $S[j]=i$ if j is in S_i.

Note: On-the-fly array initialization gives an $O(1)$ bound for initializing S in the analysis above. In practice, the time for initialization in the straightforward way may not be significant.

Union-Find Problem

General problem: Given sets $S_1...S_n$ where initially $S_i=\{i\}$, perform a sequence of $O(n)$ of the following two operations:

FIND(i):   Return the index of the set that contains i.

UNION(i,j,k):   Replace S_k by the union of S_i and S_j.
                    if $i{\neq}k$ **then** $S_i := \{\}$
                    if $j{\neq}k$ **then** $S_j := \{\}$

Standard problem: Like the general problem except that the name of a union is the name of the larger of the two sets:

UNION(i,j):   **if** $|S_i| \geq |S_j|$
                         then $S_i := S_i \cup S_j$ and $S_j := \{\}$
                         else $S_j := S_i \cup S_j$ and $S_i := \{\}$

Standard problem is equivalent to the general problem:

If we have an implementation for the standard version of the union-find problem, with only a constant factor overhead we can maintain a map between "internal" and "external" names to obtain an implementation of the general problem:

- Initialize an array EXT[1]...EXT[n] so that EXT[i]=i, $1 \leq i \leq n$.

- Initialize an array INT[1]...INT[n] so that INT[i]=i, $1 \leq i \leq n$.

- Replace FIND(i) by FIND(INT[i]) and when a value j is returned, replace it by EXT[j].

- Replace UNION(i,j,k) by UNION(INT[i],INT[j]) and for the case that SIZE[i] $\geq$ SIZE[j] ≥ 1 do:

        INT[k] := i; INT[i] := 0; INT[j] := 0
        EXT[i] := k; EXT[j] := 0;

* From this point on we consider only the standard problem.

Linked List Implementation of Union-Find

- Initialize a bit vector $S[1]...S[n]$ so that $S[i]=i$, $1 \leq i \leq n$.

- Initialize an array of linked lists $L[1]...L[n]$ so that for $1 \leq i \leq n$, $L[i]=i$, and $NEXT[i]=nil$ (0 can be used to represent the nil pointer).

- Initialize an array $SIZE[1]...SIZE[n]$ so that $SIZE[i]=1$, $1 \leq i \leq n$.

- FIND(i) simply returns $S[i]$.

- UNION(i,j) when $SIZE[i] \geq SIZE[j] \geq 1$ does:

    Traverse $L[j]$ and for each value k, do $S[k] := i$ and
    let x denote the last element of $L[j]$.

    Insert the list $L[j]$ at the beginning of $L[i]$:
    $$NEXT[x] := L[i]$$
    $$L[i] := L[j]$$

    Make $L[j]$ empty:
    $$L[j] := nil$$

    Update the sizes:
    $$SIZE[i] := SIZE[i]+SIZE[j]$$
    $$SIZE[j] := 0$$

- UNION(i,j) when $1 \leq SIZE[i]<SIZE[j]$ is symmetric.

- UNION(i,j) when SIZE[i]=0 or SIZE[j]=0 does nothing.

Time for $O(n)$ operations is $O(n\log(n))$:

    finds: Each FIND is $O(1)$, for a total of $O(n)$ time for all FIND operations.

    unions: If an element participates in a UNION and is in the larger set, it is not moved. If it is in the smaller set, $O(1)$ time is spent to move it to become part of a set that is at least twice as big. Hence, an element can be moved at most $\log_2(n)$ times, for a total of at most $O(n\log(n))$ time for all UNIONS.

Tree Implementation of Union-Find

- Initialize n trees so that for $1 \le i \le n$, PARENT[i]=*nil* (again, 0 can be used to represent the nil pointer); that is, a vertex of a tree is simply a parent pointer, and initially, each tree is just a root vertex with a *nil* parent pointer.

- Initialize an array $SIZE[1]...SIZE[n]$ so that $SIZE[i]=1$, $1 \le i \le n$.

- FIND(i):
    $p := i$
    while PARENT[p]$\ne$*nil* **do** $p := $ PARENT[p]
    return p

- UNION(i,j) when $SIZE[i] \ge SIZE[j] \ge 1$ does:
    PARENT[j] := i
    $SIZE[i] := SIZE[i]+SIZE[j]$
    $SIZE[j] := 0$

- UNION(i,j) when $1 \le SIZE[i] < SIZE[j]$ is symmetric.

- UNION(i,j) when SIZE[i]=0 or SIZE[j]=0 does nothing.

Time for $O(n)$ operations: $O(n\log(n))$

Each UNION is $O(1)$. Now consider the total time spent for each element over all FINDS. Since the height of a tree representing a set increases by one only when a UNION causes the elements of a set to become part of a set with at least twice as many elements, the maximum depth of any vertex in any tree is at most $\log_2(n)$, and the total time for all finds is $O(n\log(n))$.

Tree Implementation of Union-Find with Path Compression

Idea: After each FIND, attach each vertex that was visited directly to the root (so that time for future FIND operations may be reduced):

function FIND(i):

Initialize an empty stack.

$p := i$

while PARENT[p]$\neq$*nil* **do begin**
    PUSH(p)
    $p := $ PARENT[p]
    end

while stack is not empty **do begin**
    Pop an integer j.
    PARENT[j] $:= p$
    end
return p

end

Example:

INITIALIZE: 1   2   3   4   5   6   7   8

UNION(1,2), UNION(3,4), UNION(5,6), UNION(7,8):

UNION(1,3), UNION(5,7):

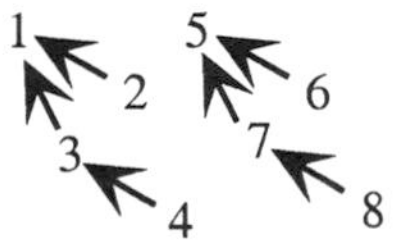

UNION(1,5):

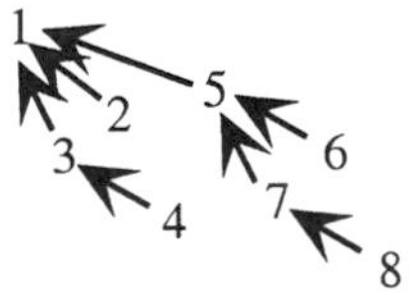

FIND(4), FIND(8):

Analysis of path compression:

Although path compression is simple, its analysis is more complicated.

- When path compression is employed in a tree implementation, a sequence of $O(n)$ UNION and FIND operations is $O(nA^{-1}(n))$ where A^{-1} denotes the inverse of a version of *Ackermann's function*; that is, if i is the smallest integer for which $A[i] \geq j$, then $A^{-1}[j]=i$.

- Ackermann's function has extremely large asymptotic growth and hence its inverse has extremely small asymptotic growth.

- For example, if $\log^2(n)$ denotes $\log_2(\log_2(n))$, $\log^3(n)$ denotes $\log_2(\log_2(\log_2(n)))$, and so on, then for any $i>0$, $A^{-1}(n)$ is $O(\log^i(n))$; in fact it is *much* smaller .

- For all practical purposes, path compression yields linear time for a sequence of $O(n)$ operations; for example, if $n<2^{1,000,000}$, then $A^{-1}(n) \leq 4$.

- The proof of this bound is due to Tarjan; see the chapter notes.

Formal definition of Ackermann's function:

For $n \geq 0$ define $A(n) = ACK(n,n)$ where:

> **function** ACK(i,n)
>   **if** $i=0$ **then return** $n+1$
>   **else if** $n=0$ **then return** ACK($i-1,1$)
>   **else return** ACK($i-1,$ACK($i,n-1$))
> **end**

If we let $ESTACK(a,n) = a^{a^{a^{\cdot^{\cdot^{\cdot^{a}}}}}}$ — a stack of n a's, it can be shown by induction that:

$$A(0,n) = n+1$$
$$A(1,n) = n+2$$
$$A(2,n) = 2n+3$$
$$A(3,n) = 2^n-3$$
$$A(4,m) = ESTACK(2,n+3)-3$$

One can think of the first argument as specifying the operator (where 1 denotes adding 2, 2 roughly denotes multiplying by 2, 3 roughly denotes exponentiation, 4 roughly denotes the ESTACK function, and 5 is more complicated). The $i+1^{th}$ operation on n is expressed recursively as the ith operation applied to the $i+1^{th}$ operation on $n-1$.

For all practical purposes, $A^{-1}(n) \leq 4$, since ESTACK(2,4)=65,536 and hence:

$$ACK(4,4) = ESTACK(2,7)-3 > 2^{2^{2^{65,535}}}$$

Example: Off-Line Processing of Heap Operations

Problem: Starting with an empty set S, process a sequence of $O(n)$ instructions of the following form *off-line* (all instructions may be seen before any output is produced):

INSERT(i):   Insert the integer i; i must be in the range $0 \leq i \leq n$ and no integer may be inserted more than once.

DELETEMIN: Delete and print the smallest integer currently in S.

Idea: We could use a heap to do this in $O(n\log(n))$ time. However, because the problem is off-line (and we can see all of the instructions in advance before having to produce any output), and because the integers inserted are unique and in the range 0 to n, we can produce the same output in $O(nA^{-1}(n))$ time employing the fast union-find data structure. Since we only care about printing the correct output, we do not bother deleting from S. However, this could be done (see the example on the following page).

1. Input the instructions and form a doubly linked list of sets where set i contains all of the elements that are inserted between the $i{-}1^{th}$ and i^{th} DELETEMIN instruction (set 1 is all elements inserted before the first instruction and set $i{+}1$ is all elements inserted after the last instruction).

2. Union sets to create a single set that is the result of executing the instructions, and keep track of the set into which each item goes:

    Q := an empty queue
    for i := 1 to n **do begin**
        j := FIND(i)
        if j is not the last set **then begin**
            Enqueue to Q the pair (i,j).
            Union set j with $j{+}1$ and name it $j{+}1$, and patch up the linked list.
        end
    end

3. Bucket sort the items deleted into the order that they would have been deleted if the instructions were processed on-line:

    Bucket sort Q using the second components of the pairs as the key.
    while Q is not empty **do begin**
        Dequeue from Q a pair (i,j).
        Output "i is deleted by the jth DELETEMIN instruction".
    end

Analysis: The *for* loop of Step 2 considers each integer i in *increasing* order, and finds the set into which i is merged, which is the same as the index of the instruction that deletes i. By bucket sorting on the second component of the pairs produced in Step 2, Step 3 correctly reorders the deletions to be in the order they would have happened on-line. Space is $O(n)$. Steps 1 and 3 are $O(n)$ time, and Step 2 is $O(nA^{-1}(n))$ time.

Other Operations that can be Added to Union-Find

Idea: MEMBER, INSERT, and DELETE can all be implemented in a way that does not affect the asymptotic running time of the fast union-find algorithm. Here i denotes an integer in the range 1 to n, and S denotes the name of a set, which is also an integer in the range 1 to n.

MEMBER(i,j): Do a FIND(i) and check if the result is j.

INSERT(i,j): Do a FIND(i) to check that i is in a set of size 1 with name i (i.e., since we are assuming disjoint sets, it would be an error if i was in any set other than $\{i\}$). Then do UNION(j,i). Note that even if SIZE[j]=1, the way we have defined UNION insures that the resulting set will be named j.

DELETE(i): Like deleting from a binary search tree, we can replace a vertex by one of the leaves of the tree that it is in:

For each set keep a linked list of its leaves.

For each vertex keep a variable *count* of the number of vertices that point to it (including itself).

The UNION operation is still $O(1)$ time since the only new thing that must be done is to concatenate the leaf lists (for the new root leaf list) and update the count field of the root (to be the sum of the two old count fields).

The time for a find operation is only increased by the constant factor needed to update the *count* fields of the vertices traversed. That is, each time a vertex is moved to point to the root, the count of the vertex it used to point to is decremented, and if that count becomes 0, that vertex is added to the leaf list of the root.

To perform a DELETE(i), first do a FIND(i) to determine the root of i and then in $O(1)$ time replace i by a leaf of the root (or simply delete i if it is the first leaf on the roots leaf list).

Example: The algorithm for processing off-line heap operations that was presented on the previous page produces the correct output, but does not delete any elements from S. If we want to have the algorithm terminate with S correctly computed as well, without increasing the $O(nA^{-1}(n))$ time, we can delete i from the set j just before Step 2 places the pair (i,j) in the queue.

Exercises

196. The list L in the on-the-fly initialization algorithm stores pointers back into B. Suppose that we were using a language that did not have pointer variables and instead, we had to store in L the indices into the array A. For example, if A was a two dimensional array, we would have to store pairs of integers in L. How would this affect the complexity of on-the-fly initialization?

197. Prove that a permutation always consists of a disjoint set of cycles (a goes to b goes to c ... goes to a).

Hint: Use a proof by contradiction that shows that if this was not the case, then two items must go to the same location.

198. Method 1 for in-place permutation "cheats" by marking elements of A. That is, we do not really use $O(1)$ space in addition to A because the marking constitutes an additional n bits. In practice, there are many circumstances when these bits might be available for "free". Suppose, for example, that positive integers were stored in A but A was defined in such a way that it was legal to store a negative integer in A. Rewrite Method 1 for in-place permutation to mark elements by negating them.

199. Consider Method 1 for in-place permutation.

    A. Give a formal proof of correctness.

    B. Give a formal proof that it is $O(n)$ time.

200. Consider Method 2 for in-place permutation.

    A. Give a formal proof of correctness.

    B. Give a formal proof that it is $O(n^2)$ time.

201. Suppose that Method 1 for in-place permutation was changed slightly by making the exchange statement be:

    Exchange $A[j]$ and $A[f(j)]$.

Explain why the algorithm now computes the inverse permutation. That is, the algorithm performs a sequence of exchange operations so that the element in the $f(i)^{th}$ location ends up in the ith location.

202. The following algorithm uses f^1, the inverse of f, to perform in-place permutation:

    for $i := 1$ to n **do begin**

        $j := f^1(i)$

        while $j < i$ **do** $j := f^1(j)$

        Exchange $A[i]$ and $A[j]$.

        end

A. Give a proof of correctness.

   Hint: Correctness follows by induction. For the ith iteration, if $j < i$ then

   $$A[j] = A_{\text{original}}[f(j)]$$

   and if $j \geq i$ then

   $$A[f^k(j)] = A_{\text{original}}[f(j)]$$

   where $f^k(j)$ denotes

   $$\underbrace{f(f(f(\cdots(j)\cdots)))}_{k \text{ applications of } f}$$

   for the least k such that $f^k(j) \geq i$.

B. Explain why the time is $O(n^2)$.

203. Consider the time and space complexity of the generic bucket sorting algorithm:

   A. Explain why Steps 1, 2, and 3 of the generic bucket sorting algorithm are $O(m)$, $O(n)$, and $O(m)$ time, respectively.

   B. Although it would not change the total asymptotic time of the algorithm as a whole, explain how on-the-fly array initialization can be used to reduce Step 1 to $O(1)$ time.

   C. Explain why the algorithm uses $O(n+m)$ space.

   D. Assuming that an item and its key consume 1 unit of space, compare the exact amount of space used when on-the-fly array initialization is or is not used.

204. Given n integers each represented with d digits in base $b \geq 2$:

   A. Describe how to sort them in $O(dn)$ time.

   Hint: Do d bucket sorts working from right to left, being sure to keep equal elements in the same relative order when buckets are collected.

   B. Suppose that all of these integers are unique. Show that dn is $\Omega(n\log(n))$.

205. Prove that the fast union-find algorithm with path compression is linear time when all the UNIONs occur before any FINDs.

Hint: Show that a vertex is visited at most once when it has depth greater than 1; then add the total time spent at vertices of depth greater than one to the total time spent at vertices of depth 0 or 1.

206. We have already described how DELETE can be implemented in a way that does not affect the asymptotic running time of the fast union-find algorithm.

   A. Present pseudo-code for that method.

B. Present pseudo-code for a different method that, for each vertex, keeps a list of its children, and to perform DELETE(i), walks down from i until a leaf is encountered, performing path compression on the way. Explain why the method does not increase the asymptotic running time of the fast union-find algorithm.

207. For Ackermann's function, give formal proofs by induction that:

A. $ACK(1,n) = n+2$

B. $ACK(2,n) = 2n+3$

C. $ACK(3,n) = 2^{n+3}-3$

D. $ACK(4,n) = ESTACK(2,n+3)-3$

208. For Ackermann's function, describe the function $A(5,n)$.

209. Our definition of Ackermann's is the "traditional" one (see the chapter notes). Suppose the initial cases are changed to create a variation $ACK2(i,n)$ defined for $i,n \geq 1$:

> **function** ACK2(i,n)
>   **if** $n=1$ **then return** 2
>   **else if** $i=1$ **then return** $n+2$
>   **else return** ACK($i-1$, ACK($i,n-1$))
>   **end**

Give formal proofs by induction that we of the following equalities for $n>1$:

A. $ACK2(1,n) = n+2$

B. $ACK2(2,n) = 2n$

C. $ACK2(3,n) = 2^n$

D. $ACK2(4,n) = ESTACK(2,n)$

210. For Ackermann's function, suppose the initial cases are changed to create a variation $ACK3(i,n)$ defined for $i,n \geq 0$:

> **function** ACK3(i,n)
>   **if** $i=0$ **then return** $n+2$
>   **else if** $n=0$ **then return** 1
>   **else return** ACK($i-1$, ACK($i,n-1$))
>   **end**

Give formal proofs by induction that $ACK3(i,n)$ behaves like $ACK(i+1,n)$; that is:

A. $ACK(0,n) = n+2$

B. $ACK(1,n) = 2n+1$

C. $ACK(2,n) = 2^{n+1}-1$

D. $ACK(3,n) = ESTACK(2,n+1)-1$

Also, although $ACK3(n,n)$ is clearly not $O(ACK(n,n))$, explain why $ACK3^{-1}(n,n)$ is $O(ACK^{-1}(n,n))$, and hence reasonable variations in the initial conditions will not affect the analysis of the fast union-find algorithm.

211. Consider the straightforward implementation of $ACK(i,n)$ by using mechanical recursion elimination to produce a non-recursive program that manipulates a stack.

    A. Write the pseudo-code that results.

    B. Show that the computation of $ACK(i,n)$ uses $\Omega(ACK(i,n))$ space.

       Hint: First show that $A(1,n)$ computes $n+2$ via a sequence of n recursive calls (using tail recursion); then argue that the only way ACK can return a value $j>2$ is by ultimately calling $ACK(1,j-2)$.

    B. How does the space increase if instead of charging unit-cost to store an integer, we charge one unit of space for each bit of data stored.

212. The straightforward implementation of $ACK(i,n)$ with a stack (by employing mechanical recursion elimination) can be improved by observing that it is really only the first argument that must be pushed, a single integer containing a running total can be maintained instead of stacking the second argument.

    A. Show that the following program computes $ACK(i,n)$:

> Input i and n.
> Initialize an empty stack.
> PUSH(i)
> *total* := n
> **while** the stack is not empty **do begin**
> x := POP
>   **if** $x=0$ **then** *total* := *total*+1
>   **else if** *total*=0 **then begin** PUSH($x-1$); *total* := 1 **end**
>   **else begin** PUSH($x-1$); PUSH(x); *total* := *total*−1 **end**
> **end**
> Output *total*.

    B. By letting the variable k be the index of the item on top of the stack in an array implementation of a stack, show that the code of Part A can be simplified to:

> Input i and n.
> $S[0]$:= i
> k := 0
> *total* := n
> **while** $k \geq 0$ **do begin**
>   **if** $S[k]=0$ **then** *total* := *total*+1
>   **else if** *total*=0 **then begin** $S[k]$:= $S[k]-1$; *total* := 1 **end**
>   **else begin** $S[k]$:= $S[k]-1$; k := $k+1$; $S[k]$:= $S[k-1]+1$; *total* := *total*−1 **end**
> **end**
> **end**
> Output *total*.

17. The following program computes $ACK(i,n)$ using only $O(i)$ space for the two arrays $I[0]...I[i]$ and $V[0]...V[i]$:

> Input i and n.
> Initialize $I[0]=0$ and $I[j]=-1$, $1 \leq j \leq i$.
> Initialize $V[j]=1$, $0 \leq j \leq i$.
> **while** $I[i] \neq n$ **do begin**
> > $I[0] := I[0]+1$
> > $V[0] := V[0]+1$
> > $k := 0$
> > **while** $k<i$ **and** $I[k]=V[k+1]$ **do begin**
> > > $I[k+1] := I[k+1]+1$
> > > $V[k+1] := V[k]$
> > > $k := k+1$
> > > **end**
> > **end**

A. Prove that for $0 \leq j \leq i$ the program terminates with $V[j]=ACK(j,n)$ by showing that at the start of each iteration of either *while* loop, for $0 \leq j \leq i$, if $I[j] \geq 0$, then:
$$V[j] = ACK(j,I[j])$$
Hint: Consider the test $I[k]=V[k+1]$ made by the inner *while* loop:

> If $I[k]=V[k+1]=1$, then since we have already computed $V[k]=A(k,1)$, it is correct to fill in $V[k+1]=V[k]$ and $I[k+1]=-1+1=0$; by the second condition of ACK that states $ACK(i,0) := ACK(i-1,1)$.

> If $I[k]=V[k+1]>1$, then since $A(k+1,I[k+1])=V[k+1]=I[k]$, again it is correct to fill in $V[k+1]=V[k]$ since by the third condition of ACK that states $ACK(i,n)=ACK(i-1,ACK(i,n-1))$:
> $$ACK(k+1,I[k+1]+1) = ACK(k,A(k,A(k+1,I[k+1]))) = A(k,I[k]) = V[k]$$

B. Show that after the program terminates, $V[j]=A(i,n)$, $0 \leq j \leq i$, and in fact, the second statement of the inner *while* loop can be replaced by:
$$V[k+1] := V[0]$$

C. Show that this program uses $\Omega(A(i,n))$ time.

D. How does the $O(i)$ space used change if instead of charging unit-cost to store an integer, we charge one unit of space for each bit of data stored?

E. Modify this program for $ACK2$ and $ACK3$ as defined in the preceding exercises.

213. The standard definition of ACK with two variables applies the ith level operation to a single value. With a third argument, we can define a series of more "natural" functions:

> **function** ACKM(i,n,m)
> > **if** $i=1$ **then return** $n+m$
> > **else if** $m=1$ **then return** n
> > **else return** ACKM$(i-1,n,$ACKM$(i,n,m-1))$
> > **end**

Prove by induction that:

$$ACKM(1,n,m) = n+m$$
$$ACKM(2,n,m) = n*m$$
$$ACKM(3,n,m) = n^m$$
$$ACK(4,n,m) = \text{ESTACK}(n,m)$$

214. In the spirit of Ackermann's function, consider how to neatly define a level $i+1^{th}$ operation in terms i^{th} level one. Define incrementing an integer with the function:

> **function** INC
>   **begin**
>   **return** $i+1$
>   **end**

Now, using pseudo-code, define recursive functions for:

A. $ADD(x,y)$: Computes $x+y$ using only recursive calls and calls to *INC*.

B. $MULT(x,y)$: Computes $x*y$ using only recursive calls and calls to *ADD*.

C. $E(x,y)$: Computes x^y using only recursive calls and calls to *MULT*.

D. $ES(x,y)$: Computes $\text{ESTACK}(x,y)$ using only recursive calls and calls to *E*.

215. Give a more detailed proof of correctness for the algorithm presented for off-line processing of heap operations.

Chapter Notes

On-the-fly array initialization was presented in Hopcroft [1974].

The exercise that uses the inverse of f to perform in-place permutation is from Exercise 3-22 of the book of Manna [1974], which references a proof by Duijvestijn [1972]; the proof hint here is based on a different proof of Storer and Keller [1974].

Bucket sorting is presented in virtually any text on data structures and algorithms; in the chapter on strings we examine how to generalize it to sort vectors and strings.

The idea of path compression for the tree implementation of the union-find problem was known for some time before it was completely analyzed; the book of Aho, Hopcroft, and Ullman, upon which the presentation of the union-find algorithm here is based, credits the idea to M. D. McIlroy and R. Morris in the 1960's at Bell Laboratories and the work of A. Tritter, and presents a proof that it is $O(nG(n))$ where $G(n)$ is the inverse of $ESTACK(2,n)$. A proof that it is $O(nA^{-1}(n))$, and that this bound is tight, is due to Tarjan [1975]. A nice presentation of this proof and related concepts appears in the book of Tarjan [1983]. He uses a slightly more aggressive definition than the traditional Ackermann's function for $i,n \geq 1$, which we denote here by $ACKT$:

> **function** $ACKT(i,n)$
>   **if** $i=1$ **then return** 2^i
>   **else if** $n=1$ **then return** $ACKT(i-1,2)$
>   **else return** $ACKT(i-1,ACKT(i,n-1))$
>   **end**

Since Ackermann's function grows so fast, changes in the initial conditions like this only affect the inverse by an additive constant. For example, $ACKT(i,n)$ behaves like $ACK(i+2,n)$. Tarjan also carefully defines the inverse of $ACKT$ with two arguments, $\alpha(i,n)$, and gives a more precise bound of $O(i\alpha(i,n))$ for a sequence of i operations on n elements.

Ackermann's function was presented by W. Ackermann as an example of a function that is not *primitive recursive* (definable with basic arithmetic and simple recursion on one variable). It can be shown that for any primitive recursive function $f(n)$, there exists a fixed value of i such that $f(n)$ is $O(ACK(i,n))$. See the book of Hennie [1977] for a discussion of primitive recursive functions and Ackermann's function, as well as references on the subject. The exercise that gives a definition of Ackerman's function that results in "neat" functions of 2 ($n+2$, $2n$, 2^n, etc.) is from the book of Aho, Hopcroft, and Ullman [1983]. The exercise showing how to compute $ACK(i,n)$, using only $O(i)$ space for the two arrays $I[0]...I[i]$ and $V[0]...V[i]$, is from Teitelbaum [1974]. See also McBeth [1990].

10. Graphs

Introduction

A *graph* $G=(V,E)$ consists of a set V of *vertices* (also called *nodes*) and a set E of *edges*. In a *directed* graph, edges are ordered pairs of vertices and in an *undirected* graph, edges are unordered pairs of vertices.

Example of an undirected graph:

$V = \{1, 2, ..., 10\}$

$E = \{(1,8), (1,9), (2,3), (2,5), (3,4), (3,6), (4,10), (5,6), (6,10), (7,7), (7,10)\}$

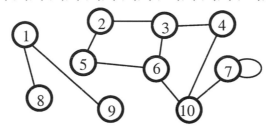

Example of a directed graph:

$V = \{1,2, ..., 10\}$

$E = \{(1,2), (2,3), (2,4), (3,6), (4,1), (4,5), (4,6), (5,4), (6,8), (8,1), (10,7), (10,10)\}$

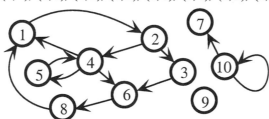

Examples of graphs that we have already seen:

- Lists.
- Trees.

Graph Terms

Adjacent vertex: A vertex w is *adjacent* to a vertex v in a directed or undirected graph if there is an edge (v,w).

Incident edge: The edges *incident* to a vertex v in a directed or undirected graph are those having v as at least one of their end points. In a directed graph, an *outgoing* edge of v is one of the form (v,w) and an *incoming* edge of v is one of the form (w,v).

Degree: In an undirected graph, the *degree* of a vertex v is the number vertices adjacent to v (or the number of edges incident to v). In a directed graph, the *in-degree* of a vertex is its number of incoming edges and the *out-degree* of a vertex is its number of outgoing edges (an edge from a vertex to itself counts in both the in and out degrees).

Path, cycle, self-loop, tour: A path is a sequence of vertices $v_0...v_k$, such that (v_i,v_{i+1}) is an edge, $0 \le i < k$; its *length* is k (the number of edges on the path). A path is a *cycle* if its first and last vertices are the same. A path is *simple* if no vertex is repeated, except possibly for the first vertex being the same as the last (in which case it is a simple cycle). A cycle that is a single edge from a vertex to itself is called a *self-loop*. A cycle is a *tour* if it visits every vertex and a *Hamilton tour* if it visits every vertex exactly once.

Clique: A subset of the vertices in an undirected graph for which there is a edge between every pair.

Subgraph: A graph obtained by deleting some edges and some vertices (if a vertex is deleted, then so are all of its incident edges). When a graph G is a subgraph of a graph H, we say H contains G.

Spanning tree, forest: A subgraph that is a set of disjoint trees and contains all vertices. If the forest is a single tree, it is called a *spanning tree*.

Connected graph: A directed or undirected graph for which it is not possible to partition its vertices into two disjoint subsets such that there are no edges from a vertex in one subset to a vertex in the other. Partitioning a graph into the minimum number of connected subgraphs defines its *connected components*.

Bi-connected graph: A connected undirected graph that is a single edge, or such that for every pair of vertices, there is a simple cycle that contains them. Equivalently, a connected undirected graph that does not contain an *articulation vertex*, a vertex such that its removal (along with its incident edges) yields two or more disconnected non-empty graphs. Partitioning the edges into the minimum number of bi-connected subgraphs defines its *bi-connected components* (we shall see that the partition is unique).

Strongly-connected graph: A directed graph where for every pair of vertices, there is a cycle containing them (i.e., there is a path between any two vertices). Partitioning the vertices into the minimum number of strongly-connected subgraphs defines its *strongly connected components* (we shall see that the partition is unique).

Acyclic graph: A directed or undirected graph that does not have a cycle.

Bipartite graph: A directed or undirected graph such that the vertices can be partitioned into two sets so that every edge has one of its end points in each set.

Isomorphic graphs: Two directed or undirected graphs are *isomorphic* if the vertices of one can be renumbered to make it identical to the other.

Representing Graphs

Idea: A list of all edges (*edge list* representation) suffices to specify a directed or undirected graph, but is not easy to work with for most applications. We usually want additional structure that allows easy access to the edges adjacent to a given vertex.

Adjacency matrix: For a graph with n vertices and m edges, form a n by n matrix, where the entry in row i and column j is 1 if there is an edge from vertex i to vertex j, and 0 otherwise. Edges can be accessed, added, and deleted in $O(1)$ time, but it may be wasteful of space because the space for the matrix is always $O(n^2)$, but m could be anywhere between $O(n)$ and $O(n^2)$ depending on the application. For undirected graphs, except for self-loops, each edge (i,j) appears twice (once in the row for i and once in the row for j) and so the matrix is symmetric about the main diagonal from $A[1,1]$ down to $A[n,n]$.

Adjacency lists: For a graph with n vertices and m edges, form an array $A[1]...A[n]$, where $A[i]$ points to a linked list of all vertices j such that (i,j) is an edge. Uses $O(n+m)$ space (i.e., space linear in the size of the graph) but may increase the time of algorithms that need the ability to access edges in an arbitrary fashion. Like adjacency matrices, except for self-loops, edges in an undirected graph appear twice.

Example: Below the undirected and directed graphs from the introduction to this chapter are shown with their adjacency matrix and list representations to the right:

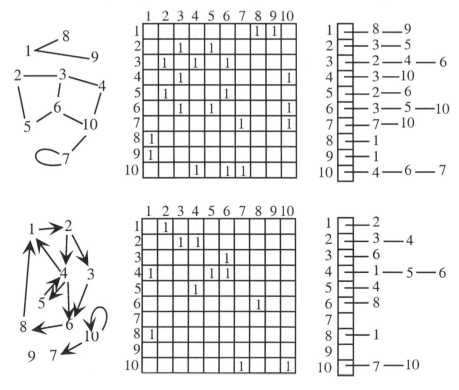

Depth-First Search

Idea: Like pre-order traversal of a tree, visit all vertices reachable from an adjacent vertex before visiting another adjacent vertex. To be sure that no vertex is missed (e.g., when the graph is disconnected), DFS is called on every vertex. To be sure that no vertex is visited twice, DFS begins by checking if the vertex has already been visited.

> **procedure** DFS(v, G):
> > **if** v is not marked **then begin**
> > > {visit v}
> > > Mark v.
> > > **for** each vertex w adjacent to v **do** DFS(w, G)
> > > **end**
> >
> > **end**
>
> Initialize each vertex in the graph G to be "unmarked".
>
> **for** each vertex v **do** DFS(v, G)

Unique ordering: Because the *for* loop does not specify in what order vertices in the adjacency list are visited, different traversals may satisfy the DFS definition. In practice, a particular representation of the graph in memory will give rise to a natural "standard" order. For example, if we are using the adjacency list representation, a unique traversal of the graph results from replacing the *for* loop by a *while* loop that traverses an adjacency list from first to last:

> **procedure** first-to-lastDFS(v, G):
> > **if** v is not marked **then begin**
> > > {visit v}
> > > Mark v.
> > > $w :=$ first vertex in the adjacency list for v
> > > **while** (w is not *nil*) **do begin** first-to-lastDFS(w, G); $w :=$ NEXT(w) **end**
> > > **end**
> >
> > **end**

Example: Assume that the adjacency lists are ordered so that lower numbered vertices come before higher numbered ones. Then depth-first search of the following graph visits the vertices in the order 1, 2, 5, 4, 6, 7, 8, 9, 10, 11, 12, 3:

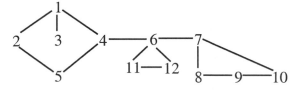

Note: Since with both adjacency matrix and adjacency list representations the vertices are in an array, it is trivial to visit them all; the real advantage of DFS is that all edges are traversed as well; this "exploration" of the graph will be useful for many applications.

Breadth-First Search

Idea: Like level-order traversal of a tree, visit all adjacent vertices before going deeper:

> **procedure** BFS(v, G)
>> Initialize a queue to contain v.
>> **while** queue is not empty **do begin**
>>> DEQUEUE a vertex v
>>> **if** (v is unmarked) **then begin**
>>>> {visit v}
>>>> Mark v.
>>>> **for** each vertex w adjacent to v **do** ENQUEUE(w)
>>>> **end**
>>> **end**
>> **end**

Initialize each vertex in the graph G to be "unmarked".
for each vertex v **do** BFS(v, G)

Unique ordering: Like DFS, because the *for* loop does not specify in what order adjacent vertices are visited, different traversals may satisfy the BFS definition. Again, if we are using the adjacency list representation, a natural unique traversal of the tree results from replacing the *for* loop by a *while* loop that traverses the list from first to last:

> **procedure** first-to-lastBFS(v, G)
>> Initialize a queue to contain v.
>> **while** queue is not empty **do begin**
>>> DEQUEUE a vertex v
>>> **if** (v is unmarked) **then begin**
>>>> {visit v}
>>>> Mark v.
>>>> w := first vertex in the adjacency list for v
>>>> **while** (w is not *nil*) **do begin** ENQUEUE(w); w := NEXT(w) **end**
>>>> **end**
>>> **end**
>> **end**

Example: Consider again the graph in the example on the previous page for DFS and again assume that the adjacency lists are ordered so that lower numbered vertices come before higher numbered ones. Breadth-first search visits the vertices of this graph in the order 1, 2, 3, 4, 5, 6, 7, 11, 12, 8, 10, 9:

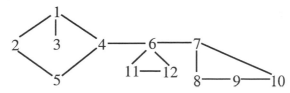

Depth-First Spanning Trees

Idea: As we perform DFS, we can remove the edges encountered and place them in one of two sets:

solid edges: Edges that led to an unmarked vertex.

dashed edges: Edges that led to a marked vertex.

The solid edges form a *DFS spanning forest* for the graph.

Note: The DFS spanning forest for a given graph is not unique; it depends on which vertex is chosen first, the order of the adjacency lists, etc.

Undirected graphs: All dashed edges must go from a vertex to one of its ancestors in the DFS spanning forest, and hence we refer to dashed edges as *back edges*. An undirected graph has a cycle if and only if DFS produces at least one back edge.

Directed graphs: The dashed edges are one of three types:

back edges: From a vertex to an ancestor in the DFS spanning tree.

forward edges: From a vertex to an descendant in the DFS spanning tree.

cross edges: From a vertex to one that was visited earlier in the DFS, but is not a descendant or ancestor.

Depth-first numbering: When constructing a depth-first spanning forest, it will often be convenient to assume that instead of just marking vertices when they are visited, we number them with a variable that is incremented each time a vertex is visited; that is, for a graph of n vertices, the first vertex to be visited is numbered 1 and the last vertex to be visited is numbered n. We refer to this numbering as a *depth-first numbering (DF numbering)*.

Breadth-first spanning trees: Like DFS, we partition edges into tree edges and dashed edges. In an undirected graph, only cross edges are possible. For directed graphs, back edges are also possible. A breadth-first numbering can also be defined. See the exercises.

Bi-Connected and Strongly-Connected Components

Bi-connected components: Recall that *bi-connected* graphs are connected undirected graphs that are a single edge, or such that for every pair of vertices, there exists a simple cycle that contains them. Equivalently, they are connected undirected graphs that do not contain an *articulation vertex*, a vertex such that its removal (along with all of its incident edges) results in two or more disconnected non-empty graphs. A connected undirected graph may be uniquely partitioned into a minimal number of *bi-connected components*, by using the equivalence relation that two edges are in the same component if and only if they lie on a common simple cycle (see the exercises). In the figure below, the bi-connected components are enclosed with the dashed lines, and the articulation vertices are shown as dark squares.

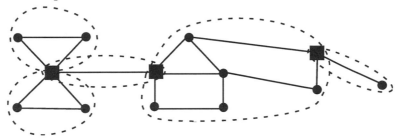

Strongly-connected components: Recall that a directed graph is *strongly-connected* if for every pair of vertices, there exists a cycle that contains them, or equivalently, there is a path between any two vertices. A directed graph may be uniquely partitioned into a minimal number of *strongly-connected components* by using the equivalence relation that two vertices are in the same component if and only if they lie on a common cycle. In the figure below, the strongly-connected components are enclosed with the dashed lines.

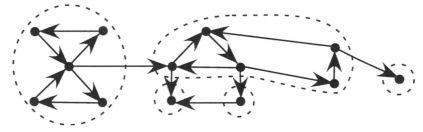

Using DFS to find components: As presented on the following two pages, edges (bi-connected components) or vertices (strongly-connected components) can be pushed on a stack as we go down and popped when returning from a vertex *v*. With strongly-connected components, we pop vertices that can't reach higher than *v*. With bi-connected components, we pop edges when we find a child *w* of *v* that can't reach higher than *v* (if all vertices below *w* have been popped, we pop only the single edge (*v*,*w*)).

Bi-Connected Components of an Undirected Graph

Use the basic depth-first search algorithm on an undirected graph G to store a depth-first numbering in the array DFN, and modify it to compute:

$bcLOW(v) =$ The vertex w of lowest depth-first number such that there is a back-edge from a (not necessarily proper) descendant of v to w.

To compute $bcLOW(v)$ as we perform the DFS, observe that $bcLOW[v]$ is the minimum of the following three quantities (see the exercises):

$DFN[v]$

$bcLOW[c]$ for all children c of v

$DFN[w]$ for all back edges (v,w); i.e., w has been visited but is not the parent of v

Because a single edge can be a bi-connected component, we check for a component when we are at a vertex v and discover a child w such that $bcLOW(w) \geq DFN(w)$.

> **procedure** bcDFS(v,G)
> $count := count+1$
> $bcLOW[v] := DFN[v] := count$
>   **for** each vertex w that is adjacent to v **do begin**
>   **if** w has not been visited (i.e., $DFN[w]=0$) **then begin**
> $PARENTDFN[w] := DFN[v]$
>   Push the edge (v,w) onto S.
>   bcDFS(w)
> $bcLOW[v] := \text{MINIMUM}\{bcLOW[v],bcLOW[w]\}$
>   **if** $bcLOW(w) \geq DFN[v]$ **then**
>   Output a component by popping all edges up to and including (v,w).
>   **end**
>   **else if** $DFN(w) \neq PARENTDFN[v]$ **then** $bcLOW[v] := \text{MIN}\{bcLOW[v],DFN[w]\}$
>   **end**
>   **end**
> $count := 0$
> **for** each vertex v **do** $bcLOW[v] := PARENTDFN[v] := DFN[v] := 0$
> Initialize an empty stack S.
> **for** each vertex v **do if** $DFNUM[v]=0$ **then** bcDFS(v,G)

Complexity: For an undirected graph of n vertices and m edges, the only addition that has been made to the $O(n+m)$ DFS algorithm that is not $O(1)$ time is the popping of edges from the stack when a bi-connected component is discovered. Like the analysis of the $POP(i)$ operation in the chapter on lists, popping an edge can be "charged" to the operation that pushed it there, and so the total cost of popping edges is $O(m)$. Hence, the entire algorithm is $O(n+m)$ time and space.

Strongly-Connected Components of a Directed Graph

Similar to bi-connected components, we store a depth-first numbering in the array DFN and push vertices onto a stack as they are encountered (as opposed to pushing edges for bi-connected components). Roots of subtrees that correspond to strongly-connected components play the analogous role to articulation vertices. When discovering one upon returning from a depth-first search of a vertex, we have discovered a strongly-connected component that can be popped off the stack. Similar bi-connected components, define:

$scLOW(v) =$ The vertex w of lowest depth-first number such that there is a cross or back edge from a (not necessarily proper) descendant of v to w, and w is in the strongly connected component that contains v.

In general, a vertex x with a back-edge that goes to a vertex w numbered lower than v in the same component may be reached from v by traversing a sequence of edges that includes cross and back edges in addition to tree edges. Because vertices numbered lower than a vertex x that are in the same component will already be on the stack, when a new dashed edge (x,w) is discovered, we simply check if it is in the stack using a bit vector $INSTACK$.

> **procedure** scDFS(v,G)
>> $count := count+1$
>> $scLOW[v] := DFN[v] := count$
>> Push v onto S and set INSTACK[v]:=1.
>>
>> **for** each vertex w that is adjacent to v **do begin**
>>> **if** w has not been visited (i.e., $DFN[w]=0$) **then begin**
>>>> scDFS(w)
>>>> $scLOW[v] := \text{MINIMUM}\{scLOW[v],scLOW[w]\}$
>>> **end**
>>>
>>> **else if** $DFN[w]<DFN[v]$ and INSTACK[w]=1 **then**
>>>> $scLOW[v] := \text{MINIMUM}\{scLOW[v],DFN[w]\}$
>>
>> **end**
>>
>> **if** $scLOW(v)=DFN[v]$ **then**
>>> Output a component by popping all vertices up to and including v, and for each vertex w that is popped set INSTACK[w]:=0.
>
> **end**
> $count := 0$
> **for** each vertex v **do** INSTACK[v] := $scLOW[v]$:= $DFNUM[v]$:= 0
> Initialize an empty stack S.
> **for** each vertex v **do if** $DFNUM[v]=0$ **then** scDFS(v,G)

Complexity: For a directed graph of n vertices and m edges, time and space is $O(n+m)$ by an argument virtually identical to that for the bi-connected components algorithm.

Minimum Weight Spanning Trees

Definition: Suppose weights are associated with each edge of a connected undirected graph G of n vertices and m edges (since G is connected, $m \geq n-1$). The *weight* of a spanning tree T for G is the sum of the weights of its edges. T is a *minimum spanning tree* for G if there is no other spanning tree for G of lower weight.

Prim's algorithm:

> $T :=$ an arbitrary vertex of G
> **while** T has less than $n-1$ edges **do**
> > Add to T the lowest cost edge (v,w) such that v is in T and w is not in T.

Kruskal's algorithm:

> $F :=$ a forest of n one-vertex trees (one for each vertex of G)
> **while** F has less than $n-1$ edges **do**
> > Add to F the lowest cost edge (v,w) such that v and w are in different trees.

Proof of Correctness of Prim and Kruskal Algorithms

Lemma: Let T be any tree in a spanning forest F, and let e be a minimum cost edge that connects T to a different tree in F. If there exists a minimum spanning tree that contains F, then there exists a minimum spanning tree that contains both F and e.

Proof: Suppose the contrary; that is, that there exists a minimum spanning tree U that contains F but there does not exist a minimum spanning tree that contains both F and e:

> The edges of T are in U (since U contains F), and hence there is an edge f in U that connects T to the remainder of U.

> By definition of e, $cost(e) \leq cost(f)$, and we can replace f by e in U to obtain a new spanning tree V.

> We now have a contradiction: Either V is minimal (and it contains both F and e) or it is not, in which case U cannot be either (since $cost(e) \leq cost(f)$ implies $cost(V) \leq cost(U)$).

The correctness of both algorithms now follows by induction on the number of iterations of the *while* loop (the current number of edges i in the spanning forest being built), since both algorithms can be viewed as starting with a forest of single vertex trees and adding one edge at a time until the forest has only a single tree:

> For $i=0$, the forest is a set of single vertex trees and is contained by any minimal weight spanning tree.

> For $0 < i \leq m$, assume the $i-1$ edges are contained by a minimal weight spanning tree. Then so must the i edges, since the edge added by the ith iteration satisfies the lemma.

Implementation of Prim's Algorithm

We use a heap to store the edges of G, so that they can be examined in order of increasing cost. Initially, the edges adjacent to the first vertex chosen are placed in the heap. Each time a new vertex is visited, all of its adjacent edges are added to the heap.

$T :=$ an arbitrary vertex r of G

$S := \{r\}$

$H :=$ a heap containing all edges incident to r

while S contains less than n vertices and H is not empty **do begin**

    $(u,v) :=$ DELETEMIN(H)

    if v is not in S **then begin**

        Add (u,v) to T.

        Add v to S.

        for each edge (v,w) such that w is not in S **do** INSERT$((v,w),H)$

        end

    end

if S contains less than n vertices **then** *ERROR — G is not connected*

Output T.

Time: $O(m\log(n))$

Any standard linear space representation for T (e.g., adjacency list) for which edges can be added in $O(1)$ time suffices. S can be represented with a bit-vector $S[1]...S[n]$ where $S[i]$ is 1 if the ith vertex is in S and 0 otherwise. So that the *while* loop can easily test if S contains n vertices, a counter can be initialized to 1 and then incremented each time a vertex is added to S.

If we exclude the time spent to insert items into H, $O(\log(m))$ time is used by each iteration of the *while* loop to perform a DELETEMIN and $O(1)$ additional work, for a total of $O(m\log(m))$ time.

H can be initialized in $O(m)$ time with the linear time heap construction algorithm. The total time spent by all executions of the *for* loop is $O(m\log(m))$, since each edge can be inserted into H at most two times (an edge (v, w) appears once on the adjacency list for v and once on the adjacency list for w).

Since $\log(m) \leq \log(n^2) = 2\log(n) = O(\log(n))$ the time is $O(m\log(n))$.

Space: $O(m)$

S uses $O(n)$ space, T uses $O(m)$ space, and H uses $O(m)$ space, for a total of $O(n+m)$ space, which is $O(m)$ if the graph is connected.

Implementation of Kruskal's Algorithm

We again employ a heap to store the edges of G, so that they can be examined in order of increasing cost. Each time a new edge is examined, a UNION-FIND data structure is used to determine whether its endpoints are in different trees.

> $T :=$ an empty tree
>
> $S := \{\{v\}: v \text{ is in } G\}$
>
> $H :=$ a heap containing the edges of G
>
> **while** S contains more than one set and H is not empty **do begin**
>> $(u,v) :=$ DELETEMIN(H)
>>
>> $A :=$ FIND(u)
>>
>> $B :=$ FIND(v)
>>
>> **if** $(A{\neq}B)$ **then begin**
>>> Add (u,v) to T.
>>> UNION(A,B)
>>> **end**
>>
>> **end**
>
> **if** S contains more than one set **then** *ERROR — G is not connected*
>
> Output T.

Time: $O(m\log(n))$

Any standard linear space representation for T (e.g., adjacency list) for which edges can be added in $O(1)$ time suffices.

H can be initialized in $O(m)$ time with the linear time heap construction algorithm.

Like Prim's algorithm, the while loop performs at most m DELETEMIN heap operations for a total of $O(m\log(m)) = O(m\log(n))$ time.

The UNION and FIND operations can be implemented with the fast UNION-FIND data structure in time $O(mA^{-1}(n))$, which is asymptotically smaller than $O(m\log(n))$.

Hence, the total time is $O(m\log(n))$.

Space: $O(m)$

Like Prim's algorithm, S uses $O(n)$ space, T uses $O(m)$ space, and H uses $O(m)$ space, for a total of $O(n+m)$ space, which is $O(m)$ if the graph is connected.

Topological Sort of a Directed Graph

Problem: Given a directed graph G of n vertices and m edges, either determine that a cycle exists or label each vertex so that if there is an edge from v to w then $label(v)>label(w)$.

Example: In a mathematics textbook, number the theorems so that a theorem only depends on theorems with lower numbers. In a dictionary, check for circular definitions (A refers to B refers to C refers to A, etc.).

Method 1: Do depth-first search and number a vertex when it is exited. A cycle exists if and only if there are back edges.

Method 2: Use a modified breadth-first search. Work from the highest numbered vertex (that has no incoming edges) down to the lowest numbered vertex (that has no outgoing edges). All vertices are initialized with a count equal to the number of incoming edges to that vertex. The queue is initialized to contain all vertices with count 0 (no incoming edges). When a vertex is removed from the queue, the counts of all of its adjacent vertices are decremented, and those whose counts become 0 are added to the queue.

Set counts of vertices to the number of incoming edges:
  for each vertex v **do** $count[v] := 0$
  for each edge (v,w) **do** increment $count[w]$

Initialize a queue with all vertices with no incoming edges:
  for each vertex v **do if** $count[v]=0$ **then** ENQUEUE(v)

Process the queue:
  $index := n$
  while *queue* is not empty **do begin**
      DEQUEUE a vertex v
      $label[v] := index$
      $index := index-1$
      for each vertex w adjacent to v **do begin**
          $count[w] := count[w]-1$
          if $count[w]=0$ **then** ENQUEUE(w)
          end
      end

A cycle exists if not all vertices have been labeled:
  if $index>0$ **then** {a cycle exists}

Complexity: Both methods are $O(n+m)$ time and space. Method 1 is a minor variation of standard depth-first search. Although Method 2 modifies standard breadth-first search by incorporating the counts, its behavior is essentially the same; initialization of the counts is $O(n)$ for the first *for* loop and $O(m)$ for the second *for* loop, initialization of the queue is $O(n)$, and the *while* loop is $O(m)$ since it visits each edge at most once.

Euler Paths

Definition: An *Euler path* for a directed or undirected graph visits each edge once.

Euler path theorem for undirected graphs: An undirected connected graph G has an Euler path if and only if it is connected and has either two vertices of odd degree or no vertices of odd degree (in which case it has an Euler cycle).

Proof: If G is not connected, or has more than two odd-degree vertices, or exactly one odd-degree vertex, it cannot have an Euler path (since, except for its endpoints, the path consumes two incident edges each time it passes through a vertex). Conversely, let P be a path between two odd-degree vertices s and t of G and let H be the subgraph of G consisting of the edges not in P (or let P be a single arbitrary vertex $s=t$ if G has no odd-degree vertices). Since P goes between s and t, H cannot have an odd-degree vertex (i.e., s and t might be in H, but since the edges of P are not, s and t have even degree in H). We can now repeatedly delete cycles from H and *splice* them into P until P contains all edges of G (and H is empty). A splice operation starts at a vertex v of P that has an incident edge not in P, and deletes edges from H until it gets stuck at a vertex w. Since all vertices of H have even degree, it must be that $w=v$, and P can be modified to follow the edges of this cycle the first time it visits v.

Euler path algorithm for undirected graphs: For a graph G of n vertices and m edges, we follow the construction of the Euler path theorem, although the initialization of P can be eliminated (see the exercises).

Verify that G is connected and has at most two odd-degree vertices.

Initialize P to be a path between two odd-degree vertices s and t (or let P be any single vertex if there are no odd-degree vertices).

Initialize a queue to contain the vertices of P.

while queue is not empty **do begin**
    Dequeue a vertex v.
    $C :=$ empty cycle
    while v has a non-empty adjacency list **do begin**
        Delete an edge (v,w) from G and add it to C.
        $v := w$
    end
    Splice C into P.
end

Complexity: $O(n+m)$ time and space. A simple pass over G can verify that it is connected and has at most two odd-degree vertices. Initialization of P can be done with a depth-first search starting at s. The *while* loop is $O(m)$ since each edge is visited once.

Directed graphs: The same algorithm works if initialization of P is changed to a path from a vertex s with out-degree one greater than in-degree to a vertex t with in-degree one greater than out-degree (see the exercises).

Single-Source Minimum Cost Paths

Problem: Given a directed graph of n vertices and m edges, a *source* vertex s, and for each edge (v,w) a non-negative *cost* $c(v,w)$, defined to be ∞ if the edge does not exist, label each vertex v with $D[v]$, the minimum cost (sum of costs of its edges) of a path from s to v. We often refer to $D[v]$ as the *distance* from the source to v.

Dijkstra's Algorithm

> Initialize $D[s] := 0$ and $D[v] := c[s,v]$ for all other vertices.
> Initialize all vertices to be unmarked.
> **while** all vertices are not marked **do begin**
> > $v :=$ an unmarked vertex for which $D[v]$ is minimum
> > Label v "marked".
> > **for** each unmarked vertex w that is adjacent to v **do**
> > > $D[w] := \text{MINIMUM}\{D[w], D[v]+c(v,w)\}$
> >
> > **end**

Note: Store a back-pointer field with each vertex that points to a vertex on a minimum cost path back to s (so that the minimum cost path from any vertex v to s can be output in time proportional to its length).

Correctness of Dijkstra's algorithm (induction on the number of marked vertices):

(*hypothesis*)

> For marked vertices, $D[v]$ is the length of a shortest path; otherwise, $D[v]$ is the length of a shortest path for which all vertices except v are marked.

(*basis*)

> Correctness when there is only one marked vertex follows from the initialization.

(*inductive step*)

> Assume the hypothesis true for $k \geq 1$ marked vertices and consider when the $k+1^{th}$ vertex v is marked.

> Suppose that there is a path P from s to v such that $cost(P)<D[v]$, where $cost(P)$ denotes the sum of the costs of the edges of P. Since by the inductive hypothesis $D[v]$ is the length of a shortest path from s to v for which all vertices except v are marked, P must contain at least one unmarked vertex besides v; let w be the first unmarked vertex encountered when traversing P from s to v, and let Q be the portion of P going from s to w. By the inductive hypothesis, $D[w] \leq cost(Q)$, and since $cost(Q) \leq cost(P)$ $< D[v]$, it follows that w would have been selected instead of v to be marked (so it is not possible that P exists). Hence, the inductive hypothesis for marked vertices has been maintained.

> Now consider the unmarked vertices. Since v is the only additional vertex to be marked at this step, and the *for* loop checks for each vertex adjacent to v if it is shorter to go through v, the inductive hypothesis for unmarked vertices is maintained.

Adjacency Matrix Implementation of Dijkstra's Algorithm

Use of an adjacency matrix is straightforward. Each iteration of the *while* loop can in $O(n)$ time find the v for which $D[v]$ is minimum and then in $O(n)$ time the *for* loop can sequence through a row of the adjacency matrix.

Time: $O(n^2)$

Space: $O(n^2)$

Adjacency List Implementation of Dijkstra's Algorithm

When m is relatively small compared to n^2, adjacency lists provide the basis for an algorithm that is linear or near linear in the number of edges.

Time:

Keep all of the unmarked vertices in a dictionary with the operations INSERT, DELETEMIN, and DECREASE (make an item smaller and fix up the data structure).

Initialization takes $O(n)$ time to initialize the array D, $O(n)$ time to mark all vertices, and $O(n\log(n))$ time to INSERT all vertices into the dictionary.

Each of the n iterations of the *while* loop, excluding the time needed for the inner *for* loop, is dominated by the DELETEMIN operation used to obtain v.

The inner for loop is executed at most once for each edge, for a total of $O(m)$ DECREASE operations over all iterations of the *while* loop (that is, constant time to compute $D[v]$ plus the DECREASE operation to adjust v's position in the dictionary).

By using a standard balanced tree data structure, the INSERT, DELETEMIN, and DECREASE operations are $O(\log(n))$ time (a DECREASE can be implemented with a DELETE and an INSERT); yielding a time of $O(m\log(n))$.

With more complex heap data structures (e.g., "Fibonacci heaps"), the amortized time for all of the DECREASE operations can be reduced to $O(m)$, yielding a $O(n\log(n)+m)$ time (see the chapter notes).

Space:

$O(n+m)$ for the adjacency list representation of the graph.

All Pairs Minimum Cost Paths

Problem: Given a graph G of n vertices where associated with each edge (i,j) is a non-negative cost $c(i,j)$, compute a n by n matrix A where $A[i,j]$ is the cost of a minimum length path from vertex i to vertex j (or $+\infty$ if there is no edge from i to j).

Idea: Use dynamic programming to compute least-cost paths with intermediate vertices numbered $\leq k$ from least-cost paths with intermediate vertices numbered $<k$.

Floyd's Algorithm for Shortest Paths

> **for** $1 \leq i,j \leq n$ **do if** there is an edge from i to j **then** $A[i,j] := c(i,j)$ **else** $A[i,j] := +\infty$
> **for** $i := 1$ **to** n **do** $A[i,i] := 0$
> **for** $k := 1$ **to** n **do**
> > **for** $1 \leq i,j \leq n$ **do**
> > > $A[i,j] := \text{MINIMUM}(A[i,j], A[i,k]+A[k,j])$

Time: $O(n^3)$

The initialization of A by the first *for* loop is $O(n^2)$. The setting of $A[i,i] := 0$ by the second for loop is $O(n)$. For each of the n iterations of the third *for* loop, the inner *for* loop is $O(n^2)$, for a total of $O(n^3)$ time.

Space: $O(1)$ in addition to the $O(n^2)$ used by the input.

If the graph is provided in matrix form (i.e., the matrix A is the input), then the first *for* loop that initializes A can be eliminated and the space used by the algorithm is $O(1)$ in addition to the space used by A.

Correctness of Floyd's algorithm (by induction on k):

(*hypothesis*)
After the kth iteration, $A[i,j]$ is the cost of the shortest path from vertex i to vertex j using intermediate vertices $\leq k$.

(*basis*)
For $k=0$, the initialization correctly sets $A[i,j]$ to $c(i,j)$ or 0 if $i=j$.

(*inductive step*)
At the start of the kth iteration of the outer loop, assume that $A[i,j]$ is the cost of the minimum length path from i to j using vertices numbered $<k$. Then the shortest path from i to j using vertices numbered $\leq k$ either does not use k or goes to k (using vertices numbered less than k) and then to j; this is exactly reflected by the MIN statement.

Transitive Closure

Definition: The *transitive closure* of a directed or undirected graph G is a directed graph that has the same vertices as G and an edge (v,w) if and only if there is a path from v to w in G.

Problem: Given a graph G of n vertices, compute a n by n matrix A that is the adjacency matrix for the transitive closure of G.

Warshall's Algorithm for Transitive Closure

Similar to the all-pairs shortest path problem, use dynamic programming to compute the set of all paths using vertices numbered $\leq k$ as intermediate vertices from the paths that only use vertices numbered $<k$ as intermediate vertices (if there is an edge from i to k and an edge from k to j, then $A[i,k]*A[k,j]=1$):

> **for** $1 \leq i,j \leq n$ **do if** there is an edge from i to j **then** $A[i,j] := 1$ **else** $A[i,j] := 0$
> **for** $i := 1$ **to** n **do** $A[i,i] := 1$
> **for** $k := 1$ **to** n **do**
> > **for** $1 \leq i,j \leq n$ **do**
> > > $A[i,j] := \text{MAXIMUM}(A[i,j], A[i,k]*A[k,j])$

Time: $O(n^3)$

> Analysis is identical to Floyd's Algorithm.

Space: $O(1)$ in addition to the $O(n^2)$ used by the input.

> Analysis is identical to Floyd's Algorithm.

Correctness of Warshall's algorithm:

> Similar to Floyd's Algorithm.

Generic Path Finding Framework

Idea: By changing initialization from 0 to 1 and operations from *MINIMUM* and addition to *MAXIMUM* and multiplication, Floyd's becomes Warshall's algorithm. Here, we present a framework that includes both the Floyd and Warshall algorithms as well as problems that may need self-loops and cycles in paths.

Generic framework: Initially we only have paths between two vertices i and j that are a single edge from i to j. Then we work our way up from $k=1$ to n, where at each stage, we know about all paths from i to j that use intermediate vertices numbered $<k$, and we must choose between not using vertex k (the current value of $A[i,j]$) or going from i to k using vertices numbered $<k$, possibly milling around and coming back to k one or more times using vertices numbered $<k$ (not needed with the Floyd and Warshall algorithms), and then going from k to j using vertices numbered $<k$.

> **for** $1 \le i,j \le n$ **do if** there is an edge from i to j **then** $A[i,j] := c(i,j)$ **else** $A[i,j] :=$ UNDEF
>
> **for** $i := 1$ **to** n **do** $A[i,i] :=$ CHOOSE($A[i,i]$, INIT)
>
> **for** $k := 1$ **to** n **do**
>> **for** $1 \le i,j \le n$ **do**
>>> $A[i,j] :=$ CHOOSE($A[i,j]$, COMBINE($A[i,k]$,CYCLE($A[k,k]$),$A[k,j]$))

Example, Floyd's algorithm:

>   UNDEF $= +\infty$
>   INIT $= 0$
>   CYCLE(x) $= 0$
>   COMBINE(x,y,z) $= x+y+z$
>   CHOOSE(x,y) $=$ MINIMUM(x,y)

Example, Warshall's algorithm:

>   UNDEF $= 0$
>   INIT $= 1$
>   CYCLE(x) $= 1$
>   COMBINE(x,y,z) $= xyz$
>   CHOOSE(x,y) $=$ MAXIMUM(x,y)

Example, conversion of a pattern diagram to a regular expression: Floyd's and Warshall's algorithms do not make use of the full generality of the framework. In the chapter on strings we shall see how a pattern such as "an *a* followed by any number of *b*'s and *c*'s followed by a *d*" can be represented as a graph or as an expression like $a(b+c)*d$. To convert a graph to an expression, we will need to use paths that include cycles (see the exercises to that chapter).

Conditions under which the general framework is correct: Algebraic conditions can characterize the values of *UNDEF*, *INIT*, *CHOOSE*, *COMBINE*, and *CYCLE* that can be used to correctly "sum" path "labels" (see the exercises).

Maximum Flow

A *transport network* is a connected directed graph G such that:
- G contains a vertex s called the *source* that has no incoming edges.
- G contains a vertex t called the *sink* that has no outgoing edges.
- Associated with each edge (v,w) is a capacity $C(v,w) \geq 0$ (when there is no edge from v to w, $C(v,w)$ is defined to be 0).

A *flow* in a transport network is an assignment $F(v,w) \geq 0$ to each edge (v,w) such that:
1. $F(v,w) \leq C(v,w)$
2. For all vertices v that are not the source or sink, the flow into v is equal to the flow out of v; that is:

$$\sum_{w \in G} F(v,w) = \sum_{w \in G} F(w,v)$$

The *value* of a flow is the flow into the sink, which by Conditions 1 and 2 is the same as the flow out of the source; that is:

$$value \text{ of the flow } F = \sum_{v \in G} F(s,v) = \sum_{v \in G} F(v,t)$$

A flow is *maximum* if there is no flow with a greater value.

A *cut* in a transport network G is a set of edges X whose removal divides G into two graphs G_s and G_t such that G_s contains s and G_t contains t. An edge (v,w) such that v is in G_s and w is in G_t is called a forward edge of X; the other edges of X are its backward edges. The *capacity* of X is the sum of the capacities of all of its forward edges.

Maximum flow theorem: If we think of flow as water going through pipes, after picking any cut, the water can only get from s to t by flowing through one of the forward edges in the cut. That is, given any cut X that divides G into two graphs G_s and G_t, since for all vertices but s and t the flow in must be the same as the flow out, the net flow out of G_s is at most the sum of the flow along the forward edges of X (since flow on backward edges, which must be ≥ 0, can only reduce the net flow from G_s to G_t). Hence:

The value of any flow is at most the capacity of any cut.

Example: Below, edges are labeled with their capacities and a flow assignment in parentheses. Not minimal cuts include $\{(s,a),(s,b)\}$ and $\{(a,c),(c,b),(b,d)\}$. The edges (s,b), (a,b), (c,b), (c,d), and (c,t) form a cut of capacity 14, equal to the flow from s to t.

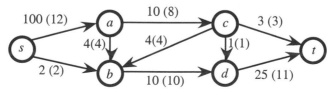

Undirected Paths

A sequence of edges P is an *undirected path* from v to w in G if P is a path from v to w in G or it is possible to change the direction of some or all of the edges in P to obtain a path from v to w in G. When traversing P from v to w in G, if an edge (x,y) is traversed by first visiting x and then y, then it is called a *forward* edge in P (it points in the direction of w); otherwise it is called a *backward* edge in P (it points in the direction of v). For example, in the graph presented on the previous page, the sequence of edges (s,b), (a,b), (a,c), (c,t) is an undirected path from s to b to a to c to t, where (a,b) is a backward edge.

Augmenting Paths

An undirected path P from v to w in a transport network G is an *augmenting path* from v to w with respect to a flow F if:

1. For each forward edge (v,w) in P, $F(v,w)<C(v,w)$.

2. For each backward edge (v,w) in P, $F(v,w)>0$.

The *residual capacity* of an augmenting path P from s to t with respect to a flow F is the minimum of the flow on any backward edge of P, and for each forward edge, the difference between its capacity and its flow.

Key Observation: Given an augmenting path P from s to t with residual capacity z, the flow from s to t can be increased by subtracting z from each backward edge on P and adding z to each forward edge on P.

Augmenting Path Theorem for Flow

A flow in a transport network is maximum if and only if there is no augmenting path from the source to the sink.

Proof: We have already observed that if there is an augmenting path P from s to t, then the flow from s to t can be increased by the residual capacity of P. For the converse, suppose that F is a flow for which there is no augmenting path from s to t. Let S be the set of all vertices v, including s, for which there is an augmenting path from s to v. Since S cannot include t (or there would be an augmenting path from s to t), the set X of all edges with one endpoint in S and one endpoint not in S forms a cut. Furthermore, by the definition of S, it must be that for all forward edges (v,w) in X, $F(v,w)=C(v,w)$, and for all backward edges in X, $F(v,w)=0$ (if not, there must be an augmenting path from S to w, and w should have been in S). Hence, the value of F is equal to the capacity of X, and by the maximum flow theorem, F is maximum.

Max-Flow = Min-Cut Theorem

The value of a maximum flow in a transport network is equal to the minimum possible capacity of a cut.

Proof: Let F be a maximum flow. By the proof of the augmenting path theorem that we can construct a cut X with a capacity equal to the value of F. Since by the maximum flow theorem, the value of F is $\leq$ the capacity of any cut, X must be a cut of minimum capacity, and the theorem follows.

Generic Approach to Computing Maximum Flow

Idea: Using a procedure "UPFLOW", repeatedly increase the flow along an augmenting path in a transport network until there are no augmenting paths from s to t.

    for each edge (v,w) of G do $F(v,w) := 0$

    while there exists an augmenting path P **do** UPFLOW(P)

Increasing flow along an augmenting path: The following procedure assumes that its argument P is an augmenting path from s to t, and increases the flow along P.

    procedure UPFLOW(P)

        $z := 0$

        for each edge e on P **do**
            if e is a forward edge (v,w) **then** $z := \text{MINIMUM}\{z, C(v,w) - F(v,w)\}$
            else if e is a backward edge (w,v) **then** $z := \text{MINIMUM}\{z, C(w,v)\}$

        for each edge e on P **do**
            if e is a forward edge (v,w) **then** $F(v,w) := C(v,w) + z$
            else if e is a backward edge (w,v) **then** $F(w,v) := C(w,v) - z$

    end

A worst-case example: Consider the following graph:

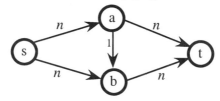

Starting with a flow of 0, if we augment with the path (s,a,t) and then with the path (s,b,t), we find the maximum flow of $2n$ in only 2 steps. However, by alternately choosing the paths (s,a,b,t) and (s,b,a,t), it takes $2n$ steps. Since the size of the input is constant in addition to the $O(\log(n))$ bits used to represent n, the worst case running time uses an exponential number of steps as a function of the length of the input.

Idea: In the example above, choosing augmenting paths in order of increasing residual capacity was highly inefficient. Although it is a reasonable idea is to choose paths in order of decreasing residual capacity, it is computationally problematic (see the exercises). Instead, we will organize the generic algorithm to process paths from s to t in order of increasing path *length*; that is, in order of increasing number of edges on the path (regardless of the residual capacity of the path). We first present a very simple implementation of this idea (the Edmonds–Karp algorithm) that has a relatively poor worst case running time (but may work well in practice), and then "refine" this idea to achieve better asymptotic performance.

Edmonds–Karp Algorithm

Idea: Repeatedly find an augmenting path of non-decreasing length; do this by performing a breadth-first search from the source and placing on each vertex w that is encountered a label of the form $<c, v, +/->$ that indicates:

c: There is an augmenting path from s to w with capacity c.

w: The vertex before w on this path is v.

$+/-$: $+$ indicates that the last edge is a forward edge (it goes from v to w); $-$ indicates that the last edge is a backward edge (it goes from w to v).

The Edmonds–Karp maximum flow algorithm:

> **for** each edge (v,w) **do** $F(v,w) := 0$
> **repeat**
>> Initialize $z := 0$ and all vertices to be not labeled.
>> Enqueue s.
>> **while** (the queue is not empty) and (t is not labeled) **do begin**
>>> Dequeue a vertex v.
>>> For each edge (v,w) such that w is not labeled and $F(v,w)<C(v,w)$ **do**
>>>> Enqueue w with label $<\text{MINIMUM}\{z,C(v,w)-F(v,w)\},v,+>$
>>> For each edge (w,v) such that w is not labeled and $F(w,v)>0$ **do**
>>>> Enqueue w with label $<\text{MINIMUM}\{z,F(v,w)\},v,->$
>> **end**
>> **if** t is labeled **then**
>>> Perform the UPFLOW operation on the new augmenting path obtained by following second components of labels back to s.
>> **until** t is not labeled

Complexity: Given a transport network G of n vertices and m edges, we use a modified adjacency list representation that stores for each vertex a list of all edges that start or end at that vertex (so (v,w) is stored on the list for v and the list for w); the values of C and F are stored with one of the copies of each edge. Hence the space is $O(m)$. Each iteration of the repeat loop uses $O(m)$ time for the *while* loop that performs the breadth-first search, and $O(m)$ time to perform the UPFLOW operation. In practice, we do not need to actually construct the path and call UPFLOW; instead, we simply trace back using the labels:

> $w := t$
> **while** $w \neq s$ **do**
>> $v :=$ the second component of the label of v
>> **if** the third component of the label of v is $+$
>>> **then** $F(v,w) := C(v,w)+z$
>>> **else** $F(w,v) := C(w,v)-z$
>> $w := v$
> **end**

Hence, it follows that the algorithm is $O(nm^2)$ time, because it can be shown that the number of iterations is $O(nm)$; the following page presents a "direct proof", and later, after developing some additional concepts, we will see a simpler proof.

"Direct Proof" of the number of iterations used by the Edmonds–Karp algorithm:

Idea: If z is the residual capacity of an augmenting path P with respect to a flow F, an edge (v,w) of P is a *critical* edge of P with respect to F if (v,w) is a forward edge such that $z = C(v,w)-F(v,w)$ or a backward edge such that $z = F(w,v)$. Each iteration of the Edmonds–Karp algorithm increases flow along an augmenting path by an amount that is limited by its critical edges. If an edge is critical in an augmenting path, then the next time it is used in an augmenting path, it must go in the opposite direction; that is, if a forward edge is filled to capacity, it can only be useful the next time as a back edge, and if a back edge has flow 0, it can only be useful the next time as a forward edge. We shall use this principle to show that each subsequent time an edge is critical, it is on a path that is at least 2 edges longer. We make use of the following quantity:

$M_x(v,w) = $ Minimum number of edges in an augmenting path from v to w at the point just after the xth iteration has been completed.

Lemma: For all w, $M_x(s,w) \le M_{x+1}(s,w)$ and $M_x(w,t) \le M_{x+1}(w,t)$.

Proof: We consider the proof for $M_x(s,w)$; the proof for $M_x(w,t)$ is similar. Assume the contrary; let w be such that $M_{x+1}(s,w)<M_x(s,w)$ and $M_{x+1}(s,w)$ is as small as possible, and assume that, after the $(x+1)^{st}$ iteration, the last edge on an augmenting path P of $M_{x+1}(s,w)$ edges from s to w is a forward edge (v,w) where $F(v,w)<C(v,w)$; the case when it is a back edge is similar. Since $M_{x+1}(s,w)$ was chosen to be as small as possible, it must be that $M_x(s,v) \le M_{x+1}(s,v)$ (otherwise, $M_{x+1}(s,v)<M_x(s,v)$, and $M_{x+1}(s,v)$ would have been chosen instead of $M_{x+1}(s,w)$). Hence it must be that $F(v,w)=C(v,w)$ after the xth iteration or our assumption that $M_x(s,w)>M_{x+1}(s,w)$ is contradicted as follows:

$M_x(s,w) \le M_x(s,v)+1$ — because an augmenting path to v can be extended to w

$\le M_{x+1}(s,v)+1$ — because we have already seen that $M_x(s,v) \le M_{x+1}(s,v)$

$= M_{x+1}(s,w)$ — by definition of the edge (v,w)

However, if $F(v,w)=C(v,w)$ after the xth iteration and $F(v,w)<C(v,w)$ after iteration $(x+1)$, then it must be that (w,v) is a backward edge of the augmenting path P computed by iteration $(x+1)$, and our assumption is again contradicted as follows:

$M_x(s,w) = M_x(s,v)-1$ — since P contains a minimal number of edges

$\le M_{x+1}(s,w)-2$ — because we have already seen that $M_x(s,v)+1 \le M_{x+1}(s,w)$

$O(mn)$ **bound on the maximum number of iterations**: If there is a critical edge (v,w) in the $(x+1)^{st}$ augmenting path, the next time (v,w) is in an augmenting path (whether it is critical or not), suppose it is in the yth augmenting path (where it must be that $y \ge x+2$), it will be in the opposite direction. If (v,w) is a forward edge for the xth iteration then

$M_y(s,t) = M_y(s,v)+M_y(v,t)$ — by definition of M

$= 1+M_y(s,w)+M_y(v,t)$ — since (v,w) is a backward edge for the yth iteration

$\ge 1+M_x(s,w)+M_x(v,t)$ — by the lemma

$= 2+M_x(s,v)+M_x(v,t)$ — since (v,w) is a forward edge for the xth iteration

$= 2+M_x(s,t)$ — by the definition of M

and similar reasoning when (v,w) is backward. Hence each succeeding augmenting path for which (v,w) is critical has at least two more edges. Since no edge may be critical more than $n/2$ times and there are m edges, we have at most $O(mn)$ iterations.

The Residual and Level Graphs

Residual graph: Given a transport network G with a flow F, the *residual graph* with respect to F is a directed graph R with the same set of vertices as G and for each edge (v,w) of G for which $C(v,w)>0$:

- If $F(v,w)>0$, then (w,v) is an edge in R and we define $C_R(w,v)=F(v,w)$.
- If $F(v,w)<C(v,w)$, then (v,w) is an edge in R and we define $C_R(v,w)=C(v,w)-F(v,w)$.

R defines the places in G where additional flow can be added to F; that is, augmenting paths from s to t with respect to F in G correspond to paths from s to t in R. For example, the figure below shows a transport network and below its residual graph.

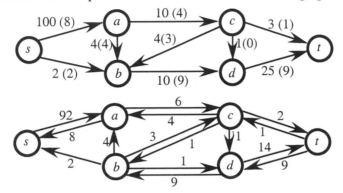

Level graph: Let G be a transport network with a flow F, and R the residual graph with respect to F. The *level* of a vertex v in R is the length of a shortest path from s to v in R (or ∞ if there is no path). The *level graph* for F is the subgraph L of R that can be constructed in linear time by performing a BFS from s in R and keeping only those edges (v,w) such that $level(w) = level(v)+1$ (actually, T is a notational convenience — L can be computed directly from G and F). L contains every shortest path from s to t in R. Continuing the previous example, we obtain the level graph shown below. Notice that although the flow of 10 from s to t in G can be increased to 14 with the path (s,a,c,t) to add 2, and the path (s,a,c,d,t) to add 1, and the path (s,a,c,b,d,t) to add 1, only the path (s,a,c,t) is present in L (because the other two paths are longer).

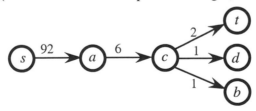

Pruned level graph: By performing BFS starting at t on reversed edges of L, in $O(m)$ time all edges that are not on a path from s to t in L can be deleted to obtain the *pruned level graph*. For example in the level graph shown in the figure above, pruning removes vertices b and d (and edges (c,b) and (c,d)).

Blocking Flows

Given a transport network G with a flow F, an edge (v,w) is *saturated* if $F(v,w)=C(v,w)$, and F is *blocking* if every path from s to t contains at least one *saturated edge*. It is important to remember the difference between an augmenting path (an undirected path defined from a directed graph) and a normal path. For example, consider the figure below, where edges are labeled with their capacities and the corresponding flow is shown in parentheses. All three paths from s to t, (s,a,t), (s,b,t), and (s,a,b,t), contain a saturated edge, and hence the flow is blocking. However, the flow of 10 from s to t can be increased to 12 with the augmenting path (s,b,a,t), which contains no saturated edges.

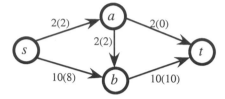

Dinic's Algorithm

Dinic's algorithm provides a general approach, which for a graph G with n vertices and m edges, repeatedly finds a blocking flow, and effectively increases flow along all augmenting paths for the corresponding level graph simultaneously:

> **for** each edge (v,w) of G do $F(v,w) := 0$
>
> **while** the pruned level graph L for F is not empty **do begin**
>
> > Compute a blocking flow F_L for L.
> >
> > **for** each edge (v,w) of L do $F(v,w) := F(v,w)+F_L(v,w)$
>
> **end**

To bound the number of iterations of the *while* loop, consider how the residual graph R for G changes from one iteration to the next. Since L is constructed by BFS of R, all shortest paths from s to t in R are also in L, and hence all of these shortest paths contain at least one edge (a saturated edge) that will not be in R in the next iteration. Hence, the level of t increases with each iteration, and there are at most $n-1$ iterations. Hence, Dinic's algorithm is $O(n\beta(n,m))$, where $\beta(n,m)$ denotes the time to construct a blocking flow.

Note: Another Analysis of the Edmonds–Karp Algorithm

Before considering efficient ways to find a blocking flow, we note here that an analysis very similar to that for Dinic's algorithm can be used to bound the number of iterations of the Edmonds–Karp algorithm. We can again argue that the level of t cannot decrease, and then observe that the level of t can remain the same for at most m iterations. That is, we can view the Edmonds–Karp algorithm as having phases of length at most m, such that after a phase is completed, the level of t increases by at least 1.

MKM Algorithm

The MKM algorithm finds a blocking flow for Dinic's algorithm by repeatedly identifying a "bottleneck" vertex and pushing the maximum flow that can go through it forward to t and back to s:

for each edge (v,w) of G do $F(v,w) := 0$

while the pruned level graph L for F is not empty **do begin**

    1. Compute for each vertex v of L its capacity $C_L(v)$ defined as the minimum of the sum of the residual capacity into v and the residual capacity out of v in L:

$$C_L(v) = \text{minimum}\left\{ \left(\sum_{(u,v)\ in\ R} C_R(u,v) \right), \left(\sum_{(v,w)\ in\ R} C_R(v,w) \right) \right\}$$

    2. Create a queue of the vertices of L sorted in order of increasing value of C_L.

    3. **while** L is not empty **do begin**

        A. $v :=$ next vertex in the queue (the vertex of L of least capacity)

        B. Modify F by "pushing" a flow of $C_L(v)$ forward to t by using as many outgoing edges as necessary (saturating all but at most one of those used) to get the flow from v to the next level, and then from each of the vertices on the next level, using as many edges as necessary to push the flow forward to the following level, and so on.

        C. Modify F by "pulling" a flow of $C_L(v)$ backward to s in a fashion symmetric to the pushing of the previous step.

        D. Delete all saturated edges from L.

        E. Delete from L all vertices and edges not on a path from s to t in L.

    end

  end

Analysis: Initialization of F is $O(m)$. Computing L at the start of each iteration of the main *while* loop is $O(m)$. Step 1 is $O(m)$. Step 2 is $O(n\log(n))$. Step 3A identifies a bottleneck (maximum possible flow through any vertex is at least as great as that through v), and so Steps 3B and 3C are well defined (for example, even if all the flow out of v went along a single edge to a vertex w, it must be that $C_L(w) \geq C_L(v)$). Since Step 3D must delete all the incoming (and perhaps some outgoing) or all the outgoing (and perhaps some incoming) edges of v, Step 3E must at a minimum delete v and all of its incident edges. Hence the *while* loop of Step 3 is executed at most n times. The time for Step 3 can be divided into two parts, the total of $O(m)$ time over all iterations devoted to edges that are saturated and deleted, and the $O(n)$ time per iteration to handle the $O(1)$ edges per vertex that are visited but not saturated. Hence, the total time for each iteration of the main *while* loop is $O(m+n\log(n)+n^2) = O(n^2)$, and the algorithm is $O(n^3)$. Space is $O(m)$. Note that asymptotically faster methods are known; see the chapter notes.

Bounded Flow

With the *bounded* flow problem, each edge (v,w) of a transport network is labeled with a lower bound $B(v,w)$ and an upper bound $C(v,w)$. In order for a flow to be legal, for every edge (v,w) it must be that $B(v,w) \leq F(v,w) \leq C(v,w)$.

Idea: If we had a legal flow with value z from s to t in a transport network G with lower and upper bounds on each edge, we could think of z as the sum of x and y, where x is the amount that z is reduced by subtracting $B(v,w)$ from $F(v,w)$ for each edge (v,w), and y is the value of the flow that remains. We form a new transport network H from G by creating a new source and sink, s_H and t_H, and adding some additional edges that effectively route a flow of value x directly from s_H to t_H. We then can define the capacity of an edge (v,w) to be $C(v,w)–B(v,w)$ and apply a standard maximum flow algorithm to find y (or determine that a legal flow is not possible).

Legal flow algorithm:

1. Compute the arrays *IN* and *OUT* where for each vertex v, $IN[v]$ is the sum of the lower bounds of all incoming edges of v, and $OUT[v]$ is the sum of the lower bounds of all outgoing edges of v.

2. Create two new vertices s_H and t_H and add the following new edges:
    A. For each vertex v except s add the edge (s_H, v) and set $C(s_H, v) := IN[v]$.
    B. For each vertex v except t add the edge (v, t_H) and set $C(v, t_H) := OUT[v]$.
    C. Add the edge (t, s) and set $C(t, s) := \infty$.

3. Compute a maximum flow in H from s_H to t_H where the maximum capacity of an edge (v,w) that was originally in G is defined to be $C(v,w)–B(v,w)$.

4. If there is an edge (s_H, v) that is not saturated, that is $F(s_H, v)<IN(v)$, then a legal flow in G is not possible. Otherwise, delete all vertices and edges added in Step 2, and for each edge (v,w) of G set $F(v,w) := F(v,w)+B(v,w)$.

Correctness: If a legal flow in G has value f, then a flow of value f in H for which all edges emanating from s_H are saturated can be constructed by setting $F(t,s) := f$ and for each edge (v,w) of G, subtracting $B(v,w)$ from $F(v,w)$ and adding $B(v,w)$ to $F(s_H, w)$ and $F(v, t_H)$. Conversely, if a flow of value f in H saturates all edges leaving s_H, then a legal flow of value f in G is achieved by adding $B(v,w)$ to $F(v,w)$ for each edge (v,w) of G.

Complexity: Step 1 can be done in a single pass over the edges, which for each edge (v,w), adds $B(v,w)$ to $IN[w]$ and $OUT[v]$, and the number of new edges created in Step 2 is proportional to the number of vertices. Hence, the entire algorithm uses linear time in addition to whatever time is used in Step 3 to compute a maximum flow.

Maximizing or minimizing a legal flow: The precise value of a flow produced by the legal flow algorithm depends on the maximum flow algorithm used in Step 3. It is possible to maximize or minimize flow that remains within the bounds; see the exercises.

Example of the bounded flow algorithm:

The first figure shows a transport network G with each edge labeled with its lower and upper bounds. The second figure shows the graph H that could have been produced by Steps 2 and 3 of the legal flow algorithm (depending on the maximum flow algorithm that was used), where the dashed edges are the new edges and the solid edges are the original edges of G labeled by $C(v,w)-B(v,w)$; the flow for all dashed edges except (t,s) is the same as the capacity and the flow for all other edges is shown in parentheses. In this example, the flow from s_H to t_H is 36 and the flow from s to t (which is the same as the flow on the edge (t,s)) is 12. However, the flow from s to t could be decreased to 11 by subtracting a unit of flow along the cycle (s,a,c,t,s) or increased to 14 by adding a unit of flow along the cycles (s,a,c,t,s) and (s,a,c,d,t,s). The third figure shows the flow for G produced by Step 4.

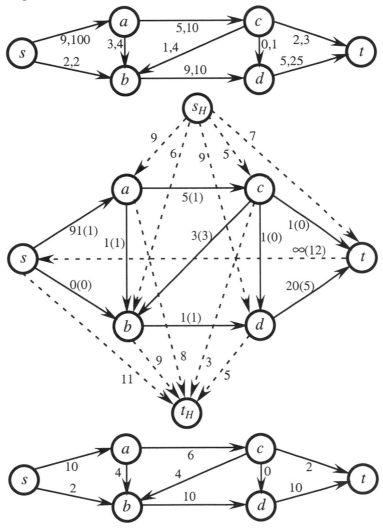

Maximum Matching

A *matching* for an undirected graph G is a subset M of the edges of G, called *matched edges*, such that no two edges in M have a vertex in common. A vertex in G is *matched* if one of its incident edges is matched. M is a *maximum* matching if there does not exist a larger matching for G.

Example: A college dormitory has rooms that can each hold 1 or 2 students. Given a graph that has one vertex per student and an edge between two vertices if those two students are willing to share a room, a maximum matching maximizes the number of shared rooms (and hence allows a minimum number of rooms to be used).

Augmenting paths: In a matching M for an undirected graph $G=(V,E)$, an *augmenting path* with respect to M is a path in G between two unmatched vertices with edges that alternate between M and $E-M$ (i.e., between M and the set of edges that are not in M).

Idea: By reversing the classification of the edges along an augmenting path P, a matching that is one edge larger can be constructed; that is, for each edge on P that is in M, remove it from M, and for each edge on P that is not in M, add it to M. For example, in the figure below, on the top is an augmenting path of 7 edges that contains 3 matched edges (as indicated by the thick lines), and on the bottom the classification of the edges has been reversed to increase the number of matched edges.

Augmenting Path Theorem for Matching

A matching M for an undirected graph $G=(V,E)$ is maximum if and only if there are no augmenting paths with respect to M.

Proof:

(*maximum $\Rightarrow$ no augmenting paths*) Given an augmenting path P with respect to M, a larger matching can be constructed for G by reversing the classification of the edges on P as described above.

(*no augmenting paths $\Rightarrow$ maximum*) Suppose there are no augmenting paths with respect to M in G but there is a matching N such that $|N|>|M|$ (and we now show that this leads to a contradiction). Consider the subgraph H of G that has all vertices of G together with the set F of all edges of G that are in M or N but not in both. All vertices in H must have degree at most 2 because, since M and N are both matchings, no two edges in M share a vertex in common and no two edges in N share a vertex in common, and so in F a vertex can have at most two incident edges. Hence all connected components in F must be simple paths or circuits. Furthermore, all of the circuits must have the same number of edges in M as in N (since their edges must alternate between M and N). Hence, since $|N|>|M|$, there must be a path P in H with more edges in N than M. But P must be an augmenting path with respect to M, which contradicts our assumption that M has no augmenting paths.

CHAPTER 10

Matching in Bipartite Graphs

We limit out attention from this point on to connected undirected graphs (since we can always do matching on each of the connected components). Recall that an undirected graph is *bipartite* if its vertices can be partitioned into two sets so that every edge has one of its end points in each set.

Reduction to a flow problem: As depicted below, two new vertices s and t can be added, directed edges placed from s to all vertices of one half of the bipartite graph, and directed edges placed from all vertices in the other half to t. If all edges have capacity 1, then the amount of flow that is possible from s to t corresponds to how many vertices are matched. We assume that edges are assigned an integral amount of flow (i.e., we do not want 1/2 unit leaving a vertex on two different edges); all of the algorithms that we considered will produce an integral flow if the capacities are all integers.

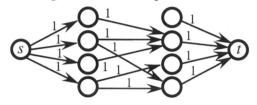

Idea: A flow algorithm may have better worst case complexity for this special case where all edges have capacity 1, and every vertex besides s and t has either in-degree 1 or out-degree 1, which we will refer to here as a *matching* transport network. For example, although the Edmonds–Karp Algorithm uses $O(nm)$ iterations in general (using time $O(m)$ per iteration), it can use at most n iterations for a matching transport network (the flow is $\leq n$ and each iteration increases it by 1), and hence is $O(nm)$ time.

An $O(n^{1/2}m)$ maximum matching algorithm for connected bipartite graphs:

Convert to a flow problem and use Dinic's algorithm. The total time spent to find blocking flows (e.g., using MKM) is $O(m)$, since all edges visited are saturated and deleted. We can argue that the number of basic steps is $\leq 2\lceil \sqrt{n-2} \rceil$ as follows:

> Suppose that we just completed the $\lceil \sqrt{n-2} \rceil^{th}$ basic step. Let f denote the current value of F and g the value of a maximum flow. Since all capacities are 1, there must be $g-f$ paths from s to t in the level graph L that have no intermediate vertices or edges in common, and so at least one of these paths has $\leq (n-2)/(g-f)$ intermediate vertices, and $\leq (n-2)(g-f)+1$ edges. Hence, since the shortest of these paths must have at least $\lceil \sqrt{n-2} \rceil+1$ edges (since Dinic's Algorithm increases the shortest path from s to t in L by at least one after each basic step), we have $\lceil \sqrt{n-2} \rceil+1 \leq (n-2)/(g-f)+1$, from which it follows that $g-f \leq \sqrt{n-2}$. Hence, after at most $\lfloor \sqrt{n-2} \rfloor$ additional basic steps the maximum flow is found, and the bound follows.

A simple $O(nm)$ "direct" algorithm for matching in connected bipartite graphs:

Although flow can be used for matching, in practice there is a very simple $O(mn)$ time algorithm that employs breadth-first search.

Generic algorithm: By the augmenting path theorem, the following generic algorithm finds a maximum matching by repeatedly complementing the labels of edges on an augmenting path (matched becomes unmatched and unmatched becomes matched):

Initialize M to be empty by labeling all edges of G "unmatched".

while G has an augmenting path P with respect to M **do**

Complement the label of each of the edges of P.

Matched vertices: Complementing the labels of the edges on an augmenting path simply changes the endpoints from unmatched to matched (leaving the other vertices on the path matched). Hence, once a vertex becomes matched, it stays matched, and the generic algorithm can be implemented as:

Initialize M to be empty by labeling all edges of G "unmatched".

for each vertex v of G **do**

if there is an augmenting path P starting at v **then**

Complement the label of each of the edges of P.

Using BFS to find augmenting paths in a bipartite graph: We can find an augmenting path starting at an unmatched vertex v by performing a BFS starting at v. At each vertex w of an even level, we stop if an edge is found going to an unmatched vertex (the path back to the root is an augmenting path); otherwise, all edges going down from w to the next odd level are added, and then extended down to the next even level by the matching edge (so vertices at odd levels have only one child). If the entire graph is searched without finding an unmatched vertex at an odd level, then there is no augmenting path. In the figure below, the root is unmatched and may have many children, the vertices at level one can have only one child (since two matched edges cannot have a common endpoint), the vertices at level two can have many children, etc.

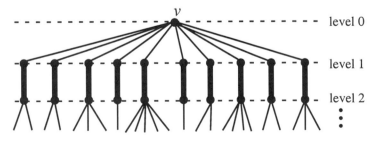

Complexity: For a graph of n vertices and m edges, a BFS of $O(m)$ time and space is performed at each of the n vertices, for a total of $O(nm)$ time and $O(m)$ space.

CHAPTER 10

Matching in Undirected Graphs

Idea: We generalize the BFS algorithm for bipartite graphs. Again, assume that the graph is connected. An undirected graph is bipartite if and only if it has no odd-length cycles (see the exercises), and indeed, it is a particular type of odd length cycle that causes the bipartite algorithm to fail. If we define levels so that matched edges always go from level i to level $i+1$ for some odd integer i, then a problem occurs when at an even level, we find a matched edge to another vertex in that level or the next even level. The figure below shows four of these situations with the edges (v,e), (s,t), (x,z), and (b,c).

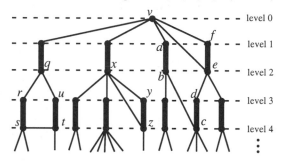

Blossoms: A *blossom* with respect to a matching M is a cycle of length $2k+1$, $k \geq 1$, that contains k matched edges. The *base* of the blossom is its highest vertex in the BFS tree. For example, the figure above shows the blossoms, (v,f,e,v) with base v, (q,r,s,t,u,q) with base q, (x,y,z) with base x, and (v,a,b,c,d,e,f,v) with base v.

Blossom lemma: For a graph G that contains a blossom B, consider the new graph $B(G)$ obtained shrinking B into its base x; that is, delete all the edges incident to vertices of B, delete all vertices of B except x, and for each former edge (u,w) where u was not in B and w was in B, add the edge (u,x). The following two pages show that G has an augmenting path if and only if $B(G)$ does.

Straightforward generalization of the bipartite algorithm: For each vertex of G repeatedly perform BFS starting at v until there is no augmenting path that can be complemented and no blossom that can be shrunk. Since a blossom can be shrunk or a path complemented in $O(m)$ time, each BFS starting at v is $O(m)$. Since shrinking a blossom reduces the number of vertices in G by at least 2, at most $nm/2$ BFS's from v must be performed for a total of $O(nm)$ time per vertex (including the $O(nm)$ time to expand the blossoms at the end). Hence, the total time for all n vertices is $O(n^2m)$. $O(nm)$ space is used if we simply make a copy of G each time a blossom is shrunk.

More efficient implementations: Instead of actually shrinking blossoms, the union-find data structure can be used to maintain sets of vertices that comprise blossoms to achieve $O(nmA^{-1}(m))$ time and $O(m)$ space. With more complex techniques, $O(n^{1/2}m)$ asymptotic time is possible (the same as the bipartite case); see the chapter notes.

Proof of the blossom lemma:

Let G be an undirected graph and consider a blossom B with base x in a BFS tree with root v (where v is an unmatched vertex). The *stem* of B is the path from v to x in the BFS tree; either the stem is empty (i.e., $v=x$) or it ends in a matched edge. For a vertex w in B, the *base path* from x to w, which we denote by $\beta(w)$ is the path that that goes from x to w in a way that ends with a matched edge; that is, there are two ways to get from x to w, depending on which way B is traversed, and since B is an odd length cycle, one way ends at w with an unmatched edge, and one with a matched edge. For example, in the figure below, $\beta(w) = (x,a,b,u,w)$ and $\beta(a) = (x,c,d,w,u,b,a)$.

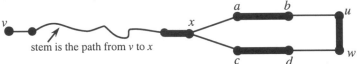

stem is the path from v to x

If $B(G)$ has an augmenting path P starting at v, then so does G: If P does not pass through x, then P is also an augmenting path in G. Otherwise, P is the combination of two paths Q and R, where Q is the stem of B, and R goes from x to the end of P. If (x,z) is the first edge of R, then there is an edge (y,z) in G for some vertex y of B. Hence, Q followed by $\beta(y)$ followed by R must be an augmenting path in G.

If G has an augmenting path P starting at v, then so does $B(G)$: If P does not pass through any vertex of B, then P is also an augmenting path in $B(G)$. Otherwise, without loss of generality, we can assume that P enters B, travels around a portion of B, leaves B, and then continues to its end without ever re-entering B; this is because any subsequent portions of P that leave B and return to B can simply be removed from P to obtain a shorter augmenting path.

CASE 1: *P enters B at x with a matched edge and leaves B with an unmatched edge or P enters B with an unmatched edge and leaves from x with an unmatched edge:*
The portion of P in B can simply be deleted to produce an augmenting path in $B(G)$.

CASE 2A: *P enters and leaves B with an unmatched edge:*
Then P is the path Q, followed by the path R, followed by the edge (z,a), followed by the path S, where Q goes from v to some vertex $y\neq x$ of B, R is a portion of B that goes from y to another vertex z of B, and S is the portion of P from vertex a to where P ends at a vertex w. Let H be the stem of B. If P does not intersect H, then H followed by the edge (z,a) followed by S is an augmenting path in $B(G)$, as depicted below:

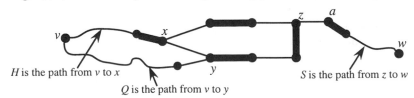

H is the path from v to x S is the path from z to w

Q is the path from v to y

It does not matter if Q intersects H if S does not (we not use Q), but if S intersects H, we need to cut and paste paths, as described on the next page.

Proof of the blossom lemma continued, when S intersects H:

We have two cases, because the cutting an splicing that we use to take care of S intersecting H will work differently depending on whether or not Q intersects H.

CASE 2B: S intersects H but Q does not: To make the figure below simpler, except for (z,a) which denotes the same single edge as in the previous figure, we have shown paths as single edges; that is, except for (z,a) all unmatched edges could be any path that starts and ends with an unmatched edge, and all matched edges could be any path that starts and ends with a matched edge. The only difference between what this figure depicts and what the previous one depicted in CASE 2A is that after going from z to a, P joins H for a while before going on to a path to w; in fact, it could join and rejoin H any number of times before leaving H at vertex b for the last time and going on to w. In any case, we know that in $B(G)$, (v,y) becomes (v,x), and hence in $B(G)$ there is an augmenting path from v to x to u to a to b to w.

CASE 2C: Both Q and S intersect H: Again, to make the figure below simpler, except for (z,a) we have shown paths as single edges. The only difference between what this figure depicts and what the previous one depicted in CASE 2B is that on the way from v to y, P joins H for a while, starting at vertex c; in fact, it could join and rejoin H any number of times before leaving H at vertex d for the last time before going on to y. In any case, we know that in $B(G)$, (z,a) becomes (x,a), and hence in $B(G)$ there is an augmenting path from v to c to d to u to x to a to b to w. Finally, we observe that the figure below covers all the relevant possibilities because Q cannot intersect the stem between vertices b and a (or P would fold back on itself), and since we made no use of the portion of the stem from v to b, it makes no difference if Q intersects H there.

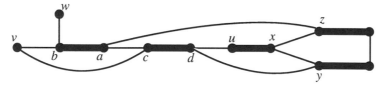

Complexity of blossom shrinking and expansion: The first part of the proof of the blossom lemma says that given an augmenting path in $B(G)$ that goes through a shrunken blossom, we can simply expand the blossom and then fill in the missing portion of the augmenting path in G by using the blossom's base path. The second half guarantees that we will not miss any paths by looking in $B(G)$. Although Cases 2B and 2C constructed augmenting paths that did not follow the stem of the blossom in question, they must follow the stem of some other blossom. In a more efficient implementation that uses the union-find data structure to keep track of blossoms without actually shrinking them, we can patch up the BFS tree and continue; see the chapter notes.

Stable Marriage

Stable marriage problem: A group of n girls and n boys are to be paired off and married. Each boy has a list of the n girls in order of his preference, and each girl has a list of the n boys in order of her preference. A pairing is *stable* if there does not exist a boy and a girl who are not married to each other but prefer each other to their spouses.

Formulation as a graph problem: Given a directed bipartite graph G of n vertices such that every vertex has n outgoing edges and every edge (v,w) has a distinct label $R(v,w)$ in the range 1 to n, find a matching such that every vertex is matched and for any two edges (b_1,g_1) and (b_2,g_2) in the matching, the following two conditions hold:

1. $R(b_1,g_1)>R(b_1,g_2)$ or $R(g_2,b_2)>R(g_2,b_1)$; that is, either b_1 ranks g_1 higher than g_2, or g_2 ranks b_2 higher than b_1, so b_1 and g_2 cannot both prefer each other to their spouses.

2. $R(b_2,g_2)>R(b_2,g_1)$ or $R(g_1,b_1)>R(g_1,b_2)$ that is, either b_2 ranks g_2 higher than g_1, or g_1 ranks b_1 higher than b_2, so b_2 and g_1 cannot both prefer each other to their spouses.

Gale–Shapely algorithm: A greedy algorithm from the boys point of view works in rounds where boys without spouses propose to girls, and girls take their best offer, until a stable marriage is found. A symmetric version of the algorithm that is greedy from the girls point of view has girls proposing to boys.

- In round 1, each boy proposes to his highest ranked girl; when a girl receives a better proposal than she has thus far, she accepts it (and breaks her current match).
- In round i, each boy who was rejected in round $i-1$ proposes to the next higher ranked girl on his list, and the girls update their proposals.

Correctness: Since no boy proposes to a girl twice, there are at most n^2 proposals made, and so the process must terminate (at least one proposal is made per round). To see that the resulting marriage is stable, observe that if boy b and girl g are not married to each other, they cannot prefer each other to their spouses, because if b prefers g to his spouse, then he must have proposed to g and been rejected (so g must prefer her spouse to b).

Implementation: For a problem of n boys and girls, the edges of the graph can be input and their labels placed in two n by n adjacency matrices B and G, where the label of edge (b_i,g_j) is stored in $B[i,j]$, and the label of edge (g_i,b_j) is stored in $G[i,j]$. A bucket sort can be used for each row of B to create for each boy b_i a linked list $L[i]$ of the girls in order of his preference (that is, $L[1]...L[n]$ is an array where each entry is a pointer to a linked list of n items). The array $X[1]...X[n]$ can be used to keep track for each girl of her current spouse (or $X[i]=0$ if girl i does not currently have a spouse). A queue can be maintained of all the boys who do not have spouses; initially, all boys are placed in the queue. Each proposal can be executed $O(1)$ time; take the next boy of the queue, go to the next item on his linked list, check Conditions 1 and 2 by consulting the arrays B and G, update X, and if a boy is rejected in the process, place him on the queue.

Complexity: Creating B and G is $O(n^2)$, the bucket sorting is $O(n^2)$, and initializing X and the queue is $O(n)$. There are at most n^2 proposals made, that take $O(1)$ time each. Hence, since the input contains n^2 edges, the algorithm is linear in the size of the input. Space is also linear in the size of the input.

NP-Complete Graph Problems

Idea: Although problems of arbitrary complexity exist, the class of NP-complete problems are well known because they include many problems of practical interest and seem to be at the boundary of what might be computable (e.g., there are algorithms that have a running time that is bounded by a polynomial in the input size) and what is probably not for reasonable size inputs (e.g., a running time that is exponential in the size of the input). Because of the generality of the graph data structure, many "classic" NP-complete problems are stated as a graph problem. Thousands of NP-complete problems have been discovered. They all have the property that a polynomial time algorithm is not known, but if a polynomial time algorithm exists for any one of them, then all of them can be solved in polynomial time.

Examples of NP-Complete graph problems:

Clique: Given an undirected graph G and an integer k, is there a subset S of k or more vertices of G such that there is an edge between every pair of vertices in S?

Independent set: Given an undirected graph G and an integer k, is there a subset S of k or more vertices of G such that there is not an edge between any two vertices in S?

Vertex cover: Given an undirected graph G and an integer k, is there a subset S of k or fewer vertices of G such that every edge in G has at least one of its endpoints in S?

Dominating set: Given an undirected graph G and an integer k, is there a subset S of k or fewer vertices of G such that for every vertex in G there is an edge between it and a vertex in S?

Subgraph isomorphism: Given two connected directed or undirected graphs G and H, is H isomorphic to a subgraph of G (is there a one-to-one map between the edges of H and the edges of a subgraph of G)?

Hamilton tour: Given a directed or undirected graph, is there a simple cycle that visits all vertices.

Hamilton path: Given a directed or undirected graph and two vertices s and t, is there a simple path from s to t that visits all vertices.

Graph partition: Given an undirected graph G and an integer k, is there a set of k or fewer edges of G whose removal divides G into two disjoint graphs with an equal number of vertices.

Yes-no problems: Like the problems listed above, NP-complete problems are traditionally posed as simple yes-no problems, since the yes-no versions are at least as hard as the computational version of the problem (e.g., if you could actually compute a maximum size clique, it would be easy to then output a "yes" if it has k or more vertices).

Polynomial-Time Reductions

A problem can be shown *NP-hard* by showing that if it has a polynomial time solution, then that solution could be used to solve in polynomial time a problem that is already known to be NP-hard (we defer the question of the "first" NP-hard problem until the next page). This is often called a *polynomial-time reduction*, because it shows how in polynomial time to reduce solving the original NP-hard problem to solving the new problem. As an example, consider the *single-source longest path problem* (SSLP):

> Given a directed graph of n vertices and m edges, a *source* vertex s, and for each edge (v,w) a non-negative *cost* $c(v,w)$, defined to be 0 if the edge does not exist, label each vertex v with $D[v]$, the cost of a *maximum* cost *simple* path from s to v.

Since we have specified that the path must be simple, the SSLP problem is well defined (i.e., you cannot keep going around a cycle to make a path longer), and it is tempting to think that it must be possible to adapt Dijkstra's algorithm to this problem. However, if we could solve SSLP quickly, then we could compute Hamilton tours quickly, since for the special case that all edge costs are 1, the maximum cost path from s to itself in a graph of n vertices is n if and only if the graph has a Hamilton tour.

NP-Complete Problems

A problem is *NP-complete* when a reduction is provided in both directions. Since NP-complete problems are "equivalent" (in the sense that if any one can be solved in polynomial time, they all can), for the reverse direction, it suffices to pick any problem that is known to be NP-complete, and show that a polynomial time algorithm for it implies a polynomial time algorithm for the problem at hand. Once this has been done, the new problem has been shown to be equivalent to any member of the class, and can hence be added to the list of known NP-complete problems.

The Class NP

The name "NP" comes from the class of *non-deterministic* algorithms that read an input w and produce an output of "yes" or "no", where we allow statements of the form:

> **if** guess **then** statement 1 **else** statement 2

A non-deterministic algorithm is *defined* to compute "yes" on a given input w if there *exist* choices for all guesses that causes a "yes" to be output. To determine this, it appears to require exponential time to try all possible combinations of guesses. For example, a non-deterministic algorithm for the directed or undirected Hamilton tour problem is:

> Guess a sequence of $n-1$ edges (where n is the number of vertices in the graph) and output "yes" if they form a simple cycle, and "no" otherwise.

This non-deterministic algorithm is well defined in the sense that if there is a Hamilton tour, then there does exist a correct choice for the guesses that pick a set of $n-1$ edges, and if not, no choice for the guesses will allow the algorithm to output "yes". However, it gives us no clue how to efficiently compute Hamilton tours in practice.

We define NP as the set of all yes-no problems that can be solved in $P(n)$ non-deterministic time for some polynomial P (i.e., there always exists a sequence of guesses so that the question can be answered in $P(n)$ steps).

CHAPTER 10

The "first" NP-Complete Problem

Historically, the "first" NP-complete problem was:

ND Acceptance: Given a non-deterministic algorithm A that always answers yes or no to its inputs and an input string s, does A answer yes on input s?

Here it is critical that we do not "cheat" with the RAM model and, for example, neglect to charge for arithmetic operations that involve large integers; see the chapter notes. Because the ND acceptance problem is so general, when showing that a new problem is NP-complete, it often makes one direction of the proof straightforward (construct a non-deterministic program that guesses a solution and checks that it is correct). For the other direction, one typically tries to give a reduction from a known NP-complete problem that is as "close" as possible to the problem in question.

Example: If we already know that the Hamilton tour problem was NP-complete, to show that the subgraph isomorphism problem (SUBISO) is NP-complete, we can:

1. *Reduce SUBISO to ND acceptance (i.e., show SUBISO is in NP)*:

   To determine if a graph H of m vertices is isomorphic to a subgraph of a graph G of n vertices (where $m \leq n$), a non-deterministic algorithm can guess a set of S of m vertices of G and delete all other vertices and their incident edges to obtain a graph I of m vertices. Then it can guess a correspondence between the vertices of H and I, guess a correspondence between the edges of H and I (or immediately output "no" if H and I do not have the same number of edges), and then check for each edge of H if the corresponding edge in I connects the corresponding vertices.

2. *Reduce Hamilton Tour to SUBISO*:

   To determine if an undirected graph G of n vertices has a Hamilton tour, construct a new graph H that is just a simple cycle of n vertices, and now H is isomorphic to a subgraph of G if and only if G has a Hamilton tour.

3. *Conclude that SUBISO is NP-complete.*

   Assuming that Hamilton tour is NP-complete, we have completed a "circle". By definition of NP-completeness, if Hamilton tour could be solved in polynomial time, then so could any problem in NP, including NP-acceptance. We have just shown that SUBISO is in NP and if you could solve SUBISO in polynomial time, then you could solve Hamilton tour in polynomial time.

The "second" NP-Complete Problem

The proof of the "second" NP-complete problem uses a reduction from ND Acceptance. Traditionally this problem is to determine if a Boolean expression is satisfiable (see the exercises). Given a non-deterministic algorithm A, for each input w, the proof shows how to construct a Boolean expression with a length that is polynomial in the length of w and is satisfiable if and only if A outputs "yes" for input w (i.e., a long Boolean expression describes the actions of A on input w). Traditionally, the third NP-complete problem is a graph problem that can model the constraints represented by the Boolean expression (see the chapter notes, and the exercise that describes how do this with the clique problem).

Dealing with NP-Complete Graph Problems

Idea: It can be useful to show that a problem is NP-complete because, under the assumption that NP-compete problems cannot be solved in polynomial time, it prevents one from wasting time trying to apply standard techniques to the problem that would lead to a polynomial time algorithm. However, the question is what to do next? Often, there are algorithms that are guaranteed to lead to a reasonable solution or at least ones that work well in practice.

Example: The travelling salesman problem is:

> TSP: Given a connected undirected graph G with a cost $C(v,w) > 0$ associated with each edge (v,w), find a tour (a cycle that visits every vertex) of minimum cost.

Clearly TSP is NP-hard, since the special case where all costs are 1 has a solution of cost n if and only if G has a Hamilton tour. However, we can employ the following algorithm to get a reasonable solution for a graph of n vertices and m edges:

1. Construct a minimum spanning tree S for G.

2. Traverse S in pre-order to form a tour T that visits each edge of S exactly twice (i.e., each edge gets visited once on the way down and once on the way back up).

3. **while** there is a vertex v such that T visits v more than once and there are edges (u,v) and (v,w) in S and an edge (u,w) in G such that $C(u,w) < C(u,v)+C(v,w)$ **do** Replace the edges (u,v) and (v,w) in T with the edge (u,w).

Step 2 approximates TSP by constructing a cycle that is less than twice as long as a minimum traveling salesman cycle (since any cycle that visits all vertices, less its longest edge, is a spanning tree). Step 3 is a heuristic that attempts to further reduce the cost of the cycle by replacing adjacent edges on the cycle with a "shortcut" edge. Steps 1 and 2 are $O(m)$. Step 3 can be implemented in $O(m)$ time by a careful traversal of T.

Note: This algorithm uses both the idea of an *approximation algorithm* and a *heuristic*. Steps 1 and 2 are an approximation algorithm that guarantees a worst a factor of 2 worse than optimal. Step 3 is reasonable heuristic that is likely to yield improvements for many practical problems, but we have provided no guarantee that it will yield any improvement in the worst case.

Exercises

216. For many of the algorithms we have considered, such as shortest paths, we have used graphs with labels associated with the edges. Describe how to incorporated labels into adjacency matrix and adjacency list representations of a graph.

217. Let G be a bipartite graph of N vertices that are partitioned into two sets of $n \geq 1$ and $m \geq 1$ vertices ($n+m = N$ and all edges have one vertex in each set). Depending on whether G is undirected or directed, explain how one or smaller matrices can be used to save space over a single N by N adjacency matrix.

218. Prove that a connected acyclic undirected graph must be a tree; that is, any of its vertices can be designated as the root.

219. A graph is *planar* if it can be drawn on the plane in such a way that no edges cross (the edges do not have to be drawn as straight lines, although it can be shown that if the graph can be drawn with curved lines, then it can also be drawn with straight lines — see the chapter notes). Drawing a graph on the plane defines *faces*, the way the plane is divided up into regions by the drawing; the infinite region around the graph is also defined to be a face. For example, in the figure below, the graph shown on the left is planar because it can be redrawn as shown on the right. It has 5 vertices, 9 edges, and 6 faces (including the infinite face that surrounds it).

In contrast, it is possible to prove that neither of the following two graphs are planar:

These two graphs are called the *Kuratowski graphs*. It can be shown that if the vertices of degree 2 in a graph G (which do not affect planarity) are first eliminated (by repeatedly replacing a path (u,v,w) for a vertex v of degree 2 by the edge (u,w)), then G is planar if and only if the resulting graph does not contain one of these two graphs as a subgraph. It is also possible to determine in linear time if a graph is planar. (See the chapter notes).

A. If n denotes the number of vertices, m the number of edges, and f the number of faces in a planar graph G, prove the classic *Euler's formula* for planar graphs:
$$n - m + f = 2$$
Hint: Use a proof by induction of the number of edges in G.

B. By applying Euler's formula for planar graphs, prove that:
$$m \leq 3n - 6$$

220. Prove that a graph is planar if and only if its bi-connected components are planar.

221. A graph is *outer-planar* if it is a planar graph that can be embedded on the plane so that every vertex lies on the exterior infinite face. Prove that a graph is outer-planar if and only if all of its bi-connected components are outer-planar.

222. Any planar graph of n vertices has a *separator* of size $O(\sqrt{n})$; that is, there is always a set of $O(\sqrt{n})$ vertices whose removal (along with all of their incident edges) divides the graph into two disconnected subgraphs of approximately equal size, where one is at most twice the size of the other (see the chapter notes). As a consequence, if every vertex has a degree that is bounded by some constant independent of n, then there is always a set of $O(\sqrt{n})$ edges whose removal divides the graph into two disconnected subgraphs of approximately equal size. Suppose that you are given a procedure DIVIDE(G) that in linear time divides an undirected graph G, where all vertices have degree 4 or less, into two disconnected graphs of approximately equal size (one is at most twice the size of the other) by removing at most $O(\sqrt{n})$ edges.

A. Describe an algorithm that employs DIVIDE to draw G on the planar grid. That is, place the vertices on line intersections and draw the edges to follow the grid lines. For example, in the figure below, on the left is a planar graph and on the right is a drawing of it on the planar grid.

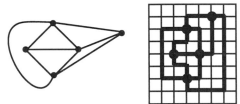

Define the area of the resulting layout to be the number of grid squares in the smallest rectangle that encloses the layout; for example, the layout in the figure above has area 42 (7 squares high by 6 squares wide). Your algorithm should produce an $O(n(\log(n))^2)$ area layout.

Hint: Use divide an conquer by taking out $O(\sqrt{n})$ edges, recursively laying out the two subgraphs, and then by brute force, "stretch" the layout to form channels into which the edges between the two halves can be routed. Analyze the area by expressing the length of the longest side of the layout with a recurrence relation; then square the solution to get the area.

B. Analyze the running time of your algorithm for Part A.

C. Now suppose that you are asked to consider some other class of graphs and given a procedure DIVIDE takes an arbitrary graph in this class, where each vertex has degree 4 or less, and produces a separator of size $O(f(n))$ using time $O(g(n))$. Generalize your construction developed in Part A.

223. Describe how depth-first search can be used to divide an undirected graph into its connected components.

224. Consider the problem of finding a path between two vertices.

    A. Describe how depth-first search can be used to find a path from a vertex s to a vertex t in an undirected graph G.

    B. Suppose that we are given a depth-first spanning forest for an undirected graph G. Describe how to find a path from a vertex s to a vertex t by tracing paths in the tree.

    C. Generalize Parts A and B to directed graphs.

    D. Generalize Parts A, B, and C to use breadth-first search.

225. Describe how depth-first search can be used to find a path in an undirected connected graph that passes through each edge exactly once in each direction.

Hint: Think of visiting a vertex as "walking down the tree edge", of finding a back edge as "walking up the edge and then walking back", and of returning from a vertex as "walking back up the tree edge".

226. Describe how depth-first search can be used to determine whether the edges of a connected undirected graph can have arrows put on them so that the resulting directed graph has a path between every pair of vertices (i.e., so that it is strongly-connected).

Hint: Show that this can be done if and only if removing any edge from the graph leaves it connected.

227. Consider the following four types of graphs:
- directed graphs
- undirected graphs
- bi-connected graphs
- trees (where edges are directed from a vertex to its children)

Assuming that a depth-first search may start at an arbitrary vertex of the graph, for each of these types of graphs, say which of the following things can occur in a forest constructed by depth-first search:
- back edges
- forward edges
- cross edges
- more than one child of the vertex used to start the depth-first search
- more than one tree in the forest

Hint: Remember that except for bi-connected graphs the graph could consist of several disconnected graphs, and that search of a tree may not start at the root.

228. The following program visits the vertices of a graph:

procedure non-recursiveDFS

    while *stack* is not empty **do begin**

        POP a vertex *v*

        if *v* is not marked **then begin**

            {visit *v*}

            mark *v*

            w := first vertex in the adjacency list for *v*

            while (*w* is not *nil*) **do begin**

                PUSH(*w*)

                w := NEXT(*w*)

            end

        end

    end

end

Initialize each vertex to be "unmarked".

for each vertex *v* **do begin**
    Initialize a *stack* to contain *v*.
    DFS
end

A. Explain why non-recursive DFS performs a depth-first search.

    Hint: View POP as entering the procedure and PUSH as making a recursive call.

B. We have already observed that DFS traversal does not necessarily visit the vertices of the graph in a unique order because the order in which adjacency vertices are visited is not specified. If *G* is represented by adjacency lists, how does the ordering of *non-recursiveDFS* compare to that of *first-to-lastDFS(v,G)* that was presented in this chapter?

C. Explain why *non-recursiveDFS* becomes a breadth-first search if *stack* is replaced by *queue*, POP is replaced by DEQUEUE, and PUSH is replaced by ENQUEUE.

229. The presentations of depth-first search assumed that the graph stored in the adjacency list representation. Write pseudo-code for an adjacency matrix representation and analyze the time and space used.

230. The presentations of breadth-first search assumed that the graph stored in the adjacency list representation. Write pseudo-code for an adjacency matrix representation and analyze the time and space used.

231. Characterize exactly the class of dashed edges that can appear in a breadth-first spanning forest (under what circumstances can they go to the left or right, from a vertex to a deeper one less deep one, between trees in the forest, etc.).

232. A binary relation $\equiv$ defined on a set S is an *equivalence relation* if it satisfies the following three properties:

        reflexive: For all elements x, $x \equiv x$.

        symmetric: If $x \equiv y$, then $y \equiv x$.

        transitive: If $x \equiv y$ and $y \equiv z$, then $x \equiv z$.

Prove that any equivalence relation over a non-empty set S partitions it into disjoint non-empty subsets $\{S_1, S_2, ...\}$ called *equivalence classes*; that is: if $i \neq j$, then S_i and S_j have no elements in common and for all elements x and y in S:

If both x and y are in S_i, then x is equivalent to y.

If x is in S_i and y is in S_j, then x is not equivalent to y.

Hint: For two elements x and y in S, let X be the set of all elements that are equivalent to x and let Y be the set of all elements that are equivalent to y. If X and Y have an element in common, use the symmetric and transitive properties to show that it must be that X is a subset of Y and Y is a subset of X (and hence X and Y must be the same set).

233. Recall the definitions of bi-connected and strongly-connected graphs.

A. For a connected undirected graph, define an edge to be equivalent to itself and two different edges to be equivalent if and only if there is at least one simple cycle that contains both of them. Prove that this is an equivalence relation on the edges (as defined in the previous problem).

B. Prove that a connected undirected graph of ≥ 3 vertices has an articulation point if and only if there is a pair of vertices that do not lie on a common cycle.

C. Prove that the following algorithm for a connected undirected graph G partitions its edges into a unique minimal number of disjoint bi-connected subgraphs.

    Initialize a set for each edge in G.
    while there are two equivalent edges in different sets **do**
       Merge those two sets.

D. For a directed graph, define a vertex to be equivalent to itself and two different vertices v and w to be equivalent if and only if they lie on a common cycle (or equivalently, there is a path from v to w and a path from w to v). Prove that this is an equivalence relation on the vertices.

E. Prove that the following algorithm for a directed graph G partitions its vertices into a unique minimal number of disjoint strongly-connected subgraphs.

    Initialize a set for each vertex in G.
    while there are two equivalent vertices in different sets **do**
       Merge those two sets.

234. Prove that a vertex of a connected undirected graph is an articulation point if and only if it is contained in more than one bi-connected component.

235. Given a connected undirected graph G, suppose that a new undirected graph H was constructed that had one vertex for each bi-connected component of G and an edge between two vertices if their corresponding bi-connected components share an articulation vertex.
- A. Prove that H is always well defined; that is, it is not possible that two bi-connected components can share more than one articulation vertex.
- B. Prove that H must be a tree.

236. Let T be a depth-first spanning tree for a connected undirected graph G of n vertices, and assume that the depth-first search algorithm that produced T also computed a depth-first numbering (by numbering the vertices of G in the order they were visited) and stored these numbers in the array $DFN[1]...DFN[n]$. Prove that a vertex v is an articulation point if and only if one of the following two conditions hold:
1. v is the root and has more than one child.
2. v is not the root and for some child w of v there is no back edge between any descendant of w (including w) and a proper ancestor of v.

237. Bi-connected components have been defined only for connected undirected graphs and strongly-connected components only for directed graphs.
- A. Explain why, if the concept of strongly-connected is extended to undirected graphs, it simply says that the graph is connected.
- B. Generalize the concept of bi-connected components to directed graphs, and give a bi-connected components algorithm for directed graphs.

238. Define a path to be *edge-simple* if no edge is repeated twice (but vertices may be repeated more than once). Define a *very strongly-connected* graph to be:

> "A directed graph such that for every pair of vertices, there exists an edge-simple cycle that contains them."

- A. Give a graph that is strongly-connected but not very strongly-connected.
- B. Give an algorithm to compute very-strongly-connected components.

239. A classic child's problem is to draw a house without taking your pencil off the paper; for example, as depicted in the figure below:

Explain why the Euler path theorem proof implies that all ways of drawing this figure must start at one of the bottom corners and end at the other bottom corner.

240. Recall that the Euler path algorithm for undirected graphs worked by first finding a path between the two odd-degree vertices (or starting with a path that is a single arbitrary vertex if there are no odd-degree vertices).

A. Modify the Euler path algorithm of undirected graphs to:

- Eliminate the check that G is connected.
- Eliminate the initialization of P and the identification of t. (So now this step just needs to find an odd vertex s or let s to an arbitrary vertex if G has no odd-degree vertex.)
- Eliminate the use of a separate data structure for C.

Hint: Let algorithm to get stuck at a different vertex from which it started at most once. Assuming that P is represented by a linked list of edges, by keeping appropriate pointers into P, edges can be placed directly into P as they are found.

B. Suppose that the undirected graph has no odd-degree vertices (so it has an Euler cycle). Give simplified pseudo-code for this case (i.e., the code can assume that the graph has no odd-degree vertices and does not need to check for them).

C. In contrast to the approach of Part A, give pseudo-code for finding an Euler path for an undirected graph that works as follows:

Find the two odd-degree vertices s and t (or let $s=t$ be an arbitrary vertex if there are no odd-degree vertices).

Add the edge (s,t), or if there is already an edge (s,t), add the path (s,d,t) where d is a new vertex.

Apply Part B.

Remove the extra path from s to t to change the cycle to a path.

D. Revise your pseudo-code of Part C to not actually add and delete the edge (s,t) or the path (s,d,t).

241. Consider Euler paths for directed graphs.

A. Prove the *Euler path theorem for directed graphs*:

A directed graph has an Euler path if and only if it is connected and all of its vertices have the same in and out degrees except possibly for one that has an out-degree 1 greater than its in-degree and one that has an in-degree 1 greater than its out-degree.

B. Prove that if a directed graph has an Euler cycle (an Euler path that starts and ends at the same vertex), then it is a strongly-connected graph for which all vertices have equal in and out degrees.

C. Prove that if a directed graph has an Euler path, then a vertex that does not have equal in and out degrees must be an endpoint of the path.

D. Explain what changes must be made to the Euler path algorithm for undirected graphs to work for directed graphs.

242. Given an n by n matrix M of integers, describe an $O(n^2)$ algorithm for determining whether there exists a permutation P such that permuting the rows and columns of M by P results in a matrix in upper triangular form (i.e., the entry for the ith row and j^{th} column is 0 if $i>j$).

Hint: Think of the matrix as an adjacency matrix for a graph (where an entry is zero if there is no edge and non-zero otherwise) and do a topological sort.

243. Suppose that all edges in a directed graph had weight 1. Give a linear time algorithm to find the shortest path between two vertices s and t.

Hint: Use breadth-first search.

244. Given a n by n matrix of positive integers, describe an algorithm that finds a sequence of adjacent entries that starts with entry $(1,1)$ and ends with entry (n,n) such that the sum of the absolute values of the differences between adjacent entries is minimized. Two entries are adjacent if their first coordinates are the same and the second coordinates differ by 1 or their second coordinates are the same and their first coordinates differ by 1.

Hint: Think of the entries of the matrix as vertices of a graph where there is an edge between adjacent entries and use entry $(1,1)$ as the source vertex for Dijkstra's Algorithm.

245. The algebraic structure that corresponds to path finding is a *closed semiring*, defined by a five-tuple $(S, \alpha, \beta, \oplus, \otimes)$ where S is a set, α and β are members of S, and $\oplus$ and $\otimes$ are binary operations defined on elements of S that satisfy:

1. For all x in S:
   α is an identity element for $\oplus$; that is: $x \oplus \alpha = \alpha \oplus x = x$
   β is an identity element for $\otimes$; that is: $x \otimes \beta = \beta \otimes x = x$
   α is an annihilator for $\otimes$; that is: $x \otimes \alpha = \alpha \otimes x = \alpha$

2. $\oplus$ is associative (i.e., $(x \oplus y) \oplus z = x \oplus (y \oplus z)$), commutative (i.e., $x \oplus y = y \oplus x$), and idempotent (i.e., $x \oplus x = x$).

3. $\otimes$ is associative and distributes over $\oplus$ (i.e., $x \otimes (y \oplus z) = (x \otimes y) \oplus (x \otimes z)$ and $(x \oplus y) \otimes z = (x \otimes z) \oplus (y \otimes z)$).

4. If $x_1, x_2, x_3, \ldots$ is a countably infinite sequence in S, then its sum exists and is unique. Also, associativity, commutativity, and idempotence apply.

5. $\otimes$ distributes over countably infinite sums as well; that is:

$$\left(\sum_i x_i\right)\left(\sum_j x_j\right) = \sum_i \left(\sum_j x_i x_j\right) = \sum_j \left(\sum_i x_i x_j\right)$$

For any closed semiring, we define the closure operation, denoted by $*$, as

$$x^* = \beta \oplus x \oplus x^2 \oplus \cdots = \sum_{i=0}^{\infty} x^i$$

where x^0 denotes β and x^i denotes $x \otimes x \otimes x \ldots \otimes x$ — i times.

A. At first it might appear that rules 4 and 5 for a closed semiring are redundant, but they are necessary in general. One has to be careful about reasoning about an infinite number of operations. For example, show that using normal addition on the following table, adding the sums of the columns gives 0, but adding the sums of the rows gives -2:

$$
\begin{array}{ccccccc|c}
-1 & 0 & 0 & 0 & 0 & \cdots & & -1 \\
\frac{1}{2} & -1 & 0 & 0 & 0 & \cdots & & -\frac{1}{2} \\
\frac{1}{4} & \frac{1}{2} & -1 & 0 & 0 & \cdots & & -\frac{1}{4} \\
\frac{1}{8} & \frac{1}{4} & \frac{1}{2} & -1 & 0 & \cdots & & -\frac{1}{8} \\
\frac{1}{16} & \frac{1}{8} & \frac{1}{4} & \frac{1}{2} & -1 & \cdots & & -\frac{1}{16} \\
\vdots & \vdots & \vdots & \vdots & \vdots & & & \vdots \\
\hline
0 & 0 & 0 & 0 & 0 & \cdots & & ?? \\
\end{array}
$$

B. Suppose that each edge of a graph is labeled by a element from a closed semiring. Define the *label* of a path as the product (using the $\otimes$ operation) of the labels on the edges in the path (taken in order), where the label of a path of length 0 is defined to be β. If we define $UNDEF=\alpha$, $INIT=\beta$, $CHOOSE=\oplus$, and $COMBINE=\otimes$, generalize the proof by induction for Floyd's Algorithm to prove that for each pair of vertices i and j, the path finding algorithm computes $A[i,j]$ to be the sum (using the $\oplus$ operation) of the labels of all paths from i to j, where the sum over an empty set of paths is defined to be α.

C. Show that:

Floyd's Algorithm is defined by the closed semiring
$$(I, +\infty, 0, MINIMUM, +)$$
where I denotes the non-negative numbers (it does not matter whether number are taken as integers or real numbers) together with the value $+\infty$.

Warshall's Algorithm is defined by the closed semiring
$$(\{0,1\}, 0, 1, MAXIMUM, \bullet)$$
where $\bullet$ denotes multiplication ($0\bullet0=0$, $0\bullet1=0$, $1\bullet1=1$).

Note: We shall see in the chapter on strings that conversion of a pattern diagram to a regular expression is defined by the closed semiring
$$(S, \{\}, \{\varepsilon\}, \cup, \bullet)$$
where S denotes the family of sets of finite strings over some alphabet (including the empty string ε), $\cup$ denotes set union, and $\bullet$ denotes the operation of appending one string to another.

246. We saw that the algorithms for all-pairs shortest paths and transitive closure were $O(n^3)$, where n denotes the number of vertices in the graph.

A. Let m be the number of edges in the graph. Assuming that there is a cost defined for most or all pairs of vertices (i.e., $m = \Omega(n^2)$), explain why the time is $O(m^{3/2})$ and the space is $O(m)$.

B. Making the same assumptions as Part A, suppose that for an integer k, each vertex is encoded as a binary number using k bits, and each edge is encoded as a pair of binary numbers. Give an expression for the minimum value of k, as a function of m, that is needed. Then, using this expression and the answer to Part A, give the time and space used as a function of number of bits to represent the input

247. Give an example of a graph with only $O(n)$ edges such that the output from the all-pairs shortest paths algorithm has $\Omega(n^2)$ pairs for which a finite cost has been computed. Give a similar example for transitive closure.

248. A *transitive reduction* for a directed graph is any directed graph with as few edges as possible that has the same transitive closure.

A. Explain why, in general, a transitive reduction is not necessarily unique.

B. Give an expression for how many different transitive reductions there are of a clique of n vertices.

C. Prove that a directed acyclic graph has a unique transitive reduction.

Hint: Assume the contrary (i.e., an acyclic graph G has two distinct transitive reductions G_1 and G_2) and derive a contradiction (either G has a cycle or G_1 and G_2 are not minimal size reductions).

249. Recall the maximum flow theorem that stated that the value of any flow is at most the capacity of any cut. We informally argued that conservation of flow in and out of a vertex implied that the net flow out of s must pass along the forward edges of the cut to get to t. Give a more formal proof that sums the flow in and out of all vertices on the side of the cut containing s, and shows that everything cancels except flow along edges of the cut.

250. Consider the generic approach to finding maximal flow by successively finding augmenting paths. Explain why choosing paths in order of decreasing residual capacity is computationally problematic. In particular, explain the problems that arise in trying to adapt Dijkstra's Algorithm to find the augmenting path of greatest residual capacity, and discuss how you might approach the analysis of the number of iterations needed.

251. In the analysis of the number of iterations used by the Edmonds–Karp algorithm, there were a number of symmetric / similar cases that were left out. Fill in the details.

252. For a graph of n vertices and m edges, we saw that the Edmonds–Karp algorithm for maximum flow ran in $O(nm^2)$ worst case time, which is $O(n^5)$ for the case that m is $O(n^2)$.

- A. Although this sounds pretty bad, explain why if N denotes the size of the input, the running time is $O(N^3)$ in the worst case, and for the case that m is $O(n^2)$, it is $O(N^{2.5})$ in the worst case; assume that the size of the input is counted by charging 1 unit of space for each vertex or edge that is listed.

- B. If $O(\log(n))$ units of space are charged for each vertex or edge listed in the input, how do the bounds of Part A change?

253. With the Goldberg–Tarjan algorithm (see the chapter notes) it is possible to find a maximum flow for a graph G with n vertices and m edges in $O(mn\log(n^2/m))$ time. We have seen how the MKM algorithm computes a flow in $O(n^3)$ time, and the chapter notes mention the Sleator and Tarjan algorithm that runs in $O(nm\log(n))$ time. Explain why the bound of $O(mn\log(n^2/m))$ is no worse asymptotically than either of these other two bounds.

254. Recall the legal flow algorithm for bounded flow that reduced the problem to a standard flow problem, and assume the Edmonds–Karp algorithm is used.

- A. Give an example of a graph for which the legal flow algorithm produces a minimum flow.

- B. Give an example of a graph for which the legal flow algorithm produces a maximum flow.

- C. Give an example of a graph for which the legal flow algorithm produces a flow that is neither minimum or maximum.

255. Assuming that a bounded flow F has been computed for a transport network G.

- A. Suppose that all of the bounds on edges are integers and describe how binary search can be used to find a flow that is legal and such that the total flow from s to t is closest to a particular integer k (so k could be chosen to be 0 to minimize the flow or to a large number to maximize the flow). Analyze the complexity of your algorithm.

  Hint: After finding a legal flow, look for smaller flows by adjusting the capacity of the edge (t,s) and look for larger flows by adjusting the capacity of a new edge from t to a new sink vertex.

- B. Describe how to compute a maximum possible legal flow for G by starting with F and employing Dinic's algorithm.

- C. Describe how to compute a minimum possible legal flow for G by starting with F and employing Dinic's algorithm.

  Hint: Reverse the roles of s and t.

256. Let G be a transport network with just a lower bound $B(v,w)$ on each edge (v,w). We can construct a legal flow by simply assigning a flow of x to each edge where x is the maximum over all bounds on edges. To get a minimum legal flow, we could optimize a solution to the bounded flow problem (with and upper bound of ∞ on each edge) as outlined in the preceding exercise. However, suppose that instead we used the following simpler algorithm:

1. Perform a simple DFS on G starting at s. Delete any zero bounded edges and isolated vertices that are not reachable from s and quit if there are any non-zero bounded edges that are not reachable from s.

2. Temporarily reverse all edges and repeat Step 1 starting at t.

3. Let H be a depth-first forest for G starting at s, remove all back-edges from H and place them in a queue, rename each vertex with its DFS number (so now, if (i,j) is an edge, then it must be that $i<j$).

4. Place the minimum possible flow on H:
   for each edge (i,j) of H **do** $F(i,j) := B(i,j)$

5. Add flow to fix each vertex in H with too much flow in:
   for $i := 2$ **to** $n-1$ **do if** (flow into i) > (flow out of i) **then**
   Increase the flow of one of i's outgoing edges by the difference.

6. Add flow to fix each vertex in H with too much flow in:
   for $i := n-1$ **downto** 2 **do if** (flow out of i) > (flow into i) **then**
   Increase the flow of one of i's incoming edges by the difference.

7. Return each back edge to H by increasing flow around a cycle:
   for each back edge (i,j) in the queue **do begin**
   Add (i,j) to H with flow $B(i,j)$ and add $B(i,j)$ to the flow of all edges on the path of tree edges from j to i (which together with (i,j) forms a simple cycle).
   end

A. Explain why this approach always produces a legal flow, but not necessarily a minimum one.

B. Explain how Steps 1 through 6 can be implemented in linear time.

C. Give an example of a class of graphs for which Step 7 uses quadratic time when implemented in the straightforward fashion (i.e., repeatedly process a back edge and increase flow along the cycle consisting of it and the tree edges that connect its endpoints).

D. Present pseudo-code for a linear time implementation of Step 7.

   Hint: Traverse G to construct for each vertex a linked list of all back edges that point to it. Simultaneously propagate the extra flow for all back edges up the spanning tree by maintaining for each vertex i the quantity $E(i)$ that keeps track of the extra flow that must go i, where for each back-edge (i,j), $E(i)$ is initialized to $B(i,j)$, and vertices i without back edges pointing to them have $E(i)$ initialized to 0. Work from $k := t$ down to s in the reverse of the DFS numbering, where the flow from the parent p of k to k is increased by $E(k)$, $E(k)$ is added to $E(p)$, and for each back edge (j,k) the flow from j to k is subtracted from $E(k)$.

257. Prove that an undirected graph is bipartite if and only if it has no odd length cycles.

Hint: Explain why if the graph is bipartite then all cycles must end on the same side that they started, and conversely, if the graph has no odd length cycles, then in any breadth-first search tree for the graph, all non-tree edges must go between an even and odd numbered level.

258. When we considered how network flow could be used to compute a maximum matching on a bipartite graph, we assumed that edges are assigned an integral amount of flow (i.e., we do not want 1/2 unit leaving a vertex on two different edges). Verify that all of the algorithms that we considered will produce an integral flow if the capacities are all integers.

259. A natural approach to the stable marriage problem is "random" spouse swapping. That is, repeatedly find a boy and girl who both prefer each other to their spouses (i.e., where one of Conditions 1 and 2 fails), break their marriages, and marry them. Consider the following B and G adjacency matrices for 4 boys a, b, c, and d, and 4 girls w, x, y, and z; the blank rows can be filled in arbitrarily:

$$B = \begin{array}{c|cccc} & w & x & y & z \\ \hline a & 1 & 3 & 2 & 4 \\ b & 2 & 3 & 4 & 1 \\ c & 4 & 3 & 1 & 2 \\ d & & & & \end{array} \qquad G = \begin{array}{c|cccc} & a & b & c & d \\ \hline w & 2 & 1 & 4 & 3 \\ x & & & & \\ y & 1 & 4 & 2 & 3 \\ z & 4 & 2 & 1 & 3 \end{array}$$

A. Explain why random spouse swapping could produce the following infinite loop:

$$\begin{array}{ccccccc} (a,w) & (a,x) & (a,y) & (a,y) & (a,w) & (a,w) & (a,w) \\ (b,x) & (b,w) & (b,w) & (b,z) & (b,z) & (b,x) & (b,x) \\ (c,y) & \Rightarrow (c,y) \Rightarrow & (c,x) & \Rightarrow (c,x) \Rightarrow & (c,x) & \Rightarrow (c,z) \Rightarrow & (c,y) \\ (d,z) & (d,z) & (d,z) & (d,w) & (d,y) & (d,y) & (d,z) \end{array}$$

B. Explain why there always exists a sequence of swaps that leads to a stable set of pairings. Discuss the effectiveness of an algorithm that made truly random swaps (i.e., would situations like that described in Part A be typical, or would the algorithm converge).

260. Let G be an undirected graph where each edge is labeled by a cost >0. Prove that a minimal cost tour of G cannot pass through any given edge more than twice. What can you say about directed graphs?

261. For bipartite graphs prove that:

A. The number of vertices in a minimal size vertex cover is equal to the number of edges in a maximal size matching (recall that a vertex cover is a subset of the vertices such that every edge is incident to at least one of them). Give an example of an undirected graph that is not bipartite for which this is not true.

B. The number of vertices in an independent set is equal to the number of vertices minus the number of edges in a maximal size matching (recall that an independent set is a set of vertices for which no two have an edge between them). Give an example of an undirected graph that is not bipartite for which this is not true.

C. Explain why Parts A and B imply that although the vertex cover and independent set are NP-complete in general, they are solvable in polynomial time for bipartite graphs.

262. An *edge cover* for an undirected graph a subset of its edges such that every vertex is the endpoint of at least one of these edges.

A. Prove that if an undirected graph (that is not necessarily bipartite) contains no isolated vertices (vertices with no incident edges), then the number of edges in a minimum size edge cover is equal to the number of vertices minus the number of edges in a maximal size matching.

B. Even though the edge cover problem sounds very much like the vertex cover problem (similar to how the Hamilton Tour problem sounds similar to the Euler Circuit problem), explain why Part A implies that the edge cover problem is not NP-complete.

263. Step 3 of the algorithm presented for the travelling salesman problem attempted to improve the cost of the tour by repeatedly replacing two adjacent edges (u,v) and (v,w) of the tour by a "short-cut edge" (u,w) such that the cost of (u,w) is less than the cost of (u,v) plus the cost of (v,w). Give an algorithm to do this in linear time.

264. For step 3 of the algorithm presented for the travelling salesman problem, we can generalize the notion of a short-cut edge by defining a *short-cut path* with respect to a tour T as a path P from a vertex v to a vertex w such that:

- There is a portion of T going from v to w, call it the path Q, such that every vertex of Q except possibly v and w is passed through more than once by T.

- P has fewer edges than Q.

- P has lower cost than Q.

We now can consider more powerful versions of Step 3 of the travelling salesman approximation algorithm based on short-cut paths:

A. Describe how to find short-cut paths that start at a vertex v, and how this can be used for an algorithm that repeatedly reduces the cost (and number of edges) of T until there are no short-cut paths with respect to T. Analyze the running time of your algorithm.

B. Refine your solution to Part A (and its analysis) under the assumption that P is restricted to have at most x edges for a fixed integer $x \geq 1$ (but Q can still have arbitrary many edges).

C. Refine your solution to Part B (and its analysis) under the assumption that Q is restricted to have at most y edges for a fixed integer $y \geq 1$.

D. Given an integer n, give an example of a graph of n vertices such that when the algorithm of Part A is used for Step 3, a minimum tour is not produced.

E. Suppose that we allow P to have any length, including a length that is greater than that of Q (but the cost of P must still be lower than Q). How does this affect the running time of Part A? How about if we simply allow the length of P to be $\leq$ that of Q instead of strictly less?

265. Let G be an undirected graph. Let H be the graph obtained from G by using the same set of vertices but the complement of the edges; that is, if (v,w) is an edge in G, then it is not an edge in H, and if (v,w) is not an edge in G, then it is an edge in H.

A. Prove that a set of vertices is a clique for G if and only if it is an independent set for H.

B. Prove that a set of vertices S is a clique for G if and only if its complement (the set of vertices that are not in S) is a vertex cover set for H.

C. Under the assumption that the clique problem has already been proved to be NP-complete, explain why Part A implies that the vertex cover problem is NP-complete.

266. Consider the following transformation that converts a directed graph G to an undirected graph H:

- For each vertex v of G add two new vertices vin and $vout$, and place an undirected edge between vin and v and an undirected edge between v and $vout$ (so each vertex of G has been replaced by a path of three vertices).

- Replace each directed edge (v,w) of G by an undirected edge between $vout$ and win.

A. Prove that H has a Hamilton tour if and only if G does.

B. We have already seen a non-deterministic polynomial time algorithm for the undirected Hamilton tour problem. Under the assumption that the Hamilton tour problem for directed graphs has already been shown to be NP-complete, explain why this in combination with Part A implies that the Hamilton tour problem for undirected graphs is NP-complete.

267. Recall the dominating set problem.

A. Give a non-deterministic polynomial time algorithm for the dominating set problem.

B. Show how in polynomial time to convert an instance of the vertex cover problem to the dominating set problem.

C. Under the assumption that the vertex cover problem has already been proved to be NP-complete, explain why Parts A and B imply that the dominating set problem is NP-complete.

268. Show that determining whether two rooted directed acyclic graphs are isomorphic is as hard as the isomorphism problem for arbitrary undirected graphs.

269. A *Boolean expression* is formed from Boolean variables (variables with value *true* or *false*) combined with the operators:

logical OR: $A+B$ is true if either of A or B (or both) is true, and false otherwise.

logical AND: $A \bullet B$ is true if both A and B are true, and false otherwise. By convention, we often write AB to mean $A \bullet B$.

logical NOT: $\neg A$ is true if A is false and false if A is true. By convention, we often write $\overline{A}$ to mean $\neg A$.

Parentheses may be used to group sub-expressions. For example, $\overline{A}(B+C)$ is true only when A is false and at least one of B and C are true. A Boolean expression is *satisfiable* if there exists an assignment to its variables that makes it true. For example, $\overline{A}(B+C)$ is satisfiable in three ways (A=*false* / B=*true* / C=*false*, A=*false* / B=*false* / C=*true*, or A=*false* / B=*true* / C=*true*), but expressions like $A\overline{A}$ are not.

A. Prove that $(\overline{A}+B+\overline{C})(\overline{A}+\overline{B}+C)(A+B+\overline{C})$ is satisfiable.

B. Prove that $(\overline{A}+\overline{B})(A+\overline{B}+\overline{C})(\overline{C}+B)(C+D)(C+\overline{D})$ is not satisfiable.

C. Give a non-deterministic polynomial time algorithm to determine whether a Boolean expression is satisfiable.

D. A Boolean expression is in *conjunctive normal form* (CNF) if it is the product of sums (called *clauses*) of *literals*, where a literal is a variable that may be complemented or uncomplemented. For example, the expressions in Parts A and B are both in conjunctive normal form, where Part A has 3 clauses of three literals each and Part B has 5 clauses where the second has three literals and the others have two literals. Given a polynomial time algorithm that takes as input a CNF Boolean expression B with k clauses and produces an undirected graph G that has a clique of size k if and only if B is satisfiable.

Hint: Create one vertex of G for every literal of every clause and place an edge between two vertices in different clauses if they do not conflict with each other;

that is either they correspond to literals that use different variables (complemented or uncomplemented) or they correspond to literals that use the same variable, and it is either complemented in both places or uncomplemented in both places.

E. Under the assumption that it has already been shown NP-complete to determine if a CNF Boolean expression is satisfiable, prove that the problem remains NP-complete when CNF expressions are restricted to have at most 3 literals per clause.

Hint: For $m \geq 4$,, given a clause

$$(x_1 + x_2 + \cdots + x_m)$$

where x_i denotes a literal (i.e., a variable that may be complemented or uncomplemented), introduce $m-3$ new variables $v_1 \ldots v_{m-3}$, and replace this clause with the sequence of clauses:

$$(x_1 + x_2 + y_1)(x_3 + \overline{y}_1 + y_2)(x_4 + \overline{y}_2 + y_3)\cdots(x_{m-2} + \overline{y}_{k-4} + y_{k-3})(x_{m-1}x_m + \overline{y}_{m-3})$$

For example, for $k=5$ we have:

$$(x_1 + x_2 + y_1)(x_3 + \overline{y}_1 + y_2)(x_4 + x_5 + \overline{y}_2)$$

270. Without techniques like simulated annealing (even with them), iterative improvement algorithms (local search algorithms, hill-climbing algorithms, etc.) can get "stuck" at a local minimum.

A. Consider the following approach to solving the travelling salesman problem:

> To find the best possible traveling salesman tour of a graph, start with an arbitrary tour and repeatedly look for a pair of edges (a,b) and (c,d) in the tour such that the cost of the tour is improved by replacing them by the edges (a,c) and (b,d).

Give an example of a solution resulting from this approach that is not optimal but for which no further improvement is possible.

B. The *graph partitioning problem* is to partition the vertices of an undirected graph into two equal size sets such that the number of edges with an endpoint in different sets is minimized. Consider the following approach to solving the graph partitioning problem:

> To find the best possible graph partition, start with an arbitrary partition and successively try to find a pair of vertices, one in each half of the partition, such that exchanging them decreases the number of edges between the two halves of the partition.

Give an example of a solution resulting from this approach that is not optimal but for which no further improvement is possible.

271. The *Traffic Jam* puzzle starts with pieces arranged as shown on the left in the figure below and the object is to slide them around to form the arrangement shown on the right (the two white squares are empty).

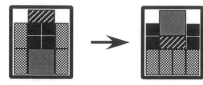

A. A seemingly simpler problem is the *sliding piece set-up problem* where you are not given the initial and final positions; the problem is simply:

Can the pieces be placed in the puzzle rectangle?

Define the sliding piece set-up problem for an infinite class of sliding piece puzzles given by a parameter n, and prove that the problem is NP-complete.

Hint: Make use of the knapsack or bin packing problems.

B. Define the Traffic Jam puzzle for an infinite class of inputs and show that it is in NP.

C. What can you say about the complexity of the Traffic Jam puzzle?

272. When solutions or even approximate solutions are not known for a problem, one can always resort to exhaustive search, aided by heuristics. Puzzles can provide a simple framework in which to study heuristic search.

The *Six piece rectangular burr* puzzle has six rectilinear shaped pieces of dimension 2 by 2 by k units for some k (typically $k=6$ or $k=8$) with a number of unit sized cubes removed to form "notches" cut along unit boundaries; the assembled shape is symmetric in all three dimensions. Some have a "key" piece that slides out. However, more complicated ones have a number of internal voids, and removing pieces may require sliding operations; the number of moves it takes to remove the first piece is the *level* of the burr (as high as 12 for a six piece rectangular burr). For example, *Coffin's Burr*, shown in the following figure, has level 3 (letters show how pieces fit together, numbers indicate an order in which they can be disassembled). Note that in general it can be possible for there to be many legal configurations, but only one that can be disassembled.

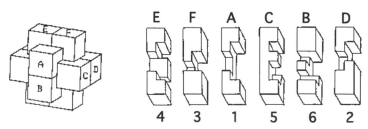

A. Describe how you would represent a position with a data structure.

B. Describe how you would generate all positions that are reachable from a given position by one move.

C. Given a representation for positions and a move generation procedure, describe how breadth-first search could be used to find the shortest solution to the puzzle.

D. Discuss heuristics that could be used to speed up the search for practical inputs.

E. Give an upper bound on the number of distinct reachable positions. How close a lower bound can you prove? If your upper an lower bounds do not match, describe an algorithm to determine the exact number. Consider the case where pieces are not allowed to twist and moves must be an integral number of units. Then discuss how to generalize your algorithm to allow twists and movements by a fraction of a unit (you should also explain why such movements can be possible and useful). Next, consider the case where notches are allowed to show when the puzzle is assembled. Finally, consider the case where pieces may have different lengths.

F. Generalize the puzzle to an input where there are x, y, and z pieces going in each of the three dimensions. Note that even for $x = y = z = 1$, quite complex puzzles exist. For example, the following figure shows the assembled shape and below it two different 3-piece rectangular burr puzzles that assemble to this shape (called Cutler's "GigaBurr" and "GigaBurr-2"); both require 8 moves to remove the first piece. Given this generalization, repeat Parts A through E. For Part E, also consider the possibility that in addition to pieces having different lengths, they may also have different horizontal and vertical thickness (i.e., there may be different height, width, and length parameters for each piece of the puzzle).

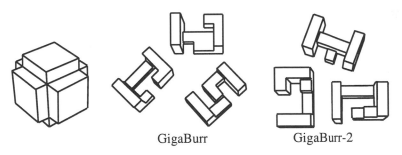

GigaBurr   GigaBurr-2

Chapter Notes

In addition to the general texts on data structures and algorithms listed in the chapter notes to the first chapter, additional textbooks on graph algorithms and optimization algorithms include: Even [1979], Lawler [1976], Papadimitriou and Steiglitz [1982], Polak [1977], Reingold, Nievergelt, and Deo [1977]. There are many general texts on combinatorics and graph theory, including the classic text of Berge [1962], and the books of Bogart [1983], Bondy and Murty [1976], Brualdi [1977], Busacker and Saaty [1965], Christofides [1975], Gibbons [1985], Golumbic [1980], Harary [1969], Liu [1968], Lovasz [1979], Ore 1963], Polya, Tarjan, and Woods [1983], Roberts [1984], Swamy and Thulasiraman [1981], Tucker [1980], and Tutte [1984]. Biggs [1993] considers algebraic graph theory.

Depth-first and breadth first search are presented in virtually any algorithms text, the presentation here, including that of bi-connected and strongly-connected components, is based on that of the book of Aho, Hopcroft, and Ullman [1974].

Armoni, Ta-Shma, Wigderson, and Zhou [2000] present an algorithm to test connectivity between two points using only $O(\log(n)^{3/4})$ space, and give a history of the work in this area.

There is a great deal of work on the minimum spanning tree problem that dates to before 1930; work before 1985 is reviewed in Graham and Hell [1985]. Prim's algorithm for minimum spanning trees is presented in Prim [1957]; the name has become standard, although the idea is much older. Kruskal's algorithm is presented in Kruskal [1956]. Karger, Klein, and Tarjan [1999] present a randomized linear-time minimum weight spanning tree algorithm; the introduction to their paper reviews the history of algorithms for the minimum weight spanning tree algorithm problem after 1985, including asymptotically faster algorithms than those presented here. Chazelle [2001] presents a spanning tree algorithm that runs in roughly $O(nACK^{-1}(n,m))$ time for a graph of n vertices and m edges where $ACK^{-1}(n,m)$ denotes the inverse of Ackermann's function (see the chapter on sets over a small universe for a presentation of Ackermann's function). Finding spanning trees that have special properties can make the problem much harder. For example, if you could find a spanning tree with a minimum number of leaves, you could compute Hamilton paths, which is NP-hard; Storer [1981] shows that it is also hard to find one with a maximum number of leaves, and gives an approximation algorithm for the problem.

See Knuth [1973] (Volume 1) for detailed discussion of topological sorting.

Euler circuits and paths are addressed in most of the texts on graphs and graph algorithms listed earlier.

Dijkstra's single-source shortest path algorithm is presented in Dijkstra [1959]. Floyd's all-pairs shortest paths algorithm is presented in Floyd [1962] and Hu [1968]. Warshall's transitive closure algorithm is presented in Warshall [1962]. The conversion of a pattern diagram to a regular expression is due to Kleene [1956], and captures the full generality of the path finding approach; the formulation of path finding based on a closed semiring, as presented in the exercises, is based on the presentation in Aho, Hopcroft, and Ullman [1974], and the example in the exercise showing infinite sums adding differently is due to G. Smith [1985]. Zwick [1999] considers the "all pairs lightest shortest paths" problem where it is desired to find a minimal length path among all paths of minimum cost. Storer and Reif [1994] and Reif and Storer [1994] consider shortest paths between polygonal obstacles, and Reif and Storer [1987] consider paths between obstacles that minimize turns.

There has been much work on the maximum flow problem, with bounds stated in terms of n (the number of vertices), m (the number of edges), and U (the maximum value of an edge capacity; that is, edge capacities are in the range 1 ... U). Note that U can be exponential in the input size, but the quantity $\log(U)$ that appears in some bounds is polynomial in the input size. Ford and Fulkerson [1956], present the general framework of augmenting paths that uses $O(nmU)$ time; also see Dantzig [1951] who presents an $O(n^2mU)$ time algorithm. Edmonds and Karp [1972] present the $O(nm^2)$ algorithm that was presented here (see also Dinic [1970]) as well as an $O(m^2\log(U))$ algorithm, Dinic [1970] presents the general approach that was presented here and gives an $O(n^2m)$ algorithm, Gabow [1985] presents an $O(nm\log(U))$ algorithm (see also Dinic [1973]), Karzanov [1974], Malhotra, Kumar, Maheshwari [1978] (for which the MKM algorithm is named), and Goldberg [1987] give a $O(n^3)$ algorithms, Cherkassky [1977] gives an $O(n^2m^{1/2})$ algorithm, Galil [1978] gives an $O(n^{5/3}m^{2/3})$ algorithm, Shiloach [1978] and Galil and Naamad [1980] give $O(nm\log(m)^2)$ algorithms, Sleator and Tarjan [1980] present an $O(mn\log(n))$ algorithm, which is also presented in the book of Tarjan [1983], Goldberg and Tarjan [1986] give an $O(nm\log(n^2/m))$ algorithm, and Ahuja and Orlin [1989] give an $O(nm+n^2\log(U))$ algorithm. Further work on network flow algorithms is presented in Ahuja, Orlin, and Tarjan [1989], Alon [1990], Phillips and Westbrook [1993], King, Rao, and Tarjan [1994], Cheriyan, Hagerup, and Mehlhorn [1996], Cherassky and Goldberg [1997] and Goldberg and Rao [1998] (which presents both an $O(n^{2/3}m\log(n^2/m)\log(U)$ and an $O(m^{3/2}\log(n^2/m)\log(U)$ algorithm, and has a table that summarizes most of the work mentioned here). The classic book of Ford and Fulkerson [1962] is devoted to network flow and many variations of the problem. The presentation here of the Edmonds-Karp maximum flow algorithm is motivated by the presentation in the book of Lawler [1976]. The presentation of Dinic's general approach is motivated by the general presentation in the book of Tarjan [1983]. The book of Denenberg and Lewis [1991] presents the MKM algorithm.

Although the MKM algorithm is relatively simple to describe, in the book of Tarjan [1983], he presents a number of alternate approaches (see also, for example, the presentation of *preflow-push* and *lift-to-front* algorithms in the book of Cormen, Leiserson, Rivest, and Stein [2001]), and points out that a potential practical drawback of the MKM algorithm is that it works by identifying bottlenecks, and may as a result in practice use a number of basic steps that comes closer to the worst case. The book of Johnson and Mcgeoch [1992] presents implementations and experiments with network flow algorithms.

There are many variations of the network flow problem, including the minimum cost flow problem, where there are costs on the edges and the goal is to find a maximum flow of minimum cost; see for example the books of Tarjan [1983], Papadimitriou and Steiglitz [1982], and Lawler [1976]. Leighton and Rao [1999] address multi-commodity max-flow min-cut theorems.

Graph matching has been studied as early as Hall [1948] and Berge [1957]. The book of Papadimitriou and Steiglitz [1982] present matching on arbitrary graphs and the blossom construction. The book of Tarjan [1983] describes how to compute maximum matchings on arbitrary graphs in $O(nm)$ time. Hopcroft and Karp [1973] present an $O(n^{1/2}m)$ algorithm for matching in bipartite graphs, and Even and Kariv [1975] generalize their construction to an $O(n^{2.5})$ algorithm for matching in general graphs; in fact, as noted in the book of Tarjan [1983], their algorithm is $O(minimum\{n^{2.5}, n^{1/2}m\log\log(n)\})$. Michali and Vazirani [1980] give an $O(n^{1/2}m)$ algorithm for matching in general graphs. Gabow, Kaplan, and Tarjan [1999] address the testing of the uniqueness of matchings. The book of Karpinski and Rytter [1998] is devoted to parallel algorithms for graph matching.

The Gale-Shapely algorithm is presented in Gale and Shapely [1962]. The book of Gusfield [1989] is devoted to the Stable Marriage problem and the wide diversity of related and generalized versions of the problem (including the one here, and solutions that are more balanced from the perspective of both the boys and the girls).

The concept of NP-completeness is clearly not limited to graph problems. We have introduced the notion in this chapter because due to the general nature of the graph data structure, it is natural to encounter practical problems that are not tractable in the worst case. Define P to be the class of all algorithms (with yes-no outputs) that run in a time that is bounded by a polynomial of the input size and NP to be the class of all non-deterministic algorithms (with yes-no outputs) that run in a time that is bounded by a polynomial of the input size. Then if P=NP, any problem that can be solved by a non-deterministic poly-time algorithm can be solved by a normal (deterministic) polynomial time algorithm; in particular, all NP-complete problems would have polynomial time algorithms. In our definition of non-deterministic algorithms, it is critical not to "cheat" with the RAM model and, for example, neglect to charge for arithmetic operations that involve large integers. For this reason, the *Turing machine* model is often used when formally defining the class NP (and then it is shown how Turing machines can simulate normal pseudo-code with only a polynomial-time slow down when there is no cheating in the RAM model). The book of Garey and Johnson [1979] is devoted to the P=NP question, the study of NP-complete problems, and related issues. A nice introduction to the subject, including the Turing machine model, is contained in the books of Aho, Hopcroft, and Ullman [1974] and Sipser [1997]. The book of Hochbaum [1997] is devoted to approximation algorithms for NP-hard problems. Vempala [1999] gives lower bounds on the approximability of the traveling salesman problem.

Many real life problems that are "hard" do not fall in the class of NP-hard problems unless they are generalized to have inputs that can be of arbitrary size. For example, the game of chess is not an NP-hard problem because there are only a finite number (a large finite number) of distinct possible positions, and so in principle, optimal moves can be made in $O(1)$ time with table lookup. However, the basic intuition that chess is hard is correct. For example, Storer [1979] shows that playing chess on a n by n board is NP-hard; in fact, *PSPACE-hard*. A problem is PSPACE-hard if a polynomial time algorithm to solve it implies that any problem that can be solved by using a polynomial amount of space can also be solved in (deterministic) polynomial time. Since, with only polynomial space, an algorithm can use exponential time (without going into an infinite loop), the class of PSPACE-complete problems (ones that are PSPACE-hard and can be solved with polynomial space) contains very complex algorithms. For example, all problems in NP are in PSPACE, since in polynomial time an algorithm can consume only a polynomial amount of space (i.e., $O(1)$ space per time unit). In fact, it is possible to convert a non-deterministic algorithm that uses polynomial space (and possibly exponential time) to a deterministic one that still uses polynomial space (i.e., NPSPACE = PSPACE). The construction, due to Savitch [1970], goes roughly as follows:

> Suppose that we have a non-deterministic algorithm A that uses polynomial space $P(n)$ for an input of length n (assumptions are needed that prevent A from "abusing" the unit-cost RAM model). Then the state of A at any point in time (the value of all variables and the instruction to be executed next) can be represented with $P(n)$ space. We can think of a computation of A on an input of length n as moving from state to state, and the question is whether appropriate choices of guesses can be made to get it from its starting state to the state where it stops successfully (e.g., in the case of a yes / no problem, answers "yes"). It is not hard to show that A can go for at most an exponential number of steps before it must repeat a state; that is at most $c^{P(n)}$ steps for some constant c. We try all possible $c^{P(n)}$ sequences of guesses to determine if any one of them causes A to accept its input. To avoid using too much space to keep track of which guesses have been tried, employ a divide and conquer strategy that guesses the middle configuration and then recursively tests if the program can get from the starting configuration to the middle one and then from the middle one to the final one. It takes exponential time to try all possible middle configurations, but since the space for the first recursive call is reused by the second recursive call, the total space used is only $O(P(n))$ per stack frame times the maximum depth of the stack, which is $P(n)$, the logarithm of the $c^{P(n)}$ time used by A. That is, A can be simulated with a deterministic algorithm that uses exponential time but only $O(P(n)^2)$ space.

See for example the books of Garey and Johnson [1979] or Sipser [1997] for a presentation of the PSPACE=NPSPACE construction and additional references.

Many general texts on algorithms such as those listed at the beginning of the notes to the first chapter contain material on planar graphs (e.g., Even [1979]); in addition, the reader can refer to texts on discrete mathematics and graph theory such as Wallis [2000], Liu [1985], or Berge [1962]. Kuratowski [1930] characterized planar graphs in terms of the Kuratowski subgraphs presented in the exercises (i.e., a graph is planar if and only if after degree two vertices are eliminated it does not contain one of those two graphs as a subgraph). Auslander [1961] (with a correction noted by Goldstein [1963], and implemented in the Ph.D. Dissertation of Shirey [1969] in $O(|V|^3)$ time) presents a basic approach to graph planarity testing (and construction of the planar embedding if the graph is planar) that removes a cycle and recursively embeds the resulting disconnected pieces; Hopcroft and Tarjan [1974] employ a depth-first search forest representation of the graph to achieve linear time, and Gries and Xue [1988] present some practical improvements. Williamson [1980, 1984] presents a linear time algorithm to find a Kuratowski subgraph. Mitchell [1979] presents linear time algorithms to recognize outer-planar graphs (planar graphs that can be embedded on the plane so that every vertex lies on the exterior infinite face) and maximal outer-planar graphs (outer-planar graphs such that the addition of any new edge causes the graph to no longer be outer-planar). Lengauer [1989] gives a linear time algorithm for testing the planarity of hierarchically described graphs. Battista and Tamassia [1989] present incremental planarity testing using linear space, $O(\log(n))$ worst case time for queries and vertex insertions, and $O(\log(n))$ amortized time for edge insertions. Ramachandran and Reif [1989] give a $O(\log(n))$ parallel algorithm for planarity testing using the CRCW PRAM model (and give further references on the subject).

The planar separator theorem that was stated in the exercises can be proved using the idea of "peeling and onion". By repeatedly traversing the perimeter of a planar representation of the planar graph, it can be decomposed into a set of concentric cycles (layers of the onion). Either a cycle of $O(\sqrt{n})$ vertices can be removed to divide the graph into approximately equal sized parts, or there cannot be too many concentric cycles in the decomposition, and it is possible to "slice" the graph (onion) into approximately equal sized parts by finding two paths from the outside to the middle that cut each of the cycles into two pieces. The material on planar graphs, the planar separator theorem, and graph layout that is presented in the exercises is based on the presentation in the book of Ullman [1984], which contains additional references on the subject. The problem of drawing graphs on the plane to minimize the number of bends in embedded edges is consider in Storer [1980].

Burr puzzles are interesting in that only the very easy "Level 1" versions of them are generally available in stores. The *level* of a burr refers to the minimum number of moves required to remove the first piece. Level 1 burrs simply have a "key" piece that slides out first (or perhaps a pair of pieces that slide together). Higher levels exist when there are internal voids that require several sliding motions to get the first piece out. Cutler [1977],

Gardner [1978], and Dewdney [1985] have written about this type of burr. Cutler [1986,1994] has enumerated all possible 6-piece rectangular burrs where all six pieces have size 2 by 2 by k units for some k, twisting and fractional movements are not allowed, and no notches show when the puzzle is assembled. He implemented a program that attempts to disassemble each potential puzzle (it is possible to have many solved states that cannot be disassembled), and discovered that the highest possible level is 12, there is no level 11 puzzle, and the highest level for a unique solution is 10. Subsequent to this work Cutler enumerated all of the approximately 250 billion possible configurations for a class of 3-piece burrs. The highest level (number of moves to remove the first piece) is 8, of which there are 80 different puzzles; out of these 80, three have just 9 internal voids. The Cutler "GigaBurr" is one of these three, the other two come apart in a different way but are similar to each other, and one of them he called the "GigaBurr-2".

11. Strings

Introduction

We use the term *string* to refer to a sequence of characters over some *alphabet Σ*. For example, Σ might be just the 26 lower-case characters of the English alphabet, the 128 characters of the standard ASCII character set, or the integers 0 through 255 (that represent all possible values of a byte). We use the term *lexicographic order* to denote the generalization of alphabetical order to any character set. We have already seen the sorting of fixed-length vectors as a special case of this notion (a vector can be viewed as a string where each component is a character).

Definition: Assuming that we have defined $\leq$ between any two characters, given two strings $s = x_1 x_2 \ldots x_m$ and $t = y_1 y_2 \ldots y_n$, $s \leq t$ to means either s is a prefix of t or for some $1 \leq i \leq k$, where k denotes the minimum of m and n, the following two conditions hold:

1. $x_j = y_j$ for $1 \leq j < i$
2. $x_i < y_i$

A sequence of strings $s_1, s_2, \ldots s_n$ is in lexicographic order if $s_i \leq s_j$, for all $1 \leq i < j \leq n$.

Of course, any computer file can be viewed as a string, but we are specifically interested in data that is organized in a linear, sequential fashion. For example, the text of this book is essentially a long string (with formatting and figures added).

A basic problem that occurs in many applications is searching for data within a string:

> Given a string $S = S[1]\ldots S[n]$, find the first occurrence in S of a string
> $T = T[1]\ldots T[m]$ (or determine that T is not a substring of S).

A simple approach is, for each position i of S, go from left to right in T to determine if T matches at position i. If a position i is ever successfully identified as the left end of a match of T to S (i.e., $T[1]\ldots T[m] = S[i]..S[i+m-1]$), then i is returned. If we get all the way to the end of S and find that T is not a substring of S, 0 is returned.

```
function left-to-rightMATCH
    for i := 1 to n−m+1 do begin
        j := 1
        while j ≤ m and S[i+j−1]=T[j] do j := j+1
        if j>m then return i
        end
    return 0
    end
```

In practice we would hope that relatively short prefixes of T are enough to exclude a possible match for most positions of S (and so the *while* loop typically does not go through many iterations). If not, another approach is to hope that relatively short suffixes of T are enough to exclude a possible match for most positions of S, and work from right to left in T:

> **function** right-to-leftMATCH
>   **for** $i := 1$ to $n-m+1$ n **do**
> $j := m$
>   **while** $j>0$ and $S[i+j-1]=T[j]$ **do** $j := j-1$
>   **if** $j=0$ **then return** $i-m+1$
>   **end**
>   **return** 0
> **end**

Although both of these approaches are $\Omega(mn)$ in the worst case (see the exercises), each can be improved to $O(m+n)$ time by employing knowledge of how previous matches and mismatches occurred to advance more quickly through S. The *Knuth–Morris–Pratt algorithm* improves upon *left_to_rightMATCH* and the *Boyer–Moore algorithm* improves upon *right-to-leftMATCH*. We will also look at a randomized approach (the *Karp–Rabin algorithm*) and one specialized to short patterns (the *shift–and algorithm*).

A more general problem than searching for a substring is to search for a *pattern* in S of length m. For example, a reference in the chapter notes, such as "Knuth, Morris, and Pratt [1977]", follows a pattern that typically consists of a sequence of names (appropriately formatted with commas and the word "and"), a left bracket, a sequence of 4 digits, and a right bracket. We shall see how to find such a pattern in $O(mn)$ time.

We then consider the *trie* data structure for storing a set of strings, where each root-to-leaf path corresponds to a string. For an alphabet of size k, tries are essentially ordered trees where a vertex may have any subset of its k possible children present (binary trees are essentially tries for the case of binary alphabets). We shall also study *suffix tries*, an augmented trie that can be constructed in $O(n)$ time for a string s of n characters and in some sense store information about all substrings of s in only a $O(n)$ space.

Next we consider the *edit distance* between two strings, which is the minimum cost sequence of *insert character*, *delete character*, and *change character* operations required to convert one string to the other; it is often used as a measure of string similarity.

We then present arithmetic codes, a tool for compact representation of strings.

Finally, we present the *Burrows–Wheeler transform* to convert a string to a new one that in some sense displays its information content in a more explicit way. We also present *move-to-front* data compression, an application of the Burrows–Wheeler transform.

Lexicographic Sorting of Strings

Notation:

$S[1]...S[n]$: Set of strings.

$L[1]...L[n]$: Lengths of the strings; i.e., $S[i]$ has $L[i]$ characters.

MaxLength: Maximum string length.

N: Sum of all string lengths (so it is always true that $n \leq N$).

k: The alphabet size; characters are integers in the range 0 to $k-1$.

$S[i,1]...S[i,L[i]]$: The characters of $S[i]$.

Idea: Like the algorithm presented to sort fixed length vectors, work from $i=MaxLength$ down to $i=1$ and bucket sort on the ith component at the ith stage. To avoid wasting time on short strings, include only those strings at the ith stage of length $\geq i$, and check only non-empty buckets when performing the bucket sort.

Variable length string sorting algorithm:

1. Input the strings and form the array L of lengths and the array of strings S; that is $S[i]$ points to the list $S[i,1] ... S[i,L[i]]$.

2. Make a single pass through L to form, for $1 \leq j \leq MaxLength$, the lists:
    INDICES$[j]$ = List of all indices of strings of length j.

3. Form the list of pairs $(j,S[i,j])$ and bucket sort it; and then from the contiguous blocks that have the same first component, form for $1 \leq j \leq MaxLength$ the list:
    CHARS$[j]$ = Sorted list of characters that occur in position j of some string.

4. $Q :=$ an empty queue
    for $0 \leq i < k$ **do** BUCKET$[i] :=$ an empty queue

5. **for** $j := MaxLength$ **downto** 1 **do begin**
    A. Place the items of the list *INDICES*$[j]$ at the front of Q.
    B. Dequeue each index i in Q and enqueue it to BUCKET$[S[i,j]]$.
    C. **for** each character c in CHARS$[j]$ **do begin**
        Dequeue each index i in BUCKET$[c]$ and enqueue i to Q.
        Make BUCKET$[c]$ empty.
        end
    end

Time: $O(N+k)$ — Step 1 is $O(N)$. Step 2 is $O(n)$. Step 3 is $O(N+k)$. Step 4 is $O(k)$. Step 5A can be done in $O(1)$ time with the appropriate list implementation, or it suffices to use $O(|INDICES[j]|)$ time to move the indices one at a time; Step 5B uses time proportional to the current number of indices in Q (the number of strings of length $\geq j$); Step 5C collects strings only from non-empty buckets. Hence, each iteration of Step 5 takes time proportional to the number of strings of length $\geq j$, and Step 5 is $O(N)$.

Space: $O(N+k)$ — S and the *CHARS* lists use $O(N)$ space, the *INDICES* lists use $O(n) = O(N)$ space, and the *BUCKET* queues use $O(n+k)$ space.

Knuth–Morris–Pratt (KMP) String Matching

Idea: The function *left-to-rightMATCH* presented in the introduction finds the position of the string $T=T[1]...T[m]$ in the string $S=S[1]...S[n]$ by, for each position i of S, moving from left to right in T to determine if T matches at position i (i.e., if $S[i]...S[i+m-1] = T[1]...T[m]$). The time is $\Omega(mn)$ in the worst case because when a mismatch occurs at position j in T and we have already progressed to position $i+j-1$ in S, we start all over again at the beginning of T and position $i+1$ in S. To improve performance, before examining S, the KMP Algorithm computes from T a *back-up* array, so that when a mismatch occurs, we can adjust where we are in T, and continue forward in S.

Back-up array:

   $B[j]$ = largest $x<j$ such that $T[1]...T[x]$ is a suffix of $T[1]...T[j]$
       ($B[j]$ is defined to be 0 if there is no such x)

KMP Algorithm Using Back-Up Links

Unlike *left-to-leftMATCH*, the index i will now track the right end of a match. If $B[j-1]=0$ and $S[i]\neq T[j]$, then there is no possibility that a match would have been found starting at any of the positions $i-j+1$ through i, and we might as well go back to the beginning of T and continue matching in S at position $i+1$. However, if $B[j-1]>0$ and $S[i]=T[j]$, then we can simply proceed ahead at position $i+1$ in S and position $j+1$ in T (since by the definition of B and the fact that we have matched up to position j of T, it must be that $T[1]...T[j] = S[i-j+1]...S[i]$). If we succeed in finding a position i that is the right end of a match, we return $i-m+1$ (the position of the left end).

```
function kmpMATCH
    j := 0
    for i := 1 to n do begin
        j := j+1
        while j ≥ 1 and S[i]≠T[j] do j := B[j-1]
        if j=m then return i-m+1
        end
    return 0
    end
```

Time (excluding the time to compute the back-up array): We use an amortized analysis similar to that used for the *POP(i)* operation for stacks where one "credit" pays for $O(1)$ computing time. For each iteration of the kmpMATCH *for* loop, allocate two new credits, one to pay for all time used by that iteration of the *for* loop, except for multiple iterations of the *while* loop (that is, the first iteration of the *while* loop is covered by this credit, but not additional ones); the other credit is placed on j. Since each iteration of the *while* loop reduces j by at least 1, the number of credits currently on j is always $\geq j$ and hence suffice to pay for the time used by the *while* loop. Since only a total of $2n$ credits are used, the total time is $O(n)$.

Back-Up Diagrams

A convenient way to picture the back-up array B for a search string T is to draw a diagram with a start vertex along with one vertex for each character of T, where we simply move forward to the next vertex when there is a match, and drop back according to the back-up array when there is not. For example, suppose that we were searching a large text on elephants for the string T = "elephant." (places where elephant is at the end of a sentence). Then the figure below shows the back-up diagram. To make it easier to draw, arrows pointing up to 0 denote an edge back to vertex 0; the edge to 0 from vertex 9 is not needed unless we want to go on to find additional occurrences of the string.

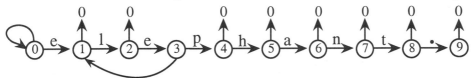

This back-up diagram represents the following rule:

> To find the string T = "elephant." in a string S, start at vertex 0 and move forward a vertex in the back-up diagram and forward a character in S whenever we match, and go back to vertex 0 (and do not move in S) whenever we mismatch. The only exception is when we mismatch after seeing "ele"; in this case, we go back to vertex 1 (and stay where we are in S). If we get to vertex 9, we have found the string (and are at its right end in S).

The pattern diagram for "elephant." is not very interesting because English is a relatively concise language and it is typical that the first few characters of a string more or less uniquely determine it. However, electronic files in practice can be quite varied (highly formatted data like spread sheets, science data, long genetic sequences, etc.). For a more interesting example, suppose that we were searching a long string of a's and b's for the string *ababbabaa*. The figure below shows the back-up diagram. Again, an arrow going up to a number denotes an edge back to the corresponding vertex.

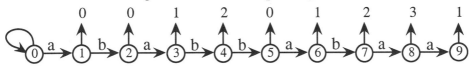

This back-up diagram saves time when searching in S. For example, if we get to vertex 8 and then discover that the next character in S is b, we follow the edge back to vertex 3 and continue forward both in the diagram and in S.

It is natural to consider "shortcuts". For example, in the figure above, if at vertex 8 and the next character in S is b, go directly to vertex 4. For a binary string, there are the same number of edges either way. In general, however, we need one these "direct" edges leaving a vertex for every possible character in the alphabet. After considering how to efficiently compute the back-up array, we shall see how a "direct" array can be computed in a similar fashion at the expense of more space for the array.

Efficient Computation of the KMP Back-Up Array

Idea: For $j>1$, if $T[j] = T[B[j-1]+1]$ then set $B[j] = B[j-1]+1$; if not, check if $T[j] = T[B[B[j-1]]]+1$, and so on. In the worst case we go all the way back and set $B[j]=0$.

> $B[0] := B[1] := 0$
> **for** $j := 2$ **to** m **do begin**
> > $B[j] := B[j-1]$
> > **while** $B[j]>0$ and $T[j] \neq T[B[j]+1]$ **do** $B[j] := B[B[j]]$
> > **if** $B[j]>0$ or $T[1]=T[j]$ **then** $B[j] := B[j]+1$
> **end**

Example: The figure below shows an array T and its back-up array B.

T:	a	a	c	a	a	c	a	c	a	a	c	a	a	a	c
	0	1	2	3	4	5	6	7	8	9	10	11	12	13	14	15

B:	0	0	1	0	1	2	3	4	0	1	2	3	4	5	2	3

$B[0] = B[1] = 0$

The $j=2$ iteration of the *for* loop sets $B[2]=B[1]$ (i.e., $B[2]=0$) since $B[2]=0$ the *while* loop is not entered, and then since $T[1]=T[2]$, the *if* statement sets $B[2]=1$.

The $j=3$ iteration of the *for* loop sets $B[3]=B[2]$ (i.e., $B[3]=1$), since $T[3] \neq T[2]$ the *while* loop sets $B[3]=B[B[3]]$ (i.e., $B[3]=0$), and then the *if* statement condition fails leaving $B[3]=0$.

The $j=4$ iteration of the *for* loop sets $B[4]=B[3]$ (i.e., $B[4]=0$), since $B[4]=0$ the *while* loop is not entered, and then since $T[1]=T[4]$ the *if* statement sets $B[4]=1$.

The $j=5$ iteration of the *for* loop sets $B[5]=B[4]$ (i.e., $B[5]=1$), since $T[5]=T[B[5]+1]$ (i.e., $T[5]=T[2]$) the *while* loop is not entered, and then since $T[5]>0$ the *if* statement sets $B[5]=2$.

Etc.

Complexity: To analyze the time used to compute B, we again use an amortized analysis where one "credit" pays for $O(1)$ computing time. For each iteration of the *for* loop, we allocate two new credits, one to pay for all time used by that iteration exclusive of the time used by the multiple iterations of the *while* loop (but the first iteration of the *while* loop is covered by the first credit) and one that is placed on j. The credits placed on j are used to pay for the iterations of the *while* loop. Consider a single iteration of the *for* loop that sets $B[j]$ to x for some $0 \leq x < j$. The only way that values of $B[y]$ for $y>j$ can be greater than x is by repeated iterations of the *for* loop that successfully perform the assignment statement $B[j] := B[j]+1$, and each of these iterations places a credit on j. Hence, j always has $\geq B[j]$ credits, which suffice to pay for multiple iterations of the *while* loop. Since only a total of $2m$ credits are used, the total time to compute B is $O(m)$. Space is $O(m)$.

Converting the KMP Back-Up Array to Direct Links

Notation: k = the number of characters in the input alphabet

Direct Links: The *while* loop of *kmpTEST* simply follows the back-up array until it finds a place in T where the next character matches the current position in S. Although we used an amortized analysis to show that the total time consumed by this process is only $O(n)$, we can improve practical performance by jumping directly to this position in T. For $0 \le j < m$ we define $D[T[j],j-1]=j$ and for a character $c \ne T[j]$:

$$D[c,j] = \text{largest } x \text{ such that } T[1]...T[x] \text{ is a suffix of } T[1]...T[j]c \text{ (or 0 if no such } x)$$

Computation of the direct array from the back-up array:

1. Put in the values for successful matches:
   for $j := 1$ **to** m **do** $D[T[j],j-1] := j$

2. Put in the remaining values for position 0:
   for all characters $c \ne T[1]$ **do** $D[c,0] := 0$

3. Put in the direct backward values:
   for $j := 1$ **to** $m-1$ **do**
       for all characters $c \ne T[j+1]$ **do** $D[c,j] := D[c,B[j]]$

Example: The figure below shows the same string T as the previous example along with the corresponding direct array D, assuming that the input alphabet is $\{a,b,c,d\}$ (i.e., $k=4$).

T:	a	a	c	a	a	c	a	c	a	a	c	a	a	a	c
	0	1	2	3	4	5	6	7	8	9	10	11	12	13	14	15

D:	0	1	2	3	4	5	6	7	8	9	10	11	12	13	14
a	1	2	2	4	5	2	7	5	9	10	2	12	13	14	2
b	0	0	0	0	0	0	0	0	0	0	0	0	0	0	0
c	0	0	3	0	0	6	0	8	0	0	11	0	0	6	15
d	0	0	0	0	0	0	0	0	0	0	0	0	0	0	0

KMP Algorithm Using Direct Links

```
function fastkmpMATCH
    j := 0
    for i := 1 to n do begin
        j := D[S[i],j]
        if j=m then return i−m+1
    end
    return 0
end
```

Complexity: $O(m)$ to compute B and $O(km)$ to compute D. Space is $O(km)$.

Reducing the Space for the KMP Direct Array

Problem: Often in practice T has relatively few distinct characters. For example T might be an English word containing less than 10 distinct characters but the alphabet size is 256 because characters are represented by bytes, and hence most rows of D are all 0's.

Map array: We assume that characters are represented by the integers 0 through $k-1$. For $0 \le c < k$, the map array translates between characters in the input alphabet and the indices 0 through $h-1$, where index 0 represents all characters not appearing in T and indices 1 through $h-1$ represent the characters appearing in T in sorted order.

$$M[c] = \begin{cases} 0 \text{ if } c \text{ does not appear in } T \\ i \text{ if } c \text{ is } i^{th} \text{ in the sorted list of characters appearing in } T \end{cases}$$

To compute M, we essentially do a bucket sort to compute M in $O(k)$ time:

1. Initialize the buckets:
   for $c := 0$ **to** $k-1$ **do** $M[c] := 0$

2. Place each character of T into its bucket (duplicates are discarded):
   for $j := 1$ **to** m **do** $M[T[j]] := 1$

3. Number the non-empty buckets:
   $i := 0$
   for $c := 0$ **to** $k-1$ **do if** $M[c]=1$ **then begin**
   $\quad i := i+1$
   $\quad M[c] := i$
   end

Representing D with less space: The rows of D are now indexed from 0 to $h-1$ and accessed by going through M (i.e., a reference $D[c,j]$ now becomes $D[M[c],j]$). For the previous example we have:

T:	a	a	c	a	a	c	a	c	a	a	c	a	a	a	c
	0	1	2	3	4	5	6	7	8	9	10	11	12	13	14	15

D:	0	0	0	0	0	0	0	0	0	0	0	0	0	0	0	0
1	1	2	2	4	5	2	7	5	9	10	2	12	13	14	2
2	0	0	3	0	0	6	0	8	0	0	11	0	0	6	15

M:	1	0	2	0
	a	b	c	d

Revised *fastkmpMATCH*: Replace $j := D[S[i],j]$ by $j := D[M[S[i]],j]$.

Complexity: $O(k+hm+n)$ time. $O(k+hm)$ space in addition to the space for S and T.

Boyer–Moore String Matching

Idea: The function *right-to-leftMATCH* presented in the introduction is a brute force approach to finding the position of the string $T=T[1]...T[m]$ in the string $S=S[1]...S[n]$. For each position i of S, it goes from right to left in T to see if T matches at position i (i.e., if $S[i]...S[i+m-1] = T[1]...T[m]$). The time is $\Omega(mn)$ in the worst case, because when a mismatch occurs in T, we start all over again at the right end of T and position $i+1$ in S. In a spirit similar to the KMP algorithm, before examining S, the Boyer–Moore algorithm computes from T an *advance-forward* array that will be used to speed up matching. However, instead of being used to readjust where we are in T, the advance forward array is used to determine how far forward we can safely advance in S without any possibility of missing a match (after a mismatch, the Boyer–Moore algorithm will always start again at the right end of T).

The Advance-Forward Array:

> $A[j]$ = smallest $x>0$ such that $T[j]\neq T[j-x]$ but for $j < k \leq m$ $T[k]=T[k-x]$
>
> > ($A[j]$ is defined to be 1 if there is no such x)

Boyer–Moore Algorithm

Unlike *right-to-leftMATCH*, the index i will now track the right end of a match. Suppose that starting at a position i in S, we started at the right end of T and worked to the left in until we found a mismatch; that is, for some $j \geq 1$ we found that $S[i]=T[m]$, $S[i-1]=T[m-1]$, $S[i-2]=T[m-2]$, ..., $S[i-m+j+1]=T[j+1]$, but $S[i+m-j]\neq T[j]$. If $A[j]=x$, it is not possible that any of the positions i, $i+1$, ..., $i+x-1$ in S are right ends of a match with T, and hence we can skip these positions.

```
function bmMATCH
    i := m
    while i ≤ n do begin
        j := m
        while j>0 and S[i−m+j]=T[j] do j := j−1
        if j=0 then return i−m+1
        i := i+A[j]
    end
    return 0
end
```

Complexity: It is possible to compute the advance-forward array in $O(m)$ time and an amortized analysis can be used to show that the number of comparisons used by bmMATCH is bounded by $3n$ in the worst case; in fact, bmMATCH can be modified (without changing A) so that the number of comparisons is less than $2n$ in the worst case (see the chapter notes).

Karp–Rabin Randomized "Finger Print" String Matching

Idea: To find the first position of $T=T[1]...T[m]$ in $S=S[1]...S[n]$, compare the hash of T (its "finger print") to that of substrings of S. Use a hash function that can be computed in $O(1)$ time as each character of S is processed, by modifying the hash value from the previous step.

Definition: A hash function h that maps strings to integers is *incremental* if:
1. For any string w, $h(w)$ can be computed in $O(|w|)$ time.
2. For any string w and any characters a and b, $h(wb)$ can be computed in $O(1)$ time from the value of $h(aw)$.

Generic matching algorithm: Let h denote an incremental hash function.

> **function** krMATCH
> > $x := h(T)$
> > **for** $i := m$ **to** n **do begin**
> > > $y := h(S[i-m+1]...S[i])$
> > > **if** $x=y$ and $T = S[i-m+1]...S[i]$) **then return** $i-m+1$
> > > **end**
> > **return** 0
> > **end**

Complexity: Since h is incremental, x can be computed in $O(m)$ time, on the first iteration of the *for* loop ($i=m$) y can be computed in $O(m)$ time, and on subsequent iterations of the *for* loop y can be computed in $O(1)$ time. Thus, assuming that the *if* statement tests $x=y$ first and immediately fails if $x \neq y$, the time is $O(m+n)$ excluding the total time spent when the *if* statement finds a "false match" ($x=y$ but $T \neq S[i-m+1]...S[i]$). Although checking for a false match takes $\Theta(m)$ time, if h is equally likely to produce any value in the range 0 to p, and $p>m$, then on the average, less than $1/m$ of the iterations have to check a false match, and hence expected time for false matches is $O(n)$, yielding overall expected time of $O(m+n)$. $O(1)$ space is used in addition to the space for S and T.

The hash function: A good choice for h is something we have already seen, the polynomial hash function; that is:

$$h\big(S[i-m+1]\cdots S[i]\big) = \big(S[i-m+1]p^{m-1} + \cdots + S[i-1]p + S[i]\big) \text{ MOD } q$$

This hash function is incremental because the first iteration of the *for* loop ($i=m$) can compute h in $O(m)$ time (using Horner's method) and then each subsequent iteration ($i>m$) can subtract $s[i-m]p^{m-1}$ from y (p^{m-1} can be pre-computed, and since the *for* loop has just incremented i, we use the index $i-m$ instead of $i-m+1$), multiply y by p, and then add on $S[i]$ (all arithmetic is MOD q).

Practical considerations: Choose p to be a prime $>k$ and q as large as can be easily accommodated on your computer. Depending on the speed of arithmetic operations on the machine being used and the choice of p and q, this approach may be competitive with *fastkmpMATCH*, and no preprocessing of T is required.

Shift-And String Matching

Idea: By packing the bits representing the characters of $T=T[1]...T[m]$ into a single integer, we can check if T matches at a given position of $S=S[1]...S[n]$ in $O(1)$ time by computing a logical AND with the corresponding packed characters of S (here we assume all integers are non-negative and represented in binary). We assume that characters are integers in the range 0 to $k-1$ and use the following notation:

$ones =$ one less than the least power of 2 greater $\geq k$ (i.e., the binary representation of *ones* is $\lceil \log_2(k) \rceil$ 1's)

$charsize =$ the number of bits to represent a character ($charsize = \log_2(ones+1)$)

x OR $y =$ bitwise logical OR (OR of two bits = 1 if either is 1, 0 otherwise)

x AND $y =$ bitwise logical AND (AND of two bits = 1 if both are 1, 0 otherwise)

$ShiftLeft(x,i) =$ shift the bits of x to the left by i positions and place 0's in the rightmost i positions that have been vacated (i.e., $x = x*2^i$)

Pre-processing to pack T into an integer and create the match mask:

procedure PACK

       $Tpack := MatchMask := 0$

       for $i := 1$ **to** m **do begin**

             $Tpack := \text{ShiftLeft}(Tpack, charsize)$ OR $T[i]$

             $MatchMask := \text{ShiftLeft}(MatchMask, charsize)$ OR $ones$

             end

       end

Finding T in S:

function saMATCH

       $Spack := 0$

       for $i := 1$ **to** n **do begin**

             $Spack := (\text{ShiftLeft}(Spack, charsize)$ OR $S[i])$ AND $MatchMask$

             if $i \geq m$ and $Spack=Tpack$ **then return** $i-m+1$

             end

       return 0

       end

Complexity: In practice, T must be small enough to packed into the machine's maximum word size. $O(1)$ space is used in addition to that used by S and T. The time is $O(m+n)$ assuming that the standard OR, AND, and SHIFT operations are $O(1)$ time. On some machines, only shifting by one position is available, but a loop to shift *charsize* bits is $O(1)$ assuming that the alphabet size is constant with respect to m and n (also, in practice, this loop is very fast since it can all done in the processor's registers).

Shift-and with don't-care positions

Positions of T that do not matter can be indicated by use of a special character or by an auxiliary array of m true/false values. To incorporate don't-cares, modify PACK to perform its two OR operations only when i is not a don't-care position. Because this trivial change is all that is needed, don't-care capability is essentially "free".

Shift-And with Anything-But Positions

Anything-but positions: Like don't-care positions, *anything-but* positions are special positions of T that can be indicated by an auxiliary array of m true/false values. We define T to *match* at a position i of S if for each position j of T at least one of the following conditions holds:

- j is a don't-care position
- j is an anything-but position and $T[j] \neq S[i+j-1]$
- j is not an anything-but position and $T[j] = S[i+j-1]$

Idea: In order for an anything-but position j in T to satisfy $T[j] \neq S[i+j-1]$, at least one of the bits in the block of *charsize* bits corresponding to $T[j]$ in *Tpack* must differ from the corresponding bit in *Spack*. To test this, we create an integer where blocks of *charsize* bits are all 0's except possibly for the leading bit, which is set to 1 if the corresponding blocks of *Tpack* and *Spack* differ in at least one bit position. We make use of three new constants:

> *HighBit* = The largest power of two $\leq k$; i.e., the integer whose binary representation is a 1 followed by *charsize*–1 0's.

> *HighMask* = The integer whose binary representation is all 0's except that for $1 \leq i \leq m$ the $(charsize * i)^{th}$ bit from the right is 1 (i.e., only the high order bit in each block of *charsize* bits is 1).

> *ABMask* = The integer whose binary representation is all 0's except that for $1 \leq i \leq m$ the $(charsize * i)^{th}$ bit from the right is 1 if and only if i is an anything-but position.

Combining bits: Identify all mis-matches between Spack and *Tpack* by producing an integer whose bits are all 0 except possibly for the high order positions of the blocks of *charsize* bits where a mis-match occurs. It makes use of the standard bit-wise logical exclusive OR function (the XOR of two bits is 1 if they differ and 0 otherwise).

> **function** CombineBits
> $x := y := Spack$ XOR *Tpack*
>   **for** $i := 2$ **to** *charsize* **do begin**
> $x := \text{ShiftLeft}(x,1)$
> $y := x$ OR y
>   **end**
>   **return** (y AND *HighMask*)
>   **end**

Complexity: *CombineBits* runs in $O(charsize)$ time, which is $O(1)$ assuming that *charsize* is a constant independent of m and n. In practice, it is very fast since all operations are performed on the two variables x and y, which can be stored in the processors registers. Also, *CombineBits* only shifts by one bit at a time, which on some machines can be faster than shifting by more than one bit at a time.

Revised *PACK* Procedure for Anything-But Positions (includes don't-cares)

Like *PACK*, *revisedPACK* processes *T* one character at a time and uses shift operations to build up the integers *Tpack*, *MatchMask*, *ABMask*, and *HighMask*.

> **procedure** revisedPACK
>> *Tpack* := *MatchMask* := *ABMask* := *HighMask* := 0
>> **for** i := 1 **to** m **do begin**
>>> *Tpack* := ShiftLeft(*Tpack*,*charsize*)
>>> *MatchMask* := ShiftLeft(*MatchMask*,*charsize*)
>>> *ABMask* := ShiftLeft(*ABMask*,*charsize*)
>>> *HighMask* := ShiftLeft(*HighMask*,*charsize*) OR *HighBit*
>>> **if** i is an anything-but position **then** *ABMask* := *ABMask* OR *HighBit*
>>> **else if** i is not a don't-care position **then begin**
>>>> *Tpack* := *Tpack* OR *T*[i]
>>>> *MatchMask* := *MatchMask* OR *ones*
>>>> **end**
>>> **end**
>> **end**

Revised *saMATCH* Function for Anything-But Positions (includes don't-cares):

The revised function *sa+dc+abMATCH* is the same as *saMATCH* except that the *if* statement must also check that the anything-but positions are all mis-matches. To do this, *CombineBits* is used to determine all positions where mis-matches occur and the integer *test* is set to the logical AND of this result and *ABMask* (so only the subset of the mis-match positions that are also anything-but positions is retained).

> **function** sa+dc+abMATCH
>> *Spack* := 0
>> **for** i := 1 **to** n **do begin**
>>> *Spack* := (ShiftLeft(*Spack*,*charsize*) OR *S*[i]) AND *MatchMask*
>>> *test* := *ABMask* AND CombineBits(*Spack*,*Tpack*)
>>> **if** $i \geq m$ and *Spack*=*Tpack* and *ABMask*=*test* **then return** $i-m+1$
>>> **end**
>> **return** 0
>> **end**

Complexity: Since *revisedPACK* remains $O(m)$ time, assuming that *charsize* is a constant independent of m and n (and hence *CombineBits* is $O(1)$ time), the algorithm remains $O(m+n)$. The space is still $O(1)$ in addition to the space used by *S* and *T*.

Shift-And with Minimum Mis-Matches

Mis-matches: T matches S at position i with at most q *mis-matches* if there are at most q values of j for which j is not a don't-care or anything-but position but $T[j] \neq S[i+j-1]$.

Minimum mis-match problem: For an integer $0 \leq q < m$, find the first position i in S for which the number of mis-matches with T at position i is $\leq q$ and for no other position is the number of mis-matches with T fewer (or determine that no such position exists).

Idea: We employ the *CombineBits* function and appropriate masking to create an integer x with 1's in the high-order bit of the charsize blocks corresponding to mis-match positions. We then employ the operation *OneCount(x)* to determine the number of 1's appearing in the binary representation of x.

Revised algorithm: We use the procedure *revisedPACK* that was presented on the preceding page. Matching is further revised to keep track of the position of S for which the least number of mis-matches have occurred thus far; the position with the least number is returned if that number is $\leq q$, otherwise 0 is returned.

> **function** sa+dc+ab+mmMATCH(q)
>> *bestposition* := $m-1$
>> *bestvalue* := $q+1$
>> *Spack* := 0
>> **for** i := 1 **to** n **do begin**
>>> *Spack* := (ShiftLeft(*Spack,charsize*) OR $S[i]$) AND *MatchMask*
>>> *test* := CombineBits(*Spack,Tpack*)
>>> **if** $i \geq m$ and *ABMask*=(*ABMask* AND *test*) **then begin**
>>>> *newvalue* := OneCount(*MatchMask* AND *test*)
>>>> **if** *newvalue*<*bestvalue* **then begin**
>>>>> *bestposition* := i
>>>>> *bestvalue* := *newvalue*
>>>>> **end**
>>> **end**
>> **end**
>> **return** *bestposition*$-m+1$
> **end**

Complexity: Assuming that *OneCount* is available as an $O(1)$ time machine instruction, asymptotic complexity is unchanged and the algorithm remains $O(m+n)$ time and $O(1)$ space in addition to the space used by S and T. Unfortunately, historically *OneCount* has only been available on special purpose machines and some "mainframe" machines with larger instruction sets. Alternately, *OneCount* can be implemented in $O(\log(m))$ time using just *ShiftRight*, AND, and binary addition operations (see the exercises) for a total of $O(m+n\log(m))$ time; in practice, since *OneCount* is computed by the processor using just its registers (and no memory accesses to S or T), it is relatively fast in any case.

Shift-And with Character Mapping

Idea: It is typical in practice for the characters that appear in T to be a small subset of the input alphabet. For example, for English text, input characters might be represented with a 7-bit ASCII code but T might have only 25 characters total ($m=25$), and hence at most 5 bits are need to represent the number of distinct characters in T.

The map array: We can proceed exactly as was done to reduce space for the KMP direct array. We assume that characters are represented by the integers 0 through $k-1$. For $0 \leq c < k$, the map array translates between characters in the input alphabet and the indices 0 through $h-1$, where index 0 represents all characters not appearing in T and indices 1 through $h-1$ represent the characters appearing in T in sorted order.

$$M[c] = \begin{cases} 0 \text{ if } c \text{ does not appear in } T \\ i \text{ if } c \text{ is } i^{th} \text{ in the sorted list of characters appearing in } T \end{cases}$$

Reduced versions of the character related constants:

Rones = One less than the least power of 2 greater $\geq h$. That is, the binary representation of *ones* is $\lceil \log_2(h) \rceil$ 1's.

Rcharsize = The number of bits to represent a character in T (including the special value of 0 that represents a character not in T). That is, the binary representation of *charsize* is $\log_2(ones+1)$.

Modifying the pack procedure and match function to incorporate the map array:

In the pack procedure, replace the reference $T[i]$ by $M[T[i]]$. In the search function replace the reference $S[i]$ by $M[S[i]]$.

Complexity: The running times of the pack procedure and search function are only changed by a (small) constant factor. An additional $O(k)$ space is used for M, for a total of $O(k+m+n)$ space, which is $O(m+n)$ assuming that the alphabet size is a constant with respect to m and n (or at least that k is $O(m+n)$).

Shift-And with Character Bit-Vectors

Idea: Instead of keeping m blocks of *charsize* bits in *Spack*, we employ a different shift-and computation that requires only one bit in *Spack* for each position of T. For example, if 7 bit ASCII codes are used to represent characters, then the maximum length of the search string T that can be used without overflowing a given machine word size (i.e., the maximum practical value for m) is increased by a factor of 7.

New definitions of Tpack and Spack:

$Tpack[c]$: Instead of a single integer *Tpack*, we now have one for each character of the input alphabet (again we assume that characters are represented by the integers 0 through $k–1$); for $0 \leq c < k$, $Tpack[c]$ is the integer whose j^{th} bit from the right, $1 \leq j \leq m$, is 1 if and only if $T[j]=c$.

$Spack$: At position i of S, the jth bit from the right in *Spack* is set to 1 if and only if the first j characters of T match ending at position i; that is for $1 \leq j \leq m$, the jth bit from the right of *Spack* is 1 if and only if $T[1]...T[j] = S[i–j+1]...S[i]$.

The new packing procedure: In $O(k+m)$ time compute *Tpack* and the constant *LeftOne* (a 1 followed by $m–1$ 0's that will be used to test for a successful match). Theoretically, the time can be reduced to $O(m)$ with on-the-fly array initialization; in any case, it is $O(m)$ assuming that k is a constant independent of m.

```
procedure sacbvPACK
    for c := 0 to k–1 do Tpack[c] := 0
    for j := 1 to m do begin
        if j=1 then LeftOne := 1 else LeftOne := ShiftLeft(LeftOne,1)
        Tpack[T[j]] := Tpack[T[j]] OR LeftOne
        end
    end
```

New matching function: For each position i of S we shift *Spack* left and then retain a 1 only in those positions that match $S[i]$ (so if the xth bit of *Spack* was 1 on the previous iteration, it becomes the $x+1^{th}$ bit and remains 1 if the $x+1^{th}$ bit of $Tpack[S[i]]$ is 1). A match is been found when the mth of *Spack* is 1. Including *sacbvPACK*, $O(m+n)$ time is used, assuming that m bits can be manipulated by a machine instruction, and $O(k)$ space is used in addition to that used by S and T.

```
function sacbvMATCH
    Spack :=  0
    for i := 1 to n do begin
        Spack := (ShiftLeft(Spack,1) OR 1) AND Tpack[S[i]]
        if (Spack AND LeftOne) ≠ 0 then return i–m+1
        end
    return 0
    end
```

Don't-Care and Anything-But Positions: If j is a don't-care position, set position j to 1 for all items of *Tpack* and if it is an anything-but position, set position j to 1 for all items of *Tpack* except $Tpack[T[j]]$. Preprocessing is $O(km) = O(m)$ assuming k is constant.

Minimum Mis-Matches with Character Bit-Vectors

Idea: When q is small, it may be practical to maintain a version of *Spack* for each integer in the range 0 to q, where *Spack*[x] keeps track of prefixes of T that match with x mismatches (the same preprocessing as before is used to compute *LeftOne* and *Tpack*[0]...*Tpack*[k–1]). The *for* loop of the match function uses an inner loop from 0 to q to update each of *Spack*[0]...*Spack*[q].

Generalization to an array *Spack*[x]: At position i of S, for $1 \le j \le m$, the jth bit from the right of *Spack*[x] is 1 if and only if $T[1]...T[j]$ matches $S[i-j+1]...S[i]$ with at most x mismatches (after taking into account don't-cares and anything-buts).

Matching: As illustrated by the example on the following page, *Spack*[0] is computed as before. For $x>0$, the jth bit from the right of *Spack*[x] is initialized to 1 if $j \le q$; that is, we imagine that S is preceded by a sequence of special characters $S[-q]...S[-1]$, none of which appear in T. Then, at each position i of S, there are two reasons for setting the jth bit from the right in *Spack*[x] to 1, $1 < j \le q$ (the first bit from the right is always 1):

1. If $T[j]=S[i]$, then it suffices that there be x or fewer mis-matches between $T[1]...T[j-1]$ and $S[i-j+1]...S[i-1]$; that is the jth bit of *Spack*[x] should be 1 if and only if the $j-1^{th}$ bit of *Spack*[x] was 1 at position $i-1$.

2. If $T[j] \ne S[i]$ (we have a new mis-match), it suffices that there be $x-1$ or fewer mis-matches between $T[1]...T[j-1]$ and $S[i-j+1]...S[i-1]$; that is the jth bit of *Spack*[x] should be 1 if and only if the $j-1^{th}$ bit of *Spack*[$x-1$] was 1 at position $i-1$.

Save[x] saves the value that *Spack*[x] had at position $i-1$, shifted left by one bit.

```
function sacbv+dc+ab+mmMATCH(q)
    bestposition := m−1
    bestvalue := q+1
    for x := 0 to q do Spack[x] := 0
    for x := 1 to q do Spack[x] := ShiftLeft(Spack[x−1],1) OR 1
    for i := 1 to n do begin
        for x := 0 to q do begin
            Save[x] := Spack[x] := ShiftLeft(Spack[x],1) OR 1
            Spack[x] := Spack[x] AND Tpack[S[i]]
            if x>0 then Spack[x] := Spack[x] OR Save[x−1]
            if i ≥ m and ((Spack[x] AND LeftOne) ≠ 0) and x<bestvalue then begin
                bestposition := i
                bestvalue := x
            end
        end
    end
    return bestposition−m+1
end
```

Complexity: $O(m+qn)$ time assuming that m bits can be manipulated by a machine instruction. $O(q+k)$ space in addition to that used by S and T.

Example of running sacbv+dc+ab+mmMATCH(q):

Let $S = abacd$ ($n=5$), $T = adac$ ($m=4$), and $q = 3$. The best that can be done is to match with one mis-match at position– 1 of S:

Pre-processing computes:

```
LeftBit  = 1000
Tpack[a] = 0101
Tpack[b] = 0000
Tpack[c] = 1000
Tpack[d] = 0010
```

This column shows the value of $Spack[x]$ after the AND with the bit vector for $S[i]$.

Initialize:

```
Spack[0] = 0000
Spack[1] = 0001
Spack[2] = 0011
Spack[3] = 0111
```

This column shows the value of $Save[x]$.

This column shows the final value of $Spack[x]$ after the OR with $Save[x-1]$.

i=1:

Spack[0]=0000	shift left	0001	AND	0001		0001
Spack[1]=0001	and OR	0011	with	0001	OR with	0001
Spack[2]=0011	with 1	0111	0101	0101	$Save[x–1]$	0111
Spack[3]=0111	→	1111	→	0101	→	0111

i=2 (finds that the left bit is one, but does nothing since i<m):

Spack[0]=0001	shift left	0011	AND	0000		0000
Spack[1]=0001	and OR	0011	with	0000	OR with	0011
Spack[2]=0111	with 1	1111	0000	0000	$Save[x–1]$	0011
Spack[3]=0111	→	1111	→	0000	→	1111

i=3:

Spack[0]=0000	shift left	0001	AND	0001		0001
Spack[1]=0011	and OR	0111	with	0101	OR with	0101
Spack[2]=0011	with 1	0111	0101	0101	$Save[x–1]$	0111
Spack[3]=1111	→	1111	→	0101	→	0111

i=4 (sets bestposition=1, bestvalue=1):

Spack[0]=0001	shift left	0011	AND	0000		0000
Spack[1]=0101	and OR	1011	with	1000	OR with	1011
Spack[2]=0111	with 1	1111	1000	1000	$Save[x–1]$	1011
Spack[3]=0111	→	1111	→	1000	→	1111

i=5:

Spack[0]=0000	shift left	0001	AND	0000		0000
Spack[1]=1011	and OR	0111	with	0010	OR with	0011
Spack[2]=1011	with 1	0111	0010	0010	$Save[x–1]$	0111
Spack[3]=1111	→	1111	→	0010	→	0111

Comparison of String Matching Methods

Idea: String matching has been extensively studied; the chapter notes provide references to a wealth of work that goes beyond what has been presented here. Even among the methods that are presented here, there are many tradeoffs that arise in practice when we consider the constants that can be expected under different assumptions.

KMP using the direct array: In terms of worst case performance, it is difficult to imagine a faster method; for each character of S, only a simple assignment statement (with one memory reference to the D array and one into S) is made. For on-line applications where the characters of S must be scanned in any case, this may be the best choice, especially when S is large relative to T (so the preprocessing is insignificant).

KMP using the back-up array: The worst case performance is only about a factor of two worse than when using the direct array and the pre-processing is simpler, using an amount of space that does not depend on the alphabet size. For on-line applications where the characters of S must be scanned in any case, this may be the best choice, especially when the alphabet size is relatively large.

Boyer–Moore: In the worst case, Boyer–Moore is likely to have a constant that is a bit larger than KMP with the back-up array since the modifications that get the number of comparisons down to $2n$ complicate the code somewhat. The real advantage of the Boyer–Moore algorithm is expected performance. Because it scans the pattern from right to left and skips portions of S when a mismatch is found, there may be a fraction of S that is never even examined. In fact, under some assumptions about the input, we might even hope that the expected number of comparisons to positions in S is $O(n/m)$. So, when S can be read in a random access fashion, and especially when m is large, Boyer–Moore can be the best choice. In practice, the presence of caching hardware on the machine in question may diminish the increase in speed of Boyer–Moore over KMP since with long patterns the cache hardware is likely to benefit KMP more than Boyer–Moore.

Karp–Rabin: Although theoretically one can argue that expected time is less desirable than worst case, hashing techniques are well established. Karp–Rabin is simple, requires little preprocessing, and performs a quick computation for each position of S. In addition it generalizes easily to multi-dimensional problems (see the chapter notes).

Shift–and: For the special situation that the search string can be packed into a word that can be manipulated with single machine instructions, this approach requires little pre-processing and performs only a quick computation for each position of S. For example, by making use of a *MAP* array, if less than 32 distinct characters are used in T and the machine being used has 128 bit arithmetic, then search strings up to length 25 can be used; and by using character bit-vectors, 128 character search strings can be used. In addition, don't-care positions come essentially free and the approach is easily adapted to anything-but positions. Minimizing mis-matches may also be practical.

Pattern Matching

Idea: We use a formal syntax for describing simple patterns that denote a set of strings over an alphabet Σ.

The empty string: ε denotes the *empty string*; a string of length zero. It is different than the empty set; for example, $\{\varepsilon\}$ is a non-empty set containing the string of length zero.

Regular expressions: A *regular expression* is a string that describes a set of strings:

ϕ denotes the empty set.

ε denotes $\{\varepsilon\}$.

For a character c in Σ, c denotes $\{c\}$.

For two regular expressions r and s that denote sets X and Y, respectively:

$$r + s = X \cup Y$$

$$rs = \{xy: x \text{ is a string in } X \text{ and } y \text{ is a string in } Y\}$$

$$r^* = \bigcup_{i=0}^{\infty} r^i, \text{ where } r^i = \begin{cases} \varepsilon & \text{if } i = 0 \\ rr^{i-1} & \text{if } i > 0 \end{cases}$$

Note: We assume that * takes precedence over • which takes precedence over + and allow the use of parentheses. We also allow the • symbol to be omitted (so that rs denotes $r{\bullet}s$).

Examples of regular expressions:

$ac+bc = (a+b)c = \{ac, bc\}$

$(a+b)(c+d)a = \{aca, ada, bca, bda\}$

$(a+b)(\varepsilon+c)d = \{ad, bd, acd, bcd\}$

$(a+b)^* = $ all possible strings of a's and b's (including ε)

$(a+b)^*a(a+b)^* = $ strings of a's and b's with at least one a

$(aa+bbb)^* = $ strings of a's and b's with an even number of a's and a multiple of 3 b's

$(S+s)t(u+ew)art = $ two different ways to spell the name Stuart capitalized or not

The basic pattern matching problem: Given a string $S = S[1]...S[n]$ and a regular expression $R = R[1]...R[m]$:

Find the least j such that for some i, $S[i]...S[j]$ is denoted by R.

Common variations of the basic pattern matching problem (see the exercises):

1. Find the least j, and given this j, the least i such that $S[i]...S[j]$ is described by R.
2. Find the least j, and given this j, the greatest i such that $S[i]...S[j]$ is described by R.
3. Find the least i, and given this i, the greatest j such that $S[i]...S[j]$ is described by R.
4. Find all positions of S that are left or right ends of patterns described by R.

Pattern Diagrams

Idea: Regular expression provide a nice language for users to describe patterns. We convert regular expressions to a directed graph where paths can be followed according to the characters of the string being searched.

Definition: A *pattern diagram* representing a regular expression *r* consists of:

- A set of vertices numbered starting at 1.
- A set of directed edges that connect the vertices; an edge from a vertex x to a vertex y is represented by the ordered pair (x,y).
- The lowest numbered vertex is the *start vertex* and the highest numbered vertex is the *final vertex* (in the case that there is only one vertex, it is both the start and the final vertex).
- Each edge is labeled by a character or by the empty string symbol ε.
- A string s is one of the strings denoted by R if and only if there is a path from the start to the final vertex whose labels form s.

Representing a pattern diagram: For n vertices, we use an adjacency list representation consisting of an array $V[1]...V[n]$ where $V[i]$ points to a list of pairs of the form c, j to represent an edge from vertex i to vertex j that is labeled by c (where c is a character or ε). We always assume that vertex 1 is the start vertex and vertex n is the final vertex.

Example: Below are three of the regular expressions from the previous example along with a corresponding pattern diagram and its linked list representation.

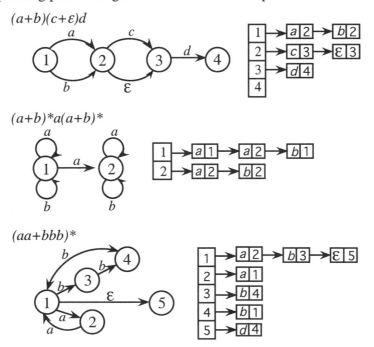

McNaughton–Yamada Algorithm

Idea: Recursively make diagrams for sub-expressions and then put them together.

$R = \phi$: Create a disconnected start and final vertex.

$R = \varepsilon$: Create a single edge labeled ε from the start to final vertex:

$R = c$: Create a single edge labeled by the character c from start to final vertex:

$R = A+B$: Create new start and final and "fork" with ε to the patterns for A and B:

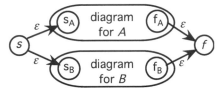

$R = A \bullet B$ or AB: Connect the final vertex of A to the start vertex of B:

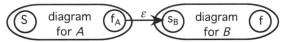

$R = A^*$: Create new start and final and put in ε edges to skip and loop:

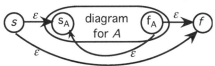

$R = (A)$: Ignore the parentheses and use the diagram for A.

Properties of the diagram constructed from a regular expression of length m:

- $\leq 2m$ vertices

  This follows by induction: The first three conditions introduce two vertices for the symbols ε and ϕ and for any single character c, at most two vertices are introduced for each +, no new vertices are introduced for $\bullet$ (so it is ok if it is omitted), two new vertices are introduced for *, and no new vertices are introduced for (and).

- $< 4m$ edges (final vertex has no outgoing edges, all others have out-degree ≤ 2)

  Again, induction can be used to show that the final vertex has no outgoing edges and all others have out-degree ≤ 2, from which $\leq 4m$ edges follows.

Linear Time Implementation of the McNaughton–Yamada Construction

Idea: Given a regular expression $R[1]...R[m]$, on the first pass convert R to prefix notation in the stack P where the operator precedes its arguments and parentheses are not needed; for example, $(a+bc)*$ is rewritten as $*+a \bullet bc$ and placed in P with $*$ on top and c on the bottom. By viewing R as a stack where $R[1]$ is the top and $R[m]$ is the bottom, we can use the linear time infix to prefix conversion algorithm presented in the chapter on induction and recursion by taking $\%$ to be the $*$ operation. On the second pass, pop off an operator and recursively pop off its argument(s) to apply McNaughton–Yamada.

Pattern diagram construction algorithm: $CONSTRUCT(1)$ forms an adjacency list representation for the pattern diagram for R, where $CONSTRUCT(i)$ takes the index i of the start vertex for a sub-expression, works recursively to construct its portions of the adjacency lists, and returns the final vertex for that sub-expression.

> **function** CONSTRUCT(i)
>> $c := POP(P)$
>> **if** c is φ **then** return $i+1$
>> **else if** c is ε or a character in Σ **then begin**
>>> Place an edge labeled c from i to $i+1$.
>>> **return** $i+1$
>>> **end**
>>
>> **else if** c is $\bullet$ **then begin**
>>> $x := CONSTRUCT(i)$
>>> $y := CONSTURUCT(x+1)$
>>> Place an edge labeled ε from x to $x+1$.
>>> **return** y
>>> **end**
>>
>> **else if** c is $+$ **then begin**
>>> $x := CONSTRUCT(i+1)$
>>> $y := CONSTRUCT(x+1)$
>>> Place edges labeled ε from i to $i+1$ and from i to $x+1$.
>>> Place edges labeled ε from x to $y+1$ and from y to $y+1$.
>>> **return** $y+1$
>>> **end**
>>
>> **else if** c is $*$ **then begin**
>>> $x := CONSTRUCT(i+1)$
>>> Place an edge labeled ε from i to $i+1$ and from i to $x+1$.
>>> Place an edge labeled ε from x to $i+1$ and from x to $x+1$.
>>> **return** $x+1$
>>> **end**
>>
>> **else** $ERROR$ — *illegal pattern syntax*
>> **end**

Complexity: Each call to $CONSTRUCT$ pops one item, and so time is proportional to the size of P, which is $O(m)$. Space is proportional to the adjacency lists constructed, $O(m)$.

Matching with Pattern Diagrams

Idea: Given a pattern diagram and a string $S=S[1]..S[n]$, for $i=0$ to n maintain the set of the vertices that are reachable by following a legal path from the start vertex according to the characters of S processed thus far. The case $i=0$ initializes the set to contain 1 (the start vertex) together with all vertices reachable by ε edges. At each stage i, we create a new set of reachable vertices by following all edges labeled by $S[i]$ from the current set and then adding all vertices reachable by ε edges. The bit vector A holds the vertices reachable at step $i-1$ and B is used to compute the vertices reachable at step i (and then to start the next iteration we set $A=B$). To find all vertices reachable by ε edges, a "breadth-first" search finds all that are reachable by a path of length 1, then all that are reachable by a path of length 2, and so on, being careful not to loop to vertices already discovered. The procedure returns i (the right end of a match) if m (the final vertex) is added at the ith stage, or goes through S and returns 0 to indicate that there was no match.

function basicpatternMATCH
>**for** $i := 0$ **to** n **do begin**
>> 1. Place in B all vertices reachable from a vertex in A on $S[i]$:
>>>**for** $j := 1$ **to** m **do** $B[j] := 0$
>>>**if** $i=0$ **then** $B[1] := 1$
>>>**else for** $j := 1$ **to** m **do if** $A[j]=1$ **then**
>>>> **for** each edge (j,x) on the list of edges for vertex j **do**
>>>>> **if** (j,x) has label $S[i]$ **then** $B[x] := 1$
>>
>> 2. Add to B all vertices reachable from B by edges labeled ε:
>>> Initialize a *queue* to contain each vertex in B.
>>> **while** *queue* is not empty **do begin**
>>>> Dequeue a vertex x.
>>>> **for** each y such that $B[y]=0$ and there is an ε edge (x,y) **do begin**
>>>>> $B[y] := 1$
>>>>> Enqueue y.
>>>> **end**
>>> **end**
>>
>> 3. Return i if the final vertex is in B, otherwise set $A := B$
>>> **if** $B[n]=1$ **then return** i **else** $A := B$
>
>**end**
>**return** 0
>**end**

Complexity: If we let v and e denote the number of vertices and edges in the pattern diagram, then both Steps 1 and 2 are $O(v+e)$ since they sequence through the vertices and visit each edge at most once. Hence the time is $O((v+e)n)$ and the space is $O(v+e)$ in addition to the space used by S, which implies $O(kn)$ time and $O(k)$ space if the pattern diagram was produced by McNaughton–Yamada from a pattern of length k.

Tries

Idea: Store a set of strings written over an alphabet of size k as root-to-leaf paths in an ordered tree that has at most k children per vertex.

Example: A trie where the leaves represent the set of strings {*aaccaa, aaccab, abc,bb, bca, bcbb, c*} and the non-leaf vertices represent the prefixes of these strings:

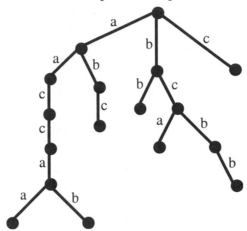

Trie representations: There is a tradeoff between the space used and efficient computation of CHILD(v,c), the child of vertex v that corresponds to character c.

> **Arrays:** Store at each vertex v and array $v[0]...v[k-1]$ that corresponds to the characters of the alphabet in lexicographic order, where $v[i]$ is a pointer to CHILD(v,i) (or *nil* if there is no child of v corresponding to the ith character of the alphabet). $O(kn)$ space and $O(1)$ time for CHILD. The arrays can be initialized in $O(k)$ time per vertex (also, theoretically, on-the-fly initialization could be used).
>
> **Linked lists:** Store the children of a vertex in lexicographic order in a linked list. $O(n)$ space and $O(k)$ time for CHILD.
>
> **Balanced trees:** Store the children of a vertex in a balanced tree (according to lexicographic order of characters). $O(n)$ space and $O(\log(k))$ time for CHILD.
>
> **Hash table:** To find a child c of a vertex v, hash on the pair v,c. $O(n)$ space and $O(1)$ *expected* time for CHILD. Theoretically, "perfect hashing" techniques can also be employed (see the notes to the chapter on hashing).

Example: Sorting Strings with Tries

To sort a set of strings whose lengths sum to n, initialize an empty trie, insert all the strings, and traverse the trie in prefix order. The trie must be stored so that children of a vertex are accessed in lexicographic order; for example, arrays, linked list, or balanced trees all yield $O(n)$ time assuming that the alphabet size k is constant with respect to n.

Example: Aho–Corasick Multiple String Matching

Idea: Given a set of strings X whose lengths sum to m, search for an occurrence of any one of these strings in a string $S = S[1]...S[n]$ by generalizing the preprocessing for the KMP string matching algorithm with direct links to place the strings of X into a trie. Number the root 0 and the other vertices from 1 to t (e.g., use a prefix traversal — it must be that $t \leq m+1$). Also, to allow strings to be substrings of other strings, $MARK[0]...MARK[t]$ is an array of true/false values that indicate whether a vertex corresponds to a string in X (leaves are always marked *true*). For a vertex v with a child w that corresponds to character c, a trie edge (solid edge) is defined by $D[c,v]=w$; for a character c that does not correspond to a child of v, a *dashed edge* is defined by:

$D[c,v]$ = The deepest vertex w such that the string corresponding to w is a suffix of the string corresponding to v followed by c (or the root if there is no such w).

The trie, D, and $MARK$ can be constructed in $O(m)$ time (see the chapter notes). The example below shows {*arnie, isabel, lala, larsa, lawren, lisa, rene, sara, salsa*}:

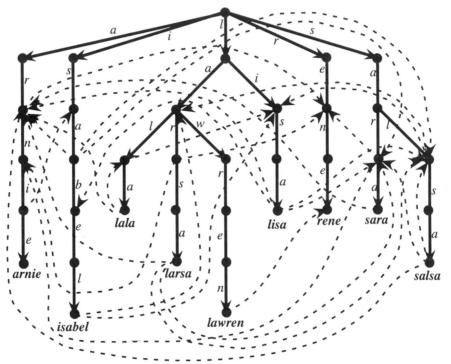

Note: Labels on dashed edges are not shown (the label is the same as the trie edge going to the parent of the vertex pointed to by the dashed edge) and dashed edges going to the root or vertices of depth one are not shown.

String matching with the augmented trie: We can use exactly the function fastkmpMATCH except change the *if* statement to check if $MARK[j]=true$. A map array can be used in exactly the same fashion as for KMP to reduce space for D.

Example: Prefix and Huffman Codes

Idea: Usually an alphabet of k characters is encoded into binary by associating with each character a string of $\lceil \log_2(k) \rceil$ bits. For example, the standard ASCII code associates with each of the 128 characters a string of 7 bits. However, any method for encoding characters as *variable length* bit strings may be acceptable in practice as long as strings are always *uniquely decodable*; that is, it is always possible to determine where the code for one character ends and the code for the next one begins.

Prefix codes: A sufficient (but not necessary) condition for an encoding method to be uniquely decodable is that the code for one character is never a prefix of a code for another; such codes are called *binary prefix codes*, and are naturally represented as root to leaf paths in a trie. Below are algorithms for encoding and decoding a single character using a binary prefix trie; they read from an input stream and write to an output stream.

Binary prefix code encoding algorithm:
>**initialize** a stack to be empty
>Read a character c.
>p = the leaf corresponding to c
>**while** p is not the root **do begin**
>>Push on to the stack the character that connects p to its parent.
>>$p = \text{PARENT}(p)$
>>**end**
>
>**while** the stack is not empty **do** write POP

Binary prefix code decoding algorithm:
>p = the root
>**while** p is not a leaf **do begin**
>>Read a bit b.
>>p = the child of p corresponding to b
>>**end**
>
>**write** the character corresponding to p

Huffman tries: The Huffman algorithm presented in the chapter on algorithm design techniques constructs a trie if we assume that tree edges are labeled with 0's and 1's (e.g., edges to left subtrees are labeled 0 and edges to right subtrees are labeled 1). If the weights are the probability distribution of characters in an input stream, then the Huffman codes are the best way to represent characters with variable length bit strings if characters are coded independently. Of course, if we take into account the pattern of characters that has occurred in the past, we may be able to do better. For example, with English text, if we have just seen the characters "Elephan", the probability that the next character is a "t" is much greater than reflected by the frequency of "t" in English (see the chapter notes).

Example: Data Compression using a Dynamic Dictionary

Idea: Maintain a dynamically changing dictionary D of strings and replace strings in the input text by their indices in D, where $|D|$ denotes the current number of strings in D.

Generic encoding algorithm:

1. Initialize the *local dictionary* D to have one entry for each character of the input alphabet (e.g., initialize entries 0 through 255 to the 256 different possible bytes).

2. **while** there is more input **do begin**

   A. Get the current match string s:

   Read as many characters as possible from the input stream to obtain a string s that is in the dictionary (i.e., reading one addition character would give a string not in the dictionary or go past the end of the input).

   Transmit $\lceil \log_2(|D|) \rceil$ bits for the index of s.

   B. Modify D:
   Add each of the strings specified by an *update method* to D (if D is full, use a *deletion method* to make space).

   end

Generic decoding algorithm:

1. Initialize D by performing Step 1 of the encoding algorithm.

2. **while** there is more input **do begin**

   A. Get the current match string s:
   Receive $\lceil \log_2(|D|) \rceil$ bits for the index of s.
   Retrieve s from D and output the characters of s.

   B. Modify D:
   Perform Step 2b of the encoding algorithm.

   end

Update and deletion methods: The encoding and decoding algorithms are well defined and the compression is lossless as long as modifications made to D depend only on the data seen thus far (which is known to both the encoder and the decoder). A simple trie data structure suffices as long as the *prefix property* is maintained (whenever a string is in the dictionary, then so are all of its prefixes). For example, the v.42bis modem compression standard uses something similar to:

FC update method: Add the single string formed by appending the first character of the current match to the previous match (provided it is not already in the dictionary).

LRU deletion method: Delete the least recently used leaf.

Example of encoding with the FC update method with the alphabet {*a,b*}:

Starting with a trie that contains only the single characters *a* and *b*, in six steps the input string is translated to six indices:

dictionary	trie for the dictionary	input string
0: a 1: b	(trie: root with branches a, b)	baaaabaaab

---

current match = b (output index 1 to the decoder)
add nothing to dictionary (no previous match)

baaaabaaab

---

current match = a (output the index 0 to the decoder)
add ba to dictionary (last match was b and the first character of the current match is a)

0: a
1: b
2: ba

(trie: root; a; b—a)

b,**a**aaabaaab

---

current match = a (output the index 0 to the decoder)
add aa to the dictionary (last match was a and the first character of the current match is a)

0: a
1: b
2: ba
3: aa

(trie: root; a—a; b—a)

b,a,**a**aabaaab

---

current match = aa (output the index 3 to the decoder)
add nothing to dictionary (last match was a and first character of the current match is a)

0: a
1: b
2: ba
3: aa

(trie: root; a—a; b—a)

b,a,a,**aa**baaab

---

current match = ba (output the index 2 to the decoder)
add aab to the dictionary (last match was aa and first character of the current match is b)

0: a
1: b
2: ba
3: aa
4: aab

(trie: root; a—a—b; b—a)

b,a,a,aa,**ba**aab

---

current match = aab (output the index 4 to the decoder)
add baa to the dictionary (last match was ba and first character of the current match is a)

0: a
1: b
2: ba
3: aa
4: aab
5: baa

(trie: root; a—a—b; b—a—a)

b,a,a,aa,ba,**aab**

Compact Tries

Idea: *Compact* tries are like regular tries except that edges from a vertex to its children are labeled by strings that start with distinct characters. CHILD(v, c) returns the child of vertex v that its attached to v with an edge labeled by a string that starts with character c.

Example: The compact trie for {*aaccaa, aaccab, abc, bb, bca, bcbb, c*} is:

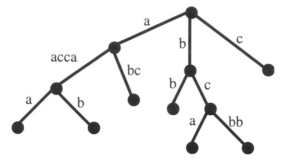

Reduced space when the strings are substrings of a master string:

If the strings are explicitly saved, then the space consumed is still proportional to the space used in the uncompacted trie.

However, if the trie is used to represent substrings of a "master" string s that is already stored, edge labels can be pairs of integers indicating a position and length in s, and hence the space per edge is constant. Since each non-leaf vertex has at least two children (so more than half the vertices are leaves), the space for the entire trie is proportional to the space consumed by the leaves.

Note: Labels on edges are not unique. Depending on s and the choice of strings to store in the trie, leaves may correspond to multiple stings in s. However, in any case, the string represented by an non-leaf vertex is a substring of each of the stings represented by its descendant leaves; the lengths of these strings are all the same, but any one of those positions can be used for the edge label. It may be appropriate for a particular application to use some standard convention (for example, when we consider sliding suffix tries, labels will always refer to the rightmost occurrence).

Definition: The *virtual depth* of a vertex v in a compact trie, denoted VDEPTH(v), is its depth in the corresponding (non-compact) trie.

Note: Vertices of a compact trie can be labeled with their virtual depth with a post-order traversal of the trie (virtual depth of a vertex is one greater than the maximum of its children).

Suffix Tries

Definition: A *suffix trie* for a string s is a compact trie that stores all suffixes of the string $s\$$ where $\$$ is a new symbol.

Example: Below is the suffix trie for the string $s=abcbbacbbab\$$:

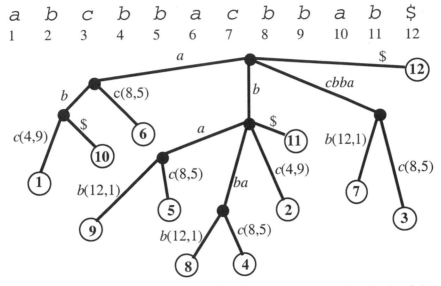

Note: For presentation, edges to a non-leaf vertex are labeled with the full string and edges to leaves are labeled with the first character of the string followed by a (position, length) reference to a suffix of s. In practice, all edges can be labeled by a pair of integers representing a position and length in $s\$$. Since $\$$ does not occur in s, the labels on edges to leaves are unique. Labels on edges to non-leaf vertices can refer to any of the occurrences in s of the corresponding string.

Example Applications of Suffix Tries

Find all occurrences of a string t in a string s: Move down from the root in the suffix trie for s according to t. If a leaf is reached before finishing t, t is not a substring of s. Otherwise, after reaching a vertex v, any of the leaves below v are suffixes of s that have t as a prefix (and so the suffix starting position is where t matches).

Find the longest repeated string in a string s: Traverse the suffix tree (e.g., preorder traversal) to find the non-leaf vertex of lowest virtual depth.

Find the longest common substring of a string s and a string t: Construct the suffix trie for $s\#t\$$. Then traverse this suffix trie in post-order and label each vertex as to whether it has descendants in only s, in only t, or in both s and t (once you know this information for the children, it is easily computed for the parent), to find the non-leaf vertex of lowest virtual depth that has both.

Sort all suffixes of a string s: Traverse the suffix trie for s in lexicographic order.

Simple Suffix Trie Construction Algorithm

Idea:

At stage i, add the leaf for position i by modifying the suffix trie already constructed for positions 1 through $i-1$ of $s\$$.

Input characters may be read more than once because input is read forward from position i to insert the new suffix, and then we back up to position $i+1$ to start the next step. In fact, the only situation where it is not necessary to back up is when the ith input character has not appeared earlier (and a new child of the root is created).

Basic brute force algorithm: We let n denote the length of $s\$$.

Initialize an empty suffix trie.

for $i := 1$ **to** n **do begin**

        cursor $:= i$

        $v := \text{SCAN}(root)$

        Label v with position i.

        end

procedure SCAN(x):

Starting at vertex x reading the input character pointed to by *cursor*, move down the suffix trie and keep reading input characters (and advancing *cursor*) to create a new leaf in one of two ways:

1. A vertex is reached that does not have a child for the current input character (and so a new leaf can be added to that vertex) — **return** this new leaf.

2. The string labeling an edge cannot be matched, the edge is "split", and a new non-leaf vertex added with one child being the edge labeled by the remainder of the string, and the other child the new leaf — **return** this new leaf.

Time: In the worst case, $O(i)$ time could be required to process the ith character (see the exercises), for a total of $O(n^2)$ time.

Space: Only $O(1)$ space is used in addition to the $O(n)$ space for the suffix trie.

Maximum match length: SCAN can be modified to limit leaf depth to some maximum depth m. As the basic algorithm is presented above, if a leaf corresponds to a string that occurs more than once in s, then each time it is visited, it can be re-labeled with the rightmost position this string has occurred thus far, and the trie will have size $O(m)$. Other conventions, such as retaining the leftmost position or keeping a linked list at each leaf of all positions (increasing the trie size to $O(n)$) can also be used; see the exercises.

McCreight's Linear Time Suffix Trie Construction

Definition: An *uncle* of a non-leaf vertex v in a suffix trie is a vertex corresponding to the same string less the first character.

Theorem: Every non-leaf vertex in a suffix trie has a unique uncle, which is a non-leaf.
Proof: Let v be a non-leaf vertex in a suffix trie for a string s. If v is the root, it is its own uncle. If v is a child of the root where the edge from the root to v is labeled by a single character, then the root is the uncle of v. Otherwise, let aB be the string represented by v, where a is a character and B is a string of length ≥ 1. Then since v is a non-leaf vertex, the substring aB must occur at least twice in s, and hence the string B must occur at least twice in s, and so there must be a non-leaf vertex that corresponds to B.

Idea: The suffix for position i is like that for position $i-1$ except that the leftmost character is removed. Instead of starting at the root each time to insert the next suffix, maintain uncle links so that "shortcuts" can be taken. If the parent v of the previous leaf added does not already have an uncle link, using v's parent's suffix link (which will always be present), you end up higher than you would like in the trie, and have to rescan the string t that labels the edge from the parent p of v to v. Add the new suffix link from v to its uncle for future use, scan down to add a new leaf for position i, and then move up to its parent, which becomes the new vertex v for the next iteration of the algorithm.

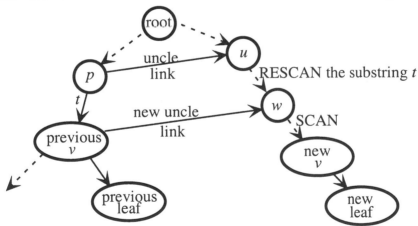

The figure above depicts the general case, where solid lines are single edges and dashed lines denote a path of one or more edges. The two special cases are when v is the root (in which case we simply scan down from the root) and when v is a child of the root, in which case we rescan t less its first character, since the uncle link from the root to itself goes to a sting of the same length (length 0) rather than to one that is one character shorter (as with all other uncle links).

Key invariant: At each stage, every non-leaf vertex, except possibly for one created at the last step, has an uncle link.

The McCreight algorithm:

We employ the function SCAN used by the basic algorithm and a new function RESCAN that moves down the trie according to positions of the input that are already in the trie.

> Initialize the trie to a single root vertex v.
> $cursor := 1$
> **for** $i := 1$ **to** n **do begin**
>   **if** v is the root **then** $w := v$
>   **else if** v already has an uncle link **then** $w := \text{UNCLE}(v)$
>   **else begin**
> $p := \text{PARENT}(v)$
> $u := \text{UNCLE}(p)$
> $k :=$ the length of the string t labeling the edge from p to v
>   **if** p is the root **then begin** $k := k\text{-}1$; $cursor := cursor+1$ **end**
> $w := \text{RESCAN}(u,k)$
>   Add an uncle link from v to w.
>   **end**
> $v := \text{SCAN}(w)$
>   Label v with position i.
> $v := \text{PARENT}(v)$
> $cursor := cursor-1$
> **end**

function RESCAN(x,k)**:**

> Starting at vertex x scanning the input symbol pointed to by the cursor, advance the cursor as we move down the trie. Each edge is traversed in constant time because the suffix trie has *already been constructed* for this portion of the input, and it suffices to match the first character of edge label to the input (and check that its length does not cause the total length traversed to exceed k). The vertex where the rescan ends is returned (a new vertex is created if it ends in the middle of an edge).

Complexity (assuming the alphabet size is constant): Space is $O(n)$; so is the time:

> **Main loop, excluding calls to SCAN and RESCAN:** Each iteration is $O(1)$ excluding the time for SCAN and RESCAN operations, for a total of $O(n)$.

> **Calls to SCAN:** Each step of SCAN reads a new input symbol or creates a new trie vertex (or both), so the total time for all calls to SCAN is $O(n)$.

> **Calls to RESCAN:** Traversal of each edge is $O(1)$ since only the first character of its label has to be matched (the position in the input can be calculated using virtual depth fields). If the RESCAN visits more than one edge, except for the last one, each corresponds to positions of the input that will never be rescanned or scanned again (since each step either follows v's uncle link or the one of parent of v, but goes no higher in the trie), and hence the time for all call to RESCAN is $O(n)$.

Example: Brute-Force Versus McCreight on $a^n\$$

Below is the suffix trie for the string $s = a^5\$$, where characters are indexed from 1 to 6. Vertices and solid edges are constructed by both the brute-force and McCreight's algorithm; the dashed edges are the suffix links constructed by McCreight's algorithm. A non-leaf vertex is labeled v_i to indicate that it was created by the ith iteration.

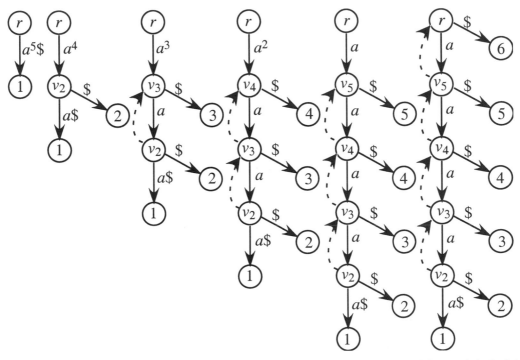

The first iteration of the brute-force algorithm places an edge for position 1 labeled by $a^n\$$. The second iteration reads from position 2 to the end of s to place v_2 that forks on $\$$ to leaf 2 and on $a\$$ to leaf 1. The third iteration reads from position 3 to the end of s to place v_3 that forks on $\$$ to leaf 3 and on a to v_2. It is straightforward to generalize this example to $a^n\$$ for any $n \geq 0$. For $i>2$ the ith iteration reads from position i to the end of s to put in a new non-leaf vertex that forks on $\$$ to leaf i and with an a to v_{i-1}. Since the time for the ith iteration is $\Omega(i)$, the time for the algorithm is given by the standard arithmetic sum (see the appendix) $\Omega(1+2+...+n) = \Omega(n^2)$.

McCreight's algorithm begins the same way as the brute force algorithm with the first two iterations constructing the edge $a^n\$$ for position 1 and then scanning to the end of s to place a new non-leaf vertex v_2 that forks on $\$$ to leaf 2 and on $a\$$ to leaf 1. However, on the third iteration, instead of scanning to the end of s, RESCAN reads a single a to jump to v_2 and creates the vertex v_3, and then SCAN places leaf 3. In general, for the string $a^n\$$, MsCreight's algorithm is $O(1)$ time for iteration 1, $O(n)$ time for iteration 2, and $O(1)$ time on each of the remaining iterations, for a total of $O(n)$ time.

Sliding Suffix Tries

Idea: When processing a string $s\$$ from left to right, we may only be interested in answering pattern matching questions about text that has been seen recently in a "window" of the preceding n characters (or the window is all characters seen thus far if less than n characters have been read).

Maximum match length: For sliding suffix tries, we assume that SCAN limits the depth of leaves to some maximum $m \leq n$.

Brute force sliding suffix trie construction algorithm: Modify the "brute force" trie construction algorithm as follows:

- Child counts are associated with each vertex (so we will know when a non-leaf vertex has only one remaining child).

- Leaves are ordered from oldest to newest in a queue.

- When SCAN moves down to add a new suffix, edge labels are updated so that they refer to the rightmost occurrence seen thus far in s of their corresponding strings; for convenience, we also label non-leaf vertices with their position, although this can be calculated from virtual depths and edge labels (see the exercises).

Each time a new suffix is added (except during the first n characters of the input, where n is the window length), delete the oldest leaf and "patch" up the trie:

> Initialize an empty suffix trie.
>
> **for** $i := 1$ **to** $|s|+1$ **do begin**
>
> > **if** the trie is full then begin
> >
> > > DEQUEUE a leaf x
> > >
> > > $y := \text{PARENT}(x)$
> > >
> > > **if** y has only one child **then begin**
> > > > Combine the labels on the edges from $\text{PARENT}(y)$ to y and from y to $\text{CHILD}(y)$ into a single edge from $\text{PARENT}(y)$ to $\text{CHILD}(y)$.
> > > > **end**
> > >
> > > **end**
> >
> > $cursor := i$
> >
> > $v := \text{SCAN}(root)$
> >
> > Label v with position i.
> >
> > ENQUEUE v
> > **end**

Time: $O(m|s|)$

Sliding Window With Two McCreight Tries

Idea: Start a new McCreight suffix trie for each block of n characters. Only the most recent two tries need to be stored; the "real" window of length n will overlap the window of $2n$ represented by these two tries, and both tries must be used when searching for strings in the window.

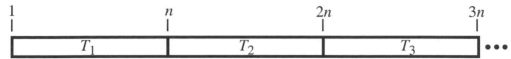

Two–trie algorithm: After a suffix trie is constructed for a block of n characters, update edge and vertex labels to indicate the rightmost occurrence that has been seen in this block; a simple post-order traversal can be employed (see the exercises).

1. Run McCreight's algorithm for n steps starting at position 1 to form T_1.
   Do a post-order traversal of T_1 to update the labels of edges and vertices.
2. Run McCreight's algorithm for n steps starting at position $n+1$ to form T_2.
   Do a post-order traversal of T_2 to update the labels of edges and vertices.
3. Reclaim the memory for T_1 and construct T_3 for positions $2n+1$ to $3n$.
   Do a post-order traversal of T_3 to update the labels of edges and vertices.

Etc.

Using the two tries: Suppose that we wish to scan the input from left to right and at each position i, be able to search for substrings of the window of the preceding n characters (we shall shortly present an example of sliding window data compression that does this). The two trie algorithm can be implemented to proceed at roughly the same pace (buffering is needed to allow it to scan and rescan ahead as necessary to insert strings into the current trie at least up to the current position i). In general, when at a position i and we wish to search for a string t in the preceding n characters, for some k, T_k represents positions j through $j+n-1$ for some position $j \leq i \leq j+n-1$, and T_{k-1} represents positions $j-n$ through $j-1$. So we can search for t by looking in T_k, and by also looking in T_{k-1} by starting at the root and going down as far as possible until we get to a leaf, or must stop at a non-leaf vertex because going down to the appropriate child goes to a vertex representing a string at a position earlier than $i-n$.

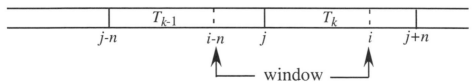

Complexity: Space and time for two suffix tries is proportional to that for one, and the $O(n)$ time for post-order traversal can be charged to the cost of the constructing that trie. So the total time spent is linear in the size of the input, and the space is $O(n)$.

Fiala–Greene Sliding Suffix Trie Algorithm

Idea: Use McCreight's algorithm, and in the spirit of the brute-force sliding suffix trie algorithm, delete leaves when they become obsolete. The brute-force algorithm is able to easily keep vertex and edge labels updated to refer to the rightmost occurrence seen thus far of the corresponding string seen because it starts each SCAN at the root. However, with McCreight's algorithm, RESCAN and SCAN start at non-leaf vertices that may be deep in the trie, and even if they update labels, the labels above them in the trie will be missed. Moving up to the root to fix labels each time a RESCAN / SCAN is done would be too time consuming. Instead we use a "lazy" method of "percolating" updates that has amortized linear time.

Additions to McCreight's data structure:

- Like the brute force version, associate child counts with each vertex and keep the leaves in a queue.

- Store a flag bit with each non-leaf vertex that will be used to determine when to spend the time to go higher in the trie to update labels to refer to the rightmost position seen thus far of the corresponding string.

Modifications to McCreight's suffix trie construction algorithm:

- Once the window is full, a leaf queue is maintained so that the oldest leaf is deleted at each stage to make room for the new one that is added.

- When a child count of a non-leaf vertex v is reduced to 1 as a result of a leaf deletion, v is deleted and the child is attached to the parent of v.

- After a new leaf v has been created and labeled with the current position, "percolate" the updating of ancestor vertices by complementing *flag* bits from 1 to 0 until a 0 flag bit is found and changed to 1. That is, if we think of a root to leaf path of flag bits as forming a binary number that ends in a 0 bit at the leaf, we set the leaf bit to 1, and then increment this number:

    flag(v) := 1
    while $v \neq$ root and *flag*(v)=1 **do begin**
        flag(v) := 0
        v := PARENT(v)
        Update the labels of v and the edge that was traversed.
    end
    flag(v) := 1

Analysis of the Percolate Update

Theorem: When a non-leaf vertex v is deleted (because a leaf deletion has caused it to have only one child), there cannot be an uncle link from some other non-leaf vertex w that points to v.

Proof: Assume the contrary. Since w is a non-leaf vertex, it must have at least two children. Hence an uncle link from w to v implies that v has at least two children, a contradiction.

Theorem: Every vertex is updated at least once during every interval of n characters, where n is the length of the window.

Proof: We show that in fact each is updated at least twice during every interval of n characters (this will allow the proof to work inductively). Assume the contrary and let:

$i =$ The first position where there is at least one vertex that has not been updated twice in the last n characters.

$v =$ The vertex farthest from the root that has not been updated during the last n characters before position i.

There are now three possibilities, all of which lead to a contradiction:

v has two children that have remained during the last n characters:

Then v must have received updates from these children, and we have a contradiction (since these children are farther from the root).

v has only one child that has remained during the last n characters:

Then v must also have a new child (since every non-leaf vertex has at least two children), and again we have a contradiction (since v must have gotten two updates — one from the child there during the last n characters and one from the new child when it was created).

v has no children that have remained during the last n characters:

Then v must have two new children (since every non-leaf vertex has at least two children), and again we have a contradiction (since v must have gotten two updates — one from each new child when it was created).

Time: The flag bits on a root to leaf path form a binary number, and a percolate update increments this number. So updates at the leaves cause at most half as many to non-leaf vertices at height 1, which in turn cause at most 1/4 as many to non-leaf vertices at height 2, and so on, which gives a geometric series (see the appendix) that adds up to at most doubling the work done at the leaves. A formal proof can use an amortized analysis with "credits"; see the exercises.

Example: Data Compression using a Sliding Window

Idea: Given a string, we represent it with a smaller string by replacing substrings by a *pointer* consisting of a pair of integers (d,l) where d is a distance back in a window of the last n characters and l is the length of an identical substring. The amount of space saved depends on how often strings are repeated in the text, how the pairs (d,l) are coded, etc.

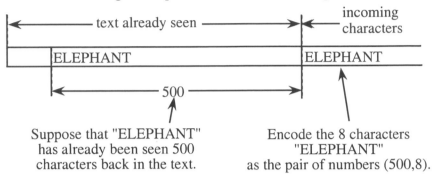

Suppose that "ELEPHANT" has already been seen 500 characters back in the text.

Encode the 8 characters "ELEPHANT" as the pair of numbers (500,8).

Greedy matching: Typically "greedy" matching is used (always take the longest match possible). However, depending on the encoding strategy used, it may be that there are better strategies (see the chapter notes).

Coding characters and pointers:

For an alphabet of size k, use $\lceil \log_2(k) \rceil$ bits per character.

Assuming that pointers use more bits than used by two characters, lengths range from 3 to some upper limit *MaxLength*, which can be represented by the *MaxLength*-2 integers 0 through *MaxLength*-3, and use $\lceil \log_2(MaxLength-2) \rceil$ bits.

A displacement can be represented with $\lceil \log_2(n) \rceil$ bits.

A flag bit can distinguish a pointer from a character, where the leading bit is a 0 if the next $\lceil \log_2(k) \rceil$ bits to follow are a character and a 1 if the next $\lceil \log_2(n) \rceil$ + $\lceil \log_2(MaxLength-2) \rceil$ bits to follow are a displacement and length.

For example, if $n=2,048$, *MaxLength*=18, and we use the standard 128 character ASCII alphabet ($k=128$), characters use one byte (a flag bit and 7 ASCII bits) and a pointers use two (a flag bit, 11 displacement bits, and 4 length bits).

Better methods that use variable length codes (sometimes capable of representing 2 bytes with less than 16 bits) can also be used (see the chapter notes).

Encoding complexity: To make the encoding linear time (in the total number of input characters) and $O(m)$ space, the encoder can use a sliding suffix trie.

Decoding complexity The decoder can simply maintain the same sliding window and look up matches when it receives a pair.

Edit Distance: A Measure of String Similarity

Definition: The edit distance from a string $s = s_1 \ldots s_m$ to a string $t = t_1 \ldots t_n$ is the minimum cost of converting s to t via the following operations:

        cost a: add a new character
        cost d: delete a character
        cost c: change one character to another

The parameters a, d, and c are positive real values; without loss of generality we assume that $c \leq a+d$. For many applications, $a=d$, and hence the edit distance from s to t is the same as that from t to s (however, the algorithm presented here does not assume this).

Dynamic programming algorithm: The edit distance from s to t is computed by working from smaller indices to larger indices in the two dimensional array ED, where $ED[i,j]$ is the edit distance from $s_1 \ldots s_i$ to $t_1 \ldots t_j$:

1. Initialize:
    for $0 \leq i \leq m$ **do** $ED[i,0] := d*i$
    for $0 \leq i \leq n$ **do** $ED[0,j] := a*j$

2. For $i,j \geq 1$ the entries of ED can be computed in any order such that when $ED[i,j]$ is computed $ED[x,y]$ has already been computed for $x<i$ and $y<j$ (for example, we can proceed a column at a time with **for** $i := 1$ **to** m **do for** $j := 1$ **to** n **do**, or a row at a time with **for** $j := 1$ **to** n **do for** $i := 1$ **to** m **do**). $ED[i,j]$ is computed by using whichever one of the following cases yields the minimum value:

    A. $ED[i,j] := ED[i,j-1]+a$ (convert $s_1 \ldots s_i$ to $t_1 \ldots t_{j-1}$ and then add t_j)

    B. $ED[i,j] := ED[i-1,j]+d$ (delete s_i and then convert $s_1 \ldots s_{i-1}$ to $t_1 \ldots t_j$)

    C. $ED[i,j] := ED[i-1,j-1]$ (convert $s_1 \ldots s_{i-1}$ to $t_1 \ldots t_{j-1}$)
       if $s_i \neq t_j$ **then** $ED[i,j] := ED[i,j]+c$ (if $s_i \neq t_j$ then change s_i to t_j)

Complexity: The straightforward implementation uses $O(mn)$ time and space to store the entire ED array. If only a particular entry such as $ED(m,n)$ is ultimately needed, the computation can use only $O(m+n)$ space by storing only the most recently computed "cross-section" of ED. For example, if we are computing ED a column at a time, then only the previous column of ED has to be saved since all of cases A, B, and C use values that are in the previous column or earlier in the same column.

Relaxed ordering: If we proceed through ED with diagonal cross-sections, then the order in which the inner loop is performed does not matter, which is desirable for a parallel implementation (see the exercises to the last chapter).

    for $k := 1$ **to** $m+n$ **do**
        for each i and j such that $i+j=k$ **do**
            Compute $ED[i,j]$ from the values of $ED[i-1,j]$, $ED[i,j-1]$, and $ED[i-1,j-1]$.

Computing the edit sequence: As ED is being computed, store with each entry the operation that was used. For any entry i,j a sequence of edit operations that produces $ED[i,j]$ (the sequence is not necessarily unique since it depends on how ties are broken) can be listed by working backwards from position i,j (see the exercises).

Example: Longest Common Sub-Sequence

Definition: A *common subsequence* of two strings s and t is a string w that can be obtained from s or t by deleting characters (the number and positions of characters deleted from s and t may be different); it is *longest* if there is not a common subsequence of s and t of greater length. $LCS(s,t)$ denotes the length of a longest common subsequence of s and t.

Example:

$s = a\,a\,b\,b\,a\,a\,a\,b\,a$

$t = a\,b\,a\,b\,b\,b\,a\,a$

$LCS(s,t)=6$ since both *aabbaa* and *aabbba* are common subsequences of length 6 but there is no common sub-sequence of length 7.

LCS Theorem

If the cost of inserting and deleting characters is the same, and the cost of changing a character is twice that cost, then:

$$ED(s,t) = ED(t,s) = |s| + |t| - 2LCS(s,t)$$

Proof: We can assume that no change operations are used since a with the same cost a change can be simulated by a delete followed by an insert, and since the cost of a delete is the same as that of an insert, $ED(s,t)=ED(t,s)$ since a sequence of operations can simply be reversed. Since s can be edited to t by first deleting the characters in s that are not in a longest common subsequence of s and t and then inserting the missing characters, $ED(s,t) \le |s|+|t|-2LCS(s,t)$. Conversely, if d is the number of deletes and i is the number of inserts in a minimal edit from s to t, then there is common subsequence of s and t of length $|s|-d = |t|-i$, and hence, $2LCS(s,t) \ge |s|+|t|-(d+i) = |s|+|t|-ED(s,t)$, which implies $ED(s,t) \ge |s|+|t|-2LCS(s,t)$.

Computation of longest common subsequence: As a consequence of the LCS theorem, we can use the edit distance algorithm to compute longest common sub-sequences (see the exercises).

Arithmetic Codes

Idea: Similar to Huffman codes, instead of encoding characters of a string with a fixed number of bits, use a number of bits that reflect the character's frequency. Unlike Huffman codes, we do not limit ourselves to an integral number of bits per character; a character's frequency is represented by effectively using a fractional numbers of bits.

Example: To simplify presentation, we use base 10 rather. Suppose that we have strings over the alphabet $\{A,B,C\}$ where 20% of their characters are A, 30% B, and 50% C. Then the real line between 0 and 1 is divided into regions that correspond to A, B, and C:

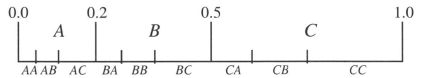

The proportions .2, .3, .5 can be used to further sub-divide these intervals for longer strings, as depicted for pairs of characters with the smaller ticks in the figure above. In general, define each such string s to correspond to a region on the real line containing all real numbers r satisfying $\text{LEFT}(s) \le r < \text{RIGHT}(s)$. The empty string corresponds to all real numbers r such that $0 \le r < 1$, and if $s = tc$ where t is a string and c is a character, then s defines a region on the real line containing all real numbers r satisfying:

$$\text{LEFT}(s) = \text{LEFT}(t) + \text{LEFT}(c)*(\text{RIGHT}(t)-\text{LEFT}(t))$$

$$\text{RIGHT}(s) = \text{LEFT}(t) + \text{RIGHT}(c)*(\text{RIGHT}(t)-\text{LEFT}(t))$$

For example:

$$
\begin{array}{rcccl}
0 & \le & A & < & .2 \\
.2 & \le & B & < & .5 \\
.5 & \le & C & < & 1 \\
0 + .5*(.2-0) = .1 & \le & AC & < & .2 = 0 + 1*(.2-0) \\
.2 + .5*(.5-.2) = .35 & \le & BC & < & .5 = .2 + 1*(.5-.2) \\
.5 + 0*(1-.5) = .5 & \le & CA & < & .6 = .5 + .2*(1-.5) \\
.1 + .2*(.2-.1) = .12 & \le & ACB & < & .15 = .1 + .5*(.2-.1) \\
.35 + .5*(.5-.35) = .425 & \le & BCC & < & .5 = .35 + 1*(.5-.35) \\
.5 + .5*(.6-.5) = .55 & \le & CAC & < & .6 = .5 + 1*(.6-.5) \\
\end{array}
$$

Applications: Like Huffman coding, arithmetic coding can be used as a compression method in itself, but is also effective as a component in a more complex system that computes a probability distribution for the next character to be processed based on the characters seen thus far; these probabilities can be given to the arithmetic coder to use to partition the real line on its next encoding step. The output is well defined because the decoder has already decoded the same set of characters and can therefore duplicate at each step the encoders partition of the real line.

Conceptual Arithmetic Encoding / Decoding Algorithm

Idea:

Both the encoder and decoder maintain a *coding interval* between the points *low* and *high*; initially *high*=1 and *low*=0.

Each time a character *c* is read by the encoder or written by the decoder, they both *REFINE* the coding interval by doing:

$length := high - low$
$low := low + (LEFT(c) * length)$
$high := low + (RIGHT(c) * length)$

As more and more characters are encoded, the leading digits of the left and right ends become the same and can be transmitted from the encoder to the decoder.

If the current digit received by the decoder is the ith digit to the right of the decimal point in the real number *r* formed by all of the digits received thus far, then the interval from *r* to $r + 10^{-i}$ is a *bounding interval*; the left and right ends of the final interval will both begin with the digits of *r*.

Conceptual Static First-Order Arithmetic Encoding Algorithm:

$high := 1$
$low := 0$
while there is input **do begin**
    read the next input character *c*
    REFINE the coding interval according to *c*
    output any new identical leading digits of the coding interval ends
    end

Conceptual Static First-Order Arithmetic Decoding Algorithm:

$high := 1$
$low := 0$
while there is input **do begin**
    read the next digit
    while bounding interval is contained in the region defined by some character *c*
    do begin
        output *c*
        REFINE the coding interval
        end
    end

Defining The End of a Finite String

Problem: For a finite string, if after writing the last character, the bounding interval happens to straddle two regions of the coding interval, then the decoder knows that it has completed its task. However, when the bounding interval is completely contained in the region defined by the character, the decoder will incorrectly continue and output another character (e.g., in the previous example, if the decoder receives a single 0, it does not know whether it represents a, aa, aaa, $aaaa$, etc.).

Solution 1: For a sequence of length n, precede the compressed sequence with an integer that specifies the length of the decompressed sequence — see the exercises.

Solution 2: Add a special character $ that appears nowhere in the input and is encoded at the end of the input. The decoder stops when it received $ (and does not output $). Assign $ a sufficiently small interval so that the "waste" is not significant (ideally, an interval corresponding to the probability $1/n$ for an input of length n — see the exercises).

Example: In the previous example, "borrow" a bit from c and add $ with probability 0.1 (in practice, we would want to make the interval for $ much smaller than this, but it makes this example easier to illustrate).

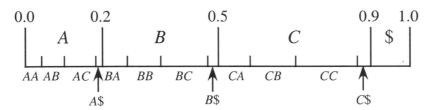

The proportions .2, .3, .4, and .1 further sub-divide these intervals a bit differently from the previous example. As depicted above by the smaller ticks, pairs of characters sub-divide each of the intervals defined by the single characters into four regions (the region for $ will never be sub-divided since $ occurs only once at the end of the input).

$$
\begin{array}{rcccl}
0 & \leq & A & < & .2 \\
.2 & \leq & B & < & .5 \\
.5 & \leq & C & < & .9 \\
.9 & \leq & \$ & < & 1 \\
.2+.5*(.5-.2) = .35 & \leq & BC & < & .47 = .2+.9*(.5-.2) \\
0+.5*(.2-0) = .1 & \leq & AC & < & .18 = 0+.9*(.2-0) \\
.5+0*(.9-.5) = .5 & \leq & CA & < & .58 = .5+.2*(.9-.5) \\
.1+.2*(.18-.1) = .116 & \leq & ACB & < & .14 = .1+.5*(.18-.1) \\
.35+.5*(.47-.35) = .41 & \leq & BCC & < & .458 = .35+.9*(.47-.35) \\
.5+.5*(.58-.5) = .54 & \leq & CAC & < & .572 = .5+.9*(.58-.5)
\end{array}
$$

Example

Consider the encoding of the string *a b c a a $* with the conceptual encoder:

> low = 0
> high = 1
>
> *a*
> length = high–low = 1
> low = low + (left(*a*)*length) = .0 + (.0*1) = .0
> high = low + (right(*a*)*length) = .0 + (.2*1) = .2
>
> *b*
> length = high–low = .2
> low = low + (left(*b*)*length) = .0 + (.2*.2) = .04
> high = low + (right(*b*)*length) = .0 + (.5*.2) = .1
>
> *c*
> length = high–low = .06
> low = low + (left(*c*)*length) = .04 + (.5*.06) = .07
> high = low + (right(*c*)*length) = .04 + (.9*.06) = .094
> **OUTPUT 0**
>
> *a*
> length = high–low = .024
> low = low + (left(*a*)*length) = .07 + (.0*.03) = .07
> high = low + (right(*a*)*length) = .07 + (.2*.024) = .0748
> **OUTPUT 7**
>
> *a*
> length = high–low = 0.0048
> low = low + (left(*a*)*length) = .07 + (0*.0048) = .07
> high = low + (right(*a*)*length) = .07 + (.2*.0048) = .07096
> **OUTPUT 0**
>
> *$*
> length = high–low = 0.00096
> low = low + (left(*$*)*length) = .07 + (.9*0.00096) = .070864
> high = low + (right(*$*)*length) = .07 + (1*0.00096) = .07096
> **OUTPUT 9**

Example continued, decoding the numbers 0 7 0 9:

low=0, high=1, interval boundaries are: 0 *a* .2 *b* .5 *c* .9 $ 1

0
bounding interval = [0,.1)
since bounding interval is contained in *a*'s region, **OUTPUT** *a*
length = high–low = 1
low = low + (left(*a*)*length) = .0 + (.0*1) = .0
high = low + (right(*a*)*length) = .0 + (.2*1) = .2
interval boundaries are: 0 *a* .04 *b* .1 *c* .18 $ .2

7
bounding interval = [07,.08)
since bounding interval is contained in *b*'s region, **OUTPUT** *b*
length = high–low = .2
low = low + (left(*b*)*length) = .0 + (.2*.2) = .04
high = low + (right(*b*)*length) = .0 + (.5*.2) = .1
interval boundaries are: .04 *a* .052 *b* .07 *c* .094 $ .1
since bounding interval is contained in *c*'s region, **OUTPUT** *c*
length = high–low = .06
low = low + (left(*c*)*length) = .04 + (.5*.06) = .07
high = low + (right(*c*)*length) = .04 + (.9*.06) = .094
interval boundaries are: .07 *a* .0748 *b* .082 *c* .0916 $ .094

0
bounding interval = [07,.071)
since bounding interval is contained in *a*'s region, **OUTPUT** *a*
length = high–low = .024
low = low + (left(*a*)*length) = .07 + (.0*.03) = .07
high = low + (right(*a*)*length) = .07 + (.2*.024) = .0748
interval boundaries are: .07 *a* .07096 *b* .0724 *c* .07432 $ .0748

9
end of input, final value = .0709
since final value is contained in *a*'s region, **OUTPUT** *a*

Although input has ended, the decoder cannot be sure that there are not other characters that result from receiving the 9, and so it continues:
length = high–low = 0.0048
low = low + (left(A)*length) = .07 + (0*.0048) = .07
high = low + (right(A)*length) = .07 + (.2*.0048) = .07096
interval boundaries are: .07 *a* .070192 *b* .07048 *c* .070864 $ .07096
since final value is contained in $'s interval, the decoder can terminate

If we continue another step, we get the same values as resulted from the encoder:
length = high–low = 0.00096
low = low + (left($)*length) = .07 + (.9*0.00096) = .070864
high = low + (right($)*length) = .07 + (1*0.00096) = .07096

On-Line Encoding and Decoding

Although the encoder and decoder work on-line, the encoder can output more than one number on a given step and the decoder can output more than one character on a given step. In fact, in the previous example, the decoder produced two characters when processing the number 7. Timing variations can be handled by input and output queues.

For an example of the encoder outputting more than one number for a given character, consider the input string a b c a b a $. Encoding of a b c a proceeds as in the previous example, and then:

b
length = high–low = 0.0048
high = low + (right(*b*)*length) = 0.07 + (0.5*0.0048) = 0.0724
low = low + (left(*b*)*length) = 0.07 + (0.2*0.0048) = 0.07096

a
length = high–low = 0.00144
high = low + (right(*a*)*length) = 0.07096 + (0.2*0.00144) = 0.071248
low = low + (left(*a*)*length) = 0.07096 + (0.0*0.0018) = 0.07096

$
length = high–low = 0.000288
high = low + (right(*$*)*length) = 0.07096 + (1.0*0.000288) = 0.071248
low = low + (left(*$*)*length) = 0.07096 + (0.9*0.000288) = 0.0712192
OUTPUT 1
OUTPUT 2
OUTPUT 4

Practical Considerations
(see the chapter notes)

> **Binary encoding:** We have used decimal notation for ease of presentation. In practice, the conceptual encoding and decoding algorithms can easily be reformulated for binary encoding.

> **Finite precision arithmetic:** The conceptual encoding and decoding algorithms use higher and higher precision arithmetic as the input gets longer, making them impractical. However, as the encoder consumes characters and narrows the interval, each time corresponding digits of the beginning and end of the interval become the same, a digit (or digits) is transmitted to the decoder. Rescaling at these points (by both the encoder and decoder) allows finite precision arithmetic.

> **Approximations:** Even with finite precision, the computations are much more time consuming than Huffman coding. Methods that are much faster and closely approximate the "pure" version have been proposed in the literature.

> **Higher order versions:** Like Huffman codes, arithmetic codes have a dynamic version and can be generalized to higher-order statistics.

> **Large alphabets:** Appropriate data structures can be employed.

The Burrows–Wheeler Transform

Idea: The suffix trie for a string $S = S[1]...S[n]$ uses $O(n)$ space and captures the "structure" of S. The *Burrows–Wheeler Transform* of S, $BWT(S)$, is a new string of exactly n characters that is a permutation of the characters of S corresponding to a pre-order traversal of the suffix trie of S (assuming children are visited in sorted order).

The $ character: Assume that $S[n]$ is a special character $\$$ that is lexicographically larger than all other characters in S (this assumption is not necessary — see the exercises).

$BWT(S)$: Let $M(S)$ be the matrix of all cyclic rotations of S, listed in lexicographic order. $BWT(S)$ is the last column of M. On the left the figure below shows $M(bratatbat\$)$.

```
    1  2  3  4  5  6  7  8  9  10
 1  a  t  a  t  b  a  t  $  b  r
 2  a  t  b  a  t  $  b  r  a  t
 3  a  t  $  b  r  a  t  a  t  b
 4  b  a  t  $  b  r  a  t  a  t
 5  b  r  a  t  a  t  b  a  t  $
 6  r  a  t  a  t  b  a  t  $  b
 7  t  a  t  b  a  t  $  b  r  a
 8  t  b  a  t  $  b  r  a  t  a
 9  t  $  b  r  a  t  a  t  b  a
10  $  b  r  a  t  a  t  b  a  t
```

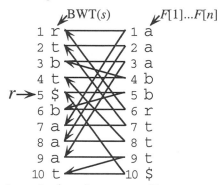

The last column of the matrix of all cyclic permutations of "bratatbat" is BWT(*bratatbat$*).

bratatbat$ can be recovered by reconstructing the first column F and visiting F in the order of the permutation links, starting at row r.

$O(n)$ computation of $BWT(S)$: Since M has n^2 entries, we do not actually compute it. Instead, since the rows of M correspond to the suffixes of S (underlined in the figure above), traverse in pre-order the leaves of the suffix trie for S (output $S[i-1]$ at the leaf for position i, except output $S[n]$ when $i=1$). The suffix trie must access vertex children in lexicographic order; provided that the alphabet size is constant with respect to n, the suffix trie construction is linear time and space.

$O(n)$ computation of the inverse BWT: Given the index r of the row in M that contains S, S can be recovered from $BWT(S)$, as depicted on the right above. Since the first column of M, denoted $F[1]...F[n]$, is the characters of S in sorted order (duplicate copies occur consecutively in "blocks") it can be reconstructed with a bucket sort. Then construct the permutation array $P[1]...P[n]$, where if $F[i]$ is the jth character in its block, then $P[i]$ is the index of the jth occurrence of $F[i]$ in $BWT(S)$; compute P by using the array $C[1]...C[k]$ (where k is the alphabet size, including $\$$) that points to the blocks in F. Follow P to produce S.

1. Use bucket sort to reconstruct the first column of M, $F[1]...F[n]$.
2. **for** $i := 1$ **to** n **do if** $i=1$ or $F[i] \neq (F[i]-1)$ **then** $C[F[i]] := i$
3. **for** $i := 1$ **to** n **do begin** $P[C[BWT[i]]] := i$; $C[BWT[i]] := (C[BWT[i]]+1)$ **end**
4. **for** $i :- 1$ **to** n **do begin** output $F[r]$; $r :- P[r]$ **end**

Inverse BWT Using Only Two Passes

Idea: To recover $S = S[1]...S[n]$ from $BWT(S) = BWT[1]...BWT[n]$ and the integer r (the index of the row of M that contains S), the algorithm presented on the previous page made four passes over $BWT(S)$, one for each of the four steps. Assuming that the input alphabet is represented by the integers 0 through $k-1$, it is possible to use only $O(k)$ time in addition to only two passes over $BWT(S)$. Construct S from right to left and use links that go in the opposite direction; also, instead of having direct permutation links, use something analogous to a base and offset. Do not actually reconstruct $F[1]...F[n]$; instead it is implicitly stored by redefining the arrays C and P as follows:

$C[0]...C[k-1]$: $C[j]$ is the number of positions in BWT that contain a character $c < j$.

$P[1]...P[n]$: $P[i]$ is the number of times character $BWT[i]$ appears in $BWT[1]...BWT[i]$.

Algorithm:

1. Begin by computing P and ending up with $C[j]$, $0 \le j < k$, equal to the number of instances of the character j in $BWT[1]...BWT[n]$:

    for $j := 0$ **to** $k-1$ **do** $C[j] := 0$
    for $i := 1$ **to** n **do begin** $C[BWT[i]] := (C[BWT[i]]+1)$; $P[i] := C[BWT[i]]$ **end**

2. Fix up C:

    $sum := 0$
    for $j := 0$ to $k-1$ **do begin** $sum := sum+C[j]$; $C[j] := sum-C[j]$ **end**

3. Reconstruct S from right to left:

    for $i := n$ **downto** 1 **do begin** $S[i] := BWT[r]$; $r := C[BWT[r]]+P[r]$ **end**

Example: Consider again the string $S = bratatbat\$$:

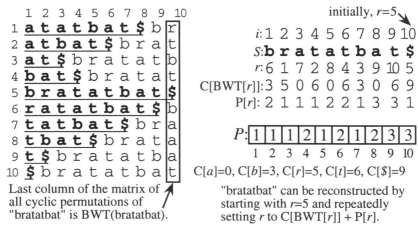

Last column of the matrix of
all cyclic permutations of
"bratatbat" is BWT(bratatbat).

initially, $r=5$

i:	1	2	3	4	5	6	7	8	9	10
S:	b	r	a	t	a	t	b	a	t	$
r:	6	1	7	2	8	4	3	9	10	5
$C[BWT[r]]$:	3	5	0	6	0	6	3	0	6	9
$P[r]$:	2	1	1	1	2	2	1	3	3	1

P:

	1	1	1	2	1	2	1	2	3	3
	1	2	3	4	5	6	7	8	9	10

$C[a]=0$, $C[b]=3$, $C[r]=5$, $C[t]=6$, $C[\$]=9$

"bratatbat" can be reconstructed by
starting with $r=5$ and repeatedly
setting r to $C[BWT[r]] + P[r]$.

Example: MTF Data Compression

Idea: Assume that we are given a basic coder (e.g., Huffman or Arithmetic):

CODE(c): Outputs a code for character c to a compressed output stream of bits.

DECODE: Reads a variable number of bits from a compressed input stream and returns the corresponding character.

To make more effective use of the coder, *move-to-front* (MTF) compression maintains an ordered list of the characters of the input alphabet, which we assume are the integers 0 through $k-1$ (e.g., if we are storing one character per byte, then $k=256$). Each time a character is processed, the MTF algorithm codes the position of the character in the list (where the front item is position 0 and the rear item is position $k-1$) and then moves the character to the front of the list. So the "character" that is being encoded by CODE, although it is still an integer in the range 0 to $k-1$, is actually the current index of the corresponding character in the character list.

Generic MTF encoding and decoding algorithms: We use $Z[1]...Z[n]$ for input and output. The encoder gets its input from Z and the decoder writes its output to Z. CODE and DECODE are assumed to write to and read from a compressed stream of bits.

Generic MTF Encoder:

Initialize *list* to contain the characters of the alphabet in positions 0 through $k-1$.
for $i := 1$ **to** n **do**
    $p :=$ the position of $Z[i]$ in *list*
    CODE(p)
    Move $Z[i]$ to the front of *list* (position 0).
end

Generic MTF Decoder:

Initialize *list* to contain the characters of the alphabet in positions 0 through $k-1$.
for $i := 1$ **to** n **do**
    $p :=$ DECODE
    $Z[i] :=$ the character at position p of *list*
    Move $Z[i]$ to the front of *list* (position 0).
end

Using MTF with Burrows–Wheeler: When a string S is highly compressible, BWT(S) will have many runs of repeated characters. An effective data compression method, used by a number of compression utilities (see the chapter notes) is to pre-process the string with the Burrows–Wheeler transform, and use a MTF coder. Since the MTF coder is a pre-processor for basic coding, this amounts to a three layered method.

Efficient implementation: We could use an array *list*$[0]...list[k-1]$. In the worst case, the encoder spends $O(k)$ time to search for the position p of a character c in list, and both the encoder and decoder take $O(k)$ time to slide *list*$[0]...list[p-1]$ back one position so that c can be placed in *list*$[0]$, for a total of $O(nk)$ time to process a sequence of n characters. For better asymptotic time when k is large, we can represent *list* with a balanced tree, as presented on the following page.

Balanced Tree Implementation of MTF

Idea: Represent the list with a balanced tree (e.g., a 2-3 tree) of *time stamps*, and employ the INDEX function for search trees, which in $O(\log(k))$ time can find the index of an element in a list in sorted order of the elements that are stored in the tree.

Data structures:

X: The array $X[0]...X[k-1]$ stores the current *time stamp* of each character. Initially, $X[c]=-c$, $0 \leq c < k$, so that all time stamps start out ≤ 0. When the ith character is processed, it is given time stamp i, $1 \leq i \leq n$.

T: A balanced tree (e.g., a 2-3 tree) stores the list of characters, where the key used is that character's current time stamp.

Q: The array $Q[0]...Q[k-1]$ contains pointers to the corresponding characters in T.

Basic operations:

POS(c): Returns the position of c by using the INDEX function for T to find the index of $X[c]$ in T in $O(\log(k))$ time; since this index is between 1 and n, POS subtracts 1 before returning the value.

CHAR(p): Returns the character stored at position p of the list by using a MEMBER operation for T.

MOVE(i): Moves $Z[i]$ to the front of the list by deleting $X[Z[i]]$ from T, setting $X[Z[i]] := i$, and then inserting $X[Z[i]]$ into T.

Encoding and decoding algorithms: We again use an input-output array $Z[1]...Z[n]$.

MTF Encoder:

> $T :=$ empty tree
> **for** $c := 0$ **to** $k-1$ insert c into T with the key $-c$
> **for** $i := 1$ **to** n **do**
> > $p := $ POS($Z(i)$)
> > CODE(p)
> > MOVE(i)
> > **end**

MTF decoder:

> $T :=$ empty tree
> **for** $c := 0$ **to** $k-1$ insert c into T with the key $-c$
> **for** $i := 1$ **to** n **do**
> > $p := $ DECODE
> > $Z[i] := $ CHAR(p)
> > MOVE(i)
> > **end**

Complexity: POS, MOVE, and CHAR all work in $O(\log(k))$ time. Hence, assuming that CODE and DECODE work in $O(1)$ time (or even in $O(\log(k))$ time), both the MTF encoder and decoder work in $O(n\log(k))$ time. The space is $O(k)$ in addition to the space used by the input-output array Z.

Exercises

273. An *anagram* of a word is another word of the same number of letters that is formed by permuting those letters. For example, "step" is an anagram of "pest". Given a list L of words, describe how sorting may be used to make a list for each word of all of its anagrams that appear in L. Assuming that L is provided as a linked list and that each string in L is represented as a linked list, if n denotes the sum of lengths of the strings in L, your algorithm should run in $O(n)$ time and space.

Hint: Associate with each word in L a key string consisting of its characters in alphabetical order, and then sort L according to these keys.

274. Define two unordered trees T_1 and T_2 to be *isomorphic* if when distinct labels are placed on the vertices of T_1, it is possible to place the same set of distinct labels on the vertices of T_2 so that T_1 and T_2 are identical (that is, if two vertices have the same label, then the sets of labels defined by their children are identical). Describe in English and give pseudo-code to determine in linear time (time proportional to the number of vertices) if two trees are isomorphic.

Hint: First label each leaf 0. Then repeatedly label each vertex that has all of its children indexed by the string of these indices. Then sort sibling labels to create indices for siblings. After both trees are labeled, in linear time work upward from the leaves to check that the labels and indices are consistent. For example, below, labels are written inside the vertices and indices are written next to them:

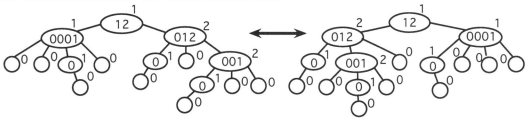

275. Give a formal proof by induction that the computation of the back-up array for the KMP string matching algorithm is correct.

276. The $O(m+n)$ worst case time of the KMP string matching algorithm using the back-up array is about the best that can be hoped for, given that in the worst case, we have to at least look at each of the characters of the search and the target strings; $O(m)$ space in addition to that used by S and T is very good as well. Thus, from the asymptotic point of view, there is no reason to compute the direct array or the more space efficient version using the map array, since both the asymptotic time and space are both increased. However, these increases are all in the preprocessing phase that constructs B and D and not in the matching phase. Discuss how the three different versions of the KMP method presented (directly using the B array, using the D array, and using the MAP array with the reduced version of the D array) compare in practice under various assumptions for the values of h, k, m, and n. In particular, consider the case that $T = pattern\ matching$ and S is 1 billion characters where characters are represented by bytes (i.e., $h=12$, $k=256$, $m=16$, $n=1,000,000,000$).

277. For string matching, suppose that S is in an area of memory where it takes time a to read a character, T is in an area of memory where it takes time b to read or write a value, arrays constructed from T (back-up array, etc.) are in an area of memory where it takes time c to read or write a value, and single integer variables are stored in registers where it takes time d to read or write a value. If only the time to read or write to memory and registers is counted (so comparisons and arithmetic operations are "free"), compare the expected and worst case time of KMP using the back-up array, KMP using the direct array, KMP using the direct array and the MAP array, Boyer–Moore, Shift-And, and Karp–Rabin, and discuss which is better for different assumptions about a, b, c, and d.

278. The basic assumption of the shift-and approach is that the search string $T[1]...T[m]$ can be packed into an integer that is small enough to be manipulated with individual arithmetic operations (OR, AND, etc.) available on the machine in question. For the case that T is too long to be packed into a single machine word, explain why the different versions of the shift-and approach presented do not cause any machine errors and explain what is actually computed (i.e., what does it mean when the search returns an index i).

279. Describe in English and present pseudo-code for a modification of basic shift-and matching where T fits into w machine words for some integer $w \geq 1$. Also describe what changes are necessary to incorporate don't-cares, exclusions, and mis-matches.

280. Recall the function sa+dc+ab+mmMATCH(q) for a match of T in S with a minimal number of mis-matches, subject to a maximum allowable number of mis-matches q. Suppose that the operation $OneCount(i)$ is not available as a basic machine instruction.

    A. Discuss how a combination of binary searching and masking on the integer i can be used instead of $OneCount(i)$ to yield an $O(m+nq\log(m))$ algorithm.

    Hint: By employing a mask, it can be determined whether the left or right half (or both) of i contains a 1, and this can be repeated in a binary search fashion to find a 1; the 1 can be masked out and the process repeated until i becomes 0.

    B. A more clever method for implementing $OneCount$ is to sum the number of ones in exponentially increasing size blocks. For example, in the language C, a program to count the number of 1's in a 32-bit number is:

```
/*Count the number of 1's in the binary representation of an integer.*/
main() {
#include <stdio.h>
int i, a;
printf("Enter an integer: "); scanf("%i",&i);
a = ((i >> 1) & 0x55555555) + (i & 0x55555555);
a = ((a >> 2) & 0x33333333) + (a & 0x33333333);
```

```
a = ((a >> 4) & 0x0F0F0F0F) + (a & 0x0F0F0F0F);
a = ((a >> 8) & 0x00FF00FF) + (a & 0x00FF00FF);
a = ((a >> 16) & 0x0000FFFF) + (a & 0x0000FFFF);
printf("There are %d ones in the binary representation of %d.\n", a, i);
}
```

After beginning the program and reading in the input integer i, in this C program, a number starting as $0x$ denotes a hexidecimal constant, where each of the characters that follows translates to 4 binary bits (so 0 -> 0000, 5 -> 0101, 3 -> 0011, F -> 1111), & denotes the AND operation, $a >> x$ means *ShiftRight(a,x)*, and + denotes the usual integer addition operation. Explain why this program computes the number of 1's in a 32-bit integer in the variable a, and give pseudo-code to count the number of ones in a number i of m bits (where i and m are inputs); assume m is a power of 2 and that you have an array $M[1]...M[\log_2(m)]$ that contains the constants you need for the bit masks at each step (i.e., the array M has been pre-computed once and for all before processing i).

Hint: After Step 1 (first assignment statement), think of a as being blocks of 2 bits, each a binary number equal to the number of ones in the corresponding block of i, after Step 2 (second assignment statement) think of a as blocks of 4 bits, each a binary number equal to the number of ones in the corresponding block of i, and so on.

C. When using the program of Part B for *OneCount* in sa+dc+ab+mmMATCH(q), it makes little difference for many practical applications how M is pre-computed, since the time is only a function of m and not n (and we are usually concerned with the case when n is much larger than m). Nevertheless, it is interesting to consider the complexity of computing M, and it may be important for special applications where m is large. Describe how to compute each entry of M in $O(\log(m))$ time (thus yielding an $O(\log(m)^2) = O(m)$ algorithm to compute M).

Hint: Use *ShiftLeft* and OR to keep doubling the length of a block of ones, and then starting with that block of ones together with the same number of zeros, keep doubling the length of the pattern until you get to length m.

281. Consider the function sacbv+dc+ab+mmMATCH(q) for matching with character bit-vectors in the presence of don't-cares, anything-buts, and at most q mis-matches.

A. Prove its correctness with a proof by induction.

B. By the definition of *Spack*, it should be true that the jth bit from the right in *Spack*[x], $1 \leq j \leq m$ and $1 \leq x \leq q$, should be 1 if the jth bit from the right in *Spack*[$x-1$] is 1. That is, if $T[1]...T[j-1]$ matches $S[i-j+1]...S[i]$ with at most $x-1$ mismatches, then it also matches with at most x mis-matches. Explain why sacbv+dc+ab+mmMATCH(q) maintains this property (which is why it was not necessary to include the statement *Spack*[x] := *Spack*[x] OR *Spack*[$x-1$]).

282. A pattern diagram is called a *non-deterministic finite automaton* (NFA) in the literature. A NFA looks like a "machine" to recognize strings, but it is difficult to actually "run" one because when you are at a given vertex, there may be a number of edges leaving that vertex that all have the label of the current input character (and there may be a number of ε-edges as well). In fact, the way that we used pattern diagrams reflected this; we had to keep track of the current *set* of vertices that we could be in as a result of reading the input thus far. A special type of pattern diagrams are *deterministic finite automata* (DFA) where there are no ε edges and for every vertex v and for every character c of the input alphabet, there is exactly one edge labeled c that leaves v. With a DFA, we can easily "run" the diagram by simply following the edges as the input is read.

A. Explain why the pattern diagram corresponding to the KMP direct array is an example of a DFA.

B. Show that any pattern diagram can be converted to an equivalent one with no ε-edges.

Hint: For each vertex, determine the set of vertices reachable by ε-edges and then put in direct edges to where these vertices can go via a non ε-edge.

C. Show that any pattern diagram without ε-edges can be converted to an equivalent DFA

Hint: Construct a new pattern diagram that has one vertex corresponding to each possible subset of the vertices of the original diagram. There is an edge from a vertex v to a vertex w in this new diagram if the set of states corresponding to w is the set of vertices reachable in the original diagram from the set of vertices corresponding to v.

283. Redo the example tree given for the Aho-Corasick algorithm to be based on the back-up array rather than the direct array (so that it could be used with kmpMATCH rather than fastkmpMATCH.

284. We have seen how to convert a regular expression to a pattern diagram. It is natural to consider how to do the reverse — convert a pattern diagram to a regular expression. A pattern diagram is just a directed graph where edges are labeled by characters (or by the symbol ε that represents the empty string). In the chapter on graphs, Floyd's and Warshall's algorithms are presented as special cases of a general path-finding framework that finds a path of least cost between all pairs of vertices. In a pattern diagram, define the "cost" of an edge to be its label. Explain why this general framework gives an algorithm to convert a pattern diagram to a regular expression if we define the basic operations of the framework in terms of the $+$, $\bullet$, and $*$ regular expression operators as follows:

$$\text{UNDEF} = \varphi$$
$$\text{INIT} = \varepsilon$$
$$\text{CYCLE}(x) = x^*$$
$$\text{COMBINE}(x,y,z) = x \bullet y \bullet z$$
$$\text{CHOOSE}(x,y) = x+y$$

285. Consider the greedy algorithm for the superstring problem that was presented as an example in the chapter on algorithm design techniques. Let n denote the number of strings in S and N the sum of lengths of the strings in S.

    A. To compute the overlap of two strings u and v, a naive algorithm slides the strings one character at a time. The naive implementation then tests all pairs to find the best overlap, and repeats this process for each iteration of the *while* loop. Analyze the running time of this approach.

    B. Show how to modify the Knuth–Morris–Pratt string matching algorithm to compute the overlap of two strings u and v in linear time.

    C. Using trie data structures to represent the strings of S and the reverses of the strings of S, show that the body of the *while* loop can be $O(N)$, and hence the total time is nN. Is there a way to do even better by combining some of the work done by different iterations of the *while* loop?

286. Given an example of a code that is uniquely decodable but not a prefix code.

287. We have defined what are usually refereed to as 0th *order Huffman codes*, where each character is coded independently of the characters that preceded it. Generalize the Huffman encoding and decoding algorithms to k^{th} order codes, where a different Huffman trie is used depending on the previous k characters encoded. Analyze the time and space used.

288. Suppose that we wished to send a trie that stored binary strings over a slow communication channel as quickly as possible. Describe a simple encoding of a binary trie that uses 2 bits per vertex and give algorithms to produce this encoding from a standard pointer-based representation, and to recover the trie from the encoding.

Hint: Store it in level order where each vertex is an integer 0 through 3 to signal whether it has no children, just a left child, just a right child, or two children.

289. Given a string $s\$$ of length n, where $\$$ is a character that does not appear in s, the *substring identifier* for position i, $1 \le i \le n$, is the shortest string which is a substring of $s\$$ starting at position i and occurs no where else in $s\$$. Describe how to in $O(n)$ time construct a trie that represents all of the substring identifiers of $s\$$.

290. Given the details of how to maintain a virtual depth field for each vertex in:

    A. The basic brute-force trie construction algorithm.

    B. McCreight's algorithm.

    C. The Fialia–Greene algorithm.

291. With sliding suffix tries, we assumed that non-leaf vertices were also labeled with a position, and that whenever SCAN or RESCAN moved down, both edge and vertex labels were updated with the rightmost occurrence seen thus far of the corresponding string. Describe how to search for a string in a suffix trie that has only the edge labels, and not labels on non-leaf vertices.

Hint: As you move down the trie, calculate virtual depth and subtract this from an edge label to determine where the substring ending with that edge label begins in the input.

292. The description of the two trie algorithm for sliding suffix trie construction was an informal one that described the first few steps of a straightforward pattern. Give pseudo-code for this algorithm.

293. The linear time and space bounds for McCreight's algorithm assume that the alphabet size is a constant, which makes the basic CHILD(v,c) operation constant. Different implementations of the CHILD operation will change the bound when the alphabet size k is taken into account. Explain why:

    A. With a hash implementation, the time becomes *expected $O(n)$*.

    B. With an array based implementation the space becomes $O(kn)$.

    C. With a balanced tree implementation, the time becomes $O(\log(k)n)$.

294. An approach to analyzing the time used by the percolate update is with a proof that uses an amortized analysis with "credits", where one credit pays for the $O(1)$ time used each time a vertex is updated. Give a proof that provides two new credits whenever a leaf is created (these are the only credits provided, and they must be moved around to "pay" for the updating of vertices), and maintains the following *credit invariant*:

    All vertices with *flag*=1 contain at least one credit.

Hint: Each time a leaf v is added, the percolate loop will change a sequence of ones to zeros and then after the last iteration, the flag bit of the final vertex is changed from a zero to a one. One of the new credits provided with the creation of the leaf pays for updating at the leaf; the other iterations are paid for by the credits that are guaranteed by the credit invariant to be on vertices with ones. The second of the new credits provided by the leaf creation can be placed on the final vertex, which serves to maintain the credit invariant.

295. Consider the algorithm presented to compute the edit distance from s to t. Fill in the details of how to produce a sequence of edits from $s_1...s_i$ to $t_1...t_j$ of length $ED[i,j]$ and present pseudo-code.

296. When the algorithm to compute the edit distance from s and t was presented, it was discussed how to implement it in $O(m+n)$ space by only remembering a cross-section.

    A. Give pseudo-code that uses $O(m+n)$ space by using vertical cross-sections.

    B. Give pseudo-code that uses $O(m+n)$ space by using horizontal cross-sections.

    C. Give pseudo-code that uses $O(m+n)$ space by using diagonal cross-sections.

297. We observed that the problem of computing the longest common substring of s and t can be reduced to the problem of computing the edit distance from s to t.

    A. Give pseudo-code for a direct implementation that computes the entries $LCS[i,j]$, $0 \le i \le m, 0 \le j \le n$.

    B. Fill in the details of how to produce a subsequence of $s_1...s_i$ and $t_1...t_j$ of length $LCS[i,j]$ and present pseudo-code.

    C. We have see that for the case that the cost of inserting and deleting characters is the same, and the cost of changing a character is twice that cost, an algorithm for the longest common subsequence can be used for edit distance. However, give an example to show that when the costs of all three operations are equal, it is not necessarily true that the longest common subsequence gives the correct edit distance.

298. For arithmetic coding, we presented two ways of defining the end of a finite string. The first method was to precede it with its length. The second method was to add a special symbol $ to the end. Suppose that the input string s is a binary string of $n \ge 1$ bits.

    A. For the first method, in practice we can agree on a maximum string length, and then transmit the string length with an appropriate number of bits (e.g., 32 bits would allow strings of length up to about 4 billion); the decoder knows to read this many bits before starting to read the first bit of s. But suppose that the decoder does not know how many bits to expect for the length. We know from the first chapter that $\lfloor \log_2(n) \rfloor + 1$ bits suffice to represent n. Give a way to encode n with $2(\lfloor \log_2(n) \rfloor + 1)$ bits so that when you read its bits from left to right you know when you have read the rightmost bit.

    B. Improve your answer to Part A to encode the length with $\lfloor \log_2(n) \rfloor + 1 + \varepsilon$ bits where $\varepsilon \to 0$ as $n \to \infty$.

    Hint: Precede the string of bits used for the length with its length, precede that string with its length, and so on until you are down to $O(1)$ bits.

    C. For the second method, since $ does not occur in s, it occurs with frequency $1/(n+1)$ in s, and so it should be assigned the interval $1/(n+1)$. Explain why this means that about $\log_2(n)$ bits will be sent for $ (so the overhead is about the same as Part B).

299. Compute the Burrows-Wheeler transform of the 18 character string:

    c a n a c a t c a l l c a t t l e $

Chapter Notes

The books of Crochemore and Rytter [1994], Gusfield [1997], and Apostolico [1997] address the general subject of string matching. The book of Stephen [1994] surveys algorithms for inexact methods. For an introduction to computational biology, from which many variations of string matching algorithms are motivated, see the books of Waterman [1995] and Setubal and Meidanis [1997], as well as the chapter by Lesk in Kent and Williams [1994]; Pevzner and Waterman [1995] survey open problems in this area. The book of Sankoff and Kruskal addresses pattern matching and related problems for a variety of applications, including molecular sequences. An excellent source of work done in the area of pattern matching is the series of annual conferences of Apostolico, Crochemore, Galil, and Manber [1992–2000].

The presentation of lexicographic sorting here is based on that in Aho, Hopcroft, and Ullman [1974]. Andersson, Hagerup, Hastad, and Petersson [1994] consider the complexity of searching a sorted array of strings.

The KMP algorithm is presented in Knuth, Morris, and Pratt [1977] and appears in may algorithms texts, including the book of Aho, Hopcroft, and Ullman [1974]. See also the paper of Morris and Pratt [1970].

The Boyer–Moore algorithm was introduced by Boyer and Moore [1977]. Since that time it has been studied by many authors. Rytter [1980] presents the algorithm used here for computing the advance-forward array. Cole [1991] shows that $3n$ is a tight upper bound on the worst case number of comparisons used by the Boyer–Moore algorithm on a string of length n; earlier, Guibas and Odlyzko [1980] presented a $4n$ upper bound. Crochermore, Czumaj, Gasieniec, Jarominek, Lecroq, Plandowski, and Rytter [1992] introduced an improvement to the Boyer–Moore algorithm, called *turbo shifts*, that can be used to reduce the worst case complexity to less than $2n$ comparisons for a string of length n. Galil [1979] presents improvements for when all occurrences of the pattern must be found (the original Boyer–Moore algorithm is quadratic in the worst case for finding all occurrences). Apostolico and Giancarlo [1986] present additional improvements to the algorithm. The original algorithm and the improvements indicated above are well described in the book of Crochermore and Rytter [1994]. Sunday [1990] considers three variations that his experiments show to be faster than the Boyer–Moore Algorithm.

The Karp–Rabin algorithm is presented in Karp and Rabin [1987]. A nice aspect of this approach is that it generalizes easily to multi-dimensional patterns. The presentation here is based on the presentation in the book of Lewis and Denenberg [1991]. Bentley and McIlroy[1999] present efficient methods to detect long common substrings that may be far apart (and not captured by a practical sliding window data compression method) by employing Karp–Rabin finger prints.

Here we have viewed the term "shift-and" to generally refer to methods that exploit the ability of arithmetic operations to perform collections of bit operations in parallel when the bits can be packed into a single machine word; in the literature, the "shift-and method" usually refers to the character bit vector version that was presented here. Baeza–Yates and Gonnet [1992] and Wu and Manber [1992] present the basic shift-and method.

Galil and Sieferas [1983] consider string matching using an optimal time-space tradeoff.

Landau and Vishkin [1986] consider the general problem of string matching with errors (where m is of arbitrary size) and Galil and Giancarlo [1986] improve their bounds for a string of length n and a pattern of length m to $O(kn+m\log(m))$ time and $O(m)$ space in addition to the $O(n)$ space used by the input. Meyers [1999] presents a dynamic programming approach for approximate string matching where the algorithm's performance is independent of the number of mis-matches allowed. Ukkonen [1993] considers approximate string matching employing pre-processed suffix tries.

The C-program in the exercises for counting the number of 1's in a 32-bit integer is based on a program provided by G. Motta.

The Aho–Corasick generalization of KMP to multiple string searching with an augmented trie is due to Aho and Corasick [1975].

Hoffman and O'Donnell [1982] and Kosaraju [1989] consider pattern matching in trees.

Abrahamson [1987] presents a generalized string matching algorithm that searches for all occurrences of substrings of a string of length n that are consistent with a sequence of "pattern elements" of length m in time $O\left(n+m+m\sqrt{n}\,poly\log(n)\right)$.

There has been much work on finding patterns in data that has been compressed, particularly with lossless LZ based methods; for example see Kida, Takeda, Shinohara, and Arikawa [1999], Shibata, Takeda, Shinohara, Arikawa [1999], Navarro and Raffinot [1999], Klein and Shapira [2000,2001], Navarro and J. Tarhio [2000], Shibata, Matsumoto, Takeda, Shinohara, and Arikawa [2000], Karkkainen, Navarro, Ukkonen [2000], Navarro, Kida, Takeda, Shinohara, and Arikawa [2001], Mitarai, Hirao, Matsumoto, Shinohara, Takeda, and Arikawa [2001], and their references.

The McNaughton–Yamada Algorithm is presented in McNaughton and Yamada [1960]; see also Thompson [1968]. See also the presentation in the books of Ullman [1984], Hopcroft and Ullman [1979], and Aho, Hopcroft, and Ullman [1974], upon which the presentation here is based.

Tries were proposed by Fredkin [1960].

Huffman coding is perhaps the oldest and most well known data compression technique. We have defined 0th *order Huffman codes* where each character is encoded independent of the characters that preceded it in the input; for any $k \geq 0$, a kth order Huffman code uses a different Huffman trie depending on which string of k characters precedes the character being encoded. It is also possible to encode blocks of k characters at a time with a single large Huffman trie (and in this case, a 0th order Huffman coder is sometimes called a first order Huffman coder). For further reading and references on Huffman codes see the book of Storer [1988]. Huffman codes were introduced by Huffman [1952]. Gallager [1978] (see also Faller [1973]) describes how to implement adaptive Huffman coding, where the tree is modified as the input is processed (this algorithm is used by the UNIX compact utility); see also Cormack and Horspool [1984], Knuth [1982,1985] and Vitter [1987,1989]. Storer [1991] compares the costs of adaptive and static implementations. For any integer $k \geq 2$, the construction that was presented in the chapter on algorithm design for *binary* Huffman trees can be generalized to the case where all non-leaf vertices have k children; the only new detail is that since each step of the construction algorithm reduces the number of trees in the forest by $k-1$, if $k-1$ does not evenly divide $n-1$, then the first step of the construction algorithm is to combine the $(n-1)$ MOD $(k-1))+1$ trees of smallest weight into a single one. Karp [1961], Krause [1962], and Mehlhorn [1980] consider the case when characters have non-uniform cost. Gilbert [1971] considers coding with inaccurate source probabilities. Moffat, Turpin, and Katajainen [1995] and Moffat and Turpin [1996] consider efficient implementation of prefix code construction. Guivarch, Carlach, and Siohan [2000] consider Huffman decoding in combination with channel coding. Larmore and Hirschberg [1987] and Liddell and Moffat [2001] consider length-limited Huffman codes. Golin and Na [2001] consider optimal prefix codes that end in a specified pattern, and discuss the problem of when prefix codes must be drawn from a specified language. Hankamer [1979] considers memory requirements. Jacobsson [1978] considers Huffman coding of bit-vectors. Basu [1991] considers the use of several Huffman tries to compress a string, and Mitzenmacher [2001] shows that finding optimal preset dictionaries for Huffman coding (and some LZ-based schemes) is NP-hard. Jacobson [1992] considers random access in Huffman encoded strings. Klein and Shapira [2001] consider pattern matching in Huffman encoded strings. Parallel and hardware architectures for Huffman coding are considered by Park and Prasanna [1993], Park, Son and Cho [1995], Milidiu, Laber, and Pessoa [1999], and Klein and Y. Wiseman [2000]. McIntyre and Pechura [1985] and Nekritch [2000] consider decoding look-up tables. Analysis of Huffman coding has been addressed by many, including Golomb [1980], Johnsen [1980], Capocelli, Giancarlo, and Taneja [1986], Blumer and McEliece [1988], Capocelli and A. De Santis [1988, 1991, 1991b, 1992], Yeung [1991], Manstetten [1992]. See also Baron and Singer [2001], who compare the worst case performance of a constrained Huffman code to that of an unconstrained code, as opposed to the entropy (inherent compressibility) of the source in question. Huffman codes are used in the JPEG and MPEG standards.

We have presented a somewhat simplified version of data compression with a dynamic dictionary. For example, Step 2A of the encoding algorithm amounts to a greedy algorithm for getting the next match string; although this works well in practice, it can be that by taking a shorter match at one stage, longer matches will be possible at later stages.

See the book of Storer [1988] for an overview of dictionary based data compression employing the trie data structure; Storer [1985] presents the different update heuristics, Reif and Storer [1998] consider dynamic sources, Storer [2000] and Storer and Reif [1995, 1997] consider dictionary compression in the presence of corrupted compressed data. The book of Storer [1992] contains a chapter that presents a number of parallel systolic algorithms for sliding window compression, including that of ZitoWolf [1991], and Gonzalez and Storer [1985]; generalizations to two dimensions are addressed in Storer and Helfgott [1997].

For additional reading on compression with dictionaries (and sliding-window compression), see the books of Storer [1988], Bell, Cleary, and Witten [1990], and Witten, Moffat, and Bell [1994]; Storer and Szymanski [1978] consider theoretical tradeoffs between different off-line dictionary based methods; DeAgostino and Storer [1996] consider tradeoffs between off-line and on-line methods; Storer and Reif [1991] and Royals, Markas, Kanopoulos, Reif, and Storer [1993] describe a parallel VLSI implementation of adaptive dictionary compression. Rizzo, Carpentieri, and Storer [2000] survey dictionary based methods. Note that here we are using the term compression to refer to *lossless* compression, where the decompressed data must be bit for bit identical to the original. In contrast, *lossy* compression techniques allow the decompressed data to differ from the original according to some fidelity criterion (e.g., a decompressed image looks as good to the human eye as the original). Constantinescu and Storer [1994, 1994b], Carpentieri and Storer [1994,1996], Rizzo, Storer, and Carpentieri [1999], and Rizzo and Storer [2001] generalize adaptive lossless dictionary based data compression to lossy compression of images and video.

Early work on suffix tries, often called PATRICIA trees, is presented in Morrison [1968]; see also the paper of Szpankowski [1990,1993] and Jacquet and Szpankowski [1994] for theoretical analysis of this work and related work on suffix tries. McCreight's Algorithm for linear time suffix trie construction is presented in McCreight [1976]. In earlier work, Weiner [1973] presents a linear suffix trie construction algorithm that works from right to left and also uses an amortized analysis; his algorithm is presented in the book of Aho, Hopcroft, and Ullman [1974]. Slisenko [1981] (see also Slisenko 1980, 1978, 1977]), in a long and technical paper, claims that suffix tries can be used for a number of real time string problems, including recognizing strings of the form $\$xyz\$y$ and $\$xyz\$y^R$, where y^R denotes the reverse of y; comments about this work can be found in Chen and Sieferas 1985] and Sieferas [1977].

McCreight's algorithm is not strictly on-line because it reads some distance ahead of the current position in order to insert the leaf for that position (although for many practical applications it may be close enough). In fact, as we saw in the example that showed McCreight's algorithm on the input $a^n\$$, in the worst-case scanning ahead could go all the way to the right end of the string. Ukkonen [1995] presents a strictly on-line suffix trie construction algorithm; on-line and real-time suffix trie construction is also considered by Majster and Reiser [1980] and Kosaraju [1994]. Fiala and Greene [1989] present sliding suffix tries with the percolate update. Laarson [1996,1999] uses the techniques of Fiala and Greene to incorporation deletion to Ukkonen's algorithm to slide the trie so that it evolves like Fiala–Greene, but without any scanning ahead.

The algorithm of Rodeh, Pratt, and Even [1981] (who use sliding suffix tries for linear time sliding window data compression) can be viewed as an on-line generalization of the "two–trie" algorithm that was presented in this chapter. Imagine actually extracting a block of n characters from the input, placing a $ at its right end, and running McCreight's algorithm on that string. They use three overlapping copies, where $m \leq n/2$ is the maximum match length.

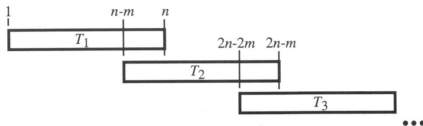

1. Start constructing T_1.
2. After $n-m$ characters: Start constructing T_2 (and continue constructing T_1).
3. After n characters: Stop constructing T_1 (but continue constructing T_2).
4. After $2n-2m$ characters: Start constructing T_3 (and continue constructing T_2).
5. After $2n-m$ characters: Stop constructing T_2 (and continue constructing T_3). Also, reclaim the memory for T_1 and start constructing T_4.

Etc.

Although there are now three tries instead of two, it is still the case that only two need to be consulted to find matches for the current position (the right trie does not become "active" until it is time to "recycle" the left one), and like the two–trie algorithm, a postorder traversal can be performed after a trie is completed to label internal vertices with the position of the rightmost occurrence seen thus far of their corresponding string.

Makinen [2000] considers compact suffix arrays, and Grossi and Vitter [2000] consider compressed suffix arrays and suffix trees. Maab [2000] considers bi-directional on-line construction of *affix trees*, a generalization of suffix tries. Rocke [2000] uses suffix tries to find similar components in sequences representing proteins. Cole and Hariharan [2000] consider suffix trie construction with missing links and present a randomized linear time algorithm for constructing suffix tries for parameterized strings.

Blumer, Blumer, Haussler, and McConnell [1984], Blumer, Blumer, and Haussler [1985], and Blumer, Ehrenfeucht, and Haussler [1984,1987] consider the DAWG (Directed Acyclic Word Graph) data structure that corresponds to a suffix trie condensed into a directed graph; Chen and Sieferas [1985] present a version of Weiner's algorithm that also constructs the smallest deterministic finite automaton recognizing reverse subwords.

Throughout our presentation of suffix tries, we have assumed that the alphabet size is a constant with respect to the length of the input string. Farach–Colton, Ferragina, and S. Muthukrishnan [2001] consider the complexity of suffix trie construction when the alphabet size is unbounded, and present a divide and conquer approach for suffix trie construction that matches the sorting lower bound. That is, since suffix trie construction can be used to sort the characters of the input string, the complexity of sorting gives a lower bound on the complexity of suffix trie construction for unbounded alphabets.

Galil [1985] and Viskin [1990] consider fast parallel string matching. Apostolico [1984], Apostolico [1985], Landau, Schieber, and Vishkin [1987], and Apostolico, Iliopoulos, Landau, Schieber, and Vishkin [1987], Hariharan [1994], and Sahinalp and Vishkin [1994] consider parallel construction of a suffix trie. Hagerup [1994] employs tries to solve several problems on strings. Gusfield, Landau, and Schieber [1991] employ suffix tries to find maximal overlaps between strings the end of one to the beginning of another, which has applications to the superstring problem and to splicing of genetic sequences.

Giancarlo [1995], Giancarlo and Grossi [1995], and Giancarlo and Guiana [1996], and Giancarlo and Guiana [1997] consider two-dimensional suffix tries.

Apostolico and Preparata [1982] use suffix tries to gather statistics on all substrings of a string. Applications of Suffix Tries to self alignments is considered in Apostolico [1992]. Applications of suffix tries to detection of rare substrings is considered in Apostolico, Bock, and Lonardi [1999]. Applications of suffix tries to data compression, including biological sequences, is considered in Apostolico and Lonardi [1998,2000].

Wagner and Fisher [1974] consider the edit distance problem, and Lowrance and Wagner [1975] consider extensions; see also the book of Sankoff and Kruskal [1983]. Ukkonen [1985] considers improvements to edit distance computations. Lipton and Lopresti [1985] present a parallel systolic algorithm to compute edit distance. Kim and Park [2000] consider dynamic edit distance tables.

The book of Abramson [1963] credits P. Elias for the idea of arithmetic coding. The book of Bell, Cleary, and Witten [1990] gives a clear presentation of practical considerations such as how to perform rescaling and use finite precision arithmetic, as does Moffat [2002]. For further material on arithmetic coding and its use in compression, see for example Cleary, Teahan, and Witten [1995], Moffat, Neal, and Witten [1995], Moffat, and Bell [1994], Cheng and Langdon [1992], Langdon [1991], Moffat [1990], Bell, Witten, and Cleary [1989], Rissanen and Mouhiuddin [1989], Witten, Neal, and Cleary [1987], Cleary and Witten [1984,1984b], Langdon [1984], Martin, Langdon, and Todd [1983], Langdon and Rissanen [1981], Rissanen and Langdon [1979]. Arithmetic coding is used by the IBM Q-coder; for example, Pennebaker, Mitchell, Langdon, and Arps [1988]. See the thesis of Howard [1992] for a presentation of fast "quasi-arithmetic" coding algorithms as well as further references. Moffat [1989] considers word-based text compression. Motta, Storer, and Carpentieri [2000] present image lossless image compression algorithms based on adaptive linear prediction, and compare these to methods employing arithmetic coding.

The Burrows–Wheeler transform, sometimes called *block sorting*, is due to Burrows and Wheeler [1994]; the description of the transform with the matrix M and the two-pass algorithm for the inverse transform that produces S from right to left is based on their presentation. The figure showing decoding with the permutation links that constructs S from left to right is based on the presentation in Sadakane [1999]. Burrows and Wheeler present a simple and effective compression algorithm that performs the transform and then uses move-to-front coding to drive a first-order coder (e.g., Huffman or arithmetic); it is the basis for the popular "bzip" and "bzip2" UNIX compression utilities. Fenwick [1996] considers a variety of ways to apply the transform to data compression and to speed up the algorithm, as does Larrson [1999]. Balkenhol and Kurtz [2000] also discuss implementations and present experiments that compare favorably with or improve slightly upon the performance of some existing utilities (including bzip2). M. Nelson [1996] overviews the transform and its application to data compression. Arnavut, Leavitt, and Abdulazizoglu [1998] consider generalizations of the transform. Helfgott [1997] examines the problem of generalizing the transform to two dimensions. Laarson [1998,1999] addresses practical connections with context trees. Sadakane [1998] considers fast practical suffix sorting for improved implementation of the Burrows–Wheeler transform. Effros [1999] considers universal lossless source coding with the transform. Other improvements to the Burrows–Wheeler transform are considered in Moffat and Wirth [2001], Isal and Moffat [2001], Seward [2001], and Chapin [2000]. Manzini [2001] gives a theoretical analysis of the performance of the Burrows–Wheeler transform.

Bentley, Sleator, Tarjan, and Wei [1986] present move-to-front compression, including analysis, data structures, and experimental results. Thomborson and Wei [1989] present (parallel) systolic implementations. Modifications of move-to-front and related ranking protocols to be used for compression with the Burrows–Wheeler transform are considered by Chapin [2000] and Schindler [1997]. Volf and F. M. J. Williams [1998,1998b] consider switching between two different ranking methods.

The anagram exercise was suggested by Bentley [1999]. The exercise to count the number of one's in a binary number was suggested by Motta [2000]. The 2-bit per vertex binary trie representation exercise was suggested by F. Mignosi.

12. Discrete Fourier Transform

Introduction

The Fourier Transform has wide applications in scientific computing and engineering. Although it has a continuous version, we will consider only the discrete version (DFT) and present what is commonly known as the Fast Fourier Transform (FFT) algorithm. The DFT can replace a set of n values with another set of n values that represents the same information in a different way that may be easier to work with in a particular application. Later, after the work has been performed, the transformation can be reversed.

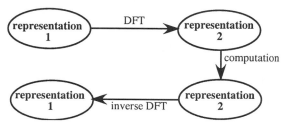

Example: A polynomial p of degree $n-1$ can be represented by storing its n coefficients $A[0]...A[n-1]$ or by storing the value of $p(x)$ at n distinct points. When represented by values, some operations are easier; for example, multiplying two polynomials is just pair-wise multiplication of the corresponding values.

Example: In the *time domain*, an audio signal can be represented dividing time into n intervals per second (e.g., in the case of a standard compact disc, $n=44,100$), and then storing for each interval the (normalized) average distance the speaker cone is in or out from its resting position. In the *frequency domain*, an audio signal can be represented as the sum of n basic tones (e.g., sine and cosine waves), where each tone can be represented by a *magnitude* (how much of that tone is used) and a *phase* (how much that tone has been shifted from being centered about 0). In both cases, larger values of n yield a more accurate representation of complicated signals. In many practical applications, computations can be simplified by using the DFT to change domains.

Although it is typical for the input to be an array of real numbers (and for the output from the inverse DFT to be an array of real numbers), it is convenient to use a "richer" domain to store the values of the transformed data. It is most common in the literature to define the DFT over the complex numbers, and we limit our attention to the complex numbers in our examples. However, we shall see that the derivation of the FFT algorithm can be generalized to appropriate classes of finite fields and rings. We shall also present the *discrete cosine transform*, a related transform that uses only real values.

Complex Numbers

Idea: In the real numbers, negative numbers do not have square roots. However, if one defines an "imaginary" number $\sqrt{-1}$ that has the property that $\sqrt{-1}^2 = -1$, then for any non-negative real number x, we can define $\sqrt{-x} = \sqrt{-1} * \sqrt{x}$.

Complex numbers: By adding an element $\sqrt{-1}$ to the real numbers (and all quantities that can now be derived with arithmetic operations), we obtain the *complex numbers*. All complex numbers c are of the form $c = a + \sqrt{-1}b$; where a and b may be any real numbers; a is called the *real part* of c, which we denote by $RE(c)$ and b is called the *imaginary part* of c, which we denote by $IM(c)$. The complex number $a - \sqrt{-1}b$ is called the *complex conjugate* of c, which we denote as c^*; for an array A, we also use the notation A^* to denote the array where the ith entry is $A[i]^*$. In the literature, an imaginary number is often referred to as "i" or "j"; we avoid this notation to avoid confusion with the common use of i and j as an index variable in pseudo-code. The notation $\sqrt{-1}$ suggest the property in which we are interested; a quantity x such that $x^2 = -1$. There must always be two complex numbers with this property; if $x^2 = -1$, then it must also be that $(-x)^2 = -1$. That is, when we extend the real numbers to include x and all resulting arithmetic combinations, we also get $-x$.

Storing pairs of real numbers: A complex number can be thought of as a way to store two real numbers at once (the real and imaginary part). For example, with an audio signal that is represented in the frequency domain, the real part can store the magnitude and the imaginary part can store the phase shift.

Polar coordinates: Instead of representing a complex number $c = a + \sqrt{-1}b$ in *Cartesian* coordinates by the two numbers (a,b), we can represent c in *polar* coordinates by the two numbers (x,d), where x is its angle from the real axis d is its distance from the origin. We measure angles in *radians*, the distance along the perimeter of the unit circle from the real axis, so that $\pi/2$ corresponds to 90 degrees, π corresponds to 180 degrees, $3\pi/2$ corresponds to 270 degrees, and 2π is the same as 0 degrees:

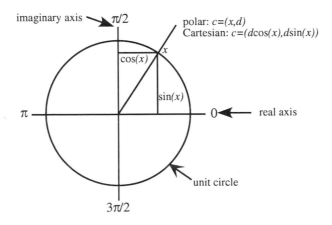

CHAPTER 12

Complex Exponentials

Notation: e denotes the natural logarithm base $e=2.71...$ (the reader may wish to review the facts about e that were presented in the chapter on hashing).

Theorem: For any real number x, $e^{\sqrt{-1}x} = \cos(x) + \sqrt{-1}\sin(x)$.

Proof: The standard *Taylor Series* for a continuously differentiable function $f(x)$ is

$$f(x) = f(0) + f'(0)x + \frac{f''(0)x^2}{2!} + \frac{f'''(0)x^3}{3!}\frac{f''''(0)x^4}{4!} + \cdots$$

where the primes denote differentiation (f' is the first derivative of f, f'' the second derivative, etc.). Then since the derivative of e^x is e^x, the derivative of $\sin(x)$ is $\cos(x)$, and the derivative of $\cos(x)$ is $-\sin(x)$, it follows that

$$e^{\sqrt{-1}x} = 1 + \sqrt{-1}x - \frac{x^2}{2} - \frac{\sqrt{-1}x^3}{3!} + \frac{x^4}{4!} + \frac{\sqrt{-1}x^5}{5!} - \cdots$$

$$\cos(x) = 1 - \frac{x^2}{2} + \frac{x^4}{4!} - \frac{x^6}{6!} + \frac{x^8}{8!} - \cdots$$

$$\sin(x) = x - \frac{x^3}{3!} + \frac{x^5}{5!} - \frac{x^7}{7!} + \frac{x^9}{9!} - \cdots$$

and hence:

$$e^{\sqrt{-1}x} = 1 + \sqrt{-1}x - \frac{x^2}{2} - \frac{\sqrt{-1}x^3}{3!} + \frac{x^4}{4!} + \frac{\sqrt{-1}x^5}{5!} - \cdots$$

$$= \left(1 - \frac{x^2}{2} + \frac{x^4}{4!} - \cdots\right) + \left(\sqrt{-1}x - \frac{\sqrt{-1}x^3}{3!} + \frac{\sqrt{-1}x^5}{5!} - \right)$$

$$= \left(1 - \frac{x^2}{2} + \frac{x^4}{4!} - \cdots\right) + \sqrt{-1}\left(x - \frac{x^3}{3!} + \frac{x^5}{5!} - \right)$$

$$= \cos(x) + \sqrt{-1}\sin(x)$$

Intuition: $e^{\sqrt{-1}x}$ is a point on the unit diameter complex circle centered at the origin of the complex plane at angle x (in radians) with real component $\cos(x)$ and imaginary component $\sin(x)$.

Idea: Complex exponentials are a nice way to represent a point on the complex plane (to multiply two complex exponentials, we can just add the exponents), and complex numbers are a nice way to store magnitude and phase information for points in the frequency domain of a digitally sampled analog signal.

Principal n^{th} Roots of Unity

Definition: For an integer $n>1$, a *principal n^{th} root of unity* over the complex numbers is any element r satisfying the following two axioms:

1. $r^n = 1$
2. $r^k \neq 1$, for all integers $1 \leq k < n$

Theorem: For $n>1$, $r = e^{\sqrt{-1}\left(\frac{2\pi}{n}\right)}$ is a principal n^{th} root of unity for the complex numbers.

Proof: Axiom 1 is satisfied because:

$$r = e^{\sqrt{-1}\left(\frac{2\pi}{n}\right)n} = e^{\sqrt{-1}2\pi} = \cos(2\pi) + \sqrt{-1}\sin(2\pi) = 1 + 0 = 1$$

Axiom 2 is satisfied because $n>1$ implies $2k\pi/n \neq 2\pi$, and hence the real part of r, $\cos(2k\pi/n)$, cannot be 1.

Facts about principal n^{th} roots of unity r that follow from the axioms:

3. For all integers $k \neq 0$ such that $-n<k<n$, $\displaystyle\sum_{j=0}^{n-1} r^{jk} = 0$.

   We have a geometric sum (see the appendix) of complex numbers which can be solved in the same way as for real numbers (see the exercises):

   $$\sum_{j=0}^{n-1} r^{jk} = \sum_{j=0}^{n-1}\left(r^k\right)^j = \frac{\left(r^k\right)^n - 1}{r^k - 1} = \frac{\left(r^n\right)^k - 1}{r^k - 1} = \frac{(1)^k - 1}{r^k - 1} = 0$$

   Axiom 1 allows us to replace r^n by 1, and Axiom 2 guarantees that the denominator is not 0 (since for $-n<k<0$, $r^k = r^n r^k = r^{n+k}$).

4. $r^{-k} \neq 1$, for all integers $1 \leq k < n$

   This fact was observed in the proof of Fact 3. Another way to see it is true is to observe that if $r-k=1$, then $r^k r^{-k}=1$ violates Axiom 1.

5. If n is even, then for $n>2$, r^2 is a principal $n/2^{th}$ root of unity.

   If $r \neq 1$, then since $n>2$, by Axiom 2, $r^2 \neq 1$. Also, $\left(r^2\right)^{n/2} = r^n = 1$.

6. If n is even, then for all integers $0 \leq k < n$, $r^{n/2}=r^{-n/2}=-1$ and $r^{k-n/2} = r^{k+n/2} = -r^k$.

   We know $(r^{n/2})^2=1$. Since the two solutions to $r^2=1$ are 1 and -1, and we know by Axiom 2 that $r^{n/2} \neq 1$, it must be that $r^{n/2}=-1$ and hence $r^{-n/2}=1$. Thus:

   $$r^{k-n/2} = r^n r^{k-n/2} = r^{k+n-n/2} = r^{k+n/2} = r^{n/2}r^k = -r^k$$

7. r^{-1} is a principal n^{th} root of unity.

   By Axiom 1 for r, $\left(r^{-1}\right)^n = \left(r^n\right)^{-1} = 1^{-1} = 1$, and hence r^{-1} satisfies Axiom 1. Axiom 2 follows from Fact 4.

Definition of the DFT

Definition: Given a principal n^{th} root of unity r over the complex numbers, the nth order *Discrete Fourier Transform* (DFT) F^n and the nth order *Inverse Discrete Fourier Transform* G^n are functions that each take an array $A[0]...A[n-1]$ of complex numbers and produce a new array of n complex numbers that has the kth entry given by:

$$F_k^n(A) = \sum_{j=0}^{n-1} A[j]r^{jk}$$

$$G_k^n(A) = \frac{1}{n}\sum_{j=0}^{n-1} A[j]r^{-jk}$$

Note: Whenever n is understood, we will omit it a superscript to F and G.

Matrix formulation: An equivalent way to view the DFT definition is to compute F by multiplying A by the matrix that has r^{jk} in its $(j,k)^{th}$ entry, $0 \le j, k < n$,

$$F(A) = \begin{pmatrix} 1 & 1 & 1 & 1 & \cdots & 1 \\ 1 & r & r^2 & r^3 & \cdots & r^{n-1} \\ 1 & r^2 & r^4 & r^6 & \cdots & r^{2n-2} \\ 1 & r^3 & r^6 & r^9 & \cdots & r^{3n-3} \\ \vdots & \vdots & \vdots & \vdots & \vdots & \vdots \\ 1 & r^{n-1} & r^{2n-2} & r^{3n-3} & \cdots & r^{(n-1)(n-1)} \end{pmatrix} \bullet A$$

and G by multiplying A by $1/n$ times the matrix that has r^{-jk} in its $(j,k)^{th}$ entry, $0 \le j, k < n$:

$$G(A) = \frac{1}{n}\begin{pmatrix} 1 & 1 & 1 & 1 & \cdots & 1 \\ 1 & r^{-1} & r^{-2} & r^{-3} & \cdots & r^{-(n-1)} \\ 1 & r^{-2} & r^{-4} & r^{-6} & \cdots & r^{-(2n-2)} \\ 1 & r^{-3} & r^{-6} & r^{-9} & \cdots & r^{-(3n-3)} \\ \vdots & \vdots & \vdots & \vdots & \vdots & \vdots \\ 1 & r^{-(n-1)} & r^{-(2n-2)} & r^{-(3n-3)} & \cdots & r^{-(n-1)(n-1)} \end{pmatrix} \bullet A$$

Simplified matrices: The $(j,k)^{th}$ entry is shown above as r^{jk} in F and r^{-jk} in G, but using identities like $r^{j(n-k)} = r^{-jk}$, the matrices can be simplified. For example, the last row and column of F are $1, r^{-1}, r^{-2}, r^{-3}, ..., r$, the last row and column of G are $1, r^1, r^2, r^3, ..., r^{-1}$, and for both F and G, the $n/2^{th}$ row is an alternating sequence of 1 and -1 (since $r^{n/2} = -1$).

Power of 2 assumption: To simplify presentation, for the remainder of this chapter we always assume that n, the size of the input (which is indexed from 0 to $n-1$), is a power of 2. See the chapter notes for a discussion of what can be done when n is not a power of 2.

F and G are Inverse Functions

Idea: Using the axioms for principal n^{th} roots of unity, one can verify that G is the inverse of F; or equivalently, multiplying the matrix for F by $1/n$ times the matrix for G yields the identity matrix, a matrix with 1's on the main diagonal and 0's everywhere else.

Theorem: For any array A of complex numbers (it could be that A consists entirely of real numbers or integers), $F(G(A))=G(F(A))=A$.

Proof: Let H be the matrix that results from multiplying the matrix for F by $1/n$ times the matrix for G, and consider the $(p,q)^{th}$ element of H, $0 \le p,q < n$. If we denote the $(p,x)^{th}$ entry of the matrix for F by $F[p,x]$ and the $(x,q)^{th}$ entry of the matrix for G by $G[x,q]$, then by definition:

$$H[p,q] = \frac{1}{n}\sum_{x=0}^{n-1} F[p,x]G[x,q]$$

$$= \frac{1}{n}\sum_{x=0}^{n-1} r^{px}r^{-xq}$$

$$= \frac{1}{n}\sum_{x=0}^{n-1} r^{(p-q)x}$$

If $p=q$ (and we have one of the elements on the main diagonal of H), then:

$$H[p,q] = \frac{1}{n}\sum_{x=0}^{n-1} r^0$$

$$= \frac{1}{n}\sum_{x=0}^{n-1} 1$$

$$= \frac{1}{n}n$$

$$= 1$$

Otherwise, if $p \ne q$, it must be that $(p-q) \ne 0$ and $-n<(p-q)<n$ (since $0 \le p,q < n$), and by Fact 3 of a principal n^{th} root of unity it follows that $H[p,q]=0$. Hence, H has one's down the main diagonal and 0's everywhere else, and the theorem follows.

Similarity of F and G

Since r^{jn}, it follows that $r^{j(n-k)} = r^{jn}r^{-jk} = r^{-jk}$, it follows that:

$$G_k^n = \begin{cases} F_0^n & \text{if } k = 0 \\ F_{n-k}^n & \text{if } 0 < k < n \end{cases}$$

Examples

Example of _n=2_: The powers of the principal second root of unity for the complex numbers are

$$r^0 = 1, \ r = -1$$

and the negative powers are:

$$r^0 = 1, \ r^{-1} = -1$$

Hence, for example:

$$F(1,2) = \begin{pmatrix} 1 & 1 \\ 1 & -1 \end{pmatrix} \bullet \begin{pmatrix} 1 \\ 2 \end{pmatrix} = \begin{pmatrix} 3 \\ -1 \end{pmatrix}$$

$$G(3,-1) = \frac{1}{2} \begin{pmatrix} 1 & 1 \\ 1 & -1 \end{pmatrix} \bullet \begin{pmatrix} 3 \\ -1 \end{pmatrix} = \begin{pmatrix} 1 \\ 2 \end{pmatrix}$$

Example of _n=4_: The powers of the principal fourth root of unity for the complex numbers are

$$r^0 = 1, \ r = \sqrt{-1}, \ r^2 = -1, \ r^3 = -\sqrt{-1}$$

and the negative powers are:

$$r^0 = 1, \ r^{-1} = -\sqrt{-1}, \ r^{-2} = -1, \ r^{-3} = \sqrt{-1}$$

Hence, for example:

$$F(1,2,3,4) = \begin{pmatrix} 1 & 1 & 1 & 1 \\ 1 & \sqrt{-1} & -1 & -\sqrt{-1} \\ 1 & -1 & 1 & -1 \\ 1 & -\sqrt{-1} & -1 & \sqrt{-1} \end{pmatrix} \bullet \begin{pmatrix} 1 \\ 2 \\ 3 \\ 4 \end{pmatrix} = \begin{pmatrix} 10 \\ -2-2\sqrt{-1} \\ -2 \\ -2+2\sqrt{-1} \end{pmatrix}$$

$$G(10,-2-2\sqrt{-1},-2,-2+2\sqrt{-1})$$

$$= \frac{1}{4} \begin{pmatrix} 1 & 1 & 1 & 1 \\ 1 & -\sqrt{-1} & -1 & \sqrt{-1} \\ 1 & -1 & 1 & -1 \\ 1 & \sqrt{-1} & -1 & -\sqrt{-1} \end{pmatrix} \bullet \begin{pmatrix} 10 \\ -2-2\sqrt{-1} \\ -2 \\ -2+2\sqrt{-1} \end{pmatrix} = \begin{pmatrix} 1 \\ 2 \\ 3 \\ 4 \end{pmatrix}$$

Example of $n=8$:

The eight powers of the principal 8th root of unity for the complex numbers involve the quantity $1/\sqrt{2}$ because they are equally spaced around the unit circle in the complex plane. That is,

$$r = e^{\sqrt{-1}\left(\frac{2\pi}{8}\right)} = \cos(\pi/4) + \sqrt{-1}\sin(\pi/4) = \frac{1}{\sqrt{2}} + \sqrt{-1}\frac{1}{\sqrt{2}}$$

and the other powers of r that land on a point 45 degrees out from the origin have the same form except for a change of sign on the real and / or imaginary portions.

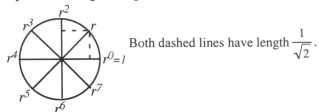 Both dashed lines have length $\dfrac{1}{\sqrt{2}}$.

If we let $x = 1/\sqrt{2}$ and $i = \sqrt{-1}$, the 8 powers of r are
$$r^0=1,\ r^1=x+ix,\ r^2=i,\ r^3=-x+ix,\ r^4=-1,\ r^5=-x-ix,\ r^6=-i,\ r^7=x-ix$$

and the 8 inverse powers are the same, but with the sign of the imaginary part changed (i.e., they are the complex conjugates of the powers of r):

$$r^0=1,\ r^1=x-ix,\ r^2=-i,\ r^3=-x-ix,\ r^4=-1,\ r^5=-x+ix,\ r^6=i,\ r^7=x+ix$$

Hence, the matrix formulations for F and G are:

$$F = \begin{pmatrix}
1 & 1 & 1 & 1 & 1 & 1 & 1 & 1 \\
1 & x+ix & i & -x+ix & -1 & -x-ix & -i & x-ix \\
1 & i & -1 & -i & 1 & i & -1 & -i \\
1 & -x+ix & -i & x+ix & -1 & x-ix & i & -x-ix \\
1 & -1 & 1 & -1 & 1 & -1 & 1 & -1 \\
1 & -x-ix & i & x-ix & -1 & x+ix & -i & -x+ix \\
1 & -i & -1 & i & 1 & -i & -1 & i \\
1 & x-ix & -i & -x-ix & -1 & -x+ix & i & x+ix
\end{pmatrix}$$

$$G = \frac{1}{8}\begin{pmatrix}
1 & 1 & 1 & 1 & 1 & 1 & 1 & 1 \\
1 & x-ix & -i & -x-ix & -1 & -x+ix & i & x+ix \\
1 & -i & -1 & i & 1 & -i & -1 & i \\
1 & -x-ix & i & x-ix & -1 & x+ix & -i & -x+ix \\
1 & -1 & 1 & -1 & 1 & -1 & 1 & -1 \\
1 & -x+ix & -i & x+ix & -1 & x-ix & i & -x-ix \\
1 & i & -1 & -i & 1 & i & -1 & -i \\
1 & x+ix & i & -x+ix & -1 & -x-ix & -i & x-ix
\end{pmatrix}$$

Example of how the DFT represents signals as sums of cosines and sines:

Continuing with the previous example of $n=8$ over the complex numbers, it is interesting to consider how the DFT represents functions as sums of cosines and sines. Perhaps the simplest experiment is to see how a simple cosine signal is transformed. If we let $x = 1/\sqrt{2}$, it is easy to check that the values of $\cos(z)$ for $z=0$, $z=\pi/8$, $z=\pi/4$, $z=3\pi/8$, $z=\pi$, $z=5\pi/8$, $z=3\pi/4$, and $z=7\pi/8$ form the vector:

$$A = (1, x, 0, -x, -1, -x, 0, x)$$

By performing the matrix multiplication, we see:

$$F*A = (0, 4, 0, 0, 0, 0, 0, 4)$$

If we think of the 8 positions of the vector $F*A$ as being indexed from 0 to 7, where position 0 stores the average value of the function (it is often called the "DC" component), position 1 stores the amount of the function $\cos(z)$ that makes up the signal, position 2 stores the amount of the signal $\cos(2z)$ that makes up the signal, etc., then we might first expect that the only non-zero entry should be the second component. However, it is not hard to check that:

$$\cos\left((n-1)k\left(\frac{2\pi}{n} \right) \right) = \cos\left(k\left(\frac{2\pi}{n} \right) \right)$$

This is because multiplying by $n-1$ causes us to wrap around the unit circle to a place that has the same cosine but the opposite sine.

Intuitively, the DFT cannot distinguish these n sample points of the $\cos(z)$ function from the same samples of the $\cos(7z)$ function, and so it puts equal weight in the second and eight components. Because of the symmetry of the cosine function, if we ignore the factor of 1/8 in G, using G gives the same result; that is:

$$F*A = 8G*A$$

As another example, consider the function $\cos(z)+\sin(z)$ which for the values $z=0$, $z=\pi/8$, $z=\pi/4$, $z=3\pi/8$, $z=\pi$, $z=5\pi/8$, $z=3\pi/4$, and $z=7\pi/8$ form the vector:

$$B = (1, 2x, 1, 0, -1, -2x, -1, 0)$$

where, again, we let $x = \frac{1}{\sqrt{2}}$ and $i = \sqrt{-1}$.

Then $F*B = (0, 4 + i4, 0, 0, 0, 0, 0, 4 - i4)$. Again, due to the identity

$$\sin\left((n-1)k\left(\frac{2\pi}{n} \right) \right) = -\sin\left(k\left(\frac{2\pi}{n} \right) \right)$$

the DFT has split the difference between the second and eight components, where the real part holds the contribution of the cosine function and the imaginary part (the phase information) holds the contribution of the sine function.

Example of representing a complicated function as sums of simple ones:

To illustrate how "simple" functions can add up to form complicated ones, consider the function $(1 + \cos(z) + \cos(3z) + \cos(15z) + \cos(35z))/10$, shown below in both Cartesian and polar coordinates. In Cartesian coordinates the x axis goes from 0 to 2π and the y axis is centered about 0 for all but the first and last graph; this is because the constant shift of 1 in the first graph shifts everything up in the last graph. In polar coordinates, all circles have radius 1 except the last, which has been scaled down from radius about 2.1 (the first circle is empty because the plot of 1 is on top of it).

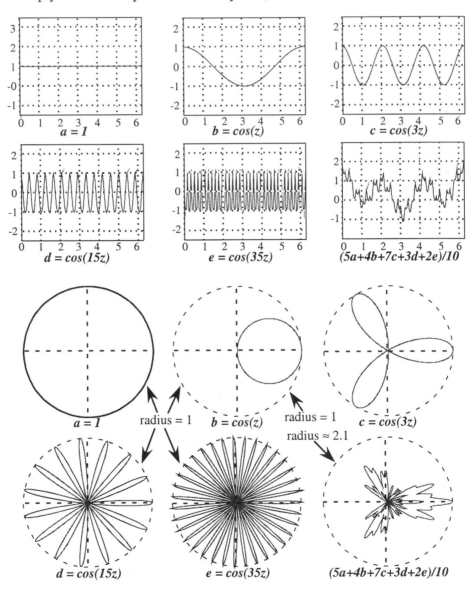

How to Split F into Two Computations of Half the Size

Recall that throughout this chapter that we assume that n is a power of 2.
For $n > 1$

$$F_k^n(A) = \sum_{j=0}^{n-1} A[j] r^{jk}$$

$$= \sum_{j=0}^{n/2-1} A[2j] r^{(2j)k} + \sum_{j=0}^{n/2-1} A[2j+1] r^{(2j+1)k}$$

$$= \sum_{j=0}^{n/2-1} A[2j] r^{2jk} + r^k \sum_{j=0}^{n/2-1} A[2j+1] r^{2jk}$$

$$= \sum_{j=0}^{n/2-1} \left(A_{even}\right)_j \left(r^2\right)^{jk} + r^k \sum_{j=0}^{n/2-1} \left(A_{odd}\right)_j \left(r^2\right)^{jk}$$

$$= \text{if } k < n/2$$
$$\text{then } F_k^{n/2}\left(A_{even}\right) + r^k F_k^{n/2}\left(A_{odd}\right)$$
$$\text{else } F_{k-n/2}^{n/2}\left(A_{even}\right) + r^k F_{k-n/2}^{n/2}\left(A_{odd}\right)$$

and similarly:

$$G_k^n(A) = \text{ if } k < n/2$$
$$\text{then } F_k^{n/2}\left(A_{even}\right) + r^{-k} F_k^{n/2}\left(A_{odd}\right)$$
$$\text{else } F_{k-n/2}^{n/2}\left(A_{even}\right) + r^{-k} F_{k-n/2}^{n/2}\left(A_{odd}\right)$$

So, $F_1^0(A) = G_1^0(A) = A$, and in general for $0 \le k < n/2$ (recall from the facts about principal n^{th} roots of unity that $r^{j+n/2} = r^{j-n/2} = -r^j$, $0 \le j < n$):

$$F_k^n(A) = F_k^{n/2}(A_{even}) + r^k F_k^{n/2}(A_{odd})$$
$$F_{n/2+k}^n(A) = F_k^{n/2}(A_{even}) - r^k F_k^{n/2}(A_{odd})$$

$$G_k^n(A) = \frac{1}{n}\left(G_k^{n/2}(A_{even}) + r^k G_k^{n/2}(A_{odd})\right)$$
$$G_{n/2+k}^n(X) = \frac{1}{n}\left(G_k^{n/2}(A_{even}) - r^{-k} G_k^{n/2}(A_{odd})\right)$$

Intuition: A_{even} and A_{odd} are uniformly spaced "samples" of the original array A, where the positions of the elements of A_{odd} are shifted to the right 1 position compared to A_{even} (or equivalently, A_{even} is shifted to the left of A_{odd}). When adding together the transforms of half the size, the scaling factor of r^k (or r^{-k} for the inverse DFT) achieves the shift.

Divide and Conquer "Butterfly"

Let:

$$F^{n/2}\left(A_{even}\right) = \left(Y[0]\ldots Y[n-1]\right)$$

$$F^{n/2}\left(A_{odd}\right) = \left(Z[0]\ldots Z[n-1]\right)$$

Then a "divide and conquer" computation gives rise to "butterflies":

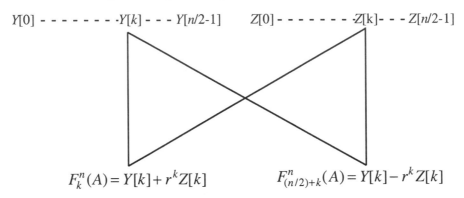

$$F_k^n(A) = Y[k] + r^k Z[k] \qquad F_{(n/2)+k}^n(A) = Y[k] - r^k Z[k]$$

Computation of G: The computation of G is nearly identical (just replace r by r^{-1} and when you are done scale all terms by $1/n$); that is, let

$$G^{n/2}\left(A_{even}\right) = \left(Y[0]\ldots Y[n-1]\right)$$

$$G^{n/2}\left(A_{odd}\right) = \left(Z[0]\ldots Z[n-1]\right)$$

and, again, a divide and conquer computation gives rise to butterflies:

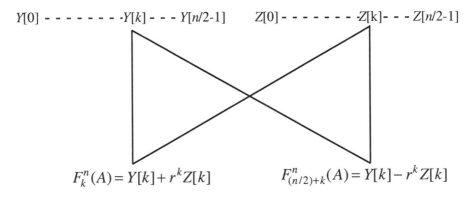

$$F_k^n(A) = Y[k] + r^k Z[k] \qquad F_{(n/2)+k}^n(A) = Y[k] - r^k Z[k]$$

* Because the computation of G is nearly identical to F, we will omit discussion of it from this point on.

CHAPTER 12

Recursive FFT Algorithm

Idea: The butterfly provides the basis for a simple recursive divide and conquer "Fast Fourier Transform" (FFT) algorithm that computes the DFT of an array X by computing two DFT's of half the size (the even and odd terms of X) and then combining the result.

> **function** FFT(X, r) **begin**
>> if $|X|=1$ then return X
>>
>> **else begin**
>>> $Y := \text{FFT}(X_{even}, r^2)$
>>>
>>> $Z := \text{FFT}(X_{odd}, r^2)$
>>>
>>> **for** $k := 0$ **to** $|X|/2-1$ **do**
>>>> $R[k] := Y[k] + r^k \bullet Z[k]$
>>>>
>>>> $R[(|X|/2)+k] := Y[k] - r^k \bullet Z[k]$
>>>>
>>>> **end**
>>>
>>> **return** R
>>>
>>> **end**
>>
>> **end**
>
> **input** an array $A[0]...A[n-1]$, n a power of 2
>
> $r :=$ a principal n^{th} root of unity
>
> FFT(A, r)

Time: If we let $T(n)$ denote the time to execute FFT on an array X of length n, then the total time used is:

$$T(n) = 2T(n/2)+O(n) = O(n\log(n))$$

Space: The space used by the first recursive call to compute Y can be re-used by the second recursive call that computes Z. The total space used excluding the recursive call is $n/2$ complex numbers for each of Y and Z and another n complex numbers for R. So the total space used is:

$$S(n) = S(n/2) + O(n) = O(n)$$

Note that this linear amount of space dominates the "hidden" cost of the stack used for recursion ($O(\log(n))$ is the maximum depth of the recursion).

In-Place Bit Reversal

The bit reversal pattern: In the straightforward recursive algorithm:

- The first recursive call puts the values with even index (the binary representation of the index ends in 0) in the array Y. Similarly, the second recursive call puts the values with odd index in the array Z.

- Thus if we think of Y and Z forming a single new array of n elements with Y forming the left half and Z forming the right half, then the indices of Y will all start with 0 in binary representation and the indices of Z will all start with 1 in binary representation.

- So, after "unwinding" the recursion, the net effect is a bit reversal:

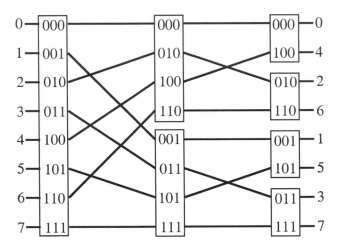

Notation: For an integer $0 \leq k < n$, $REV(k,n)$ denotes the integer whose $\lceil \log_2(n) \rceil$-bit binary representation is the reverse of the $\lceil \log_2(n) \rceil$-bit binary representation of k. For an array $A[0]...A[n-1]$ where n is a power of 2, $REV(A)$ denotes the array B that is the permutation of A given by $B[k]=A[REV(k,n)]$.

In-place bit reversal array permutation: To permute an array according to bit reversal, we exchange the appropriate pairs. Given an array $A[0]...A[n-1]$ where n, is a power of 2, the following loop computes $REV(A)$ in-place. Since when indexing from $k=0$ to $n-1$ each pair $A[j]$, $A[REV(j,n)]$ is visited twice (once when $k=j$ and once when $k=REV(j,n)$), the test $k<REV(k,n)$ checks that the exchange has not already been performed:

procedure BITREVERSE(A)
    for $k := 0$ **to** $n-1$ **do if** $k<REV(k,n)$ **then** EXCHANGE($A[k],A[REV(k,n)]$)
    end

CHAPTER 12

Recursive In-Place FFT Algorithm

Idea: By performing a bit reversal permutation beforehand, the recursive algorithm can simply call itself on the left and right halves, rather than on the odd and even indices. $FFT(a,b,r)$ computes the DFT of $A[a]...A[b]$ using the principal $(b-a+1)^{th}$ root of unity r:

> **procedure** FFT(a, b, r) **begin**
>> **if** $b-a>1$ **then begin**
>>> $$m := \left\lceil \frac{a+b}{2} \right\rceil$$
>>>
>>> FFT(a, $m-1$, r^2)
>>>
>>> FFT(m, b, r^2)
>>>
>>> **for** $k := 0$ **to** $m-1$ **do begin**
>>>> $temp := r^k \bullet A[m+k]$
>>>>
>>>> $A[m+k] := A[a+k] - temp$
>>>>
>>>> $A[a+k] := A[a+k] + temp$
>>>>
>>>> **end**
>>>
>>> **end**
>>
>> **end**

input an array $A[0]...A[n-1]$, n a power of 2

$r :=$ a principal n^{th} root of unity

BITREVERSE(A)

FFT(0, $n-1$, r)

Complexity:

Although the asymptotic time and space complexity have not changed, only $O(1)$ space is used in addition to the space used by the input array.

There is still the hidden cost of the $O(\log(n))$ space for the stack used to implement recursion.

Non-Recursive In-Place FFT Algorithm

Idea: "Unwind the in-place recursive algorithm and work bottom up from a bit-reversed input array in a "butterfly" pattern:

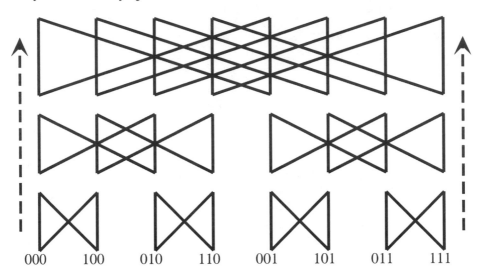

Non-recursive algorithm: The outer loop indexes the levels from bottom to top, the second loop indexes the butterflies along a given level, and the inner loop indexes the points on a given butterfly:

input an array $A[0]...A[n-1]$, n a power of 2

$r :=$ a principal n^{th} root of unity

BITREVERSE(A)

for $m := 0$ **to** $\log_2(n)-1$ **do**

    for $j := 0$ **to** $n-2^{m+1}$ **by** 2^{m+1} **do**

        for $k := 0$ **to** 2^m-1 **do begin**

$$temp := \left(r^{n/\left(2^{m+1}\right)} \right)^k \bullet A[j+k+2^m]$$

$$A[j+k+2^m] := A[j+k] - temp$$

$$A[j+k] := A[j+k] + temp$$

        end

Note: The limit of $n-2^{m+1}$ on the second loop can simply be replaced by $n-1$ since the next value of j will exceed $n-1$.

Simplified Non-Recursive In-Place FFT Algorithm

Idea: Accumulate powers of 2 as the algorithm progresses. Also, simplify the inner loop by making k start at j instead of 0.

> **input** an array $A[0]...A[n-1]$, n a power of 2
>
> $r :=$ a principal n^{th} root of unity
>
> BITREVERSE(A)
>
> $m := 1$
>
> **while** $m \leq n$ **do begin**
>> $m := 2m$
>>
>> $w := r^{n/(2m)}$
>>
>> **for** $j := 0$ **to** $n-1$ **by** $2m$ **do begin**
>>> $v := 1$
>>>
>>> **for** $k := j$ **to** $j+m-1$ **do begin**
>>>> $h := k + m$
>>>>
>>>> $temp := v \bullet A[h]$
>>>>
>>>> $A[h] := A[k] - temp$
>>>>
>>>> $A[k] := A[k] + temp$
>>>>
>>>> $v := v \bullet w$
>>>>
>>>> **end**
>>>
>>> **end**
>>
>> **end**

Further simplifications that can be made:
- Pre-compute powers of r, etc. which are constant independent of the input.
- Replace the two inner loops by a single loop from 0 to $n-1$.

The constant associated with the $O(n\log(n))$ time: Counting just the (complex) arithmetic operations involving elements of A, gives a total of:

- $\dfrac{n}{2}\log_2(n)$ multiplications

- $\dfrac{n}{2}\log_2(n)$ additions

- $\dfrac{n}{2}\log_2(n)$ subtractions

An Example Using the Simplified Algorithm

The following diagram uses butterflies to depict the operation of the simplified non-recursive in-place FFT algorithm using complex numbers on the input 1,2,3,4 (recall that the principal 4th roots of unity for the complex numbers are $1, \sqrt{-1}, -1, -\sqrt{-1}$):

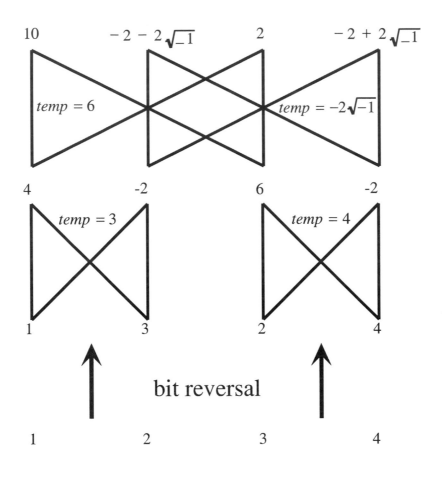

DFT over Finite Fields on an Array of Integers

Idea: In a "typical" use of the DFT, the input is an array of real or complex numbers that are represented to some degree of precision, and the transform produces an array of complex numbers represented to some precision. The complex numbers are typically used because they have principal n^{th} roots of unity for any n and the real numbers do not. However, our development thus far of the FFT algorithm makes no explicit use of the complex numbers; it is valid for any *field* that satisfies the two axioms in the definition of a principal n^{th} root of unity. Although the integers and the real numbers do not have principal n^{th} roots of unity, many finite fields that employ modular arithmetic do. Furthermore, if we add Fact 3 for principal n^{th} roots of unity as a third axiom, so do many finite rings (see the exercises for definitions of fields and rings and further information about modular arithmetic, or references in the chapter notes to basic mathematics texts).

Modular arithmetic: For an integer p, *the integers modulo p* are the set of integers 0 through $p-1$, where all arithmetic operations are defined as usual, except that they are done MOD p; that is, after every arithmetic operation, repeatedly subtract (or add) p until the integer is again in the range 0 to $p-1$. When p is a prime, it is straightforward to show that the integers modulo p form a field (see the exercises).

Example: We can verify that 2 is a principal 5th root of unity of the integers modulo 31 by checking the two axioms:

1. $2^5 = 32 = 1$ MOD 31

2. For $1 \leq k < 5$, 2^k is 2, 4, 8, or 16, and hence not equal to 1.

Commutative rings: Although there are many finite fields that have principal n^{th} roots of unity (e.g., the integers modulo 2^n-1 when 2^n-1 is prime — see the exercises), we can increase our "selection" by also allowing appropriate commutative rings. In this case, we include Fact 3 of principal n^{th} roots of unity as a third axiom that must be satisfied. It can be shown that for any $n>1$ and $w>1$ that are powers of 2, in the commutative ring of integers modulo $w^{n/2}+1$, n has a multiplicative inverse and w is a principal n^{th} root of unity (see the chapter notes). Thus, using the integers modulo $2^{n/2}+1$, we have all we need for the development of the FFT algorithm.

Computing the FFT over finite fields or rings: To compute the DFT of an array of n integers, first select a finite field or ring of integers modulo m that has a principal n^{th} root of unity and such that m is larger than any integers that are used in the computation; that is, m must be larger than any of the n input values and also larger than any values that will be computed for the transform. Then apply the standard FFT algorithm, doing all arithmetic modulo m. Note that if m is not chosen sufficiently large, the values of the transformed array will still be correct modulo m.

Example: Fast Convolutions with the DFT

Polynomial multiplication: Given two polynomials $A[0] + A[1]x + A[2]x^2 + \ldots + A[n-1]X^{n-1}$ and $B[0] + B[1]x + B[2]x^2 + \ldots + B[n-1]X^{n-1}$, each represented by the n coefficients stored in the arrays A and B, it takes $O(n^2)$ time to compute the $2n-1$ coefficients of their product in the straightforward way for $0 \le j \le 2n-2$:

$$C[j] = \sum_{u=0}^{n-1} A[u]B[j-u] \quad \text{(terms out of range are defined to be 0)}$$

This computation is called a *convolution*. For example, for $n=3$:
$$(a + bx + cx^2)(d + ex + fx^2) = ad + (ae + bd)x + (af + be + cd)x^2 + (bf + ce)x^3 + cfx^4$$

Idea: To perform the multiplication in only $O(n\log(n))$ time, transform the coefficients, perform pair-wise multiplications, and then compute the inverse transform. Since the degree of the product polynomial is $2n-2$, we first pad with zero coefficients. If the coefficients are real or complex numbers, the standard transform over the complex numbers can be used to get an answer to a high precision; for exact computations (for example with integer coefficients), a finite field can be used.

Notation: For an array A of n elements, $PAD(A)$ denotes the array of $2n-1$ elements where the first $|A|$ elements are the same as A and the remaining elements are 0 (so if A has elements 0 through $n-1$, $PAD(A)$ has elements 0 through $2n-2$).

Convolution theorem: The convolution of two arrays $A[0]...A[n-1]$ and $B[0]...B[n-1]$, denoted by $C[0]...C[2n-2]$, may be computed by doing:

$$C = G(F(PAD(A)) \bullet F(PAD(B)))$$

where $\bullet$ denotes pair-wise multiplication of the arrays $F(PAD(A))$ and $F(PAD(B))$.

Proof: Let $F(PAD(A)) = FA[0]...FA[2n-2]$ and $F(PAD(B)) = FB[0]...FB[2n-2]$, and let $F(C) = FC[0]...FC[2n-2]$. Given that we already know that G is the inverse of F, it suffices to show that for $0 \le k \le 2n-2$:

$$FC[k] = \sum_{j=0}^{2n-2} C[j]r^{jk} \quad \text{---\ by definition of } F$$

$$= \sum_{j=0}^{2n-2} \sum_{u=0}^{n-1} A[u]B[j-u]r^{jk} \quad \text{---\ by definition of } C$$

$$= \sum_{u=0}^{n-1} \sum_{j=0}^{2n-2} A[u]B[j-u]r^{jk} \quad \text{---\ rearrange sum order}$$

$$= \sum_{u=0}^{n-1} \sum_{v=-x}^{2n-2-u} A[u]B[v]r^{(u+v)k} \quad \text{---\ substitute } v = j - u$$

$$= \sum_{u=0}^{2n-2} \sum_{v=0}^{2n-2} A[u]B[v]r^{(u+v)k} \quad \text{---\ add and remove terms that are 0}$$

$$= \left(\sum_{u=0}^{2n-2} A[u]r^{uk}\right)\left(\sum_{v=0}^{2n-2} B[v]r^{vk}\right) \quad \text{---\ rearrange sum order}$$

$$= FA[k]FB[k] \quad \text{---\ by definition of } F$$

DFT On Two Arrays of Reals

Idea: Since a complex number effectively stores two real values, we can simultaneously compute the DFT of two arrays of n real values by combining them into a single array of n complex values and doing a single DFT.

DFT conjugate property: Recall that we use the notation c^* to denote the complex conjugate of a complex number c, and A^* to denote the complex conjugate of an array of complex numbers A (the ith entry of A^* is $A[i]^*$). The *DFT conjugate property* is that the kth entry of the DFT of the conjugate of A is the same as the $(n-k)^{th}$ entry of the conjugate of the DFT of A; that is, for $0<k<n$:

$$F_k^n(A^*) = F_{n-k}^n(A)^*$$

To see this, we observe that when r is a principal n^{th} root of unity over the complex numbers, for $0<k<n$:

$$r^{j(n-k)} = r^{jn-jk} = r^{jn}r^{-jk} = r^{-jk}$$

Recalling that for any real number x, $e^{\sqrt{-1}x} = \cos(x) + \sqrt{-1}\sin(x)$, and using the fact that $\cos(-\alpha)=\cos(\alpha)$ and $\sin(-\alpha)=-\sin(\alpha)$, it follows that r^{-jk}, and hence $r^{j\,(n-k)}$, is the complex conjugate of r^{jk}. Now suppose $A[j] = a + \sqrt{-1}b$ and $r^{jk} = c + \sqrt{-1}d$. Then we can see that the corresponding terms of $F_k^n(A^*)$ and $F_{n-k}^n(A)^*$ are complex conjugates:

$$A[j]^* r^{jk} = \left(a - \sqrt{-1}b\right)\left(c + \sqrt{-1}d\right) = (ac + bd) + \sqrt{-1}(ad - bc)$$

$$A[j]r^{j(n-k)} = A[j]r^{-jk} = \left(a + \sqrt{-1}b\right)\left(c - \sqrt{-1}d\right) = (ac + bd) - \sqrt{-1}(ad - bc)$$

two n-point DFT's on real arrays with n-point DFT on a complex array:

Given arrays $P[0]...A[n-1]$ and $Q[0]...B[n-1]$ of real numbers, pack them to form the complex array X where for $0 \le k < n$, $X[k] = P[k] + \sqrt{-1}Q[k]$. Observe that if $RE(X)$ denotes the array where the kth component is $RE(X[k])$ and $IM(X)$ denotes the array where the kth component is $IM(X[k])$, then $P[k]$ and $Q[k]$ can be recovered from $X[k]$ by doing:

$$P[k] = RE(X[k]) = \tfrac{1}{2}\left(X[k] + X[k]^*\right)$$

$$Q[k] = IM(X[k]) = \tfrac{1}{2\sqrt{-1}}\left(X[k] - X[k]^*\right) = \tfrac{1}{2}\sqrt{-1}\left(X[k]^* - X[k]\right)$$

Hence, since F_k^n is linear:

$$F^n(X) = F^n(RE(X)) + F^n(IM(X))$$

Thus, for $0<k<n$, the DFT conjugate property can be used to *decompose* the DFT of X:

$$F_k^n(P) = \tfrac{1}{2}\left(F_k^n(X) + F_k^n(X^*)\right) = \tfrac{1}{2}\left(F_k^n(X) + F_{n-k}^n(X)^*\right)$$

$$F_k^n(Q) = \tfrac{1}{2}\sqrt{-1}\left(F_k^n(X^*) - F_k^n(X)\right) = \tfrac{1}{2}\sqrt{-1}\left(F_{n-k}^n(X)^* - F_k^n(X)\right)$$

DFT On A Single Array of Reals

Idea: The capability of the DFT to work on an input array $A[0]...A[n-1]$ of complex numbers is being "wasted" when A consists entirely of real numbers. Since $x^* = x$ for any real value x ($*$ denotes complex conjugate), the DFT conjugate property derived on the preceding page implies that the array $F^n(A)$ is of a special form which we henceforth refer to as a *conjugate array*, where the 0th and $n/2^{th}$ components are real and for $n/2 < k < n$:

$$F_k^n(A) = \left(F_{n-k}^n(A)\right)^*$$

Hence, it suffices to compute $F_0^n(A)$ through $F_{n/2}^n(A)$, which we will do with a $n/2$-point DFT on an array of complex numbers (with respect to a principal $n/2^{th}$ root of unity). This does not change the $O(n\log(n))$ asymptotic time, but is a nice practical observation.

n-point DFT on a real array with $n/2$-point DFT on a complex array:

1. Pack the real input array A into a complex array X that is half the size:
   for $0 \leq k < n/2$ **do** $X[k] := A[2k] + \sqrt{-1}\,A[2k+1]$

2. Place $F^{n/2}(X)$ in $Y[0]...Y[n/2-1]$: $Y = F^{n/2}(X)$

3. Place $F^n(A)$ in $B[0]...B[n-1]$ by using Y to compute $B[0]...B[n/2]$, and then using the DFT conjugate property to fill in $B[n/2+1]...B[n-1]$:

   $B[0] = RE(Y[0]) + IM(Y[0])$

   for $0 < k < (n/2)$ **do** $B[k] := \frac{1}{2}\left(Y[k] + Y\left[\frac{n}{2}-k\right]^*\right) + \frac{1}{2}r^k\sqrt{-1}\left(Y\left[\frac{n}{2}-k\right]^* - Y[k]\right)$

   $B[n/2] := RE(Y[0]) - IM(Y[0])$

   for $(n/2) < k < n$ **do** $B[k] := B[n-k]^*$

Correctness: $B[0]$ follows because the DFT simply sums the elements of A (the first row of the matrix is all 1's). For $0 \leq k < n/2$, we can group terms in the definition of $F_n^n(A)$ to see that $B[k]$ must be equal to

$$B[k] = \sum_{j=0}^{(n/2)-1} \left(A[2j]r^{2jk} + A[2j+1]r^{(2j+1)k}\right)$$

and since r^2 is a $n/2^{th}$ principal root of unity, $B[k]$ follows from the decomposition developed on the preceding page:

$$\sum_{j=0}^{(n/2)-1} A[2j]r^{2jk} = \sum_{j=0}^{(n/2)-1} A[2j](r^2)^{jk} = \frac{1}{2}\left(Y[k] + Y\left[\frac{n}{2}-k\right]^*\right)$$

$$\sum_{j=0}^{(n/2)-1} A[2j+1]r^{(2j+1)k} = r^k\sum_{j=0}^{(n/2)-1} A[2j+1](r^2)^{jk} = \frac{1}{2}r^k\sqrt{-1}\left(Y\left[\frac{n}{2}-k\right]^* - Y[k]\right)$$

$B[n/2]$ follows because the $n/2^{th}$ row of the DFT matrix is always an alternating sequence of 1 and -1. For $n/2 < k < n$, $B[k]$ follows from the DFT conjugate property.

Note: An alternate approach to correctness is that the decomposition implies that the computation of $B[k]$ corresponds to the first level of the FFT butterfly (see the exercises).

Inverse DFT for Reals

Idea: There are two ways to interpret the notion of the inverse DFT for real numbers. One is that the input is an array $B[0]...B[n-1]$ of real numbers. A second is that B is a sequence of complex numbers that was produced by a DFT on an array $A[0]...A[n-1]$ of real numbers (and we now wish to recover A). We show here that both problems can be solved with a single inverse DFT on $n/2$ complex values.

The n-Point inverse DFT on an array of reals:

For $n/2 < k < n$, again we observe that when r is the nth root of unity over the complex numbers we have $r^{-j(n-k)} = r^{-jn+jk} = r^{-jn}r^{jk} = r^{jk}$. Hence, using the same reasoning as was used for F, we can verify that the DFT conjugate property also holds for G:

$$G_k^n(A^*) = G_{n-k}^n(A)^*$$

Hence, the computation is *identical* except to the computation presented on the previous page, except that F is changed to G in Step 2.

The n-Point Inverse DFT on a complex array formed by a DFT on a real array with a $n/2$ point inverse DFT on an arbitrary complex array:

If an array $B[0]...B[n-1]$ of complex numbers was produced by a DFT on a real array, then we can essentially work backwards through the algorithm presented on the previous page, using G instead of F in Step 2; the key observation is that since $r^{n/2}=-1$, we have

$$B\left[\tfrac{n}{2}-k\right]^* = \tfrac{1}{2}\left(Y[k]+Y\left[\tfrac{n}{2}-k\right]^*\right) + \tfrac{1}{2}r^k\sqrt{-1}\left(Y[k]-Y\left[\tfrac{n}{2}-k\right]^*\right)$$

and hence it follows that

$$B[k]+B\left[\tfrac{n}{2}-k\right]^* = Y[k]+Y\left[\tfrac{n}{2}-k\right]^*$$

$$B[k]-B\left[\tfrac{n}{2}-k\right]^* = \tfrac{r^k}{\sqrt{-1}}\left(Y[k]-Y\left[\tfrac{n}{2}-k\right]^*\right)$$

and now we can multiply the second equation by $\sqrt{-1}r^k$ and add the two equations together to get an expression for Step 1:

1. Compute $Y[0]...Y[n/2-1]$ from B:
   for $0 \le k < (n/2)$ **do** $Y[k] := \tfrac{1}{2}\left(B[k]+B\left[\tfrac{n}{2}-k\right]^*\right) + \tfrac{1}{2}r^{-k}\sqrt{-1}\left(B[k]-B\left[\tfrac{n}{2}-k\right]^*\right)$

2. Place $F^{n/2}(Y)$ in $X[0]...X[n/2-1]$: $X = G^{n/2}(Y)$

3. Unpack the complex array X into the real array A that is twice the size:
   for $k := 0$ **to** $n/2-1$ **do begin**
   $A[2k] := RE(X[k])$
   $A[2k+1] := IM(X[k])$
   end

Observation: We only need that B be a conjugate array; that is, we could perform a DFT on A, perform transformations on B that preserve the conjugate array property, and then compute the inverse DFT of B with this algorithm.

Discrete Cosine Transform

Definition: The nth order *Discrete Cosine Transform* (DCT) C^n and the nth order *Inverse Discrete Cosine Transform* D^n are functions that each take an array $A[0]...A[n-1]$ of real numbers and produce a new array of n real numbers that has the kth entry:

$$C_k^n(A) = \sum_{j=0}^{n-1} \sqrt{\frac{\delta(k)}{n}} A[j] \cos\left(\frac{(2j+1)k\pi}{2n}\right)$$

$$D_k^n(A) = \sum_{j=0}^{n-1} \sqrt{\frac{\delta(j)}{n}} A[j] \cos\left(\frac{(2k+1)j\pi}{2n}\right)$$

$$\text{where for an integer } m \geq 0, \delta(m) = \begin{cases} 1 \text{ if } m = 0 \\ 2 \text{ if } m > 0 \end{cases}$$

Note: We continue to assume that n is a power of 2, and whenever n is understood, we omit it as a superscript to C and D.

Matrix formulation: An equivalent way to view the DCT definition is to compute C by multiplying A by the matrix that has $\sqrt{\frac{\delta(k)}{n}} \cos\left(\frac{(2j+1)k\pi}{2n}\right)$ in its $(k,j)^{th}$ entry, $0 \leq k,j < n$; to simplify the presentation, we let $a = \sqrt{1/n}$ and $b = \sqrt{2/n}$:

$$C(A) = \begin{pmatrix} a & a & a & \cdots & a \\ b\cos\left(\frac{\pi}{2n}\right) & b\cos\left(\frac{3\pi}{2n}\right) & b\cos\left(\frac{5\pi}{2n}\right) & \cdots & b\cos\left(\frac{(2n-1)\pi}{2n}\right) \\ b\cos\left(\frac{2\pi}{2n}\right) & b\cos\left(\frac{6\pi}{2n}\right) & b\cos\left(\frac{10\pi}{2n}\right) & \cdots & b\cos\left(\frac{(2n-1)2\pi}{2n}\right) \\ b\cos\left(\frac{3\pi}{2n}\right) & b\cos\left(\frac{9\pi}{2n}\right) & b\cos\left(\frac{15\pi}{2n}\right) & \cdots & b\cos\left(\frac{(2n-1)3\pi}{2n}\right) \\ \vdots & \vdots & \vdots & \vdots & \vdots \\ b\cos\left(\frac{(n-1)\pi}{2n}\right) & b\cos\left(\frac{3(n-1)\pi}{2n}\right) & b\cos\left(\frac{5(n-1)\pi}{2n}\right) & \cdots & b\cos\left(\frac{(n-1)(n-1)\pi}{2n}\right) \end{pmatrix} \bullet \begin{pmatrix} A[0] \\ A[1] \\ A[2] \\ A[3] \\ \vdots \\ A[n] \end{pmatrix}$$

For D, multiply A by the matrix that has $\sqrt{\frac{\delta(j)}{n}} \cos\left(\frac{(2k+1)j\pi}{2n}\right)$ in its $(k,j)^{th}$ entry:

$$D(A) = \begin{pmatrix} a & b\cos\left(\frac{\pi}{2n}\right) & b\cos\left(\frac{2\pi}{2n}\right) & \cdots & b\cos\left(\frac{(n-1)\pi}{2n}\right) \\ a & b\cos\left(\frac{3\pi}{2n}\right) & b\cos\left(\frac{6\pi}{2n}\right) & \cdots & b\cos\left(\frac{3(n-1)\pi}{2n}\right) \\ a & b\cos\left(\frac{5\pi}{2n}\right) & b\cos\left(\frac{10\pi}{2n}\right) & \cdots & b\cos\left(\frac{5(n-1)\pi}{2n}\right) \\ a & b\cos\left(\frac{7\pi}{2n}\right) & b\cos\left(\frac{14\pi}{2n}\right) & \cdots & b\cos\left(\frac{7(n-1)\pi}{2n}\right) \\ \vdots & \vdots & \vdots & \vdots & \vdots \\ a & b\cos\left(\frac{(2n-1)\pi}{2n}\right) & b\cos\left(\frac{(2n-1)2\pi}{2n}\right) & \cdots & b\cos\left(\frac{(n-1)(n-1)\pi}{2n}\right) \end{pmatrix} \bullet \begin{pmatrix} A[0] \\ A[1] \\ A[2] \\ A[3] \\ \vdots \\ A[n] \end{pmatrix}$$

Note: The matrix for D is simply the transpose of the matrix for C (i.e., the matrix for C reflected about its main diagonal).

C and D are Inverse Functions

Theorem: For any array A of real numbers, $C(D(A))=D(C(A))=A$.

Proof: Let E be the matrix that results from multiplying the matrix for C by the matrix for D, and consider the $(p,q)^{\text{th}}$ element of E, $0 \le p,q < n$. By using the standard identity $\cos(\alpha)\cos(\beta) = (\cos(\alpha+\beta)+\cos(\alpha-\beta))/2$ we have:

$$E[p,q] = \sum_{x=0}^{n-1} C[p,x]D[x,q]$$

$$= \sum_{x=0}^{n-1} \sqrt{\frac{\delta(p)}{n}} \cos\left(\frac{(2x+1)p\pi}{2n}\right) \sqrt{\frac{\delta(q)}{n}} \cos\left(\frac{(2x+1)q\pi}{2n}\right)$$

$$= \frac{\sqrt{\delta(p)\delta(q)}}{2n} \sum_{x=0}^{n-1} \cos\left(\frac{(2x+1)(p+q)\pi}{2n}\right) + \frac{\sqrt{\delta(p)\delta(q)}}{2n} \sum_{x=0}^{n-1} \cos\left(\frac{(2x+1)(p-q)\pi}{2n}\right)$$

If $p=q=0$, then $E[0,0]=1$ since all of the cosines become 1. If $p=q>0$, then all of the cosines of the right sum become 1

$$= \frac{\sqrt{2*2}}{2n} \sum_{x=0}^{n-1} \cos\left(\frac{(2x+1)(2p)\pi}{2n}\right) + \frac{\sqrt{2*2}}{2n} \sum_{x=0}^{n-1} 1$$

$$= 1 + \frac{1}{n} \sum_{x=0}^{n-1} \cos\left(\frac{(2x+1)p\pi}{n}\right)$$

and now every cosine term can be matched up with one that is opposite in sign:

$$= 1 + \frac{1}{n} \sum_{x=0}^{(n/2)-1} \left(\cos\left(\frac{(2x+1)p\pi}{n}\right) + \cos\left(\frac{(2(x+n/2)+1)p\pi}{n}\right) \right)$$

$$= 1 + \frac{1}{n} \sum_{x=0}^{(n/2)-1} \left(\cos\left(\frac{(2x+1)p\pi}{n}\right) + \cos\left(\frac{(2x+1)p\pi}{n} + p\pi\right) \right)$$

If p is odd, then the sum is 0 due to the identity $\cos(\alpha) = -\cos(\alpha+p\pi)$. On the other hand, if p is even, $\cos(\alpha) = \cos(\alpha+p\pi)$ and we can split the sum again:

$$= 1 + \frac{2}{n} \sum_{x=0}^{(n/2)-1} \cos\left(\frac{(2x+1)p\pi}{n}\right)$$

$$= 1 + \frac{2}{n} \sum_{x=0}^{(n/4)-1} \left(\cos\left(\frac{(2x+1)p\pi}{n}\right) + \cos\left(\frac{(2x+1)p\pi}{n} + (p/2)\pi\right) \right)$$

If $p/2$ is odd, then again the sum must be 0, otherwise we can continue splitting the sum until we get an odd multiple of π (when p is a power of 2, we go all the way down to 1). Thus all elements of the main diagonal of E are 1. When $p \neq q$, we have two sums of cosines that can both be split in the same fashion to show that all other entries of E are 0.

DCT Basis Functions

The DCT transforms an array of values to an array of weights, each representing how much each of n basic cosine functions contributes to a periodic function that passes through these points. For the integers $0 \leq i < n$ we can compute the inverse DCT on an array that is 0 everywhere except for a 1 in the ith component, to obtain n sample points of the corresponding *basis function*. For $n=8$ the figure below shows plots of these 8 basis functions. To simplify the presentation, the constants $\sqrt{1/n}$ in the first column of D and $\sqrt{2/n}$ in the other columns have been omitted, so the ith plot simply shows the function:

$$y = \cos\left(\frac{(2x+1)i\pi}{2n}\right)$$

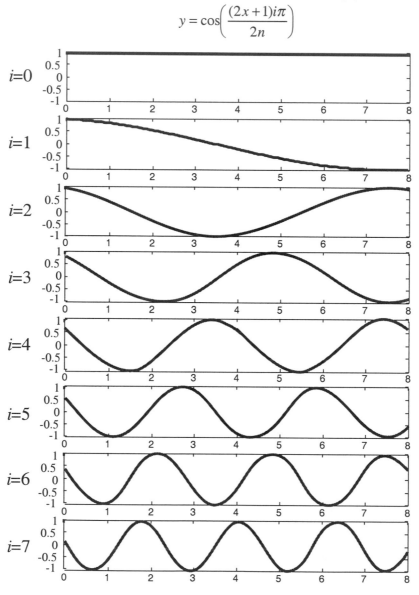

Relationship of the DCT to the DFT

Although the DCT is not simply the real part of the DFT, it is in some sense close to it. Given an array $A[0]...A[n-1]$ of real numbers, if it is "reflected" to make an array twice as long, the DCT is just the first half of the transformed values times a scale factor. Intuitively, since the reflected array is symmetric with a copy placed to the left of the origin, it looks like a portion of a periodic sequence symmetric with the origin, and phase information is not needed:

Although the next page presents a more practical approach (using a DFT of $n/2$ points), this observation leads to a very straightforward method of computing the DCT with a DFT of $2n$ points that lends intuition to the relationship between the two transforms:

1. Double the length of the input array by defining $A[j] = A[2n-j-1]$, $n \leq j < 2n$.

2. Do the DFT on A (using a principal $2n^{\text{th}}$ root of unity r over the complex numbers).

3. **for** $k := 0$ **to** $n-1$ **do** $A[k] := \frac{1}{2}\sqrt{\frac{\delta(k)}{n}} r^{k/2} A[k]$

4. output $A[0]$ through $A[n-1]$

To verify correctness, for $0 \leq k < n$ we see:

$$\frac{1}{2}\sqrt{\frac{\delta(k)}{n}} r^{k/2} F_k^{2n} = \frac{1}{2}\sqrt{\frac{\delta(k)}{n}} r^{k/2} \left(\sum_{j=0}^{n-1} A[j] r^{jk} + \sum_{j=n}^{2n-1} A[2n-j-1] r^{jk} \right)$$

$$= \frac{1}{2}\sqrt{\frac{\delta(k)}{n}} r^{k/2} \left(\sum_{j=0}^{n-1} A[j] r^{jk} + \left(A[n-1] r^{nk} + \cdots + A[1] r^{(2n-2)k} + A[0] r^{(2n-1)k} \right) \right)$$

$$= \frac{1}{2}\sqrt{\frac{\delta(k)}{n}} r^{k/2} \sum_{j=0}^{n-1} A[j] \left(r^{jk} + r^{-(j+1)k} \right) \quad \text{— since } r^{2n} = 1$$

$$= \frac{1}{2}\sqrt{\frac{\delta(k)}{n}} \sum_{j=0}^{n-1} A[j] \left(r^{\frac{(2j+1)k}{2}} + r^{-\frac{(2j+1)k}{2}} \right)$$

$$= \sqrt{\frac{\delta(k)}{n}} \sum_{j=0}^{n-1} A[j] \cos\left(\frac{(2j+1)k\pi}{2n} \right) \quad \text{— since } \cos(x) = \cos(-x) \text{ and } \sin(x) = -\sin(x)$$

Intuition: The DFT is "assuming" an "odd" extension of the sample values, whereas the DCT is "assuming" an "even" extension. For example, if for $n=4$ the values are 0,1,2,3, then the DFT is modeling the sequence 0, 1, 2, 3, 0, 1, 2, 3, 0, 1, 2, 3, 0, 1, 2, 3... whereas the DCT is modeling the sequence 0, 1, 2, 3, 3, 2, 1, 0, 0, 1, 2, 3, 3, 2, 1, 0, ... Later, when we consider practical applications of the DCT like the JPEG image compression standard, we will see that the DCT is more appropriate when one wants to discard "high frequency" information (like jumping from 3 back to 0 in this example).

Computing the DCT in $O(n\log(n))$ Time

Idea: Reflect odd entries to the right of even ones (creating a new array of length n); since we have real values, a reduction to a DFT of $n/2$ complex values can be employed.

The n-point DCT on a real array with a $n/2$-point DFT on a complex array:

1. Flip the odd numbered entries of A to the right of the even ones:

   for $0 \le k < n/2$ **do** $X[k] := A[2k]$

   for $n/2 \le k < n$ **do** $X[k] := A[2n-2k-1]$

2. Place $F^n(X)$ in $Y[0]...Y[n-1]$; since X is real, the algorithm presented earlier to compute a DFT of n real values with a DFT of $n/2$ complex values can be used:

   $Y := F^n(X)$

3. Scale Y and take its real part; r denotes a principal n^{th} root of unity:

   for $0 \le k < n$ **do** $B[k] := \sqrt{\frac{\delta(k)}{n}} RE\left(r^{k/4}Y[k]\right)$

Correctness: For $0 \le k < n$ we see:

$$B[k] = \sqrt{\tfrac{\delta(k)}{n}}RE\left(r^{k/4}Y[k]\right)$$

$$= \sqrt{\tfrac{\delta(k)}{n}}RE\left(r^{k/4}\sum_{j=0}^{n-1}X[j]r^{jk}\right)$$

$$= \sqrt{\tfrac{\delta(k)}{n}}RE\left(\sum_{j=0}^{n-1}X[j]e^{\sqrt{-1}\left(\frac{(4j+1)k\pi}{2n}\right)}\right)$$

$$= \sqrt{\tfrac{\delta(k)}{n}}\sum_{j=0}^{n-1}X[j]\cos\left(\frac{(4j+1)k\pi}{2n}\right)$$

This expression is almost correct; the re-ordering of Step 1 makes the 4 become a 2:

$$= \sqrt{\tfrac{\delta(k)}{n}}\left(\sum_{j=0}^{n/2-1} A[2j]\cos\left(\tfrac{(4j+1)k\pi}{2n}\right) + \sum_{j=n/2}^{n-1} A[2n-2j-1]\cos\left(\tfrac{(4j+1)k\pi}{2n}\right)\right)$$

$$= \sqrt{\tfrac{\delta(k)}{n}}\left(\sum_{j=0}^{n/2-1} A[2j]\cos\left(\tfrac{(4j+1)k\pi}{2n}\right) + \sum_{j=n/2}^{n-1} A[2n-(2j+1)]\cos\left(2k\pi - \tfrac{(4j+1)k\pi}{2n}\right)\right)$$

$$= \sqrt{\tfrac{\delta(k)}{n}}\left(\sum_{j=0}^{n/2-1} A[2j]\cos\left(\tfrac{(4j+1)k\pi}{2n}\right) + \sum_{j=0}^{n/2-1} A[2j+1]\cos\left(\tfrac{(4j+3)k\pi}{2n}\right)\right)$$

$$= \sqrt{\tfrac{\delta(k)}{n}}\left(\sum_{j=0}^{n/2-1} \left(A[2j]\cos\left(\tfrac{(4j+1)k\pi}{2n}\right) + A[2j+1]\cos\left(\tfrac{(4j+3)k\pi}{2n}\right)\right)\right)$$

$$= \sqrt{\tfrac{\delta(k)}{n}}\left(\sum_{j=0}^{n/2-1} \left(A[2j]\cos\left(\tfrac{(2(2j)+1)k\pi}{2n}\right) + A[2j+1]\cos\left(\tfrac{(2(2j+1)+1)k\pi}{2n}\right)\right)\right)$$

$$= \sqrt{\tfrac{\delta(k)}{n}}\sum_{j=0}^{n-1} A[2j]\cos\left(\tfrac{(2j+1)k\pi}{2n}\right)$$

$$= C_k^n(A)$$

Computing the Inverse DCT in $O(n\log(n))$ Time

Idea: We essentially reverse the computation for the DCT presented on the previous page, instead of re-order, DFT, scale, we do the reverse, and scale, DFT, re-order. However, here, Step 1 produces a complex array, and we do not have a DFT on a real array in Step 2. Instead we use a n-point DFT on a complex array (or we could just as well use a n-point inverse DFT — see the exercises); on the next page, we observe how to modify this approach so we can use a $n/2$ point inverse real DFT.

The n-point inverse DCT with a n-point DFT on a complex array:

1. Multiply A by a scale factor; r denotes a principal n^{th} root of unity:

    for $0 \le k < n$ **do** $X[k] := \sqrt{\frac{\delta(k)}{n}} r^{k/4} A[k]$

2. Place $F^n(X)$ in $Y[0]...Y[n-1]$: $Y := F^n(X)$

3. Flip the odd numbered entries of A to the right of the even ones:

    for $0 \le k < n/2$ **do** $B[2k] := RE(Y[k])$

    for $n/2 \le k < n$ **do** $B[2n-2k-1] := RE(Y[k])$

Correctness: For $0 \le k < n$ we can verify that these steps reverse the process of the DCT algorithm presented on the previous page by using more or less the same manipulations. For $0 \le k < n$, after Step 2 we have:

$$Y[k] = \sum_{j=0}^{n-1} \sqrt{\frac{\delta(j)}{n}} r^{j/4} A[k] r^{jk}$$

Step 3 can be viewed as two steps. Step 3A takes the real part and Step 3B does the re-ordering. Let $Z[k]$ denote the result after taking the real part. Then after Step 3A we have:

$$Z[k] = RE\left(\sum_{j=0}^{n-1} \sqrt{\frac{\delta(j)}{n}} r^{j/4} A[k] r^{jk} \right) = \sum_{j=0}^{n-1} \sqrt{\frac{\delta(j)}{n}} A[j] \cos\left(\frac{(4k+1)j\pi}{2n} \right)$$

To verify the re-ordering, for $0 \le k < n/2$ we have

$$B[2k] = Z[k] = \sum_{j=0}^{n-1} \sqrt{\frac{\delta(j)}{n}} A[j] \cos\left(\frac{(2(2k)+1)j\pi}{2n} \right) = D_{2k}^n(A)$$

and for $n/2 \le k < n$ we again use the identity $\cos(\alpha) = \cos(2j\pi - \alpha)$ for any integer j:

$$B[2n-2k-1] = Z[k] = \sum_{j=0}^{n-1} \sqrt{\frac{\delta(j)}{n}} A[j] \cos\left(2j\pi - \frac{(4k+1)j\pi}{2n} \right)$$

$$= \sum_{j=0}^{n-1} \sqrt{\frac{\delta(j)}{n}} A[j] \cos\left(\frac{(4n-4k-1)j\pi}{2n} \right)$$

$$= \sum_{j=0}^{n-1} \sqrt{\frac{\delta(j)}{n}} A[j] \cos\left(\frac{(2(2n-2k-1)+1)j\pi}{2n} \right)$$

$$= D_{2n-2k-1}^n(A)$$

The *n*-point inverse DCT with a *n/2*-point inverse DFT on a conjugate array:

1. Scale A; $r = e^{-\sqrt{-1}\left(\frac{2\pi}{n}\right)}$ is the standard principle n^{th} root of unity:
   $$X[0] := 2\sqrt{n}A[0]$$
   $$\textbf{for } 1 \le k < n \textbf{ do } X[k] := \sqrt{\delta(k)n}\Big(A[k] - \sqrt{-1}A[n-k]\Big)r^{k/4}$$

2. Place $G^n(X)/2$ in $Y[0]...Y[n-1]$; the algorithm presented earlier to compute an inverse real DFT with a DFT of $n/2$ complex values can be used because X is a conjugate array; for this inverse DFT, we use r^{-1} as our principal n^{th} root of unity (by Fact 7 about principal n^{th} roots of unity, if r is one, then so is r^{-1}):
   $$Y := \tfrac{1}{2}G^n(X)$$

3. Flip the odd numbered entries of A to the right of the even ones:
   $$\textbf{for } 0 \le k < n/2 \textbf{ do } B[2k] := Y[k]$$
   $$\textbf{for } n/2 \le k < n \textbf{ do } B[2n-2k-1] := Y[k]$$

Correctness: After Step 2 we have:

$$Y[k] = \frac{1}{2}\left(\frac{1}{n}2\sqrt{n}A[0]\left(r^{-1}\right)^0 + \frac{1}{n}\sum_{j=1}^{n-1}\sqrt{\delta(j)n}\Big(A[j] - \sqrt{-1}A[n-j]\Big)r^{j/4}\left(r^{-1}\right)^{-jk} \right)$$

$$= \sqrt{\frac{\delta(0)}{n}}A[0] + \frac{1}{2}\sum_{j=1}^{n-1}\sqrt{\frac{\delta(j)}{n}}\left(A[j] + e^{-\sqrt{-1}\left(\frac{\pi}{2}\right)}A[n-j] \right)e^{\sqrt{-1}\left(\frac{(4k+1)j\pi}{2n}\right)}$$

$$= \sqrt{\frac{\delta(0)}{n}}A[0] + \frac{1}{2}\sum_{j=1}^{n-1}\sqrt{\frac{\delta(j)}{n}}A[j]e^{\sqrt{-1}\left(\frac{(4k+1)j\pi}{2n}\right)} + \sum_{j=1}^{n-1}\sqrt{\frac{\delta(j)}{n}}A[n-j]e^{\sqrt{-1}\left(\frac{(4k+1)j\pi}{2n} - \frac{\pi}{2}\right)}$$

$$= \sqrt{\frac{\delta(0)}{n}}A[0] + \frac{1}{2}\sum_{j=1}^{n-1}\sqrt{\frac{\delta(j)}{n}}A[j]e^{\sqrt{-1}\left(\frac{(4k+1)j\pi}{2n}\right)} + \sum_{j=1}^{n-1}\sqrt{\frac{\delta(j)}{n}}A[j]e^{-\sqrt{-1}\left(\frac{(4k+1)j\pi}{2n}\right)}$$

$$= \sqrt{\frac{\delta(0)}{n}}A[0] + \frac{1}{2}\sum_{j=1}^{n-1}\sqrt{\frac{\delta(j)}{n}}A[j]\left(e^{\sqrt{-1}\left(\frac{(4k+1)j\pi}{2n}\right)} + e^{-\sqrt{-1}\left(\frac{(4k+1)j\pi}{2n}\right)} \right)$$

Since adding a complex number to its conjugate gives twice its real part (and so $\cos(\alpha) = (e^{\sqrt{-1}\alpha} + e^{-\sqrt{-1}\alpha})/2$), this expression simplifies to:

$$= \sum_{j=0}^{n-1}\sqrt{\frac{\delta(j)}{n}}A[j]\cos\left(\frac{(4k+1)j\pi}{2n}\right)$$

This is *exactly* the same expression as in the proof on the previous page; use exactly the same derivation as was used in that proof to show that the re-ordering of Step 3 makes the 4 become a 2 in this expression, and hence Step 3 correctly computes $B = D^n(A)$.

An inverse real DFT of $n/2$ points can be used in Step 2 because X is a conjugate array; the identities $r^{(n-k)/4} = e^{\sqrt{-1}\left(\frac{(n-k)\pi}{2n}\right)} = \sqrt{-1}e^{-\sqrt{-1}\left(\frac{k\pi}{2n}\right)}$, $\cos(-\alpha)=\cos(\alpha)$, and $\sin(-\alpha)=-\sin(\alpha)$, can be used to verify that for $n/2 < k < n$, $X[k] = X[n-k]^*$ (see the exercises), $X[0]$ is real, and so is $X[n/2]$, because it is equal to:

$$\sqrt{\delta(\tfrac{n}{2})n}\Big(A[\tfrac{n}{2}] - \sqrt{-1}A[\tfrac{n}{2}]\Big)r^{n/8} = \sqrt{2n}A[\tfrac{n}{2}]\big(1-\sqrt{-1}\big)e^{\sqrt{-1}\left(\frac{\pi}{4}\right)} = \sqrt{n}A[\tfrac{n}{2}]\big(1-\sqrt{-1}\big)\big(1+\sqrt{-1}\big) = 2\sqrt{n}A[\tfrac{n}{2}]$$

Two Dimensional DFT and DCT

A two-dimensional DFT or DCT of a n by n array of values can be obtained by first performing the transform on the rows and then on the columns (or equivalently, on the columns and then the rows — see the exercises). By using an $O(n\log(n))$ algorithm on each of the columns and rows, the time is $O(N\log(n))$ where $N=n^2$ is the size of the input. However, practical running time can be improved by a constant factor by taking advantage of 2-dimensional dependencies (see the chapter notes).

Example, the two-dimensional DCT basis functions:

The 2-dimensional DCT transforms a 2-dimensional array of values to a 2-dimensional array of "weights", each representing how much of n basic 2-dimensional cosine functions contributes to a periodic function that passes through these points. For the integers $0 \leq i,j < n$ we can compute the inverse DCT on an array that is 0 everywhere except for the $(i,j)^{th}$ component, to obtain a n by n array of sample points of the corresponding basis function. For $n=8$ the figure below shows 64 sub-plots, where the sub-plot of the ith row and j^{th} column shows an 8 by 8 array of the sample points (displayed as gray scale intensities) for the $(i,j)^{th}$ basis function.

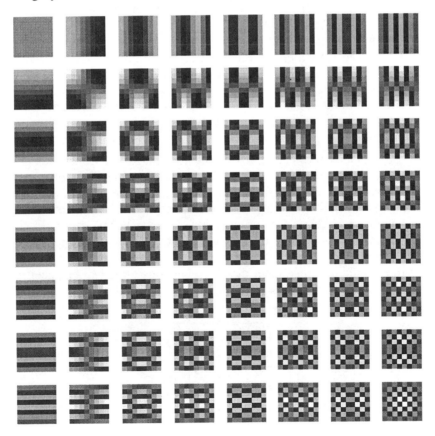

Example: JPEG Image Compression

Idea: Employ a 2-dimensional DCT to concentrate important information into fewer pixels, discard unimportant information, and do lossless compression of what is left.

Outline: For simplicity, assume each pixel is a single byte (i.e., an 8-bit gray scale image) and the dimensions of the array are multiples of 8.

1. Subtract 128 from each pixel, so now the pixel values go from -127 to $+127$ rather than from 0 to 255. Although not necessary from a mathematical point of view, this is a "natural" transformation from a signal processing point of view, and in practice reduces the precision requirements for the DCT computation.

2. Partition the image into 8 by 8 sub-images and perform independent two dimensional DCT's on each of these sub-images.

3. *Quantize* all of the coefficients according to an 8 by 8 table of *quantization step sizes*. That is, if the $(i,j)^{th}$ entry of the quantization table is q, then a value x in the $(i,j)^{th}$ position of one of the transformed sub-images is adjusted by doing:
$$x := \lfloor (x+q/2)/q \rfloor$$

4. The *DC coefficients* (the values in position $(0,0)$ of each 8 by 8 transformed sub-image, which contain the average intensity of the sub-image) are encoded by representing each as the difference between it and the one in the preceding block. These differences are then represented as (size, amplitude) pairs, where $0 \leq size \leq 11$ specifies the number of bits that are used to represent the amplitude, which uses a one's complement representation (i.e., positive values use a normal binary representation and negative values use the normal representation with each bit complemented). The sizes are then Huffman coded with the amplitudes appended (the Huffman code table can be specified).

5. The *AC coefficients* of each transformed sub-image (all values except for the one in position $(0,0)$) are traversed in a zig-zag order going from the upper left down to the lower right corner and encoded as (*run, size, amplitude*) triples, where $0 \leq run \leq 15$ is the number of zero coefficients between the previous AC coefficient and the current one; use $(15,0,-)$ as many times as necessary to represent runs of 15 or more 0's, and reserve $(0,0,-)$ to indicate that all remaining AC coefficients in the transformed sub-block are 0. The (*run, size*) values are Huffman coded with the amplitudes appended; the Huffman code table can be specified, and can be different from the one used for the DC coefficients.

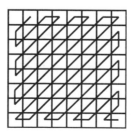

Example: MPEG Video Compression

Idea: Predict the current frame based on what was in the previous frame (or possibly what is in the frames to either side), subtract the predicted pixels from the actual values, and employ (essentially) JPEG compression on what is left.

Outline: Assume that each pixel is a color that is represented by three bytes.

1. If the input is not in this form already, perform a linear mapping of the 3 bytes of each pixel so that the first byte can be viewed as the gray scale version of the image and the other two are color components. Not only is this better for compression, but this is how broadcast TV works; color TV is backward compatible with black and white TV because the color is additional information above and beyond what is used for black and white.

2. Divide the current frame into *macroblocks* of 16 by 16 pixels.

3. "Down sample" the two color planes by a factor of two in each direction. So now we have a gray scale plane of n by n pixels and two color component planes of dimension $n/2$ by $n/2$. That is, each macroblock has four 8 by 8 blocks for the gray scale plane and one 8 by 8 block for each color plane.

4. The first frame to be encoded is always an *I-frame* ("inter-frame"), which is essentially a JPEG image. Each macroblock of a *P-frame* ("predicted frame") can be encoded like one in an I-frame or can be "predicted" as follows:

   A. For each macroblock X of the current frame, determine the macroblock Y of the previous frame that is most similar to X (e.g., least mean squared error) and send a pair of numbers to the encoder for the difference between its coordinates and that of the macroblock being encoded.

   B. Encode the difference between X and Y with (essentially) the JPEG standard. That is, each of the six 8 by 8 blocks of the macro block (four for the gray scale plane and one for each color plane) is treated in essentially the same way JPEG does 8 by 8 blocks.

   B-frames ("bidirectional" frames) are encoded by predicting from the left, the right, or averaging predictions from both the frame to the left and the right.

   B-frames cannot be predicted from other B-frames; that is, a sequence of B-frames must be preceded and followed by and I or P frame. P-frames are always predicted from the closest previous P-frame or I-frame.

   Typically I-frames appear at regular intervals in a pattern such as IBBPBBPBBI... They are sent slightly out of order so that the decoder receives the two frames surrounding a string of B frames before it has to decode them.

MPEG1 vs. MPEG2: We have described what is common to MPEG1 and MPEG2. MPEG2 adds many new features, including *interlaced video* (frames can be divided into two *fields* consisting of the odd rows and even rows of pixels), a number of new prediction modes that involve fields (and in some cases 8 by 16 blocks), prediction at the 1/2 pixel resolution (by interpolation), and a rich system level syntax that allows several video packet streams mixed together.

Exercises

300. Let $r = e^{\sqrt{-1}\left(\frac{2\pi}{n}\right)}$ be the standard principal n^{th} root of unity over the complex numbers.

   A. Compute the DFT of the array $(0,1,2,3)$ by "brute force" using the definition of the DFT (show your work).

   B. Compute the DFT of the array $(0,1,2,3)$ by applying the simplified non-recursive FFT algorithm.

301. The classic *Pythagorean Theorem* for a right triangle (a triangle where one of the angles is 90 degrees) says that the sum of the squares of the two sides is equal to the square of the hypotenuse.

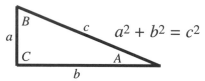

Let a, b, and c denote the lengths of the sides of the triangle, and let A, B, and C denote the angles opposite the edges of lengths a, b, and c (C is 90 degrees) then:

   A. Explain why $\sin(A) = a/c$ and $\sin(B) = b/c$.

   B. Use the following figure to prove the relationship $c^2 = a^2 + b^2$.

     Hint: Equate the sum of the areas of all the pieces to the $(a+b)^2$ area of the bounding square.

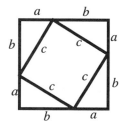

   C. Prove the basic identity $\cos(\alpha)^2 + \sin(\alpha)^2 = 1$.

   D. Show that if $x = (x_1, y_1)$ and $y = (x_2, y_2)$ are points on the plane and c is the Euclidean distance between these two points (i.e., the length of a straight line segment connecting x and y), then:

$$c^2 = \left(x_1 - x_2\right)^2 + \left(y_1 - y_2\right)^2$$

   E. Express the distance between two points in terms of the distances a and b from the origin to these two points and the angles α and β from the origin to those two points. That is:

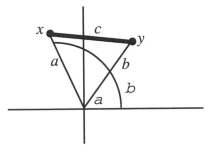

$$c^2 = a^2 + b^2 - 2ab(\cos(\alpha)\cos(\beta) + \sin(\alpha)\sin(\beta))$$

F. Prove that:

$$\cos(\alpha-\beta) = \cos(\alpha)\cos(\beta) + \sin(\alpha)\sin(\beta)$$

Hint: Use Part D with $a=b=1$ and rotate the triangle so one side lies on the *x*-axis and the angle from the *x*-axis to the other is $\alpha-\beta$.

G. Prove the *Law of Cosines*, which generalizes the Pythagorean Theorem by subtracting a correction term that depends on how much angle C differs from 90 degrees:

> In any triangle, the square of the length of any side equals the sum of the squares of the lengths of the other two sides decreased by twice the product of these two sides and the cosine of the included angle. That is, if a, b, and c denote the lengths of the sides of the triangle and C denotes the angle opposite the edge of length c (i.e., the angle between the edges of length a and length b), then:
>
> $$c^2 = a^2 + b^2 - 2ab\cos(C)$$

H. Prove the *Law of Sines*:

> For any triangle, its area is equal to one half the product of the length of any two of its sides and the angle included by those two sides; furthermore, the sines of the angles of a triangle are proportional to the lengths of their opposite sides. That is, if a, b, and c denote the lengths of the sides of a triangle of area u, A denotes the angle opposite the edge of length a, B denotes the angle opposite the edge of length b, and C denotes the angle opposite the edge of length c then
>
> $$u = \tfrac{1}{2}ab\sin(C) + \tfrac{1}{2}bc\sin(A) + ac\sin(B)$$
>
> and:
>
> $$\frac{\sin(A)}{a} + \frac{\sin(B)}{b} + \frac{\sin(C)}{c}$$

302. Prove the following trigonometric identities.

Hint: Use simple arguments about walking around the unit circle along with geometric arguments of the type used in the previous exercise.

A. $\cos(-\alpha) = \cos(\alpha)$
   $\sin(-\alpha) = -\sin(\alpha)$

B. $\cos(\alpha+2\pi) = \cos(\alpha-2\pi) = \cos(\alpha)$
   $\sin(\alpha+2\pi) = \sin(\alpha-2\pi) = \sin(\alpha)$

C. $\cos(\alpha+\pi) = \cos(\alpha-\pi) = -\cos(\alpha)$
   $\sin(\alpha+\pi) = \sin(\alpha-\pi) = -\sin(\alpha)$

D. $\cos(\alpha+\pi/2) = -\cos(\alpha-\pi/2) = -\sin(\alpha)$
   $\sin(\alpha+\pi/2) = -\sin(\alpha-\pi/2) = \cos(\alpha)$

E. $\cos(\alpha)^2 + \sin(\alpha)^2 = 1$

F. $\cos(\alpha)\cos(\beta) = \frac{1}{2}\cos(\alpha-\beta) + \frac{1}{2}\cos(\alpha+\beta)$
   $\cos(\alpha) + \cos(\beta) = 2\cos((\alpha-\beta)/2)\cos((\alpha+\beta)/2)$
   $\cos(\alpha) - \cos(\beta) = -2\sin((\alpha-\beta)/2)\sin((\alpha+\beta)/2)$

G. $\sin(\alpha)\sin(\beta) = \frac{1}{2}\cos(\alpha-\beta) - \frac{1}{2}\cos(\alpha+\beta)$
   $\sin(\alpha) + \sin(\beta) = 2\cos((\alpha-\beta)/2)\sin((\alpha+\beta)/2)$
   $\sin(\alpha) - \sin(\beta) = 2\cos((\alpha+\beta)/2)\sin((\alpha-\beta)/2)$

H. $\cos(\alpha+\beta) = \cos(\alpha)\cos(\beta) - \sin(\alpha)\sin(\beta)$
   $\cos(\alpha-\beta) = \cos(\alpha)\cos(\beta) + \sin(\alpha)\sin(\beta)$

I. $\sin(\alpha+\beta) = \sin(\alpha)\cos(\beta) + \cos(\alpha)\sin(\beta)$
   $\sin(\alpha-\beta) = \sin(\alpha)\cos(\beta) - \cos(\alpha)\sin(\beta)$

J. $\cos(\alpha) + \sin(\alpha) = \sqrt{2}\cos(\alpha-(\pi/4))$

K. $\cos(\beta)\sin(\alpha) = \frac{1}{2}\sin(\alpha-\beta) + \frac{1}{2}\sin(\alpha+\beta)$

L. $\cos(\alpha)^2 = \frac{1}{2}(1+\cos(2\alpha))$
   $\sin(\alpha)^2 = \frac{1}{2}(1-\cos(2\alpha))$

M. $\cos(2\alpha) = 2\cos(\alpha)^2 - 1 = 1 - 2\sin(\alpha)^2$
   $\sin(2\alpha) = 2\cos(\alpha)\sin(\alpha)$

N. For an integer $n \geq 2$:
   $\cos(n\alpha) = 2\cos((n-1)\alpha)\cos(\alpha) - \cos((n-2)\alpha)$
   $\sin(n\alpha) = 2\sin((n-1)\alpha)\cos(\alpha) - \sin((n-2)\alpha)$

303. Our reasoning when manipulating n^{th} roots of unity involved moving around the unit circle by angles that are some multiple of $2\pi/n$. Prove the following identities about the cosine and sine functions for any real number x and any integers $k, n \geq 1$.

Hint: Use simple geometric arguments when you place these values on the unit circle.

A. $\cos\left((n-1)k\left(\dfrac{2\pi}{n}\right)\right) = \cos\left(k\left(\dfrac{2\pi}{n}\right)\right)$

B. $\cos\left((n-1)k\left(\dfrac{2\pi}{n}\right)\right) = \cos\left(k\left(\dfrac{2\pi}{n}\right)\right)$

304. Recall that the complex conjugate of a complex number $a + \sqrt{-1}b$ is the complex number $a + \sqrt{-1}b$.

A. Prove that adding a complex number to its conjugate gives twice its real part.

B. Prove that:

$$\cos(\alpha) = \frac{e^{\sqrt{-1}\alpha} + e^{-\sqrt{-1}\alpha}}{2}$$

305. Using the fact that $\cos(0)=1$ and $\sin(0)=0$, prove that:

$$e^{\pi\sqrt{-1}} + 1 = 0$$

306. Prove that for any real number $x \neq 0$

$$\frac{1}{1 - e^{ix}} = \tfrac{1}{2} - \tfrac{1}{2}\cot(x/2)$$

where *cot* denotes the cotangent function, $\cot(\alpha) = \cos(\alpha)/\sin(\alpha)$. Give a geometric explanation of why the real part of this expression is always 1/2, independent of the value of x.

Hint: Multiply both the numerator and denominator of the expression by $e^{-ix/2}$ and then simplify.

307. Using the notation $x = \dfrac{1}{\sqrt{2}}$ and $i = \sqrt{-1}$, explain why:

A. The principal 8th root of unity for the complex number is $r = x + ix$.

B. The 8 powers of r are:

$r^0=1,\ r^1=x+ix,\ r^2=i,\ r^3=-x+ix,\ r^4=-1,\ r^5=-x-ix,\ r^6=-i,\ r^7=x-ix$

C. The 8 inverse powers are the same, but with the sign of the imaginary part changed (i.e., they are the complex conjugates of the powers of r):

$r^0=1,\ r^1=x-ix,\ r^2=-i,\ r^3=-x-ix,\ r^4=-1,\ r^5=-x+ix,\ r^6=i,\ r^7=x+ix$

308. Using the notation $x = \dfrac{1}{\sqrt{2}}$ and $i = \sqrt{-1}$, explain why:

 A. The values of $\cos(z)$ for $z=0$, $z=\pi/8$, $z=\pi/4$, $z=3\pi/8$, $z=\pi$, $z=5\pi/8$, $z=3\pi/4$, and $z=7\pi/8$ are 1, x, 0, $-x$, -1, $-x$, 0, x.

 B. The values of $\cos(z)+\sin(z)$ for $z=0$, $z=\pi/8$, $z=\pi/4$, $z=3\pi/8$, $z=\pi$, $z=5\pi/8$, $z=3\pi/4$, and $z=7\pi/8$ are 1, $2x$, 1, 0, -1, $-2x$, -1, 0.

309. The exercises to the chapter on hashing asked for proofs of a host of basic facts about logarithms and exponentials ($\log(ab) = \log(a) + \log(b)$, etc.) over the real numbers, based on the basic calculus definition of a logarithm as an integral of the function $1/x$. Generalize these facts to the complex numbers. Explain how $(-x)^y$ makes sense even when x is not an integer. Are logarithms unique?

310. Explain why the derivation for a simple geometric sum over the real numbers presented in the appendix works for the complex numbers as well.

Hint: The derivation uses only simple arithmetic. Verify that each step is just as valid over the complex numbers as it is over the reals.

311. Let: $r = e^{\sqrt{-1}\left(\frac{2\pi}{n}\right)}$

We already know that r is a principal n^{th} root of unity for the complex numbers.

 A. Prove that for any integer k that is relatively prime to n (no integer greater than 1 divides both k and n), r^k is a principal n^{th} root of unity for the complex numbers.
 Hint: If k is relatively prime to n, then the angle of r^k cannot be 2π.

 B. Prove that all principal n^{th} roots of unity for the complex numbers have this form.
 Hint: There are only n solutions to the equation $x^n = 1$, (one of which is 1).

312. Suppose that Axiom 2 of a principal n^{th} root of unity was simply $r \neq 1$.

 A. Explain why this version would be too weak for our development of the FFT.

 B. Suppose that we did take Axiom 2 to be $r \neq 1$, but also included Fact 3 of principal n^{th} roots of unity as a third axiom. Show how these three axioms can be used to prove that $r^k \neq 1$, for all integers $1 \leq k < n$.

313. Since the inverse DFT is nearly identical to the DFT (just replace r by r^{-1} and when you are done scale all terms by $1/n$), the computation of G is very similar to that of the F. Follow the development of the computation of F through the text to show the corresponding development of G, as follows:
 A. Modify the recursive FFT algorithm for G.
 B. Modify the in-place recursive FFT algorithm for G.
 C. Modify the non-recursive in-place FFT algorithm for G.
 D. Modify the simplified non-recursive in-place FFT algorithm for G.

314. For two positive real numbers x and y, define:

$$(x \bmod y) = x - \lfloor x/y \rfloor y$$

Fermat's Little Theorem states:

For an integer $n \geq 1$, if p is a prime number that does not divide n, then:

$$\left(n^{p-1} \bmod p\right) = 1$$

To prove this, we first observe that the numbers $n \bmod p$, $(2n) \bmod p$, ..., $((p-1)n) \bmod p$ are the numbers $1, 2, ..., p-1$ in some order. Hence

$$
\begin{aligned}
(p-1)! n^{p-1} \bmod p &= \left[(1n)(2n)\cdots\left((p-1)n\right)\right] \bmod p \\
&= \left[(n \bmod p)(2n \bmod p)\cdots\left(((p-1)n) \bmod p\right)\right] \bmod p \\
&= (p-1)! \bmod p
\end{aligned}
$$

and by dividing both sides by $(p-1)!$ the theorem follows.

A. The proof of this theorem uses the fact that:

$$(xy) \bmod z = (x \bmod z)(y \bmod z) \bmod z$$

Explain why this is true.

B. For an integer $n \geq 1$, the *totient function*, $\phi(n)$, denotes the number of integers in the range 1 to n that are relatively prime to n, where two non-negative integers i and j are relatively prime if their greatest common divisor is 1; that is, the only integer that divides them both is 1. For example, the first 12 values of $\phi(n)$ are:

n	1	2	3	4	5	6	7	8	9	10	11	12
$\phi(n)$	1	1	2	2	4	2	6	4	6	4	10	4

In general, for any prime p, $\phi(p)=p-1$, and for any composite integer $n>1$, $\phi(n)<n-1$. Generalize the proof of Fermat's Little Theorem to prove that if $m,n>1$ are relatively prime, then:

$$\left(n^{\phi(m)} \bmod m\right) = 1$$

315. Define two integers x and y to be *equivalent* MOD m, which we write $x =_m y$, if $(x-y)$ is a multiple of m.

A. Prove that all integers are equivalent MOD m to one of the integers in the range 0 to $m-1$.

B. Prove that: $x =_m x+m =_m x-m$

C. Prove that: $(x+y \bmod m) =_m (x \bmod m)+(y \bmod m)$

D. Prove that: $(x-y \bmod m) =_m (x \bmod m)-(y \bmod m)$

E. Prove that: $(xy \bmod m) =_m (x \bmod m)(y \bmod m)$

F. Prove that if two positive integers m and n are relatively prime (the largest integer that divides both m and n is 1), then:

$x =_{mn} y$ if and only if $x =_m y$ and $x =_n y$

G. Generalize Part F to prove the *Chinese Remainder Theorem* for positive integers, which says how (large) integers can be represented as their residues with respect to a set of relatively prime numbers.

Let m_1, m_2, ..., m_k be positive integers that are relatively prime to each other, let m denote their product, and let M be any non-negative integer. Then for any non-negative integers n_1, n_2, ..., n_k, there is exactly one integer such that:

1. $M \leq x < M+m$
2. For $1 \leq i \leq k$, x is equivalent MOD m_k to n_k.

H. Generalize your answer to Part G to all integers.

316. A *field* is specified by five items:

1. A set S.

2. An element of S called 0.

3. An element of S called 1.

4. A binary operator + defined on S that is commutative (i.e., for all a and b in S, $a+b = b+a$), associative (i.e., for all a, b, and c in S, $a+(b+c) = (a+b)+c$), for which 0 is an identity element (i.e., for all a in S, $a+0 = 0$), and for which every element has an additive inverse with respect to 0 (i.e., for all a in S there exists a unique element b such that $a+b = 0$).

5. A binary operator $*$ defined on S that is commutative, associative, for which 1 is an identity element, for which every element has a multiplicative inverse with respect to 1 (i.e., for all $a \neq 0$ in S there exists a unique element b such that $a*b = 1$), and which distributes over + (i.e., for all a, b, and c in S, $a*(b+c) = (a*b)+(a*c)$ and $(a+b)*c = (a*c)+(b*c)$.

A *commutative ring* is specified exactly like a field except that it is not required that every element have a multiplicative inverse. A *ring* is specified exactly like a commutative ring except that it is not required that multiplication be commutative.

A. Prove that the real numbers are a field.

B. Prove that the complex numbers are a field.

C. Prove that the rational numbers (all real numbers r such that there exist two integers x and y for which $r = x/y$) are a field.

D. Prove that the integers MOD p, for any prime number $p \geq 2$, are a field.

E. Prove that the integers are a commutative ring.

F. Prove that the set of polynomials over the single variable x, where coefficients are members of a field, are a commutative ring when + adds two polynomials in the usual way (e.g., $(2x^3+x^2+3x+5) + (x^2+1) = 2x^3+2x^2+3x+6$) and $*$ multiplies them in the usual way with a convolution (e.g., $(2x^3+x^2+3x+5)*(x^2+1) = 2x^5+x^4+5x^3+6x^2+3x+5$).

G. Prove that for any $n \geq 1$, the set of n by n matrices, where entries are members of a field, is a ring when $+$ adds matrices in the usual way (i.e., pair-wise multiplication of corresponding elements), $*$ is standard matrix multiplication (see the first chapter), 0 is the matrix where all entries are 0, and 1 is the identity matrix (the $(i,j)^{th}$ entry is 1 if $i=j$ and 0 otherwise).

H. For a ring R, prove that 0 and 1 must be unique and that 0 is an *annihilator*; that is, for all a in R, $a*0 = 0$.

317. For the field of integers modulo 31 prove that:

A. 2 satisfies the two axioms for a principal 5th root of unity.

B. Prove all of the facts presented for principal n^{th} roots of unity.

318. For $1 \leq k<5$, we could prove that

$$\sum_{j=0}^{4} 2^{jk} = 0 \, MOD \, 31$$

by applying the construction of the proof of Fact 3 of principal n^{th} roots of unity. Fill in the details of a different proof that goes as follows: Prove that the sum is always adding up the integers 1, 2, 4, 8, and 16 in some order by showing that if $x>31$, 2^x MOD $31 = 2^{x-5}$, and it cannot be that two terms in the sum are the same, so the sum must always add distinct powers of 2, of which there are only five in the range 1 to 31.

319. Suppose that n is a power of 2 for which $n-1$ is prime. Prove that 2 in a principal n^{th} root of unity for the integers modulo $n-1$.

320. Given two arrays $A[0]...A[n-1]$ and $B[0]...B[n-1]$, recall that their convolution was defined as:

$$C[j] = \sum_{u=0}^{n-1} A[u]B[j-u] \quad \text{(terms out of range are defined to be 0)}$$

A. The straightforward computation could be done with two simple loops:

for $j := 0$ **to** $2n-2$ **do begin**
   $C[j] := 0$
   for $u := 0$ **to** $n-1$ **do**
      if $u \leq j$ and $(j-u)<n$ **then** $C[j] := C[j] + A[u]B[j-u]$
end

Prove that this algorithm is $\Omega(n^2)$.

B. For many executions of the inner loop, the *if* statement prevents the assignment statement from being executed because it discovers that $j<u$ (and so $j-u$ is less

than 0 and $B[j-u]$ is defined to be 0) or $(j-u) \geq n$ (and so $B[j-u]$ is a 0-entry that resulted from the padding). Explain why we can eliminate this wasted testing by limiting the range of the second loop:

```
for j := 0 to 2n-2 do begin
    C[j] := 0
    if j<n then m := 0 else m := j-n+1
    for u := m to j do
        C[j] := C[j] + A[u]B[j-u]
end
```

C. Although it would certainly be a good idea to make the modification of Part B in practice, prove that the algorithm remains $\Omega(n^2)$.

D. Give an expression, as a function of j, for the number of terms in $C[j]$; that is the number of times the *for* loop is executed in the pseudo-code of Part B.

E. Consider the simplified non-recursive in-place FFT algorithm (the last version of the FFT that was presented in this chapter). Suppose that that we are only counting arithmetic operations that involve an element of one of the input arrays. That is, for the simplified non-recursive in-place FFT algorithm it is just the arithmetic operations performed by the three statements in the inner *for* loop (and similarly for its inverse), and for the algorithm of Part B, it is just the work done by the assignment statement of the inner *for* loop. Estimate for what value of n in practice it would be faster to perform a convolution by employing a simplified non-recursive in-place FFT algorithm (and its inverse) as presented in the Convolution Theorem than by using the pseudo-code of Part B.

321. Recall the presentation of how to compute the DFT of two arrays of n real values with a DFT on a single array of n complex values.

A. Give pseudo-code for this algorithm.

B. Similar to how we reversed the process of computing a DFT of n real values with a DFT of $n/2$ complex values, describe how to reverse this process.

322. Recall the algorithm to perform a n-point DFT on a real array with $n/2$-point DFT on a complex array. The proof of correctness made use of the *decomposition* of a DFT of n complex values into the DFT's on the real and imaginary parts of these values (developed in the presentation of how to compute the DFT of two arrays of n real values with a DFT on a single array of n complex values). That is, if P and Q are arrays of n real values, and we define $X[k] = P[k] + \sqrt{-1}\, Q[k]$, then:

$$F_k^n(P) = \tfrac{1}{2}\Big(F_k^n(X) + F_k^n(X^*)\Big) = \tfrac{1}{2}\Big(F_k^n(X) + F_{n-k}^n(X)^*\Big)$$

$$F_k^n(Q) = \tfrac{1}{2}\sqrt{-1}\Big(F_k^n(X^*) - F_k^n(X)\Big) = \tfrac{1}{2}\sqrt{-1}\Big(F_{n-k}^n(X)^* - F_k^n(X)\Big)$$

Give an alternate proof of correctness that makes use of this decomposition to show that the computation of $B[k]$ corresponds to the first level of the FFT butterfly.

Hint: Take P and Q to be the even and odd values of an array of size $2n$.

323. A n by n *Toeplitz matrix* is a n by n matrix T such that $T[i,j] = T[i-1,j-1]$, $2 \leq i,j \leq n$. Give an $O(n\log(n))$ algorithm for multiplying a Toeplitz matrix times a column vector.

Hint: A Toeplitz matrix can be represented by the $2n-1$ values that are along its first row and first column. Express the multiplication of a Toeplitz matrix times a column vector as a portion of the convolution of this $2n-1$ long vector with the column vector (and then use the FFT to perform the convolution).

324. Suppose that a hard-wired butterfly network to perform the FFT algorithm has exactly one faulty vertex that is outputting a 0 no matter what its two inputs are. Describe how to identify this faulty vertex by supplying appropriate inputs to the network and observing the outputs.

325. When presenting the basis functions for the DCT, we displayed graphs of continuous cosine functions which passed through the points obtained by computing the inverse DCT on an array that is 0 everywhere except in the ith component. In contrast, for the 2-dimensional DCT we simply showed a plot of an 8 by 8 array of points. Give an expression for the $(i,j)^{th}$ basis function; that is, based on the definition of the 2-dimensional DCT, an expression for a continuous function that passes through the 8 by 8 array of values obtained by computing the inverse DCT on an 8 by 8 array that is all 0's except for the $(i,j)^{th}$ position.

326. For the DCT, in the proof that D in the inverse of C, the cases $p=q$ and $p \neq q$ both made use of the identity $\cos(\alpha)\cos(\beta) = (\cos(\alpha-\beta)+\cos(\alpha+\beta))/2$ and then argued that resulting sums of cosines were 0. Another approach for the case $p=q$ is to begin with:

$$E[p,q] = \sum_{x=0}^{n-1} C[p,x]D[x,q]$$

$$= \sum_{x=0}^{n-1} \sqrt{\frac{\delta(p)}{n}} \cos\left(\frac{(2x+1)p\pi}{2n}\right) \sqrt{\frac{\delta(q)}{n}} \cos\left(\frac{(2x+1)q\pi}{2n}\right)$$

$$= \frac{\sqrt{\delta(p)\delta(q)}}{n} \sum_{x=0}^{n-1} \cos\left(\frac{(2x+1)p\pi}{2n}\right)^2$$

Again, for $p=q=0$, all cosines are 1 and it follows that $E[p,q]=1$ (since in this case the constant in front of the sum is $1/n$). For $p=q>1$, show that the expression is 1.

Hint: Using the identities $\cos(\alpha+\pi/2)=\sin(\alpha)$ and $\cos(\alpha)^2 + \sin(\alpha)^2 = 1$, pair up terms that add to 1, so you end up with the sum evaluating to $n/2$ (and since the constant in front of the sum is $2/n$ in this case, the expression evaluates to 1); try dividing the sum into four parts and argue that you can make the pairs independently of whether p is odd or even.

327. Show that computing the 2-dimensional DCT by first doing a 1-dimensional DCT on the rows and then on the columns is equivalent to doing columns followed by rows.

328. Although it is not a very practical approach to computing the DCT, we saw that given an array $A[0]...A[n-1]$ of real numbers, if it is "reflected" to make an array twice as long, the DCT is just the first half of the transformed values times a scale factor, and we verified the correctness of the following algorithm:

1. Double the length of the input array by defining $A[j] = A[2n-j-1]$, $n \leq j < 2n$.

2. Do the DFT on A (using a principal $2n^{\text{th}}$ root of unity r).

3. **for** $k := 0$ **to** $n-1$ **do** $A[k] := \frac{1}{2}\sqrt{\frac{\delta(k)}{n}}r^{k/2}A[k]$

4. output $A[0]$ through $A[n-1]$

A. Discuss the difficulties of computing the inverse DCT with this formulation.

B. Modify this approach to allow a simple symmetric computation of the inverse DCT.

  Hint: Instead of reflecting the values to the right, reflect them to the left about the origin; for the DCT of $A[0]...A[n-1]$ define

  $A[-1-k] = A[k]$, $0 \leq k < n$

  and for the inverse DCT of $C[0]...C[n-1]$ define:

  $C[-k] = C[k]$, $0 \leq k < n$, where $C[n] = C[-n] = 0$

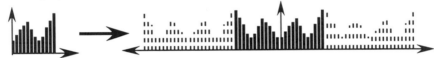

329. In this chapter's development of the inverse DCT algorithm that employs a DFT of $n/2$ complex points, we first saw how to compute an n-point inverse DCT with an n-point (forward) DFT. Explain why essentially the same algorithm can use the inverse DFT in place of where the DFT is used.

Hint: Use a scale factor of $r^{-k/4}$ and make use of the identity $\cos(-\alpha) = \cos(\alpha)$.

330. In the algorithm for the inverse DCT that employed an inverse real DFT of $n/2$ points, the first step was to scale the input array A to obtain the array X as follows:

$$X[0] := 2\sqrt{\frac{1}{n}}A[0]$$
$$\textbf{for } 1 \leq k < n \textbf{ do } X[k] := \sqrt{\frac{\delta(k)}{n}}\left(A[k] - \sqrt{-1}A[n-k]\right)r^{k/4}$$

Recall that we defined a conjugate array to be an array such that the first and $n/2^{\text{th}}$ elements are real numbers and for $n/2 < k < n$, the kth element is the complex conjugate of the $n-k^{\text{th}}$ element. The proof of correctness for the algorithm showed that $X[0]$ and $X[n/2]$ are real numbers, but left out the remaining details of the argument for why X is a conjugate array, mentioning only that it was useful to make use of the identities:

$$r^{(n-k)/4} = e^{\sqrt{-1}\left(\frac{(n-k)\pi}{2n}\right)} = \sqrt{-1}e^{-\sqrt{-1}\left(\frac{k\pi}{2n}\right)}, \quad \cos(-\alpha) = \cos(\alpha), \text{ and } \sin(-\alpha) = -\cos(\alpha)$$

Give the remainder of this argument.

Hint: Write $X[k]$ as

$$X[k] = \sqrt{\tfrac{\delta(k)}{n}}\left(A[k] - \sqrt{-1}A[n-k]\right)\left(R + \sqrt{-1}I\right)$$

where R and I denote the real and imaginary parts of $r^{k/4}$. Then use the above identities to simplify your expression for $X[n-k]$.

331. The algorithm to compute the DCT using a real DFT on $n/2$ points made use of the functions $RE()$ and $IM()$ that extract the real and imaginary parts of a complex number. However, the algorithm for the inverse DCT that made use of an inverse real DFT of $n/2$ points did not; instead it effectively used the trick of adding together a complex number and its conjugate and dividing by 2 to get the real part, by initially constructing the values $A[k] - \sqrt{-1}A[n-k]$. Using similar techniques, modify the algorithm for the DCT using a real DFT of $n/2$ points to not make use of the $RE()$ and $IM()$ functions. Note that this is not necessarily a better way of doing it, but at the very least another way to think about things and see symmetry between the forward and reverse directions.

332. Efficient computation of the DCT of n real values was presented in the form of first reducing the problem to a DFT of n real values, and then showing how to perform a DFT of n real values with a DFT of $n/2$ complex values.

- A. Combine these two algorithms to give succinct pseudo-code for a direct computation of the DCT of n real values with a DFT of $n/2$ complex values.

- B. Repeat Part A for the inverse DCT.

333. Define the *dot product* (also called the *inner product*) of two vectors of the same dimension as the sum of the products of their corresponding components; for example, the dot product of $A[1]...A[n]$ and $B[1]...B[n]$ is:

$$\sum_{i=1}^{n} A[i] * B[i]$$

Define two vectors to be *orthogonal* if they have the same dimension and their dot product is 0. Define a n by n matrix A to be *orthonormal* if the ith row is orthogonal to the jth column when $i \neq j$, and when $i = j$, the dot product of the ith row times itself is 1.

- A. Prove that the matrices for the forward and inverse DCT are orthonormal.

- B. Prove that if a matrix A is orthonormal, then its inverse is its transpose (i.e., A reflected about its main diagonal)

- C. Prove that if a n by n matrix A is orthonormal, then it is energy preserving in the sense that for any vector v of n elements, the sum of the squares of the elements of v is the same as the sum of the squares of the elements of Av.

  Hint: Make use of Part B to express the sum of the squares of v and Av.

334. The nth order *Discrete Sine Transform* (DST) S^n takes an array $A[0]...A[n-1]$ of real numbers and produces a new array of n real numbers that has the kth entry:

$$S_k^n(A) = \sum_{j=0}^{n-1} \sqrt{\frac{2}{n+1}} A[j] \sin\left(\frac{(j+1)(k+1)\pi}{n+1}\right)$$

A. The first of the DCT basis functions was the constant function. For example, with the JPEG image compression standard, in the 2-dimensional case this "DC component" plays a special role since it can be directly used to represent the average intensity of the sub-image in question. Explain why the DST does not have the constant function as one of its basis functions, and explain how a constant function is represented by a DST.

B. Describe how to compute the DST of n points with a FFT of $2n$ points in a fashion similar to what was presented in this chapter for the DCT.

C. Derive the formula for the inverse sine transform.

Chapter Notes:

The discrete Fourier transform is presented in many algorithms texts; for example, see the books of Cormen, Leiserson, Rivest, and Stein [2001] or Aho, Hopcroft, and Ullman [1974]. It is a standard topic for books on signal, speech, and image processing; for example see the books of Oppenheim, Schafer, and Buck [1999], McClellan, Schafer, and Yoder [1999], Press, Teukolsky, Vetterling, and Flannery [1999], Proakis and Manolakis [1996], Gonzalez and Woods [1992], Taylor and Mellott [1998], Gonzalez and Wintz [1987], Pratt [1991], Clarke [1985], Jayant and Noll [1984]. The books Gray and Goodman [1995], Bracewell [1986], Nussbaumer [1981], and Brigham [1974] are devoted to the Fourier transform.

The FFT algorithm is often credited to Cooley and Tukey [1965], with the basic ideas going back to Gauss [1866].

In the literature, which of F or G is considered the transform and which is the inverse transform, and which has the factor of $1/n$, may vary with the application (sometimes they are defined to both have a factor of $1/\sqrt{n}$).

When n is not a power of 2, one can pad with 0's up to the next power of 2, and the transform will work correctly in the sense that the inverse transform will reproduce the original array (which is all that is needed for applications like polynomial multiplication). However, it is not natural to do so in many signal processing applications, where the extra 0's do not reflect the signal that is being sampled. For such applications, if there is a good model of the signal available, it may be acceptable to pad up to the next power of two with points generated by the model. It is also possible to implement recursion based on the prime factorization of n. For example, if $n=24$, the first recursive decomposition could divide the problem into three parts of size 8, and then each of these three parts could be treated with the standard algorithm; see for example the book of Nussbaumer [1981].

The fact that for any $n>1$ and $w>1$ that are powers of 2, in the commutative ring of integers modulo $w^{n/2}+1$, n has a multiplicative inverse and w is a principal n^{th} root of unity, is shown in Aho, Hopcroft, and Ullman [1974], which contains a presentation of the DFT and its application to polynomial and integer multiplication (using bit operations). For an introduction to modular arithmetic, see a text on discrete mathematics such as Graham, Knuth, and Patashnik [1989] or a text on number theory such as Rosen [2000].

Cleve and Watrous [2000] and Hales and Hallgren [2000] consider algorithms for the Fourier transform using quantum computing models, and include additional references on the subject.

The book of Rao and Yip [1990] is devoted to a general presentation of the Cosine transform and its applications to signal processing. The books of Pennebaker and Mitchell [1993] and Bhaskaran and Konstantinides [1995] focus on its application to the JPEG image compression standard, discuss fast one and two dimensional implementations, and include additional references, and the book of Mitchell, Pennebaker, Fogg, and LeGall [1996] considers its application in the MPEG video compression standard.

Makhoul [1980] presents fast DCT algorithms in both 1 and two dimensions, as does Lee [1984], and Chen, Smith, and Fralick [1977] formulate a fast algorithm with matrix operations. A similar presentation of the computation of a DFT of n real numbers with a DFT of $n/2$ complex numbers appears, for example, in the book of Proakis and Manolakis [1996]. The presentation of a DCT with a DFT of $n/2$ complex numbers was suggested to the author by J. E. Storer. Instead of presenting the two steps (real n-point DFT with complex $n/2$-point DFT and then the DCT with a real DFT), the book of Haskell, Puri, and Netravali [1997] presents a "direct" implementation.

The computation of the DCT on n points by using a DFT on $2n$ points that was presented in the exercises is given in Haralick [1976]; Tseng and Miller [1978] show how to modify this construction to obtain the DCT from only the real portion of this computation.

Given that many image compression algorithms (including those employed by the JPEG and MPEG compression standards) employ two-dimensional DCT's on 8 by 8 sub-images, there has been considerable work on optimizing this special case. Ligtenberg and Vetterli [1986] present an algorithm for the 8-point one-dimensional DCT that uses 13 multiplications and 29 additions. Feig and Winograd [1992] present an algorithm for an 8 by 8 two-dimensional DCT that uses 54 multiplications, 6 arithmetic shifts (shifting by 1 bit to multiply by 1/2), and 462 additions. Note that optimized two-dimensional methods may not be "separable"; that is, they may not simply compute the 1-dimensional DCT's on rows and columns to form the two-dimensional DCT.

The figure showing the two-dimensional DCT basis functions was scaled to make the images print well, so that they are not too dark or too light. Since the DCT basis functions are scaled in any case when they are employed by the transform, this scaling factor is in some sense arbitrary. For $1 \leq i,j < 8$, the $(i,j)^{th}$ sub-plot was generated as follows:

1. Construct an 8 by 8 matrix with 256 in the $(i,j)^{th}$ position and 0's everywhere else. Any non-zero value could have been used instead of 256. The constant 256 provides a nice amount of "ink". An 8 by 8 matrix with 256 in position $(0,0)$ and 0 in all other 63 position has an inverse 2-dimensional DCT that is an 8 by 8 matrix of 1's.

2. Compute the two-dimensional inverse DCT.

3. Add the value q to each pixel, where q is magnitude of the largest positive value (or equivalently the smallest negative value, since the transform is symmetric) appearing in any sub-plot ($q \approx 3.85$). So after this step, all pixels p are in the range $0 \leq p \leq 2q$.

4. Multiply each pixel value by a scale factor s. Since $2q \approx 8$, the pixels are dark when printed unless they are scaled up a bit. The constant $s=8$ was chosen to make the lightest pixels visible on white paper while at the same time still being

able to distinguish between darker colored pixels. Anything in the range $4 \leq s \leq$ 12 worked ok with the author's printer.

The book of Pennebaker and Mitchell [1993] is devoted to the JPEG image compression standard. Compression very similar to the JPEG standard is used as a component of the MPEG1 and MPEG2 video compression standards; see for example the books of Bhaskaran and Konstantinides [1995], Mitchell, Pennebaker, Fogg, and LeGall [1996], and Haskell, Puri, and Netravali [1997]. Improved compression for a given level of quality is provided by the newer JPEG2000 standard, which works in a similar fashion but has more flexible formatting and works with higher powered components such as the wavelet transform and arithmetic coding; see the overview presented in Marcellin, Gormish, and Boliek [2000]. See the chapter on strings for an introduction to arithmetic coding. For an introduction to wavelet transforms, see the books of Debnath [2000], Grochenig [2000], Nievergelt [1999], Topiwala [1998], Suter [1997], Strang and Nguyen [1996], Vetterli and J. Kovacevic [1995], Schumaker and Webb [1993], Daubechies [1992], Akansu and Haddad [1992], Chui [1992, 1992b, 1995, 1998, 1998b], Combes and Grossman [1989], Mallat [1998].

13. Parallel Computation

Introduction

So far we have designed and analyzed algorithms for the RAM model, where there is a single processor connected to a random access memory. Most principles of algorithm design and analysis for the RAM model are useful in other settings. Here, we consider the *PRAM* (Parallel RAM) model, a natural generalization of the RAM model to having multiple processors that access a common memory.

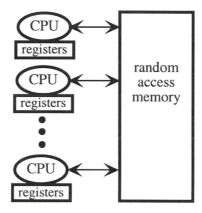

Like the RAM model, it may be that each processor has its own registers; in fact, it may be reasonable for many applications to assume that each processor has a large amount of local memory. Here, our interest is in how the processors can productively work together to solve a problem faster by making use of the common shared memory and solving parts of the problem in *parallel*. That is, we measure the total number of time steps from start to finish, but do not "charge" extra if many processors are active during a particular time step. For example, consider the simple problem of multiplying the corresponding entries of two arrays $A[1]...A[n]$ and $B[1]...B[n]$ and placing the results in a third array $C[1]...C[n]$ so that $C[i]=A[i]*B[i]$, $1 \leq i \leq n$:

> Standard serial version:
>
> **for** $i:=1$ **to** n **do** $C[i] := A[i]*B[i]$

> Parallel version:
>
> **for** $1 \leq i \leq n$ **in parallel do** $C[i] := A[i]*B[i]$

Here we have generalized our pseudo-code notation with the *in parallel do* construct to perform the operations for each value of i at the same time. With the RAM model, the *serial algorithm* is $O(n)$, whereas with the PRAM model, the *parallel algorithm* is $O(1)$ because all n multiplications are done simultaneously.

For another simple example, consider the recursive merge sort procedure presented as an example of divide and conquer, which consisted of the following basic steps:

1. Divide a list of n elements into two lists $L1$ and $L2$ of $n/2$ elements.

2. Sort $L1$ and Sort $L2$.

3. In $O(n)$ time, merge $L1$ and $L2$ into a single sorted list.

For the standard RAM model, the time is given by

$$T(n) = 2T(n/2) + O(n) = O(n\log(n))$$

where the term $2T(n/2)$ covers the two recursive calls in Step 2. However, in the PRAM model, since the two recursive calls of Step 2 can be done in parallel:

$$T(n) = T(n/2) + O(n) = O(n)$$

Clearly this is not a very impressive speed-up for the investment of n processors. The problem is that although the sorting is done in parallel, we are still using a linear time merge algorithm. We shall see later how to do better.

An issue that we shall not address in this chapter is parallel input and output. For example, in the code to multiply corresponding entries in two arrays, we did not discuss how the data was placed into the memory in the first place. If we were to simply read the elements of A and B sequentially, the $O(n)$ input time would dominate the $O(1)$ time for the parallel computation of the n multiplications, and hence, if pair-wise multiplication was the only work to be done on A and B, the complexity of the computation would depend entirely on the model of input/output used. Fortunately, in many applications, input and output is done only at the beginning and end of a very long computation involving many serial and parallel computations, and it is reasonable to consider input and output as a separate issue.

Because the power of the PRAM rests on the ability of processors to simultaneously access the same set of memory locations, the assumptions about how simultaneous access works are typically divided into three categories:

Exclusive Read, Exclusive Write Model (EREW):

- Two processors may NOT read the same location at the same time.
- Two processors may NOT write to the same location at the same time.

   The parallel code for pair-wise addition of two arrays that was presented earlier is an example of an algorithm in the EREW model.

Concurrent Read, Exclusive Write Model (CREW):

- Two processors MAY read the same location at the same time.

- Two processors may NOT write to the same location at the same time.

Concurrent Read, Concurrent Write Model (CRCW):
- Two processors MAY read the same location at the same time.
- Two processors MAY write to the same location at the same time.

For the CRCW model, a convention for what happens when two or more processors write to a given location simultaneously needs to be specified. Here are ten examples:

1. Nothing happens and the contents of the location is unchanged.

2. All of the writers must write the same value (or there is a hardware error).

3. A special non-zero value is written in that location.

4. A random non-zero value is written in that location.

5. Exactly one of the writers succeeds (but no assumptions may be made about which one).

6. The lowest numbered writer succeeds.

7. The lowest numbered writer writing a non-zero value succeeds.

8. One of the writers is randomly chosen to succeed.

9. The writer with the smallest value (ties are broken arbitrarily) succeeds.

10. All of the values simultaneously written are added together and placed in the location.

In general the ability to concurrently read and possibly write as well can be very powerful. For example, all of Conventions 2 through 10 work with the following $O(1)$ time yes-no database search algorithm:

Database yes-no search problem: Processor 0 wishes to determine if a particular value x exists in one or more of locations $A[1]$ through $A[n]$.

$O(1)$ *time CRCW PRAM algorithm* (using any of Conventions 2 through 10):

1. $z := 0$
2. **for** $1 \leq i \leq n$ **in parallel do if** $A[i]=x$ **then** $z:=1$
3. **if** $z \neq 0$ **then** output "yes" **else** output "no"

Although Convention 1 does not work correctly for this algorithm (see the exercises), it can be useful. For example, suppose the problem was to determine whether an item appears exactly once in the database; then the algorithm above works with Convention 1 but not with Convention 2. As another example, suppose the problem was to determine whether a particular value x appears two or more times in the database. Then neither of Conventions 1 and 2 work with the algorithm above, but Convention 2 can be used with a relatively simple randomized algorithm:

Duplicate database yes-no search problem: Processor 0 wishes to determine if a particular value x exists in TWO or more of locations 1 through n.

Randomized O(1) time CRCW PRAM algorithm (using Convention 2) that is always correct when it answers yes, and with probability greater than $1 - 1/2^k$ is correct when it answers no, for any specified constant k.

The algorithm uses k rounds, which each round repeats the same randomized check if x is in two or more locations. On each round, every processor flips a coin to decide if it is a "1" or a "2". If we are lucky, at least one processor becomes a 1 and one becomes a 2. All of the 1's set *test1*=1 and all of the 2's set *test2*=1. The loop is exited with a positive answer (by setting *stopflag*=1) if both *test1* and *test2* are 1. Otherwise, after k rounds, the loop exits with *stopflag*=0.

1. *rounds* := *stopflag* := 0

2. **while** *rounds* < k and *stopflag*=0 **do begin**
   test1 := *test2* := 0
   for $1 \leq i \leq n$ **in parallel do**
      if $A[i] \neq x$ **then** $R[i]:=0$
      else begin
         $R[i]$:= randomly choose 1 or 2
         if $R[i]=1$ **then** *test1* := 1 **else** *test2* := 1
         end
   if *test1* $\neq 0$ and *test2* $\neq 0$ **then** *stopflag* := 1
   rounds := *rounds*+1
   end

3. **if** *stopflag* $\neq 0$ **then** output "yes" **else** output "no"

The algorithm works better the more duplicates there are (see the exercises). But even when there are exactly two copies, for each of the k iterations of the outer *for* loop, there is an even chance that one of the copies coin flip is 1 and the other is 2, and the duplication will be discovered. As long as k is a constant independent of n, the algorithm runs in time $O(k)=O(1)$ and with probability less than $1/2^k$ incorrectly answers "no". It is also unlikely to run all k iterations when x does appear twice, since there is an even chance on each iteration that it discovers this and stops.

Because small changes in concurrent write assumptions can have significant effects on parallel complexity in the CRCW model, one must be careful to understand assumptions when developing CRCW PRAM algorithms and reasoning about the CRCW model. Fortunately, in practice, most problems that can be solved in the CRCW model can also be solved in the same asymptotic time on the CREW model. In this chapter, we take the middle ground and always assume the CREW model unless explicitly stated otherwise.

The simplicity of the PRAM model and the fact that it is a natural generalization of the standard RAM model is nice from the point of view of algorithm design. However, actually building a PRAM with n processors for large n is problematic, given that one device (the common memory) needs to be connected to n communication paths. Most approaches to date for building a PRAM have been based on building a *network* of processors where processors each have local memory and there is some pattern of communication paths (e.g., wires) that interconnects them so that in one time unit, a processor can send $O(1)$ information along a communication path to another processor. That is, any graph of n vertices represents a network of n processors, where edges represent the communication paths (edges could be directed if communication paths can go in only one direction). For example, the figure below shows on the left a network based on a tree and on the right a network based on a 2-dimensional mesh:

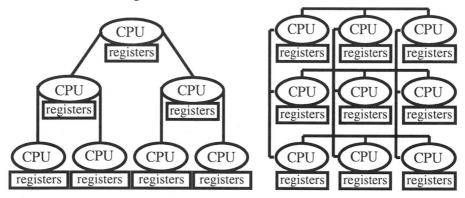

Networks can simulate the actions of a PRAM algorithm by employing *packet routing*. In an ideal simulation, each memory location of the common PRAM memory that is used is "controlled" by a distinct processor (and there may be additional processors as well). To simulate one step of the PRAM algorithm, the network first sends packets of information along the communication paths between processors so that every processor gets the data item it needs (to perform the current step) from the processor that controls that data item. Then all of the processors perform the computation for the current step, and the process is repeated; packet route, compute, packet route, compute, etc. Depending on the interconnection pattern of the network, to get from one processor to another, packets may have to travel between a number of intermediate processors. Not only is the length of the longest path traversed for a given step an issue, but congestion at intermediate points can also be the limiting factor for the time of the packet routing phase of a step. That is if many packets flow through a given processor on their way to their destination, and only one packet can be sent along a particular edge in a given time unit, then incoming packets may have to be queued at that processor to wait for a turn to be sent out along an edge.

Many networks have been proposed for fast PRAM simulation, where "fast" usually means at $O(\log(n)^k)$ for some $k \geq 1$; for example $O(\log(n))$ expected time or $O(\log(n)^2)$ worst case time are reasonable goals (although under appropriate assumptions $O(\log(n))$ worst case asymptotic worst case time is possible). Here for fast PRAM simulation we present the hypercube (HC) network (a $\log_2(n)$-dimensional cube of n processors), and the cube connected cycles (CCC) and butterfly (BF) networks that are essentially equivalent to it. The figure below depicts a three dimensional hypercube of eight vertices.

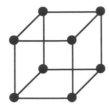

Hypercube networks have many different $O(\log(n))$ long paths between processors. By using the network to carefully collect and distribute multiple packets going to the same destination, packets to simulate a PRAM step can be handled quickly with a minimum of queuing at vertices.

The hope is that the actions of a PRAM can be efficiently simulated on a network that is practical to build. However, we shall see that there is a tradeoff between the network complexity and how efficiently it can simulate a PRAM. In particular, we shall see that, under reasonable assumptions that reflect current technology, to simulate a PRAM algorithm for a problem of size n with only a factor of $O(\log(n))$ slow-down (or even $\log(n)$ to some power), asymptotically the amount hardware for the communication paths dominates the total hardware used for the processors (because the interconnection pattern of the wire used for communication paths needs to be very complex). Although this is an asymptotic argument where the constants work in our favor, it suggests that fast PRAM simulation by a network is likely, at the very least, to be expensive (and in particular, the HC/CCC/BF networks are likely to be expensive). Indeed, this has been true of general purpose parallel machines to date that attempt to provide $O(\log(n))$ slow-down PRAM simulation, where with just a modest number of processors (e.g., on the order of 100), it is not uncommon for over half the hardware and computing time to be devoted to routing information between the processors. In contrast, to date there have been parallel machines based on a simple two-dimensional mesh that provide thousands of processors with less hardware devoted to routing information, but they have an $O(\sqrt{n})$ slowdown. Until the technology progresses to where n gets relatively large (e.g., hundreds of thousands or more), the constants and other considerations will probably be more significant that the difference between a $O(\log(n))$ and a $O(\sqrt{n})$ slowdown. However, it is clear that building a PRAM is a complex issue, and the assumptions about how it is done may greatly affect the practical efficiency of algorithms we design.

Intuitively, the reason that a PRAM can be costly to simulate with a particular network is that its power to simulate other models of parallel computation is great. Any pair of processors can easily communicate by simply writing to and reading from an agreed upon memory location. So for example, if there are n processors and $2n^2$ locations of the common RAM memory are set aside for special locations for each pair of processors to communicate (2 locations per processor for data going in both directions), then the

PRAM model can be thought of as a network with a direct communication path between every pair of processors (in addition to the communication paths from every processor to the common memory that allow processors to make use of the remainder of the common memory). Hence, in some sense, the PRAM model is at least as powerful as any reasonable model based on a network of interconnected processors. Although one can often design algorithms to work directly on the network in question, a nice aspect of the PRAM model, like the standard RAM model, is that it provides a simple general framework in which to study the design and analysis of parallel algorithms.

In the 1980's and 1990's, machines were built based on many models, including hypercube, mesh, and parallel caching, and over a host of processors scales (from relatively inexpensive systems with thousands of small processors in a mesh to systems costing millions of dollars with a few hundred very powerful processors in a hypercube). The traditionally rapid increase in processor speed from year to year and highly changing technology has made investment in such complex hardware problematic. However, it seems likely that eventually, as the limits on serial computing speed are reached, massively parallel computation will be important for high performance and complex systems (e.g., simulation of complex biological system). The goal of this chapter is only to introduce the reader to the area of parallel algorithms and provoke thought about how basic principles of algorithm design and analysis translate to other models and assumptions beyond the RAM model. Although it is unclear what models of parallel computing will prevail in the future, it seems likely that basic notions like the PRAM model and classic structures like the hypercube and mesh networks will continue to influence our thinking about the design of parallel algorithms.

Example: Summing an Array

Idea: To add up the items in an array $A[0]...A[n-1]$, in one step items one apart are added to cut the problem in half, in a second step items two apart are added to again cut the size of the problem in half, etc. This approach can be viewed as simulating a tree network with the array A stored in the leaves and the sum coming out of the root:

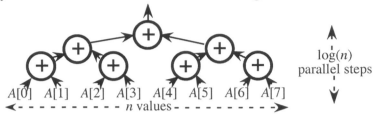

$$\log(n)$$
parallel steps

In practice, we do not really need all of the processors; we can keep overwriting A. Concurrent reads are not used, and the algorithm works in the EREW model.

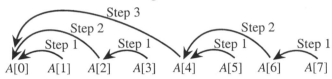

The summing algorithm:

> **function** erewSUM($A[0]...A[n-1]$)
>   **for** $k=0$ **to** $\lceil \log_2(n) \rceil - 1$ **do**
>   **for** $0 \le i < n-2^{k+1}$ **in parallel do**
>   **if** i is a multiple of 2^{k+1} **then** $A[i] := A[i]+A[i+2^k]$
>   **return** $A[0]$
>   **end**

Note: The test if i is a multiple of 2^{k+1} is not necessary (see the exercises).

An equivalent way to express the summing algorithm:

> **function** erewSUM($A[0]...A[n-1]$)
> $k := 1$
>   **while** $k<n$ **do begin**
>   **for** $0 \le i < n-2k$ **in parallel do**
>   **if** i is a multiple of $2k$ **then** $A[i] := A[i]+A[i+k]$
> $k := k*2$
>   **end**
>   **return** $A[0]$
>   **end**

Complexity: Each of the $\lceil \log_2(n) \rceil$ iterations uses $O(1)$ time (since $O(1)$ time is used for the body of the parallel *for* loop); hence the algorithm is $O(\log(n))$ time. $O(1)$ space is used in addition to the space used by A. The number of processors used is $n/2$; however, we shall see later (Brent's Lemma) that $O(n/\log(n))$ processors suffice.

Example: List Prefix-Sum / List Ranking

Notation: L is a singly-linked list represented by $A[0]...A[n-1]$, $0 \le first < n$ the index of the first item, and the array $NEXT[0]...NEXT[n-1]$ such that by starting with $i := A[first]$ and repeatedly doing $i := \text{NEXT}[i]$, we visit all positions and end up at a position i such that $NEXT[i]=nil$; for simplicity assume items are ≥ 0 and $nil = -1$.

List suffix-sum: We wish to compute for each $0 \le i < n$ the sum of all positions from position i through the end of the list. We can use the same distance-doubling idea as for array sum, emanating from every vertex; only the EREW PRAM model is needed:

> **procedure** erewListSuffixSum(L)
> > **while** $NEXT[first] \neq nil$ **do**
> > > **for** $0 \le i < n$ **in parallel do if** $NEXT[i] \neq nil$ **then begin**
> > > > $A[i] := A[i] + A[NEXT[i]]$
> > > > $NEXT[i] := NEXT[NEXT[i]]$
> > > > **end**
> >
> > **end**

List prefix-sum: We wish to compute for each $0 \le i < n$ the sum of all positions from position i through the start of the list. We can reverse the list (in parallel do $NEXT[NEXT[i]] := i$, set *first* to what used to be the last position, and set the NEXT field of what used to be the first position to *nil*) and then do a suffix sum (see the exercises).

List ranking: The special case of prefix-sum where all values of A are 1 (there could be additional data associated with each vertex), and we compute the position of each vertex.

Suffix-sum / prefix-sum / list ranking on an array: For the special case of a list in sequential positions of an array, define $NEXT[i]=i+1$, $0 \le i < n-1$, and $NEXT[n-1]=nil$. For example, suppose the array $A[0]...A[9]$ initially contained 1 in each location and consider the successive iterations of the *while* loop of erewListSuffixSum:

array *position*	0	1	2	3	4	5	6	7	8	9
starting values	1	1	1	1	1	1	1	1	1	1
values after *first* iteration	2	2	2	2	2	2	2	2	2	1
values after *second* iteration	4	4	4	4	4	4	4	3	2	1
values after *third* iteration	8	8	8	7	6	5	4	3	2	1
values after *fourth* iteration	10	9	8	7	6	5	4	3	2	1

Complexity: $O(\log(n))$ time since each iteration of the outer *while* loop for suffix-sum doubles the distance over which sums are taken, and prefix-sum adds only $O(1)$ additional time. $O(n)$ space in addition to the space for L (or $O(n)$ additional space if we cannot overwrite A and $NEXT$ and must first make copies). $O(n)$ processors are used.

Example: List Prefix-Sum on a Binary Tree

Notation: T denotes a binary tree represented by an array $A[0]...A[n-1]$ of data items, $0 \le root < n$ the index of the root, and the array $P[0]...P[n-1]$ of parent indices, the array $L[0]...L[n-1]$ of left child indices, and the array $R[0]...R[n-1]$ or right child indices. For simplicity assume items are ≥ 0 and $nil = -1$.

Problem: For each leaf of T, compute the sum of all values in leaves to its left (not necessarily physically in the array; by to the left, we mean the leaves that come earlier in a pre-order traversal of T); the values of A stored at non-leaf vertices are irrelevant.

Idea: Generalize the algorithm for summing an array, but do not overwrite A. Instead, use the whole binary tree, so that intermediate values can be collected and passed back down to the leaves. Place the prefix sum in the array B and use two additional arrays to perform the computation:

$X[i]$ = Sum of all leaves in the subtree rooted at vertex i.

$S[i]$ = Sum of all leaves to the left of i (leaves that come before i in pre-order).

Tree prefix-sum algorithm: To check if a vertex i is a leaf, check if $L[i]=R[i]=nil$. To check if a vertex i is a right child, check if $R[P[i]]=i$. For simplicity, define $X[nil]=0$. Again, only the EREW PRAM model is needed; the example on the following page illustrates the three phases of the algorithm.

> **procedure** erewTreePrefixSum(T)
>
> 1. Work up from the leaves to the root to compute the array X:
>   *height* := 0
>   **while** $X[root]=nil$ **do**
>   **for** $0 \le i < n$ **in parallel do begin**
>   **if** i is a leaf **then** $X[i] := A[i]$ **else** $X[i] := X[L[i]] + X[R[i]]$
>   *height* := *height*+1
>   **end**
>
> 2. Work back down to the leaves to compute the array S:
>   **for** $h:=0$ **to** *height* **do**
>   **for** $0 \le i < n$ **in parallel do begin**
>   **if** i is the root **then** $S[i] := 0$ **else** $S[i] := S[P[i]]$
>   **if** i is a right child **then** $S[i] := S[i] + X[L[P[i]]]$
>   **end**
>
> 3. Add A and S to place the prefix-sum for each leaf in the array B:
>   **for** $0 \le i < n$ **in parallel do** $B[i] := A[i]+S[i]$
>   **end**

Complexity: Time is proportional to the height of T, which is $O(\log(n))$ for a complete binary tree. $O(n)$ space is used in addition for the space for T. $O(n)$ processors are used.

Example of the tree prefix-sum computation:

Suppose that initially $A[i]:=1$ for every leaf i.

1. Work up from the leaves to compute the X values:

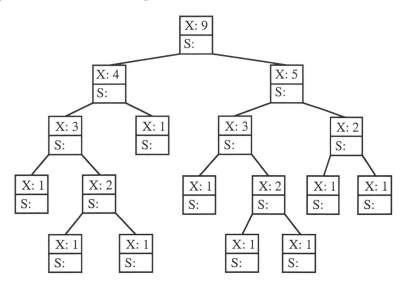

2. Work down from the root to compute the S values and compute *height*=4.

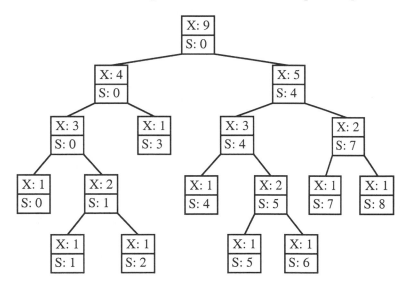

3. The values 0 through 8 are stored in the S values of the leaves; the final step computes the prefix sum for each leaf by adding 1 (since in this example, A is all 1's) to each of these S values and storing the values 0 through 9 in the leaves of the B array.

Example: $O(1)$ CRCW Array Max with $O(n^2)$ Processors

Idea: Most "collecting" operations like sum or maximum take $\Omega(\log(n))$ time on most models because of the need to fan in the data from many locations to a single location. However, the CRCW model is sufficiently powerful to solve some of these problems in $O(1)$ time.

CRCW assumption: Here, we make what is one of the most simple assumptions:

> *When concurrent writes occur, they must all write the same value.*

CRCW array maximum in $O(1)$ time with $O(n^2)$ processors: We call this function "SimpleMAX" to distinguish it from the next example that will reduce the number of processors to $O(n)$ at only a small increase in time.

function crcwSimpleMAX($A[0]...A[n-1]$)

1. Compare every pair of elements:
   for $1 \le i,j < n$ **in parallel do if** $A[i]<A[j]$ **then** $B[i,j]:=1$ **else** $B[i,j]:=0$

2. Set $X[i]:=0$ if $A[i]$ contains a maximum element:
   for $1 \le i,j < n$ **in parallel do** $X[i]:=0$
   for $1 \le i,j < n$ **in parallel do if** $B[i,j]=1$ **then** $X[i]:=1$

3. Write the maximum element to the variable m and return it:
   for $1 \le i < n$ **in parallel do if** $X[i]=0$ **then** $m:=A[i]$

4. **return** m

   end

Complexity: The argument $A[0]...A[n-1]$ to the function is just notation; in practice, parameter passing can be done in $O(1)$ time by simply passing a pointer to A and the two integers 0 and $n-1$. Steps 1 through 4 are each $O(1)$ time, for a total of $O(1)$ time. The space is dominated by the array B, which is $O(n^2)$. The number of processors is $O(n^2)$.

Computing the index of the maximum element: *crcwSimpleMAX* can be modified to return the smallest (or largest) index of a maximum element (see the exercises).

Example: $O(\log\log(n))$ CRCW Array Max with $O(n)$ Processors

Idea: Instead of using a binary tree of n leaves to compute the maximum in $O(\log(n))$ time with $O(n)$ processors (e.g., like the example that presented *erewSUM*), we employ a tree with height $O(\log\log(n))$ but with more children per vertex.

Double-exponential trees: A *double exponential tree* is one where the leaves at level 0 store $A[0]...A[n-1]$, pairs of leaves are the children of vertices on level 1 (if n is odd, then the last vertex on level 1 has only one child), groups of 4 vertices from level 1 are the children of vertices on level 2, and in general, if the vertices on level i have x children each, then the vertices on level $i+1$ have x^2 children each (except that the rightmost vertex on level i could have between 1 and x^2 children); the root always has between 2 and $\sqrt{n}$ children. For example, the double-exponential tree for 128 vertices has height 3:

14 subtrees for A[8] - A[119]

0 1 2 3 4 5 6 7   120 121 122 123 124 125 126 127

Computing the maximum of an array in $O(\log\log(n))$ time with $O(n)$ processors: We proceed in a manner like *erewSUM* to make calls on *crcwSimpleMAX*. At each stage vertices remaining are shifted left in A so calls will always be on consecutive positions:

```
function crcwMAX(A[0]...A[n−1])
    k := 2
    while k<2n do begin
        for 0 ≤ i < 2n/k in parallel do
            if i is a multiple of k then A[i/k] := crcwSimpleMAX(A[i]...A[i+k−1])
        k := k*k
    end
    return A[0]
end
```

Example: Suppose that $A[0]...A[127]$ contained the integers 0 through 127 (i.e., $A[i]=i$). Then the three iterations of the *while* loop transform A as follows:

$$
\begin{aligned}
A[0]...A[127] \quad &= 0, 1, 2, 3, ..., 127 \\
A[0]...A[63] \quad &= 1, 3, 5, 7, 9, ..., 127 \\
A[0]...A[16] \quad &= 7, 15, 23, 31, 39, 47, 55, 63, 71, 79, 87, 95, 103, 111, 119, 127 \\
A[0] \quad &= 127
\end{aligned}
$$

Complexity: Since a double exponential tree has height $O(\log\log(n))$ and the sum over each vertex at level of its number of children squared is $O(n)$ (see the exercises), each of the $O(\log\log(n))$ iterations of the *while* loop needs only $O(n)$ space to make the parallel calls to crcwSimpleMAX, all of which take only $O(1)$ time.

Example: Matrix Multiplication

Notation: Recall the "standard" matrix multiplication algorithm to multiply a m by w matrix A times a w by n matrix B to obtain a m x n matrix C; here we start indices at 0:

for $0 \leq i < m$, $0 \leq j < n$ **do begin**
    $C[i,j] := 0$
    for $k=0$ **to** $w-1$ **do** $C[i,j] = C[i,j] + A[i,k]B[k,j]$
    end

Idea: The inner *for* loop simply computes the vector product of the ith row of A times the jth column of B. Since it does not matter in which order these inner products are done (in fact the standard algorithm has been presented in a way where the outer *for* loop does not specify a particular order), they can just as easily be done simultaneously. For convenience we can first in $O(1)$ time perform all of the mwn pairwise multiplications in parallel. Then, in $O(\log(w))$ time the mn sums can be done in parallel by calling erewSUM.

function pramMATRIXMULT(A,B,C)
    for $0 \leq i < m$, $0 \leq j < w$, $0 \leq k < n$ **in parallel do** $Z[i,j,k] := A[i,j]*B[j,k]$
    for $0 \leq i < m$, $0 \leq k < n$ **in parallel do** $C[i,k] := $ erewSUM($Z[i,0,k]...Z[i,w-1,k]$)
    end

Complexity:

    Time: $O(\log(w)) = O(\log(n))$ for square matrices

    Space: $O(mwn) = O(n^3)$ for square matrices

    Number of processors: $O(mwn) = O(n^3)$ for square matrices

Reducing the number of processors by a factor of log(w):

The use of w processors by *pramSUM* is not as "cost effective" as we might like in the sense that by using w processors, the $O(w)$ time for the serial implementation was not reduced by a factor of w, but rather by a factor of only $w/\log(w)$.

This inefficiency of a factor of $\log(w)$ carries over to *crewMATRIXMULT*.

To remedy this inefficiency, we can partition an array to be summed into blocks of size $\lceil \log_2(w) \rceil$ and then use $O(w/\log(w))$ processors to each sum each block in $O(\log(w))$ time (each processor using a standard serial algorithm on its block). Then, starting with a vector of $\lceil w/\lceil \log_2(w) \rceil \rceil$ numbers, *erewSUM* can in an additional $O(\log(w))$ time sum them with $O(w/\log(w))$ processors.

Example: Merge Sort

Generic parallel merge sort procedure: We adapt the procedure *arrayMSORT* (presented as an example of divide and conquer) to perform the two recursive calls in parallel. Just as with *arrayMSORT*, the following procedure sorts $A[i]...A[j]$ with recursive calls on the two halves and then invokes a procedure (which must be supplied) to merge the two sorted halves (we assume the array name is A; a pointer to A could be passed as a third argument):

```
procedure pramMSORT(i,j)
    if i<j then begin
        m := ⌊(i+j)/2⌋   — i.e., the left side gets one more element if j–i+1 is odd
        in parallel do begin
            pramMSORT(i,m)
            pramMSORT(m+1,j)
        end
        pramMERGE(i,j,m)
    end
end
```

CREW merging: Modify the usual serial merge sort algorithm to employ binary search to determine an element's correct position in the merged list. For an element in the left half of $A[i]...A[j]$, its sorted position is its position in the sorted left half plus the number of elements in the right half that are less. Similarly, for an element in the right half of $A[i]...A[j]$, its sorted position is its position in the sorted right half minus the number of elements in the left half that are greater. This gives an $O(n\log(n)^2)$ serial algorithm instead of $O(n\log(n))$, but it will lend itself well to parallel implementation.

Notation: LESS(i,j,x) and GREATER(i,j,x) are functions that binary search $A[i]...A[j]$ to determine how many of these elements are less than or greater than x. These functions can be computed in $O(\log(j-i))$ time by a single processor.

```
procedure crewMERGE(i,j,m)
    if i<j then begin
        for i ≤ k ≤ m in parallel do P[k] := k + LESS(m+1,j,A[k])
        for m+1 ≤ k ≤ j in parallel do P[k] := k – GREATER(i,m,A[k])
        for i ≤ k ≤ j in parallel do A[P[k]] := A[k]
    end
end
```

Complexity: Time is $T(n) = T(n/2) + O(\log(n) = O((\log(n))^2)$. As with *arrayMERGE*, recursive calls of *crewMERGE* work on disjoint ranges of indices, and a single scratch "permutation" array P of the same size as A suffices, for a total of $O(n)$ space (which dominates the $O(\log(n))$ space for the recursion stack). $O(n)$ processors are used.

Example: Quick Sort

Generic parallel quick sort procedure: Similar to generic PRAM merge sorting, we adapt the procedure *arrayQSORT* with randomized 3-partitioning (presented as an example of divide and conquer) to perform the two recursive calls in parallel (we assume the array name is A; a pointer to A could be passed as a third argument):

procedure pramQSORT(i,j)
    if $i<j$ **then begin**
        a,b := PARTITION(i,j)
        in parallel do begin
            pramQSORT(i,a)
            pramQSORT(b,j)
        end
    end
end

EREW 3-partitioning: Similar to how CREW merge sort counted the elements less than a given element as a way to determine where it belonged in the rearranged array, we employ parallel prefix-sum to determine a position for each element in the partitioned array. A key improvement here is that were are using a single parallel method to do this (instead of many serial binary searches done in parallel) and can use the EREW model.

procedure erew3PARTITION(i,j)

    r := a random integer in the range $i \leq r \leq j$

    Use three parallel prefix-sum computations to compute for $i \leq k \leq j$:
        $LESS[k]$ = the number of elements x such that $x \leq k$ and $A[x]<r$
        $EQUAL[k]$ = the number of elements x such that $x \leq k$ and $A[x]=r$
        $GREATER[k]$ = the number of elements x such that $x \leq k$ and $A[x]>r$

    a := $i+LESS[n-1]-1$

    b := $a + EQUAL[n-1]+1$

    for $i \leq k \leq j$ **in parallel do**
        if $A[k]<r$ **then** $P[k]$:= $i + LESS[k] - 1$
        else if $A[k]=r$ **then** $P[k]$:= $a + EQUAL[k]$
        else $P[k]$:= $b + GREATER[k] - 1$

    for $i \leq k \leq j$ **in parallel do** $A[P[k]]$:= $A[k]$

    return a and b
    end

Complexity: Since prefix sum is $O(\log(n))$, as with parallel merging the time is given by $T(n) = T(n/2) + O(\log(n)) = O(\log(n)^2)$. $O(n)$ processors are used.

Brent's Lemma

Idea: Brent's lemma provides criteria for when a parallel algorithm is processor efficient. A parallel algorithm can be viewed as "processor efficient" if:

$$(parallel\ time) * (number\ of\ processors) = O(serial\ time)$$

Example: We have already seen how crcwSUM and pramMATRIXMULT "wasted" a factor of $\log(n)$ processors, but we noted after the presentation of PRAM matrix multiplication that this waste could be eliminated by balancing the parallel time with some serial preprocessing that first reduced the size of the vector by a factor of $\log(n)$ by summing blocks of size $\log(n)$, each block being done serially by a single processor. On the other hand, pramMSORT "efficiently" used its $O(n)$ processors to convert an $O(n\log(n)^2)$ serial algorithm to a $O(\log(n)^2)$ parallel one.

Definition: The *work* performed by a processor during a computation is the number of steps for which it is not idle (i.e., the number of steps for which it is doing work that contributes to the computation).

Example: With *parallelSUM*, on the first iteration, all $n/2$ processors are used, on the second $n/4$, and so on. Hence the total work can be expressed with a geometric sum:

$$n(1/2 + 1/4 + ...) = O(n)$$

Brent's lemma: If a problem of size n that can be solved in $s(n)$ serial time can be solved in $p(n)$ parallel time in such a way that the total work performed by all processors is $s(n)$, then this parallel time can be achieved with only $s(n)/p(n)$ processors.

Proof: Let w_i, $1 \leq i \leq p(n)$, be the total work performed by all processors at Step i; that is:

$$\sum_{i=1}^{p(n)} w_i = s(n)$$

If $w_i \leq s(n)/p(n)$, then there are enough processors at Step i. Otherwise, replace Step i with $w_i/(s(n)/p(n))$ steps. So now each step can be performed with $s(n)/p(n)$ processors, and the total time is:

$$\sum_{i=1}^{p(n)} \left\lceil \frac{w_i}{(s(n)/p(n))} \right\rceil \leq \sum_{i=1}^{p(n)} \left(\frac{p(n)w_i}{s(n)} + 1 \right)$$

$$\leq p(n) + \frac{p(n)}{s(n)} \sum_{i=1}^{p(n)} w_i$$

$$= 2p(n)$$

$$= O(p(n))$$

PRAM Simulation

Idea: Later in this chapter we shall consider weaker models of parallel computation that may be more practical to build than the CREW PRAM (the EREW PRAM, the HC/CCC/BF, and the mesh). In an "ideal" simulation there is one processor for each shared data item. For each step of the PRAM program, each processor that needs to read a shared data item located in a different processor sends a request *packet* to that processor and wait to receive a packet with the data.

Ideal PRAM simulation algorithm:
1. Each processor a that wishes to communicate a message X with processor b (send a data item, retrieve a data item, etc), constructs a *request packet* $<a,b,X>$.
2. Sort all request packets by the second component (so all packets that are destined for a particular processor are in consecutive processors).
3. Identify the leaders, where processor i is a leader if $i=0$ or the second component of i's packet differs from that of processor $i-1$. Each leader i that contains a packet $<a,b,X>$ sends a *secondary request packet* $<i,b,X>$.
4. When a secondary request packet $<i,b,X>$ is received by processor b, it returns the response Y to the message X in a secondary *acknowledgement packet* $<b,i,Y>$.
5. When the secondary acknowledgement packet $<b,i,Y>$ is received by the leader i, it sends a *distribution packet* $<i,j,Y>$ to each of its followers.
6. When a processor j receives a distribution packet $<i,j,Y>$, it sends a *return packet* $<j,b,Y>$ to the processor b that originally sent the message X to obtain Y.

Time per step of the simulation: Let:
n = number of processors
$S(n)$ = The time to *sort* n items stored 1 per processor.
$R(n)$ = Time to perform a 1-1 *routing*. That is, given a subset of the processors that each have a packet to send, and assuming that at most one packet is being sent to any given processor, the time to get all packets to their destinations.
$D(n)$ = Time to perform *data distribution*. That is, if some processors are designated as *senders* and the rest as *receivers*, and each sender needs to send a packet to all of the receivers between it and the next sender (or to all of *the higher* numbered receivers if it is the highest numbered sender), the time to get all packets to their destinations.

Step 1 is $O(1)$. Step 2 is $O(S(n))$. Steps 3, 4, and 6 are $O(R(n))$. Step 5 is $O(D(n))$. Hence, the entire algorithm uses $O(R(n)+S(n)+D(n))$ time for *each* step of the original PRAM algorithm.

CRCW PRAM Simulation: The same approach can be used, except depending on the assumptions, an additional "tool" (e.g., prefix-sum) may be needed; see the exercises.

EREW PRAM MODEL

We now focus on the EREW PRAM model, starting with some more examples, and then showing how to implement the components needed for PRAM simulation. Since 1-1 packet routing is "free" on the EREW PRAM (by writing / reading an agreed location), we need only address data distribution and sorting.

Example: Broadcast on an EREW PRAM

Problem: Copy a single data item x is to all positions of an array $A[0]...A[n]$.

Idea: Assuming that each receiver knows the index of its sender, on a standard (CREW) PRAM, in $O(1)$ time each of its receivers can simultaneously read it. However, on an EREW PRAM, we need to spread out the data to the receivers. The algorithm for summing data can essentially be reversed; that is, send the data to a receiver far away, then both send the data half as far, then all four send a quarter as far, and so on. For example, in three steps data can be copied from Position 0 to Positions 1 to 7:

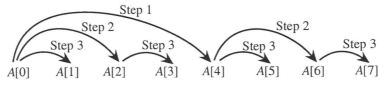

The broadcasting algorithm:

> **procedure** erewBROADCAST(x, $A[0]...A[n-1]$)
> $A[0] := x$
>   **for** $k := \lceil \log_2(n) \rceil$ **downto** 1 **do**
>   **for** $0 \le i < n-2^{k-1}$ **in parallel do if** i is a multiple of 2^k **then** $A[i+2^{k-1}] := A[i]$
>   **end**
>
> *Note*: The test if i is a multiple of 2^{k+1} is not necessary (see the exercises).

An equivalent way to express the broadcasting algorithm:

> **procedure** erewBROADCAST(x, $A[0]...A[n-1]$)
> $A[0] := x$
> $k := n$
>   **while** $k > 1$ **do begin**
>   **for** $0 \le i < n-k/2$ **in parallel do if** i is a multiple of k **then** $A[i+k/2] := A[i]$
> $k := \lceil k/2 \rceil$
>   **end**
> **end**

Complexity: $O(\log(n))$ time is used. $O(1)$ space is used in addition to the space used by A. $O(n)$ processors are used.

Example: Sum on an EREW PRAM

The first example for the PRAM, *erewSUM*, worked for the EREW PRAM model.

Example: Matrix Multiplication on an EREW PRAM

Recall the PRAM (CREW) matrix multiplication algorithm that was presented earlier::

> **function** pramMATRIXMULT(A,B,C)
> > **for** $0 \leq i < m$, $0 \leq j < w$, $0 \leq k < n$ **in parallel do** $Z[i,j,k] := A[i,j]*B[j,k]$
> > **for** $0 \leq i < m$, $0 \leq k < n$ **in parallel do** $C[i,k] := \text{erewSUM}(Z[i,0,k]...Z[i,w-1,k])$
> > **end**

Idea: The second *for* loop does not use concurrent reads because the pairwise multiplications are already distributed among the *mwn* locations of Z, and each of the calls to erewSUM employs a disjoint set of processors. The problem is that the pairwise multiplications in the first *for* loop are done by directly accessing A and B, causing each entry of A and B to be simultaneously read by many processors. Here, we first broadcast the entries of A and B to Z (by using disjoint sets of processors for each entry) so that no simultaneous reads occur:

procedure erewMATRIXMULT(A,B,C)

1. Broadcast the entries of A into X:
   > **for** $0 \leq i < m$, $0 \leq j < w$ **in parallel do** erewBROADCAST($A[i,j]$, $X[i,j,0]...X[i,j,n-1]$)

2. Broadcast the entries of B into Y:
   > **for** $0 \leq j < w$, $0 \leq k < n$ **in parallel do** erewBROADCAST($B[j,k]$, $Y[0,j,k]...X[m-1,j,k]$)

3. Perform the pair-wise multiplications into Z:
   > **for** $0 \leq i < m$, $0 \leq j < w$, $0 \leq k < n$ **in parallel do** $Z[i,j,k] := X[i,j,k]*Y[i,j,k]$

4. Compute each entry of C by summing the appropriate sub-array of Z:
   > **for** $0 \leq i < m$, $0 \leq k < n$ **in parallel do** $C[i,k] := \text{erewSUM}(Z[i,0,k]...Z[i,w-1,k])$

end

Complexity:

> **Time:** Step 1 is $O(\log(n))$, Step 2 is $O(\log(m))$, Step 3 is $O(1)$, and Step 4 is $O(\log(w))$, for a total of $O(\log(m)+\log(w)+\log(n))$, which is $O(\log(n))$ for square matrices
>
> **Space:** $O(mwn) = O(n^3)$ for square matrices
>
> **Number of processors:** $O(mwn) = O(n^3)$ for square matrices

Data Distribution on an EREW PRAM

Notation: For an array $A[0]...A[n-1]$, some positions are designated as *sender*s and the rest as *receivers*. Each sender item $A[i]$ is be copied to all receivers between it and the next sender (or to all of $A[i+1]...A[n-1]$ if $A[i]$ is the highest numbered sender).

Idea: The broadcast mechanism is initiated from each sender. It is not necessary for a receiver to know the index of its sender; we keep track of the highest numbered sender from which data has been received thus far in a second array S.

Data distribution algorithm:

> **for** $0 \leq i < n$ **in parallel do if** P_i is a sender **then** $S[i]:=i$ **else** $S[i]:=-1$
> **for** $k:=\lceil \log_2(n) \rceil$ **downto** 1 **do**
> > **for** $0 \leq i < n-2^{k-1}$ **in parallel do if** $S[i] \geq S[i+2^{k-1}]$ **then begin**
> > > $S[i+2^{k-1}] := S[i]$
> > > $A[i+2^{k-1}] := A[i]$
> > >
> > > **end**
> >
> **end**

An equivalent way of expressing the data distribution algorithm:

> **for** $0 \leq i < n$ **in parallel do if** P_i is a sender **then** $S[i]:=i$ **else** $S[i]:=-1$
>
> $k := n$
>
> **while** $k>1$ **do begin**
> > **for** $0 \leq i < n-k/2$ **in parallel do if** i is a multiple of k and $S[i] \geq S[i+k/2]$ **then begin**
> > > $S[i+k/2] := S[i]$
> > > $A[i+k/2] := A[i]$
> > > **end**
> >
> > $k := \lceil k/2 \rceil$
> >
> > **end**
>
> **end**

Complexity:

> Time: $O(\log(n))$
> Space: $O(n)$ space in addition to the space used by A
> Processors: $O(n)$

Sorting on an EREW PRAM

Idea: We again employ the procedure pramMSORT, but use a different merging method, called *odd-even merging*. Odd-even merging uses a divide and conquer strategy that allows a parallel implementation to merge in $O(\log(n))$ time without concurrent reads. Before addressing the parallel implementation, we first present a standard serial implementation and its analysis.

Notation: Let $X = x_1, x_2, ..., x_m$ and $Y = y_1, y_2, ..., y_n$ be such that $m=n$ or $m=n+1$:

$X_{odd} = x_1, x_3, x_5, ...$

$X_{even} = x_2, x_4, x_6, ...$

$\text{SHUFFLE}(X, Y) = x_1, y_1, x_2, y_2, x_3, y_3, ...$

$\text{FIX}(X) = $ **for** $i:=3$ **to** n **by** 2 **do if** $x_i < x_{i-1}$ then exchange x_i and x_{i-1}

Odd-Even merge sorting (on a standard serial RAM): Two recursive calls merge X_{odd} with Y_{odd} and X_{even} with Y_{even}. The resulting sorted lists are shuffled together, and then adjacent pairs that are out of order are fixed:

> **function** OEMERGE(X, Y)
>   **if** X is empty **then return** Y
>   **else if** Y is empty **then return** X
>   **else return** FIX(SHUFFLE(OEMERGE(X_{odd}, Y_{odd}),OEMERGE(X_{even}, Y_{even})))
>   **end**

Example:

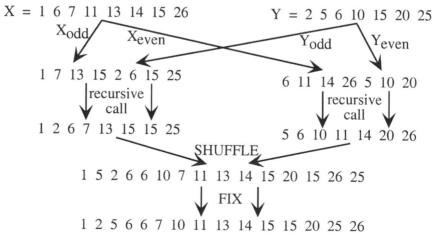

Analysis of Odd-Even Merge Sort

Correctness of odd-even merging:

Let $O = o_1...o_m$ and $E = e_1...e_n$ be the sorted lists resulting from merging X_{odd} with Y_{odd} and X_{even} with Y_{even}, and consider an item o_k, $1 \leq k \leq m$.

Suppose that o_k is from X (if it is from Y, the reasoning is identical with the roles of X and Y reversed in the argument to follow); that is, for some $i \geq 0$, $o_k = x_{2i+1}$.

Let $j \geq 0$ be the number of elements in O that are $\leq o_k$; that is:

$$y_1, y_2, ..., y_{2j-1} \leq o_k \leq y_{2j+1}$$

Hence, in O, there are $i+j$ values that are $\leq o_k$ ($x_1, x_3, ..., 2_{i-1}$ and $y_1, y_3, ..., y_{2j-1}$).

Now consider the items in E that are $\leq o_k$.

We know that $x_2, x_4, ..., x_{2i} \leq o_k$ and because $o_k < y_{2j+1} \leq y_{2j+2}$ it must be that $y_2, y_4, ..., y_{2j-2} \leq ok < y_{2j+2}$. However, we do not know whether $o_k \leq y_{2j}$ or $o_k > y_{2j}$.

Hence, in E there are at least $i+j-1$ items that are $\leq o_k$ and at most $i+j$ items that are $<o_k$.

Hence, since $k=i+j+1$, when SHUFFLE places $o_k = o_{i+j+1}$ just after E_{i+j}, this is legal, unless $o_k < y_{2j} = E_{i+j}$, in which case FIX moves o_k to just before E_{i+j}.

Array implementation (on a standard serial RAM):

We assume that initially X and Y are in sequential locations of an array A. That is, for some $i \leq j$ such that $j-i+1=m+n$, X is in $A[i]...A[i+m-1]$ and Y is in $A[i+m]...A[j]$.

As with the serial *arrayMERGE* algorithm, a single second array B of the same size of A is employed for temporary space (recursive calls use disjoint portions of B).

Division into X_{odd}, Y_{odd} and X_{even}, Y_{even} can be done by first placing X_{odd}, Y_{odd} into $B[i]...B[i+m-1]$ and X_{even}, Y_{even} into $B[i+m]...[j]$, and then copying back into A.

Next, the recursive calls simply work on the left and right halves of A.

Next, SHUFFLE can place items into B and then copy B back to A.

Finally, FIX can simply traverse $A[i]...A[j]$.

Time (on a standard serial RAM): If $M(n)$ denotes the time to merge n elements and $T(n)$ the time to sort n elements, then $M(n)=T(1)=O(1)$ and in general:

$$M(n) = 2M(\lceil n/2 \rceil) + O(n) = O(n\log(n))$$

$$T(n) = 2T(\lceil n/2 \rceil) + O(n\log(n)) = O(n(\log(n))^2)$$

Space to Sort a List of Size n: $O(n)$.

Implementation of odd-even merge sorting on an EREW PRAM

Idea:

To sort $A[i]...A[j]$ we use the generic *pramMSORT* procedure and replace pramMERGE with an array implementation of odd-even merging that employs an array B (of the same size as A) for scratch space to perform merge operations by placing items into B and then copying back into A.

The three arguments i,j,m specify that the two lists to be merged are in the left and right halves of $A[i]...A[j]$, where m is the midpoint; that is, $A[i]...A[m]$ must be merged with $A[m+1]...A[j]$ (we assume the array name is A; a pointer to A could be passed as a third argument).

It is always the case that *pramMSORT* chooses m so that the number of items in the left half (i.e., $m-j+1$) is the same as or one greater than the number in the right half (i.e., $j-m$).

procedure erewOEMERGE(i,j,m)

    if $i<j$ **then begin**

        1. DIVIDE: Rearrange A (by first copying to the temporary array B) so that the left half contains every other element that used to be in A (in the same relative order) and the right half contains the remaining elements (in the same relative order):

            for $i \leq k \leq j$ **in parallel do**
                if $k \leq m$ **then** $B[k]:=A[i+2(k-i)]$ **else** $B[k]:=A[i+2(k-m)-1]$
            for $i \leq k \leq j$ **in parallel do** $A[k]:=B[k]$

        2. RECURSE: Recursive calls to do the merging inside the left and right halves:
            pramOEMERGE(i, m, $\lfloor (i+m)/2 \rfloor$)
            pramOEMERGE($m+1$, j, $\lfloor (m+j+1)/2 \rfloor$)

        3. SHUFFLE: Shuffle the left and right halves:
            for $i \leq k \leq j$ **in parallel do**
                if $k-i$ is even **then** $B[k]:=A[i+(k-i)/2]$ **else** $B[k]:=A[m+(k-i+1)/2]$
            for $i \leq k \leq j$ **in parallel do** $A[k]:=B[k]$

        4. FIX: Fix adjacent even-odd pairs that are out of order:
            for $i<k<j$ **in parallel do**
                if $k-i$ is odd and $A[i]<A[i+1]$ **then** exchange $A[i]$ and $A[i+1]$

    end

  end

CHAPTER 13

Non-Recursive Implementation of Odd-Even Merge Sort

Idea:

*** For simplicity, we assume $A = A[0]...A[n-1]$, where n is a power of 2.

Recursive calls in both *pramMSORT* and *erewOEMERGE* are just "bookkeeping".

To merge the two halves of a range (i,j) keep dividing into odds and evens to create ranges of half the size until the range size gets down to 2. Then repeatedly use *SHUFFLE* and *FIX* to double the range size to get back to the whole range (i,j).

To sort a range (i,j), start with ranges of size 2 and keep merging the halves of ranges to create ranges of twice the size to get back to the whole range.

Step 2 of *erewOEMERGE* will be eliminated, and we shall refer to the computations of Steps 1, 3, and 4 as $DIVIDE(i,j,m)$, $SHUFFLE(i,j,m)$, and $FIX(i,j,m)$ respectively.

procedure nr-erewOEMERGE(i,j,m)

    Keep dividing until the ranges of size p (always a power of 2) have size 2:
    $p := j-i+1$
    while $p>2$ **do begin**
        for $i \le x < j$ **in parallel do if** x is a multiple of p **then** DIVIDE(x, $x+p-1$, $x+p/2$)
        $p := p/2$
    end

    Repeatedly SHUFFLE and FIX from ranges of size 2 back up to the whole range:
    $p := 2$
    while $p \le (j-i+1)$ **do begin**
        for $i \le x < j$ **in parallel do if** x is a multiple of p **then begin**
            SHUFFLE(x, $x+p-1$, $x+p/2$)
            FIX(x, $x+p-1$, $x+p/2$)
        end
        $p := p*2$
    end

    end

procedure nr-erewMSORT

    $p := 1$
    while $p \le n$ **do begin**
        for $0 \le x < n$ **in parallel do**
            if x is a multiple of p **then** nr-erewOEMERGE(x, $x+p-1$, $x+p/2$)
        $p := p*2$
    end
    end

Hypercube / CCC / Butterfly Networks

Hypercube (HC)

A *hypercube* of dimension 0 is a single vertex. For $k>0$, a hypercube of dimension k is constructed by connecting corresponding vertices of a hypercube of dimension $k-1$; these connecting edges are called *hypercube edges of dimension k*. The n vertices of a hypercube of dimension k (where $n=2^k$) are labeled from 0 to $n-1$ using k-bit binary numbers where the ith bit from the right changes when an edge of dimension k is traversed. The figure below shows hypercubes of dimensions 0 to 3:

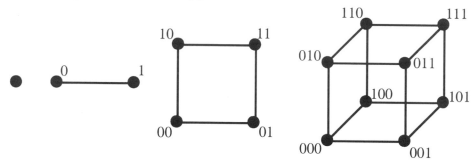

Cube Connected Cycles (CCC)

A *cube connected cycles* network of dimension k is obtained from a hypercube of dimension k by replacing each vertex i by a cycle of length k (*called the CCC cycle at vertex i*) of vertices labeled as $[1,i]$, $[2,i]$, ..., $[k,i]$ where a hypercube edge of dimension d that used to be attached to vertex i is now attached to vertex $[d,i]$. In the figure below, for each i, vertices $[1,i]$...$[k,i]$ are shown in a cycle with i in binary written in the center and $1...k$ written next to each vertex:

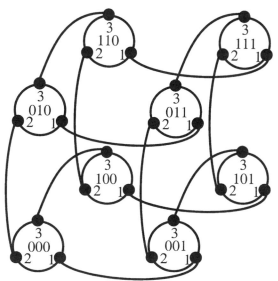

Butterfly (BF)

A *butterfly* of dimension k consists of $k+1$ *rows* of vertices numbered 0 to k, each containing 2^k vertices in *columns* 0 to 2^k-1, where a vertex $[r,i]$ in Row r, $1 \le r \le k$, and Column i, $0 \le i < n$, has a *column edge* down to $[r-1,i]$ and a *butterfly edge* to $[r-1,FLIP(r,i)]$; where $FLIP(x,y)$ denotes the integer z with the same binary representation as y except that the xth bit, counting from the right, differs (this is just notation — there is no need to compute FLIP values in practice because connections are "hard-wired" and each processor knows to which processors it is connected). We refer to the data stored at vertex $[r,i]$ as $data[r,i]$.

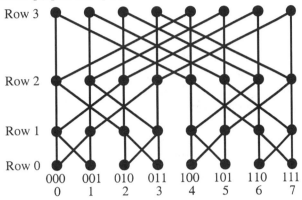

Recursive decomposition of the butterfly network: In a k-dimensional butterfly, removing the kth row (and the incident edges) leaves two networks of dimension $k-1$ consisting of the left and right halves of the columns, and removing the 0th row leaves two networks of dimension $k-1$ consisting of the even and odd numbered columns:

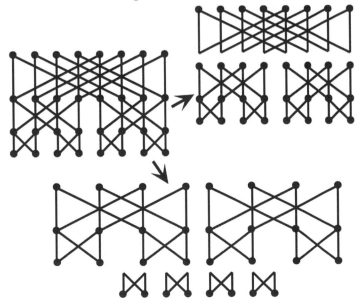

Equivalence of the CCC and Butterfly Networks

Definition: A *wrapped* butterfly of dimension k can be obtained from a butterfly of dimension k by mapping Row 0 onto Row k; that is, by eliminating the vertices of Row 0 and for each edge that used to be attached to a vertex $(0,j)$, attaching it to vertex (k,j). Thus, the wrapped butterfly has rows 1 through k and Column i in the butterfly becomes a cycle of length k in the wrapped butterfly, which we call the *column cycle* at vertex i.

Isomorphism between the CCC and the wrapped butterfly of dimension k:

The CCC cycle at vertex i, $0 \le i \le 2^k-1$, corresponds to the column cycle at vertex i in the wrapped butterfly.

We identify the dimensions of the CCC with the rows of the wrapped butterfly. For any dimension $1 \le d \le k$, a hypercube edge $([d,i],[d,j])$ in the CCC corresponds to the pair of butterfly edges $([d,i], [d-1\ MOD\ k, j])$ and $([d,j], [d-1\ MOD\ k, i])$, where we use the MOD notation here to simply mean that if $d=1$, then $d-1$ is defined to be k.

The figure below depicts edge $([d,i],\ [d,j])$ in the CCC with a solid line and shows with dashed lines the pair of edges that are in the wrapped butterfly:

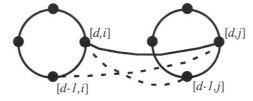

Practical equivalence of the Hypercube and the CCC/Butterfly:

An algorithm for a CCC/Butterfly of n columns can run in the same asymptotic time on a hypercube of n vertices, since a hypercube corresponds to a *collapsed butterfly* where each column maps to a single vertex with $2\log_2(n)$ incident edges.

In theory, however, there could be a slowdown of a factor of $O(\log(n))$ going in the other direction. That is, for the CCC to simulate a single step of an algorithm for the hypercube, we might have to traverse $O(\log(n))$ cycle edges for each step of the computation (e.g., to read data from one hypercube edge and send it back out along a different one on the opposite side of the cycle).

However, in practice, many algorithms for the CCC/Butterfly (and all of the ones we shall consider) have a "regular" flow in the sense that when data is being routed around the network and goes into a vertex on an edge of dimension d, it goes out on an edge of dimension $d-1$ or $d+1$ that is adjacent on the cycle. For example, passing data in parallel from one row of the butterfly to the next has this property.

For convenience, we will present all of our algorithms for the HC/CCC/BF in terms of the butterfly.

Example: Broadcast and Sum on a Butterfly

Idea: As depicted below, the butterfly contains a binary tree with root $[k,0]$ and leaves the processors of Row 0. Hence, we can proceed in essentially the same fashion as for the EREW algorithm presented earlier ($O(\log(n))$ time and $O(1)$ space per processor).

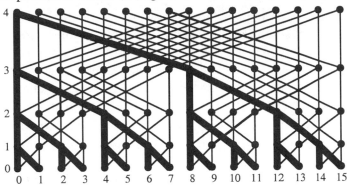

Broadcast from Row k: To simplify our presentation, we assume that the data item to be broadcast is initially in Processor $[k,0]$ in the variable $D[k,0]$; if it is in some other row l of Column 0, in $O(\log(n))$ time it can first be copied up to Row k by doing:

    for $r:=l+1$ **to** $\log_2(n)$ **do** $D[r,0] := D[r-1,0]$

Algorithm for broadcasting on the butterfly:

    procedure butterflyBroadcast
        for $r:=\log_2(n)$ **downto** 1 **do**
            for $0 \leq i < n$ **in parallel do if** i is a multiple of 2^r **then begin**
            $D[r-1,i] := D[r,i]$
            $D[r-1,FLIP(r,i)] := D[r,i]$
            end
        end

An equivalent way to express the algorithm:

    procedure butterflyBroadcast
        Label Processor $[r,0]$ "marked".
        for $r:=\log_2(n)$ **downto** 1 **do**
            for $0 \leq i < n$ **in parallel do if** Processor i is marked **then begin**
            Copy $D[r,i]$ down both the column edge and the butterfly edge going
            down from Processor $[r,i]$ and mark those processors.
            end
        end

Summing on a butterfly: We essentially reverse the broadcast algorithm with the arrows going in the reverse direction (see the exercises).

Example: Prefix-Sum on a Butterfly

Problem: For data items $A[0]...A[n-1]$ stored in Row 0 of a butterfly of n columns, we wish to compute in each Column i, $0 \leq i < n$, the sum of $A[0]$ through $A[i]$.

Idea: We could obtain an $O(\log(n))$ algorithm by simulating the connections used by an EREW list or tree prefix sum algorithm. But it is relatively simple to present a "direct" implementation. Similar to the tree prefix-sum algorithm for the EREW PRAM, we maintain an array S to store the sum and an auxiliary array X to store the corresponding portion of the butterfly; that is, for Row r and Column i, if we imagine that all rows higher than i have been removed to leave $n/(2^i)$ disjoint "sub-butterflies", then:

$X[r,i]$ = sum of all columns in the sub-butterfly containing Processor $[r,i]$

$S[r,i]$ = sum of the columns to the left in the sub-butterfly containing Processor $[r,i]$

Butterfly prefix-sum algorithm: $FLIP(r,i)$ is just notation; connections are "hard-wired" in practice. The sums are left in $S[3,i]$, $0 \leq i < n$, in the top row (if needed, they can be copied back to Row 0 on $\log_2(n)$ additional Steps).

> **for** $0 \leq i < n$ **in parallel do** $S[0,i] := X[0,i] := A[i]$
> **for** $r:=1$ **to** $\log_2(n)$ **do**
> > **for** $0 \leq i < n$ **in parallel do begin**
> > $X[r,i] := X[r-1,i] + X[r-1,FLIP(r,i)]$
> > **if** $FLIP(r,i)>i$ **then** $S[r,i] := S[r-1,i]$ **else** $S[r,i] := S[r-1,i] + X[r-1,i]$
> > **end**

Complexity: There are $\log_2(n)$ iterations, each of which takes $O(1)$ time, for a total of $O(\log(n))$ time. $O(1)$ space is used per processor.

Example: Suppose that $n=8$ and all values of A are 1:

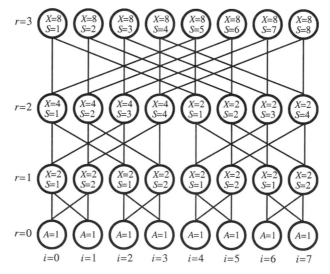

Example: Matrix Multiplication on a Butterfly

Notation: For the matrix multiplication $C=AB$, A is a m by w matrix and B is a w by n matrix, and we assume that m, w, and n are each powers of 2 (if not, "pad" the matrices with enough 0's to make the dimensions a power of 2 — see the exercises).

Idea: The matrix multiplication algorithm erewMATRIXMULT(A,B,C) for the EREW PRAM can be mapped *directly* onto the butterfly. The mwn simultaneous pairwise multiplications of Step 3 are done in Row 0 of a butterfly of mwn columns. Broadcasting of Steps 1 and 2 and the summing of Step 4 can be done on disjoint sub-butterflies.

Processor indexing: Columns of the butterfly are indexed with integers whose binary representations are comprised of three blocks of bits:

$\log_2(m)$ bits	$\log_2(w)$ bits	$\log_2(n)$ bits

For $0 \le i < m$, $0 \le j < w$, and $0 \le k < n$, $<i,j,k>$ denotes the integer with a binary representation that is the $\log_2(m)$-bit representation of i, followed by the $\log_2(w)$-bit representation of j, followed by the $\log_2(n)$-bit representation of k.

Bit substitution: The data broadcasting and summing of erewMATRIXMULT(A,B,C) is done by holding two of the three blocks constant and varying the bits in the third block. For $1 \le a \le d$ and $1 \le b \le (d-a+1)$, we let $S_{a,b,d}(i)$ denote the set of integers obtained by starting with the d-bit binary representation of i and replacing the b bits in positions a through $a+b-1$ (from the right) with all possible b-bit binary strings. For example:

$$S_{2,3,6}(000000) = \{000000, 000010, 000100, 000110, 001000, 001010, 001100, 001110\}$$

Sub-butterfly lemma: The sub-graph of the butterfly comprised of rows $a-1$ through $a+b-1$ and columns that have indices in $S_{a,b,d}(i)$ is a butterfly of dimension b.

Proof: We use induction on d that employs the two standard recursive decompositions of the butterfly. If $b=d$ (and hence $a=1$) then we have the entire butterfly. Otherwise, there are two cases. If $a>1$ then the subgraph of rows 1 through d with either the even or odd columns (depending on whether the rightmost bit of i is 0 or 1) is a butterfly. If $a=1$, then the subgraph of rows 0 through $d-1$ with either the left half or the right half of the columns (depending on whether the leftmost bit of i is 0 or 1) is a butterfly.

Butterfly matrix multiplication algorithm: Initially, $A[i,j]$, $0 \le i < m$ and $0 \le j < w$, is stored in Row 0 of Column $<i,j,0>$ and $B[j,k]$, $0 \le j < w$ and $0 \le k < n$, is stored in Row 0 of Column $<0,j,k>$. We use erewMATRIXMULT(A,B,C) where indices i,j,k in erewMATRIXMULT(A,B,C) correspond to Column $<i,j,k>$ of the butterfly. The details of the indexing necessary to perform an individual broadcast in Steps 1 or 2 or a sum in Step 3 follows from the sub-butterfly Lemma (see the exercises).

Complexity: $O(\log(n))$ time, $O(1)$ space per processor.

Data Distribution on a Butterfly

Distribution from row k: For a butterfly of dimension k, we assume that some of the columns are designated as *senders* and the rest as *receivers*. To simplify notation, we assume a sending column i initially has its data in Row k (in the variable $D[i,k]$); if data is initially in some other row x, then in $O(\log(n))$ the data can first be copied up to Row k:

> **for** $i:=x+1$ **to** k **do**
> > **for** $j:=0$ **to** $n-1$ **in parallel do**
> > > **if** j is a sender **then** $D[i,j]:=D[i-1,j]$

Thus, for $0 \le x < n$, where $n=2^k$, if Column x is a sender and Column y is the next highest numbered sender (or $y=n$ if Column x is the highest numbered sender), then the data item stored at vertex $[k,x]$ needs to be copied to vertices $[0,x]...[0,y-1]$.

Right-augmented butterfly networks: A butterfly edge that goes between $[r,i]$ and $[r-1,j]$ is a *right edge* if $i<j$ and a *left edge* if $i>j$. For a vertex $[r,i]$ that has a left edge going to $[r-1,i-2^{r-1}]$, the *right augmenting edge* for $[r,i]$ is the new edge from $[r,i]$ to $[r-1,i+2^{r-1}]$. A *right-augmented* butterfly is one where all right augmenting edges have been added. For example, the right augmenting edges are dashed in the figure below:

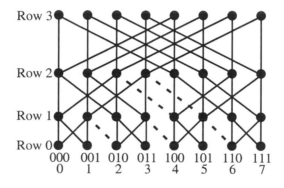

Data distribution on a right-augmented butterfly: We use essentially the same procedure as for data distribution on the EREW PRAM, where here we have a separate copy of $S[i]$ and $D[i]$ for each row r:

> **procedure** ra-butterflyDISTRIBUTE
> > **for** $0 \le i < n$ **in parallel do if** P_i is a sender **then** $S[k,i]:=i$ **else** $S[k,i]:=-1$
> > **for** $r:=\lceil \log_2(n) \rceil -1$ **downto** 0 **do**
> > > **for** $2^r \le i < n$ **in parallel do if** $S[r,i] \le S[r,i-2^r]$ **then begin**
> > > > $S[r,i] := S[r,i-2^r]$
> > > > $D[r,i] := D[r,i-2^r]$
> > > > **end**
>
> **end**

Complexity: $O(\log(n))$ time, $O(1)$ space per processor.

Detour Paths

Although half the vertices in a normal butterfly network do not have a right edge going down to the next row, there is always a path to get there that goes up the column to the first vertex with a right edge going down, along that right edge, and then back down along left edges. For example, the figure below shows with darkened edges the *detour path* from [2,2] to [1,4] and the detour path from [1,7] to [0,8]:

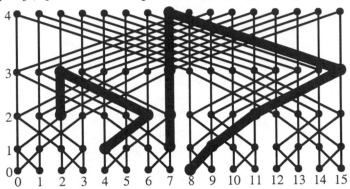

Data distribution on a (normal) butterfly: Since detour paths can be up to $O(\log(n))$ long, it is costly to simply traverse a detour path whenever the corresponding right edge is missing. Instead, use copies of the S and D arrays, LS and LD that are the values that get propagated down a sequence of left edges. Both copies are updated as usual when data comes down a right edge or when data comes down a left edge to a vertex on Row 0 or to a vertex with a right edge that goes down; otherwise, only the LS and LD arrays are updated. $FLIP(r,i)$ is just notation; connections are "hard-wired" in practice.

procedure butterflyDISTRIBUTE

    for $0 \le i < n$ **in parallel do**
        if P_i is a sender **then** $S[k,i] := LS[k,i] := i$ **else** $S[k,i] := LS[k,i] := -1$

    for $0 \le i < n$ **in parallel do** $LD[k,i] := D[k,i]$

    for $r:=k$ **downto** 0 **do**
        for $0 \le i < n$ **in parallel do**
            if $FLIP(r+1,i)<i$ or $r=0$ or $FLIP(r,i)>i$
                then begin
                    $S[r,i] := LS[r,i] := S[r,FLIP(r+1,i)]$
                    $D[r,i] := LD[d,i] := D[r,FLIP(r+1,i)]$
                end
                else begin
                    $LS[r,i] := LS[r,FLIP(r+1,i)]$
                    $LD[r,i] := LD[r,FLIP(r+1,i)]$
                end

    end

Sorting on a Butterfly

Basic approach for merging: Since removing the bottom row of a butterfly divides it into two interleaved butterflies of half the size, passing data up from one row to the next corresponds to the DIVIDE operation and passing data down from one row to the next corresponds to the SHUFFLE operation. In addition, as data is passed down from one row to the next, the butterfly edges can be used to exchange adjacent pairs.

Modified *DIVIDE*: The only problem with this basic approach is that the butterfly edges on the bottom row allow even-odd pairs to be exchanged (0-1, 2-3, 3-4, etc.) whereas the odd-even merge sort algorithm needs to exchange odd-even pairs (1-2, 3-4, 5-6, etc.). Fortunately, exchanging even-odd pairs works if we modify *DIVIDE* so that the odds of the left half are merged with the evens of the right half and the evens of the left half are merged with the odds of the right half (the proof is similar to that for normal odd-even merging). Hence, by passing data on the left side of the butterfly up along the butterfly edges instead of the column edges, the roles of odd and even can be reversed.

The sorting algorithm: Because the butterfly is so well suited, we go directly to a non-recursive implementation. Since removing the top i rows of a butterfly leaves 2^i butterflies, we think of the unsorted data initially in Row 0 as being in n one vertex butterflies and then sort the data with a loop where r goes from Row 1 to Row $\log_2(n)$. On the rth iteration, we have $n/2^r$ butterflies (Rows $r+1$ through $\log_2(n)$ are not used) that each merge the data in the left and right halves of their row 0 by passing the data up (using butterfly edges on the left half and column edges on the right half) and then passing data back down (using butterfly edges to fix even-odd pairs that are out of order).

```
procedure ModifiedDivide(d,r)
    for 0 ≤ i < n in parallel do if d<r and the r^th bit (from the right) of i is 0
        then data[d,i] := data[d−1,FLIP(d,i)]
        else data[d,i] := data[d−1,i]
    end

procedure ShuffleFix(d)
    for 0 ≤ i < n in parallel do if FLIP(d,i)>i
        then data[d−1,i] := MIN{data[d,i], data[d,FLIP(d,i)]}
        else data[d,i] := MAX{data[d,i], data[d,FLIP(d,i)]}
    end

procedure butterflyMSORT
    for r:=1 to log₂(n) do begin
        for d=1 to r do ModifiedDivide(d,r)
        for d=r downto 1 do ShuffleFix(d)
    end
end
```

Complexity: $O(\log(n)^2)$ time on a butterfly with $O(n\log(n))$ processors.

CHAPTER 13

Example of Odd-Even Merge Sorting on the Butterfly

Starting with the unsorted list g, d, a, h, e, b, c, f in Row 0, the following three pairs of figures depict the butterfly for the values of $r := 1, 2, 3$ of the outer *for* loop, where the figure on the left shows how data is moved up by the first inner *for* loop and the figure on the right shows how it is moved back down by the second inner *for* loop:

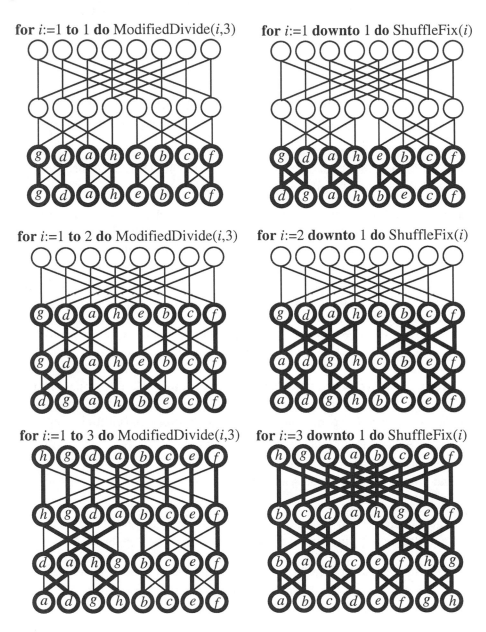

for i:=1 **to** 1 **do** ModifiedDivide(i,3)

for i:=1 **downto** 1 **do** ShuffleFix(i)

for i:=1 **to** 2 **do** ModifiedDivide(i,3)

for i:=2 **downto** 1 **do** ShuffleFix(i)

for i:=1 **to** 3 **do** ModifiedDivide(i,3)

for i:=3 **downto** 1 **do** ShuffleFix(i)

1-1 Packet Routing on a Butterfly

Problem: A subset of the n processors on Row 0 have a message ("packet") that needs to be sent to another processor on Row 0 (e.g., a data item or a request for a data item). A packet routing problem is *one-to-one* if each processor is the destination for at most one packet. As usual, we assume that in $O(1)$ time, only $O(1)$ data can be sent along an edge; so if many packets come into a vertex, they must be placed in a queue at that vertex to wait for a turn to be sent out along an outgoing edge.

Leftmost transitions:

In a hypercube, a *leftmost transition* for a processor i with respect to a processor t is a move from processor i to processor j where the address of processor j is the same as that of processor i except in the leftmost bit on which i and t differ.

For a hypercube of dimension k, the *left-right routing* of a packet from processor s to processor t is the path obtained by successively making leftmost transitions. Between any two vertices, the left-right routing uses at most k edges.

On a CCC or Butterfly of dimension k, left-right routings go from cycle to cycle or column to column corresponding to paths between vertices in the hypercube. Because transitions are leftmost, when a left-right routing comes into a cycle or column on one dimension, it goes out on an adjacent dimension. Bit positions that are initially identical correspond to traversing a portion of a cycle or column. On a standard (non-wrapped) butterfly, at most once during a left-right routing will it be necessary to traverse a column to go between Row k and Row 0. Hence, it is possible to show that between any two vertices the left-right routing has at most $2k$ edges for the CCC and butterfly and at most k edges for the wrapped butterfly (see the exercises).

Valiant–brebner one-to-one packet routing algorithm: With reasonable luck, if all packets follow a left-right routing, activity will spread evenly over the network. To eliminate dependence on the particular routing, a randomized algorithm is employed:

> **for** each processor that has a packet x **parallel begin**
>   Send x to a random processor on a left-right routing.
>   Send x to its destination on a left-right routing.
> **end**

Note: Packets are queued at vertices when necessary, with packets on their way to random destinations going ahead of packets on there way to their destinations.

Complexity: Independent of the partial routing in question, with probability less than $(0.74)^{\log_2(n)}$ that the Valiant–Brebner algorithm finishes in at most $8\log_2(n)$ steps on a hypercube or wrapped butterfly; references to its proof are in the chapter notes.

Mesh Network

Definition: A two dimensional mesh of height $h \geq 1$ and width $w \geq 1$ is a h by w array of processors, each connected only to the left, right, up, and down neighbors. For example, a mesh of height 5 and width 6 is shown below:

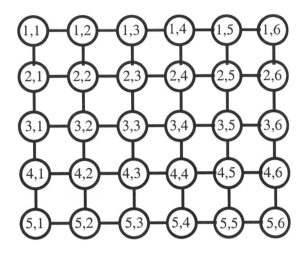

Notation, assumptions, and observations:

- For convenience we start indices of arrays and processors at 1 rather than at 0 as we did for the HC/CCC/BF because here the structure is more iterative and not based on powers of two.

- To refer to the processor in the ith row and j^{th} column we sometimes use *two-dimensional indexing* with the pair of indices $[i,j]$ and other times use *one-dimensional indexing* with the a single index $(i-1) * w+j$. For example, in the figure above, Processor [4,3] = Processor 21 because $(4-1)6+3 = 21$.

- Like the HC/CCC/BF network, we assume that each processor has local memory and that in $O(1)$ time, $O(1)$ data can be communicated along a single edge. Here it often takes longer to move data around because paths are $O(\sqrt{n})$ long.

- The mesh arrangement is much simpler than the HC/CCC/BF and our expectations for time are correspondingly lower. For $k \geq 1$, with HC/CCC/BF we typically seek algorithms that are time $O(log(n)^k)$; here we typically seek algorithms that are $O(\sqrt{n})$ or $O(\sqrt{n} \log(n)^k)$.

Example: Broadcast on a Mesh

Notation: Given a mesh of height h and width w, $n=hw$, data items $A[0]...A[n-1]$ are stored in row major order (using one-dimensional indexing).

Method 1, simulate a tree: For simplicity we assume a square mesh where $h=w$ and let $m = \lceil \sqrt{n} \rceil$. Just as the PRAM algorithm presented earlier simulated a full tree, a mesh of $O(n)$ vertices can simulate a full tree of n vertices using a recursive pattern of H's ("H-tree"), as shown below for $n=16$. However, even though there are only $O(\log(n))$ levels to the tree, the communication paths are up to $\Omega(m)$ long, and hence the computation is $\Omega(m)$. This method actually uses a grid of size $2m$ by $2m$; in $O(m)$ time the data can first be spread out by a factor of 2; the construction can be generalized to work in $O(\text{MAX}\{h,w\})$ time on a non-square (see the exercises).

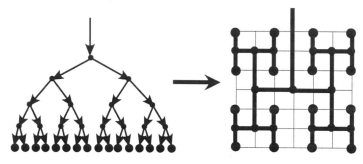

Method 2, horizontal and vertical flow: On a grid of size h by w, in $O(h+w)$ time we can propagate the sum along the rows in parallel back to the first column and then up the first column to processor $(1,1)$:

> **for** $i:=2$ **to** h **do** $A[i,1] := A[i-1,1]$
> **for** $i:=2$ **to** w **do**
>   **for** $1 \le j \le m$ **in parallel do** $A[i,j] := A[i,j-1]$

Example: Sum on a Mesh

Like the broadcast algorithm, with the arrows going in the reverse direction.

Example: Prefix-Sum on a Mesh

Problem: Given a mesh of height h and width w, $n=hw$, data items $A[0]...A[n-1]$ are stored in row major order (using one-dimensional indexing), and we wish to compute in each Processor i, $0 \leq i < n$, the sum of $A[0]$ through $A[i]$.

Idea: The prefix sum of a single row or column of the mesh can be computed in essentially the same way as summing a row or a column. The only difference is that as the sum propagates across the row, each processor saves it as it passes by. After the prefix sums have been computed for each row, they can be added to the sum of all the rows before them. We use the following arrays:

$A[i,j] =$ The value of A stored at Processor $[i,j]$; $A[i,j] = A[(i-1)*w+j]$.

$X[i,j] =$ The sum for Row i of A; $X[i,j] = A[i,1]+...+A[i,w]$.

$SR[i,j] =$ The prefix sums for Row i of A; $SR[i,j] = A[i,1]+...+A[i,j]$.

$SC[i,j] =$ The prefix sums for Column j of X; $SC[i,j] = X[i,1]+...+X[i,j]$.

$S[i,j] =$ The prefix sums for A, the sum of all $A[x,y]$'s such that $x<i$ or $x=i$ & $y \leq j$.

Mesh prefix sum algorithm:

1. Perform a prefix-sum in SR of the A's for each row:
   for $1 \leq i \leq h$ **in parallel do begin**
      $SR[i,1] := A[i,1]$
      for $j:=2$ **to** w **do** $SR[i,j] := SR[i,j-1] + A[i,j]$
   end

2. Broadcast in X the sum of the entire row across each row:
   for $1 \leq i \leq h$ **in parallel do begin**
      $X[i,w] := SR[i,w]$
      for $j:=w-1$ **downto** 1 **do** $X[i,j] := X[i,j+1]$
   end

3. Perform a prefix-sum in SC of the X's for each column:
   for $1 \leq j \leq w$ **in parallel do begin**
      $SC[1,j] := X[1,j]$
      for $j:=2$ **to** h **do** $SC[i,j] := SC[i-1,j]+X[i,j]$
   end

4. Compute the prefix sums from the adjusted value of SC (subtract X) and SR:
   for $1 \leq i \leq h$, $1 \leq j \leq w$ **in parallel do**
      $S[i,j] := SC[i,j] - X[i,j] + SR[i,j]$

Complexity: Steps 1 and 2 are $O(w)$, Step 3 is $O(h)$, and Step 4 is $O(1)$, for a total of $O(h+w)$ time. $O(1)$ space per processor is used.

Example: Matrix Multiplication on a Mesh

Idea: To compute the matrix product $C=A*B$, where A is a m by w matrix and B is a w by n matrix, use a mesh of m by n processors. The computation begins with $C[i,k]$ initialized to 0 in processor $[i,k]$, $1 \leq i \leq m$ and $1 \leq k \leq n$. The rows of A move horizontally by the columns of B that move vertically so that all terms for an entry in $C[i,j]$ meet at the processor $[i,j]$, and the computation ends with the final value of C in the mesh. The following figure depicts the computation when A has dimension 3 by 5 and B has dimension 5 by 4.

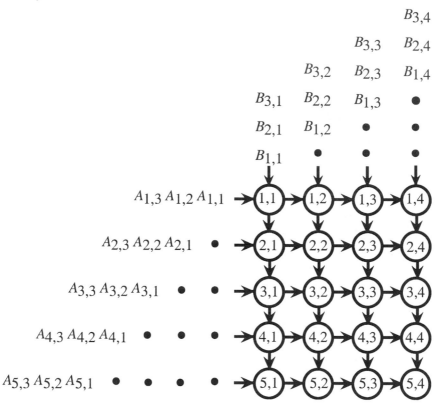

Complexity: The computation ends when in Processor $[m,n]$, $A[m,w]$ is multiplied by $B[w,n]$ and added to $C[m,n]$. If we think of the data as residing in successive positions outside the mesh before it enters, $A[m,w]$ visits $w-1$ positions initially occupied by entries of A, $n-1$ "dots", and the m processors of its row, for a total of $m+w-n-2$ steps, and similarly $B[w,n]$ visits $w-1$ positions initially occupied by entries of B, $m-1$ "dots", and the n processors of its column, also for a total of $m+w+n-2$ steps. A total of mn processors are used.

Mesh Multiplication Implementation Details

Data sequencing: Rather than attempting to model the staging of data outside the mesh, we assume there is an extra Column 0 on the left, where Processor $[i,0]$ holds the entries of Row i of A, and an extra Row 0 on the top, where Processor $[0,k]$ holds the entries of Column k of B. Alternately, if we wish to limit the space per processor to $O(1)$, we can use a $w+m$ by $w+n$ mesh with w additional columns to store the rows of A and w additional rows to store the columns of B (see the exercises).

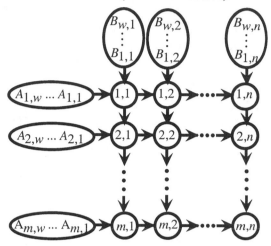

Note that A and B can be stored in a mesh in many ways and still be able to move to the edges in $O(m+w+n)$ time in order to begin the computation (see the exercises).

procedure meshMATRIXMULTIPLY

for $1 \le i \le m$, $1 \le k \le n$ **in parallel do** $X[i,k]:=Y[i,k]:=C[i,k]:=0$

for $t:=1$ **to** $m+w+n-2$ **do begin**

    for $1 \le i \le m$, $0 \le k \le n$ **in parallel do begin**

        if $k=0$ **then begin if** $i \le t \le (w+i-1)$ **then** pass $A[i,t]$ right **else** pass 0 right **end**

        else if $1 \le k < n$ **then** pass $X[i,k]$ right and receive a new value from the left

        else receive a new value of $X[i,n]$ from the left

        end

    for $0 \le i \le m$, $1 \le k \le n$ **in parallel do begin**

        if $i=0$ **then begin if** $i \le t \le (w+i-1)$ **then** pass $B[t,i]$ down **else** pass 0 down **end**

        else if $1 \le i < m$ **then** pass $Y[i,k]$ down and receive a new value from above

        else receive a new value of $Y[m,i]$ from above

        end

    end

end

Data Distribution on a Mesh

Notation: For a mesh of height h and width w, $n=hw$, we use one-dimensional indexing of Processors 1 through n. For $1 \leq i \leq n$ if Processor x is a sender and Processor y is the next highest numbered sender (or $y=n$ if Processor x is the highest numbered sender), then the data item stored at Processor x needs to be copied to vertices x through y.

Idea: Similar to our presentation for the EREW PRAM, a broadcast is initiated from each sender, and it is not necessary for a receiver to know the index of its sender; we keep track of the highest numbered sender from which data has been received thus far in a second array S. To "send" data to the right (sending down is similar), we can do:

> **procedure** SendRight
> > **for** i:=1 **to** w **do**
> > > **for** $1 \leq i \leq h$, $1 \leq j < w$ **in parallel do**
> > > > **if** $S[i,j] \geq S[i,j+1]$ **then begin** $S[i,j+1]:=S[i,j]$; $A[i,j+1]:=A[i,j]$ **end**
> > **end**

The distribution algorithm: We must be careful to address the fact that in a standard mesh, there is no "wrap-around" connection from Processor $[i,w]$ to Processor $[i+1,1]$).

1. Send data to the right for w steps.

2. In parallel, each Processor $[i,w]$, $1 \leq i < h$, in the last column sends a "message" across the mesh (in $O(w)$ time) containing $S[i,w]$ and $A[i,w]$ to Processor $[i+1,1]$, which replaces its values for these if $S[i+1,1]<S[i,w]$.

3. Send data to the right for w steps.

4. Send data down for h steps.

Complexity: $O(h+w) = O(\sqrt{n})$ for a square mesh of n vertices.

Example: The 8 by 12 mesh below has five groups shown by the different shadings. The leader of each group is marked with an X. The numbers on the followers of each group are the step of the algorithm that will first place the correct leader value in that position.

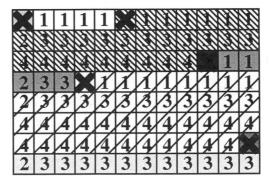

Sorting on a Mesh

Idea: A variation of odd-even merge sort where the mesh is divided into four quadrants.

Sorting: The quadrants are recursively sorted, two merges combine pairs of quadrants into halves, and an additional merge combines the two halves.

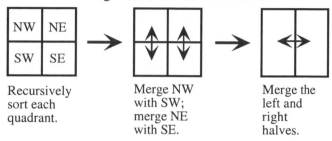

| Recursively sort each quadrant. | Merge NW with SW; merge NE with SE. | Merge the left and right halves. |

Odd-even merging: We describe how to merge two square blocks; the merging of two rectangular blocks is similar.

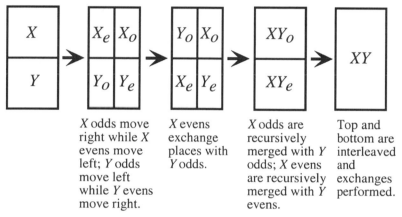

| X odds move right while X evens move left; Y odds move left while Y evens move right. | X evens exchange places with Y odds. | X odds are recursively merged with Y odds; X evens are recursively merged with Y evens. | Top and bottom are interleaved and exchanges performed. |

Complexity: Since recursive calls are done in parallel and the movement of elements at each step takes $O(\sqrt{n})$ time, for some constant a, the time to merge is given by:

$$M(n) \le M(n/2) + a\sqrt{n}$$
$$= a\sqrt{n} + a\sqrt{n/2} + a\sqrt{n/4} + \cdots$$
$$= a\sqrt{n}\left(\text{geometric sum with } r = 1/\sqrt{2}\right)$$
$$= O(\sqrt{n})$$

Hence, for a constant b, the time to sort is given by $S(n) \le S(n/4) + b\sqrt{n} = O(\sqrt{n})$.

Odd values of n and non-square meshes: The complexity analysis generalizes to $O(h+w)$ for arbitrary values of mesh height and width (see the exercises).

1-1 Packet Routing on a Mesh

Problem: A subset of the processors has a message ("packet") that needs to be sent to another processor (e.g., a data item or a request for a data item). A packet routing problem is *one-to-one* if each processor is the destination for at most one packet. As usual, we assume that in $O(1)$ time, only $O(1)$ data can be sent along an edge; so if many packets come into a vertex, they must be placed in a queue at that vertex to wait for a turn to be sent out along an outgoing edge.

Idea: First route packets vertically to form queues at each vertex of packets that are going left or right on that row. Second, flush out each row.

Vertical–horizontal algorithm:

Phase 1. All columns work in parallel. In a given column, at each step, all packets that are not in their destination row simultaneously move up or down one row closer to their destination row. After h steps, each processor has a left queue of packets destined for processors to the left in its row and a right queue of packets destined for processors to the right in its row. The sum of all queue sizes on any given row is at most w since the packet routing is 1-1.

Phase 2. All rows work in parallel. In a given row, at each time unit, all processors with non-empty left queues remove an item and send it to the left and all processors with non-empty right queues remove an item and send it to the right (when an item is received from by a Processor $[i,j]$, it is accepted if $[i,j]$ is its destination, otherwise it is added to the left or right queue (according to which direction it is going) of Processor $[i,j]$).

Complexity:

Phase 1 is $O(h)$ since there are h steps, each taking $O(1)$ time.

Let us now consider Phase 2. We limit our attention to items moving to the right in a particular row i; all rows work in parallel in the same way, and the movements of items going to the left works the same way as ones to the right. At any point in time, define the "tail processor" to be the rightmost processor that has a queue of more than one element. Since queue sizes only go down (for any queue, at each step at least one element is removed and at most one element is added), the tail processor stays the same or moves left at each step. In addition, at each step the total number of elements enqueued in the tail processor and in processors to its left goes down by one (since at each step the tail processor removes an element from its queue and sends it to the right). Hence, after at most w steps there is no longer a tail processor, and then after at most w additional steps, all items in the row will have reached their destinations. Hence, Phase 2 is $O(w)$.

Thus, the entire algorithm is $O(h+w)$.

Area-Time Tradeoffs

Computer Chips

Although many kinds of devices have been considered (optical, quantum, etc.), the standard 2-dimensional computer "chip" (and the 2-dimensional arrangement of chips on boards) is a cornerstone of computer technology. Most standard chip technologies can be modeled at a basic level as:

- There are $O(1)$ *layers* (in practice, at least 3 or 4).

- Wires on the same layer may not cross and may not come too close to each other. Here for simplicity, we assume that there is a single constant λ that is the thickness of all wires and the minimum distance between wires.

- A connection may be placed where wires on different layers cross.

- Some layers can be designated as "transistor" layers. Depending on the technology used, when current flows through a wire on a transistor layer, it can cause or prevent current to flow through a wire it crosses on a different transistor layer.

- Some layers are metal layers. Current flows faster in metal layers and has no effect when it crosses a wire on another layer.

  In practice, how much and for what distance metal wires on one layer can lie on top of wires on another layer is technology dependent (electrical effects such as capacitive coupling can reduce speed, etc.).

Boolean Functions

Any Boolean function (0-1 inputs to a 0-1 output) can be constructed from:

> AND: x AND y is 1 if both x and y are 1 and 0 otherwise.
>
> OR: x OR y is 1 if either x or y is 1 and 0 otherwise.
>
> NOT: NOT x is 1 if $x=0$ and 0 if $x=1$.

In addition, any Boolean function can be constructed from either of the following two functions (see the exercises):

> NAND: x NAND y = NOT $(x$ AND $y)$
>
> NOR: x NOR y = NOT $(x$ OR $y)$

Example: cMOS

A simple chip technology used in the 1980's and 1990's is *single-metal cMOS* where there is one layer of metal which can be used for general routing of signals and special layers, which in addition to being able to route signals (although the routing is not as fast as in the metal), can support two types of switches between special layers:

> *Negative switch layer* ("n-type" transistors): When connected to ground, a wire that it crosses on a designated layer below is effectively cut and cannot conduct current across that boundary; with positive voltage, the wire it crosses is connected.

> *Positive switch layer* ("p-type" transistors): Works in the opposite fashion to the negative switch layer.

In the figure below, x and y denote 1-bit inputs, V denotes positive voltage (i.e., a 1), G denotes ground (i.e., a 0), thick lines denote the metal layer, thin lines denote a special layer that can be switched by the dashed lines, where thick light-colored dashes are the negative switch layer and thin dark-colored dashes are the positive switch layer; the boxes are connections between layers.

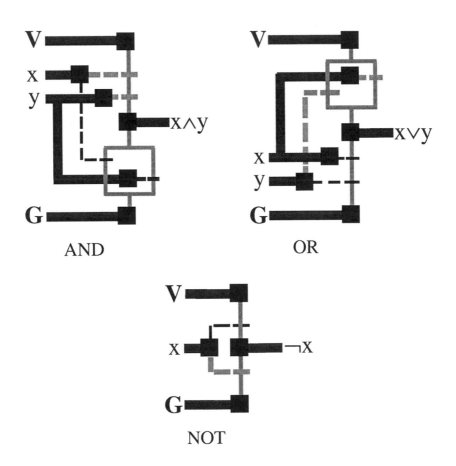

AND   OR

NOT

Example: nMOS

To avoid a short between positive voltage and ground, cMOS uses two complementary copies of a circuit, one that connects or disconnects a wire to positive voltage and the other that disconnects or connects it to ground. *nMOS* uses only the n-type transistor and instead of using complementary circuitry an, area of the chip, called a *pull-up (PU)*, can be treated to act like a resistor. The constant power drain through the pull-ups makes nMOS technology consume more power, and nMOS has been more or less abandoned in favor of cMOS. However, at least for academic exercises, it is simpler to draw small circuits in nMOS since there is in some sense half the circuitry.

Below are nMOS *NAND* and *NOR* circuits. For *NAND*, when x and y are 1 all current coming from the pull-up goes to ground, and 0 is output; but if either x or y is 0, the path to ground is cut, and current from the pull-up produces output 1. *NOR* works similarly.

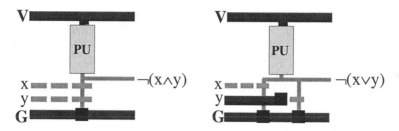

For a more complex example, below is a 1-bit nMOS adder. A carry-in bit comes into I and two bits come into x and y. The left bit of $I+x+y$ (the carry-out) is produced at O at the right bit at S. Using the identities $O = IX \lor IY \lor XY$ and $S = (\neg O)(I \lor x \lor y) \lor Ixy$, the left third (two pull-ups) computes O, the middle third (one pull-up) computes S, and the right third (two pull-ups) is one bit of memory to maintain S when $A=0$. Copies of a circuit like this can be "stacked" vertically to add any number of bits.

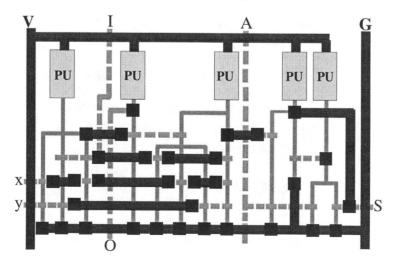

Constructing Memory With Chips

1-bit registers: A bit of memory can be constructed in $O(1)$ area using standard circuitry with a feedback loop. In the cMOS circuit shown in the figure below, when the save signal S is 0, the saved output data bit DO is locked in a positive or negative loop on the right; when $L=1$, the loops on the right are cut and an input data bit DI is let in.

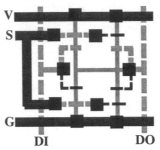

K-Bit registers: The 1-bit register shown in the figure above can form a cross section that can be replicated horizontally to store $k \geq 1$ bits in an $O(1)$ by $O(k)$ area. The circuit can be "folded" into an $O\left(\sqrt{k}\right)$ by $O\left(\sqrt{k}\right)$ area (see the exercises).

Addressable k-Bit registers: To maintain a collection of registers where only one at a time is used, a 1 can be sent down a vertical (AV) and horizontal (AH) line to activate the register where they meet, so that only it will respond to a load signal or return saved bits.

The cMOS figure below shows an enhanced version of the 1-bit storage register. The left portion AND's together AH, AV, and L so that new data is loaded only when they are all 1, the middle is a copy of the original circuit, and the right portion connects the output to the input as long as $AH=AV=1$ and $S=0$ (so a single line can both save and retrieve a bit).

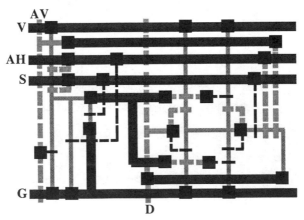

Address Selection

A regular pattern can be "programmed" to select a particular address. In the cMOS figure below, the lines $A1$, $A2$, $A3$ carry a 3-bit address; each circuit has two complementary parts connected to V and G so that the output is always connected to 1 if $A1$, $A2$, $A3$ are the specified address, and 0 otherwise.

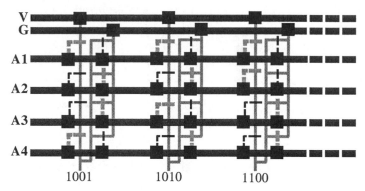

RAM Memory

Half the address bits can be sent vertically and half horizontally. In the cMOS figure below, two data bits ($D1$, $D2$) can be stored ($S=1$) or retrieved ($S=0$) from any of the 16 two-bit registers, specified by the address $A1$, $A2$, $A3$, $A4$ (power and ground to the registers is not shown). In general, the area for the kn bits is $O(kn)$, since only $O\left(k\sqrt{n}\log(n)\right)$ additional area is used for address selection (see the exercises). In practice, more specialized hardware would yield a smaller constant.

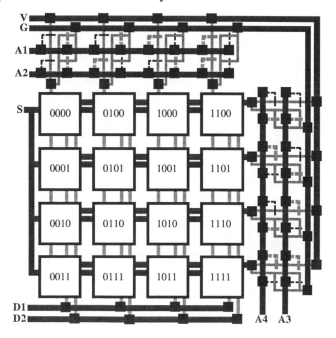

Computing with Chips

Idea: Arbitrary computations can be performed with sufficiently large logic circuits that have feedback from the outputs to the inputs that is controlled by a system clock.

Clock: For simplicity we assume a *synchronous* system controlled by a *clock* that provides a signal alternating between positive voltage and ground. In fact, it is convenient to have a *two-phase non-overlapping clock* that provides a pair of signals, ϕ_1 and ϕ_2, where only one is positive voltage at any given time.

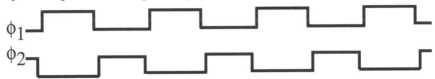

Processors: Although in practice, the design of a good processor is a major effort, the general idea is relatively straightforward once we have the ability to construct circuits for arbitrary logical functions, build memory, and control everything with a clock. A program (e.g., very low-level machine instructions) can be stored in memory, the current position in the program can be stored in a *program-counter* register, circuits can be built for each basic command (addition, logical comparisons, goto, etc.), and a circuit can be built for the flow of control (getting the next instruction, invoking the appropriate circuit to perform the command, and updating the program counter). The duration of a clock signal is made long enough to allow all computation for a step to be completed. Each time the clock "ticks", memory and computed values (such as the program counter) are held constant and are allowed to loop back to the inputs. In a simplistic computing system (serial or parallel), the clock signals ϕ_1 and ϕ_2 can perform the ticking in a lock-step fashion. When $\phi_1=1$ and $\phi_2=0$, new data is prevented from entering the buffer while new data is computed (based on the buffer and the input) and sent to the memory and the output. When $\phi_1=0$ and $\phi_2=1$, new data is prevented from entering the memory and items that are needed for the next step of the computation are copied from memory into the buffer. In a large parallel system, control may be much more complicated with local clock signals controlling different parts of the computation and communication protocols in place to go between parts.

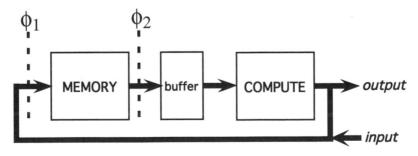

CHAPTER 13

Parallel Hardware Layout

All of our discussion to follow can be adapted to the simplest models like the single-metal cMOS described earlier, where pins may be arbitrarily positioned around the chip (see the exercises); but for convenience we assume that a parallel computer is built as follows:

- There is a basic grid of height h and width w applied to all layers that defines the notion of area. A grid square is one unit of area; the *area* of the chip is $A = hw$.

- *Pins* on to which inputs and outputs can be attached may be placed anywhere on the grid or around its perimeter.

- A *computation* consists of "reading" one value on each input pin and "writing" one value on each output pin (values of k bits take k steps to read or write).

- *Wires* must follow grid lines. Their actual thickness is irrelevant for this discussion; we assume only that wires are thin enough so that two parallel wires on adjacent grid lines do not touch (in practice, long parallel wires that are close together might have problems with capacitive coupling — see the exercises).

- On the bottom are $O(1)$ layers that supply basic hardware functions such as *power*, *ground*, and *clock* to the entire chip.

- Next there are $O(1)$ layers for the *processors* themselves; each processor uses a p by q area, which includes the local memory used by a processor (we assume that one bit of memory can be realized in $O(1)$ area — see the exercises).

- Next there is a "vertical" layer and a "horizontal" layer that are used for communication of information between processors in the vertical and horizontal directions. Where wires on the horizontal and vertical layers cross or lie on top of each other, a connection between them may be placed that takes 1 unit of area. At most one bit can be sent along a wire in one time unit.

Example: The figure below shows on the left a butterfly of 4 columns and on the right a circuit layout. Rectangles are the processors (one for each of the 12 butterfly vertices), the lines are wires in the vertical and horizontal layer, the dark squares are connections to the processors, and the dark circles are connections between the vertical and horizontal layers. Note that in practice, each of these wires might be a number of parallel wires so that a number of bits can be sent along that communication bit in a given time step.

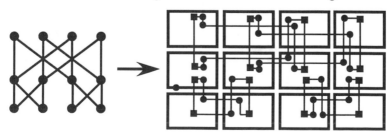

Area-Time Tradeoff for Sorting

Idea: For many problems such as sorting you cannot have it both ways. If a chip computes the problem very fast by employing a high degree of parallelism, then it must have a large area due to the inherent need for complex patterns of wire to interconnect processors. If the chip is very compact, there is insufficient room for interconnection wire required for the parallelism needed for a fast solution, and it must run relatively slow.

Simplified sorting problem: The input is the integers 0 through $n-1$, each represented with $\lceil \log_2(n) \rceil$ bits. The sort is always the same (the integers 0 through $n-1$). The problem is to output the *indices* of the sorted values. For example, for $n = 4$, if the input sequence is 2, 3, 1, 0, which is represented as 10, 11, 01, 00, then the output sequence is 3, 2, 0, 1, which is represented as 11, 10, 00, 01.

Restricted chip layouts: To begin with, we present the area-time tradeoff for sorting with respect to a restricted type of chip layout:

- The chip is square.

- Each of the n inputs arrives on a distinct pin on the left edge; there are exactly n input pins and it takes at least $\lceil \log_2(n) \rceil$ steps to read the input (since each pin can only read one of its $\lceil \log_2(n) \rceil$ bits at a time).

- Similarly, each output is produced on a distinct pin on the right edge.

Area-time tradeoff for sorting on a restricted chip layout: Imagine a vertical line dividing a chip:

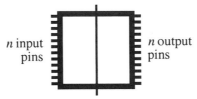

Even with the simplified sorting problem, a great deal of information must cross this line. Since the set of possible outputs consists of all 2^n possible permutations of the indices 0 to $n-1$, if computations always communicated $<n$ bits across the line, then two different inputs must produce the same output (and the chip cannot work correctly on all inputs).

Hence, there is at least one input for which the chip sends $\geq n$ bits across the line.

But since only one bit can move across a wire in one time unit and there is at most one wire per grid line, if the chip has area A, at most $\sqrt{A}$ bits can be communicated across the line in one time unit. Hence $T\sqrt{A} \geq n$ which implies $AT^2 \geq n^2$.

Generalizing the Area-time Tradeoff for Sorting

We generalize the restricted sorting area-time $AT^2 \geq n^2$ tradeoff to allow:
- Chips of height h and width w, $h, w \geq 1$.
- Inputs to arrive and outputs to be produced at arbitrary grid positions.
- The number of input and / or output pins is not necessarily n. More than one input can be read or output produced at a pin and bits of a particular value can be input or output at different pins. A particular ordering as to how the input is read and output produced is be specified with the problem (e.g., pin 1 reads the bits of the second input value, then it writes the tenth bit of the seventh output value, then it reads the fourth bit of the sixth input value, pin 2 ...).

Idea: Instead of a "perfect" partition of the chip, we find a line that has at least 1/3 of the outputs on each side (and one of the sides must have at least half of the inputs); that is, we allow at most twice as many outputs on one side of the line as the other.

Generalized partitioning construction: Assume $h \leq w$; otherwise rotate the chip. Suppose that $n/3$ of the output bits are produced on a single pin. Then $T \geq n/3$ and hence $AT^2 \geq n^2/9 = \Omega(n^2)$. Otherwise, consider a vertical line that may have at most one 1-unit "jog" in it that partitions the chip into left and right halves.

We consider the *least significant bits* (LS bits), those input and output bits that are the rightmost bits in their respective values. Consider first starting with a simple vertical line with no jog and sliding it from left to right (first between column 1 and column 2, then between column 2 and column 3, and so on) to the first position where there are $\geq n/3$ least significant bits that are output to the left. If there are $\geq n/3$ LS bits that are output to the right, we stop. Otherwise, there must be $\geq n/3$ LS bits that are output along the column just to the left of the line. Since, we are assuming that no single pin outputs $> n/3$ bits, we can move the jog down the line to place enough of this column in the right half so that $\geq n/3$ LS output bits are produced in both halves. Assume that $\geq n/3$ of the LS input bits arrive in the left half; otherwise rotate the chip (in fact, one side must have at least half the LS input bits). Let X denote the $n/3$ of the LS input bits arriving on the left half and Y the $n/3$ of the LS output bits produced on the right half. No matter what LS output bits are in Y, the values of input bits not in S can be chosen to force $Y=X$ (see the exercises). By similar reasoning as before, $wT \geq n/3$, and hence $AT^2 > n^2/9 = \Omega(n^2)$.

Sorting Area-Time Tradeoff vs. PRAM Simulation

Since a mesh of n processors can be placed on a chip of area $O(n)$, meshes are in some sense as area efficient as we can hope for. However even simulation on square meshes gives an $O(\sqrt{n})$ slow-down. In contrast the HC/CCC/BF network has only an $O(\log(n))$ slowdown but since it has the capability of sorting in $O(\log(n))$ time, the AT^2 bound says that $A = \Omega((n/\log(n))^2)$. That is, since the processors use only $O(n)$ area (assuming that each processor uses $O(1)$ area), asymptotically, the area used by the processors becomes a vanishing fraction of the almost quadratic area consumed by the interconnect wire.

Explicit recognition of the area-time tradeoff: Since the area-time tradeoff lower bound says that we must resign ourselves to nearly quadratic area for the wire if we want very fast (poly-log) PRAM simulation, networks have been proposed that explicitly allocate area for the bandwidth needed. For example, *fat trees*, which are based on the H-tree layout, use exponentially more wire as one moves up towards the root (and a switch that controls $O(k)$ wires uses $O(k^2)$ area). Although the root can be viewed as a bottleneck (all routing from any vertices in the left subtree to vertices in the right sub-tree must go through it), more bandwith has been explicitly provided to compensate. Fat trees enjoy similar PRAM simulation properties to the HC/CCC/BF and have been the basis of some parallel computer designs (see the chapter notes).

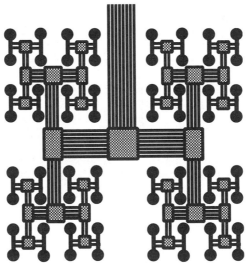

Generalizations of the Sorting Area-Time Tradeoff

> **Other Functions:** Using the notion of the inherent *information content*, I, of the problem (the amount of information that must flow from the inputs to the outputs in the worst case), $AT^2 = \Omega(I^2)$ follows; see the exercises.

> **3-Dimensions:** For 3-dimensions, a similar arguments show that $V^2T^3 = \Omega(I^3)$, and hence 3-dimensional layouts for the butterfly $V = \Omega((n/\log(n))^{3/2})$; see the exercises.

Exercises

335. Clearly the function erewSUM could be modified to perform other operations (e.g., the maximum of n elements). Describe this class of operations.

336. As presented the algorithm *erewSUM* to sum an array $A[0]...A[n-1]$ overwrites A. If the values in A will be needed later, they can be copied to an array $B[0]...B[n-1]$ first and then erewSUM can overwrite B, for a total of n additional memory locations used in addition to those used for A. However, it is possible do a bit better. Give pseudo-code that uses only $n/2$ locations in addition to those used by A by making the first iteration place the sums of pairs of elements of A in an array B of size $n/2$ and then overwriting B.

337. For a list of arrays $A_1, A_2, ..., A_m$ where each has length n give pseudo-code to compute the array $SUM[1]...SUM[m]$ where $SUM[i]$ is the sum of the n elements in array A_i. The naive approach is to re-use the *erewSUM* algorithm with n processors for each array for a total of $O(m\log(n))$ time. Your algorithm should use $O(n)$ processors and $O(m+\log(n))$ time and work for any value of n (not necessarily a power of 2).

Hint: Use the full tree configuration of processors and "pipe" the arrays in; that is on the first step, A_1 goes into the leaves, on the second step the sums of adjacent pairs moves up and A_2 goes into the leaves, and so on.

338. Broadcast on an EREW PRAM is a special case of data distribution. Give pseudo-code for broadcasting a value x to all positions of an array $A[0]...A[n-1]$ that uses $O(1)$ time in addition to the $O(\log(n))$ time for a single call to perform data distribution.

339. Explain why the following procedure computes the prefix-sum of a list and why it obeys the rules of the EREW PRAM model:

procedure pramListPrefixSum(L)
    for $0 \le i < n$ **in parallel do if** $NEXT[i]=nil$ **then** $x := i$
    for $0 \le i < n$ **in parallel do if** $NEXT[i]\ne nil$ **then** $NEXT[NEXT[i]] := i$
    $NEXT[first] := nil$
    $first := x$
    erewListSuffixSum(L)
end

340. The presentation of prefix-sum assumed that the values to be summed were ≥ 0.

    A. Modify the algorithm for list prefix-sum to work for arbitrary values.

    B. Modify the code for prefix-sum on the leaves of a binary tree to work for arbitrary values.

341. Recall that odd-even bubble sort worked by exchanging disjoint pairs on each iteration of the main *for* loop:

> **for** i:=1 **to** n **do begin**
>   **for** i:=1 **to** n **by** 2 **if** $A[i]>A[i+1]$ **then** exchange $A[i]$ and $A[i+1]$
>   **for** i:=2 **to** n **by** 2 **if** $A[i]>A[i+1]$ **then** exchange $A[i]$ and $A[i+1]$
>   **end**

- A. Give pseudo-code for the pram model that runs in $O(n)$ time with $O(n)$ processors.
- B. Explain why Brent's Lemma is no help here.

342. Give the details of the pseudo-code for the use of prefix-sum in the 3-partitioning procedure used by *erewQSORT* for parallel quick sorting. Formulate your presentation so that you set up the prefix-sum problems and then just simply call a prefix-sum procedure.

343. For simple summing of an array with $O(n)$ processors in $O(\log(n))$ time, Brent's lemma says that we can reduce the number of processors to $O(n/\log(n))$ and still perform the computation in $O(\log(n))$ time.

- A. Explain why Brent's lemma does not help with prefix-sum.
- B. Given Part A, explain why Brent's Lemma does not help us reduce the number of processors used by the procedure *erewQSORT* for parallel quick sorting.

344. Recall from the exercises to the first chapter that *displacement sort* worked, for a given element $A[i]$, by counting the number of elements that were before $A[i]$ but greater than $A[i]$ and the number that were after $A[i]$ but less than $A[i]$, and then adjusting positions accordingly.

> **for** $1 \le i \le n$ **do begin**
>   *count*$[i]$:= 0
>   **for** j:=1 **to** $i-1$ **do if** $A[j]>A[i]$ **then** *count*$[i]$:= *count*$[i]-1$
>   **for** j:=$i+1$ **to** n **do if** $A[j]<A[i]$ **then** *count*$[i]$:= *count*$[i]+1$
>   **end**
> **for** i:=1 **to** n **do** $B[i+count[i]]$:= $A[i]$

- A. Write pseudo-code for the procedure *pramDSORT* that implements displacement sort on the pram model in $O(\log(n))$ time with $O(n^2)$ processors.
- B. Apply Brent's lemma to reduce the number of processors to $O(n^2/\log(n))$.
- C. If we compare *pramDSORT* to the procedure *pramMSORT* that was presented for the pram implementation of merge sort, both are based on the idea of counting. However, *pramMSORT* uses the square of the time used by *pramDSORT* and *pramDSORT* uses the square of the number of processors use by *pramMSORT* (or slightly less in the case of Part B). Discuss which of these two algorithms would likely be more practical under reasonable assumptions.

D. The exercises for the first chapter also presented *position sort*, which calculated absolute rather than relative positions. Write pseudo-code for the procedure *pramPSORT* that implements position sort on the pram model in $O(\log(n))$ time with $O(n^2)$ processors and compare its practical performance to that of *pramDSORT* under reasonable assumptions.

345. We have presented pseudo-code for broadcasting a value to all positions of an array and have observed that summing an array is just the "reverse" of this process.

A. For an EREW PRAM, give two versions of pseudo-code for summing an array that correspond to the two versions given for broadcasting.

B. For a butterfly, give two versions of pseudo-code for summing an array that correspond to the two versions given for broadcasting.

346. The algorithms for broadcasting summing for the EREW PRAM worked by successively doubling the distance where data was sent, and all included a check so that the only positions that are "active" are the ones corresponding to the current power of two. For example, the algorithm for broadcasting on the EREW PRAM included the check. Although this check might be useful in practice to reduce unnecessary communication between processors, consider how it may be eliminated:

A. Explain why broadcasting still works without this test.

B. Explain why summing still works without this test.

C. Explain why the algorithm presented for data distribution of the EREW PRAM did not include this test. Is there anything that can be done to reduce the amount of "communication" used by this algorithm?

347. Recall Strassen's algorithm for matrix multiplication. For simplicity, assume that the problem is to multiply two n by n matrices X and Y, where n is a power of 2. An approach for a parallel implementation is to divide X into four sub-matrices A, B, C, D of size $n/2$ by $n/2$, divide Y into four sub-matrices E, F, G, H of size $n/2$ by $n/2$, and compute XY by manipulation sub-matrices in the same fashion as multiplying 2 by 2 matrices (where recursive calls to multiply sub-matrices are done in parallel). Furthermore, to make this approach work on the EREW PRAM, let A_c, B_c, ..., H_c denote copies of A, B, ..., H and compute:

$$XY = \begin{pmatrix} A & B \\ C & D \end{pmatrix}\begin{pmatrix} E & F \\ G & H \end{pmatrix} = \begin{pmatrix} AE + B_cG_c & A_cF_c + BH \\ C_cE_c + DG & CF + D_cH_c \end{pmatrix}$$

A. Present pseudo-code that in parallel makes the copies A_c, B_c, ..., H_c, in parallel computes the 8 multiplications of sub-matrices of size $n/2$ by $n/2$, and in parallel does the 4 additions of sub-matrices of size $n/2$ by $n/2$. That is, unlike Strassen's algorithm, here we do not attempt to use only 7 multiplications of sub-matrices; we just do the straightforward method.

B. Explain why the time is $O(\log(n))$ and the number of processors used is $O(n^3)$, by analyzing the appropriate recurrence relations.

C. Although for serial computing Strassen's algorithm relies critically on doing only 7 recursive calls for multiplications of sub-matrices to improve the asymptotic time to less than $O(n^3)$, explain why here using only 7 recursive calls would not improve the asymptotic time. However, explain why only 7 recursive calls would reduce the asymptotic bound on the number of processors.

D. Explain why this algorithm works on the EREW PRAM, but would not if the copies were not used.

348. Under the assumption that in $O(1)$ time, a processor can communicate with only $O(1)$ other processors, show that $O(\log(n))$ time is the best that can be hoped for to sum an array of n integers.

Hint: Since each element of the array can affect the outcome, consider the number of steps necessary for a single processor to gather *any* information about all inputs.

349. Consider the *database yes-no search problem* that was presented in the introduction. Using Convention 1 for CRCW concurrent writes listed in the introduction, describe what happens when there are no copies, 1 copy, or 2 or more copies.

350. Consider the *database yes-no search problem* that was presented in the introduction. Suppose that instead of a yes-no answer, we wanted the algorithm to answer 0 if there is no match or i if there is an i such that $A[i]=x$ (if there is more than one such i, any one of them can be returned). For each of the 10 conventions for CRCW concurrent writes listed in the introduction, explain why the algorithm presented in the introduction for the yes-no problem can or cannot be easily modified to solve this problem.

351. Consider the randomized algorithm presented in the introduction for the *duplicate database yes-no search problem*.

A. If $A[1]...A[n]$ does not contain x or contains only one copy of x, explain why the algorithm always correctly answers "no".

B. If $A[1]...A[n]$ contains $c \geq 2$ copies of x, explain why the algorithm incorrectly answers no with probability less than $\dfrac{1}{2^{(c-1)k}}$.

352. Suppose the *duplicate database yes-no search problem* that was presented in the introduction was generalized to be, given an integer constant $d \geq 1$, determine if there are d copies of x in $A[1]...A[n]$.

A. Show how to solve this problem in $O(1)$ worst case time on the CRCW PRAM, assuming that the CRCW simultaneous write convention is that one of the writers is randomly chosen to succeed.

B. Which of the other 10 CRCW simultaneous write conventions listed in the introduction can be used to solve this problem?

353. Consider the double exponential tree T with n leaves.

A. For any $k>0$ prove that if $n = \left(2^{2^k}\right)/2$, the height of T is exactly $\log_2(\log_2(2n))$.

B. For any $n \geq 2$ prove that the height of T is $\leq \left\lceil \log_2 \left\lceil \log_2(n) \right\rceil \right\rceil + 1$.

C. For any $k>0$ prove that if $n = \left(2^{2^k}\right)/2$, for any level $i>0$, the sum over each vertex at level i, of its number of children squared, is exactly $2n$.

D. For any $n \geq 2$ prove that for any level $i>0$, the sum over each vertex at level i, of its number of children squared, is $\leq 2n$.

E. Give pseudo-code for computing the maximum of an array on a double exponential tree.

354. Define an *augmented double exponential tree* as a standard one where each leaf is replaced by a vertex with two children (so the number of leaves is doubled). For example, the figure below depicts how an *augmented double exponential tree* with 16 leaves is constructed from a double exponential tree of 8 leaves:

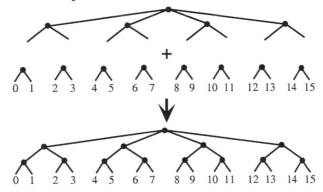

Assuming that for some $k \geq 0$, $n = 2^{2^k}$, prove that:

A. The root has $2^{2^{k-1}}$ children (i.e., $\sqrt{n}$ children), each of its children has $2^{2^{k-2}}$ children (i.e., $\sqrt[4]{n}$ children), and in general for $0 \leq i < k$ each vertex of depth i has $2^{2^{k-i-1}}$ children.

B. The height of the tree is exactly $k+1 = \log_2(\log_2(n)+1$.

C. For any depth $0 \leq i < k$ the sum over each vertex at depth i of its number of children squared is exactly n.

D. Generalize Parts A through C for arbitrary values of n.

E. Give pseudo-code for computing the maximum of an array on an augmented double exponential tree.

355. Even though the function $\log_2\log_2(n)$ is exponentially smaller than $\log_2(n)$, in practice $\log_2(n)$ is already a small function and it is useful to make a practical comparison between a double exponential tree and a straightforward computation of the maximum of an array in the CREW PRAM model using a binary tree.

A. Write pseudo-code for the procedure *pramMAX* for the CREW PRAM model that works in $O(\log(n))$ time on a standard binary tree in a similar fashion to summing an array.

B. Clearly, if we are using a double exponential tree T to compute the maximum of an array $A[0]...A[n-1]$, at the bottom of T it is probably more efficient to do a standard maximum computation at each vertex rather than call crcwMAX. Discuss the practical criteria for at what level to begin using crcwMAX.

C. A slight variation of the approach suggested in Part B is to first bunch vertices together in groups of m, apply *pramMAX* to each group of m, and then apply a double exponential tree T to the maximums produced for each group of m. For any real numbers $0<a<b$, assuming that *pramMAX* uses exactly $a\log_2(n)$ time and using T takes exactly $b\log_2(n)$ time, describe the best choice for m.

356. The procedure *crcwMAX* works in the same spirit as *erewSUM* by making parallel calls to *crcwSimpleMAX* instead of parallels calls that sum pairs of elements. A key difference however, is that *crcwMAX* at each stage shifts non-leaf vertices of the tree to the left in the array so that when it moves up to the next level of the tree, vertices will be in consecutive positions of A starting at $A[0]$. Since *erewSUM* only needs to sum pairs of elements in parallel, it was not particularly useful to do this shifting. However, show how *erewSUM* can be easily modified to do it.

357. The procedure *crcwMAX* to compute the maximum of an array $A[0]...A[n-1]$ in $O(1)$ time with $O(n^2)$ processors simply returns the maximum value. Suppose that instead, the problem was to return the index of the maximum value; that is, return an integer i such that $A[j] \leq A[i]$, $0 \leq j < n$. And suppose that as before, our assumption for the CRCW model is that when concurrent writes occur, they must all be the same value.

A. Modify *crcwMAX* to return the index; assume that all elements are distinct.

B. For the case that not all elements are distinct, the problem is that after the array X is computed by crcwMAX, it could be that for more than one value of i, $X[i]=0$, and hence, we cannot simply do

    for $1 \leq i,j < n$ **in parallel do if** $X[i]=0$ **then** $m:=i$

because we will be concurrently writing different values to m, which is not allowed under the assumptions we are using here.

Modify your pseudo-code of Part A to in parallel set $X[i]=i$ for all $X[i]$ that are 0 and then compute the largest (or smallest) index of a maximum value; you will also need to use a value other than 1 (e.g., -1) for $X's$ that are not maximums.

Hint: One approach is to use a second application of *crcwMAX*.

358. Discuss how to modify the generic (CREW) PRAM simulation algorithm for the CRCW model and for each of the 10 proposed CREW simultaneous write conventions listed in the introduction, explain which need no additional "tools" beyond sorting, data distribution, and packet routing, and which need an additional tool.

Hint: Most of those conventions do not need additional tools and for the few that do, simple tools such as array-sum or prefix-sum suffice.

359. Consider the data distribution algorithm presented for the EREW PRAM:

A. Explain why the following implementation is equivalent to the one presented for the EREW PRAM:

procedure erewDISTRIBUTE
    for $0 \le i < n$ **in parallel do if** P_i is a sender **then** $S[i]:=i$ **else** $S[i]:=-1$
    $p := n$
    repeat
        $p := \lceil p/2 \rceil$
        for $p \le i < n$ **in parallel do if** $S[i] \le S[i-p]$ **then begin**
            $S[i] := S[i-p]$
            $D[i] := D[i-p]$
        end
        until $p=1$
    end

B. Give a formal proof that erewDISTRIBUTE correctly distributes data.

C. Generalize the procedure erewDISTRIBUTE to take two arguments (a,b) and perform data distribution just for processors $P_a..P_b$.

360. The approach of merging by doing binary search that is used by pramMSORT could be used in a serial implementation. Show that this leads to an $O(n\log(n)^2)$ algorithm.

361. For $k \ge 0$ and $n=2^k$ prove that for dimension k:

A. A hypercube has $2^k = n$ vertices and $k2^{k-1} = (1/2)n\log_2(n)$ edges.

B. A CCC has $k2^k = n\log_2(n)$ vertices and $3k2^{k-1} = (3/2)n\log_2(n)$ edges.

C. A butterfly has $(k+1)2^k = n\log_2(n)+n$ vertices and $k2^{k+1} = 2n\log_2(n)$ edges.

D. A wrapped butterfly has $k2^k = n\log_2(n)$ vertices and $k2^{k+1} = 2n\log_2(n)$ edges.

E. A collapsed butterfly has $2^k = n$ vertices and $k2^{k+1} = 2n\log_2(n)$ edges.

F. A right-augmented butterfly has $(k+1)2^k = n\log_2(n)+n$ vertices and $(5k-2)2^{k-1}+1 = (5/2)n\log_2(n)-n+1$ edges.

362. Call a path in a hypercube of dimension k from a vertex x to a vertex y direct if:

1. It does not contain two edges of the same dimension.
2. It does not contain an edge of dimension i if the ith bits from the right in the binary representations of x and y are the same.

Give an expression for the number of different paths between x and y as a function of the number of bits in corresponding positions of the k-bit binary representations of x and y that differ.

363. Suppose a sequence of $n=2^k$ integers is given as input to a k-dimensional butterfly, one integer to each processor at Row k. Give an $O(\log(n))$ algorithm that leaves in Row 0 the same input but in reverse order from left to right.

   A. Solve this problem using sorting by giving the data at processor i the key $n-i$.

   B. Give a direct solution on the butterfly that inputs the list on rank k and outputs it in reverse order on Row 0 by switching the left and right halves in going from Row k to Row $k-1$ and then recursively reversing these halves on the two butterflies of dimension $k-1$;

   C. Explain why your solution to Part B could just as well work by starting with the input on Row 0 and work upward to produce the output on Row k.

364. When presenting matrix multiplication on a butterfly, we assumed that the dimensions of the matrices were powers of 2, and noted that if they are not, we can always "pad" with zeros to make dimensions that are powers of 2. Explain exactly how to do this and why it does not affect the asymptotic complexity.

365. Give pseudo-code for matrix multiplication on a butterfly that is based on the 4 steps of the algorithm for the EREW PRAM.

366. A *left-augmented* butterfly is defined in a symmetric fashion to a right-augmented butterfly, and a *fully-augmented* butterfly is defined as one that is both left and right augmented. Give pseudo-code for odd-even merge sort on a fully augmented butterfly that has the same general style as the procedure butterflyMSORT but uses the standard method of merging the odds with the odds and the evens with the evens.

367. Suppose that we have a supply of switches that take two inputs a and b and can pass them straight through to the outputs a and b, or "cross the wires", and send input a to output b and input b to output a, as shown in the figure below:

Inputs a and b can be passed straight through, or they can be switched so that input a goes to output b and input b goes to output a.

A *double butterfly* of dimension 0 is a single vertex and in general, a double butterfly of dimension k has $n = 2^k$ columns and consists of two butterflies of dimension k that are joined together by identifying the vertices at row 0 (i.e., vertices of row 0 are part of both networks). Now suppose, as depicted in the figure below, that each vertex of a double butterfly contains one of the switches, where going from left to right along the top, the inputs to the first vertex are denoted by $1a$ and $1b$, the inputs to the second vertex are denoted $2a$, and $2b$, and in general the inputs to the ith vertex from the left are denoted ia and ib, where na and nb are the inputs to the rightmost top vertex. Similarly, the outputs are labeled.

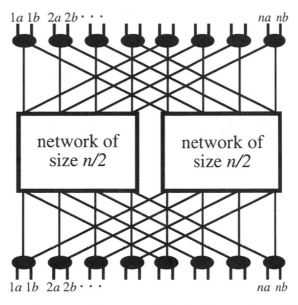

A. Give an expression as a function of k, the dimension of the network, for the number of vertices in the network.

B. Give an expression as a function of n, the number of columns of the network, for the number of vertices in the network.

C. Suppose that we are given a permutation of the integers 1 through $2n$, and we wish to route the inputs to the outputs according to this permutation. Consider the following data structure:

> Form a bipartite graph G with one "top" vertex for each of the n input vertices and one "bottom" vertex for each of the n output vertices. Place an edge between top vertex i and bottom vertex j if input ia or ib is to be connected to output ja or jb. Now consider the connected components of G. For the special case that both inputs ia and ib go to the same output vertex j, there will be a connected component consisting of just i and j connected by a single edge; label these edges with a *BOTH*. All other vertices of G have degree 2, and all other connected components of G are simple cycles. Traverse

each cycle by starting at a top vertex and alternately labeling each edge with *LEFT* or *RIGHT*.

Explain how this bipartite graph with labels of *BOTH*, *LEFT*, or *RIGHT* on the edges can be used to correctly set the input and output switches, and how to apply the construction recursively to the sub-networks.

D. Analyze the total time to set switches with the algorithm of Part C.

368. Consider the data distribution algorithm presented for the right-augmented butterfly:

A. Give a formal proof that ra-butterflyDISTRIBUTE correctly distributes data.

C. Suppose that we modified ra-butterflyDISTRIBUTE to work on a butterfly by simply traversing the corresponding detour path whenever a right edge was missing. Give a formal proof that it correctly distributes data.

B. For a vertex $[r,i]$ whose augmenting path goes to $[r-1,j]$, if p denotes the position (counting from the right) of the leftmost bit on which the binary representations of i and j differ, prove that the length of the detour path from i to j is $2(p-r)+1$.

D. Give a formal proof that the algorithm of Part C is $\Omega(\log(n)^2)$.

369. Consider a left-right routing between two vertices x and y in a HC/CCC/BF of dimension k. Prove that:

A. In a hypercube, at most k edges are traversed.

B. In a CCC, at most $2k$ edges are traversed.

C. In a wrapped butterfly, at most k edges are traversed.

D. In a standard (non-wrapped) butterfly, at most $2k$ edges are traversed.

E. Suppose that the k-bit binary representations of x and y differ in exactly i positions. Re-do Parts A through D to state your bounds in terms of i and k.

370. Taking it as a fact that, independent of the partial routing in question, with probability less than $(0.74)^{\log_2(n)}$ that the Valiant–Brebner algorithm finishes in more than $8\log_2(n)$ steps on a hypercube:

A. Our presentation of the Valiant-Brebner algorithm left out the detail as to how the network "knows" when it is done. There may be hardware in practice to detect this. But if not, describe how to periodically in $\log_2(n)$ steps do a "census" operation to detect that no processor has any more packets to send.

B. Clearly, if the census operation is done every $\log_2(n)$ steps, the number of steps is only doubled, and the asymptotic $O(\log(n))$ expected time is not affected. Suggest a better strategy that improves this constant of 2 based on the expected time for the Valiant–Brebner algorithm to complete.

C. In the worst (and extremely unlikely case) the Valiant–Brebner algorithm could take time as much as $O(n)$ time and employ queue sizes at some vertices of up to $O(n)$. Describe a strategy for using queue sizes of a maximum of $O(\log(n))$ that still has expected $O(\log(n))$ time.

Hint: Consider restarting the algorithm when you are "unlucky".

371. A 1-1 packet routing is called a permutation routing if all processors send exactly one packet and all processors receive exactly one packet. Explain how sorting can be used to perform a permutation routing.

372. Given that the butterfly network is the same as the pattern used by the non-recursive in-place FFT Algorithm, describe how to perform a FFT of n points in $O(\log(n))$ time on a butterfly network and express your algorithm in pseudo-code.

373. For a mesh of height h and width w, prove that the number of edges is $h(w-1)+w(h-1)$.

374. Consider the "H-tree" layout used by Method 1 for summing an array with a mesh.
  A. Give a formal recursive definition of the H-tree layout.
  B. Describe what to do if n is not a power of 4.
  C. Explain why a grid of size $2\lceil\sqrt{n}\rceil$ by $2\lceil\sqrt{n}\rceil$ suffices for a tree of n leaves.

375. Consider Method 1 presented for broadcasting on a mesh with the "H-tree" pattern.
  A. Write pseudo-code for this method.
  B. Write pseudo-code for summing all vertices (producing the result at Processor 1).
  C. Describe how to modify this approach for non-square meshes of height h and width w (your construction should work even when h or w are 1) and explain why the complexity is $MAX\{h,w\}$.

376. The H-tree tree layout presented for broadcasting on the mesh in nice in that it uses only $O(n)$ area. However, it may be undesirable for applications where we wish to have the leaves along the edge. Unfortunately, the obvious way of laying out a tree with the leaves on the edge uses $O(n\log(n))$ area for a complete tree of n leaves.

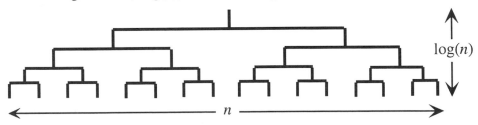

Prove that this tradeoff is unavoidable; that is, any layout of a complete binary tree of n leaves for which the leaves are on the perimeter uses $O(n\log(n))$ area.

377. The H-tree tree layout presented for broadcasting on the mesh has some edges that are $O(\sqrt{n})$ long. This can be undesirable asymptotically for models where the time for a signal to propagate down a wire depends on its length.

A. Prove that any layout for a complete binary tree with n leaves has edges that are $\Omega(\sqrt{n} / \log(n))$ long, and hence cannot be too much better in this respect.

Hint: Any path from one side to the other can only use so many edges.

B. Describe a $O(n)$ area layout for a complete binary tree with n leaves where all edges are $O(\sqrt{n} / \log(n))$ long.

378. Consider Method 1 presented for broadcasting on a mesh that sent the data item down the first column and then across the rows.

A. Write pseudo-code for this method.

B. Write pseudo-code for summing all vertices (producing the result at Processor 1).

C. Describe how to modify this approach for non-square meshes of height $h \geq 1$ and width $w \geq 1$ and explain why the complexity is $MAX\{h,w\}$.

379. The matrix multiplication algorithm for the mesh, *meshMATRIXMULT*, assumed that the rows of A were stored in the processors of a special Column 0 and that the columns of B were stored in the processors of a special Row 0. Suppose instead a mesh of $MAX\{m,w\}$ by $MAX\{n,w\}$ is used and that initially $A[i,j]$, $1 \leq i \leq m$ and $1 \leq j \leq w$, was stored in Processor $[i,j]$ and $B[j,k]$, $1 \leq j \leq w$ and $1 \leq k \leq n$, was stored in Processor $[j,k]$. Present pseudo-code that in $O(m+w+n)$ time moves A and B to the edges so that *meshMATRIXMULT* can be used.

380. Column 0 and Row 0 of the matrix multiplication algorithm for the mesh, *meshMATRIXMULT*, uses $O(w)$ space per processor. Alternately, if we wish to limit the space per processor to $O(1)$, we can use a $w+m$ by $w+n$ mesh with w additional columns to store the rows of A and w additional rows to store the columns of B (see the exercises). Modify the *meshMATRIXMULT* pseudo-code to shift the entries of A and B in from these extra Columns and Rows.

381. Consider data distribution on the mesh.

A. Pseudo-code was given for the procedure SendRight to "send" data to the right. Give similar pseudo-code for a procedure SendDown to send data down.

B. Give detailed pseudo-code for data distribution on the mesh that makes use of the procedures SendRight and SendDown.

C. Suppose that the mesh was augmented with direct connections between Processor $[i,w]$ and Processor $[i+1,1]$, $1 \leq i < h$ (i.e., from the end of one row to the start of the next). Give pseudo-code for data distribution on the mesh where processors repeatedly, for $w+h$ steps, pass data to the right, down, and also to the left when a processor is not the leader.

382. Consider sorting on the mesh.

   A. Give pseudo-code for odd-even merging of the NW with the SW quadrant, assuming that the mesh has n vertices, where $n = m^2$ and m is a power of 2.

   B. Describe how your pseudo-code of Part A can be modified for merging the NE with the SE quadrant and for merging two halves of the mesh.

   C. Give pseudo-code for sorting on the mesh that calls the merge procedures developed in Parts A and B.

   D. Describe how to generalize your solutions to Parts A, B, and C to an arbitrary mesh of height $h \geq 1$ and width $w \geq 1$, and show that the time is $O(h+w)$.

383. Consider the vertical-horizontal routing algorithm presented for a mesh of height h and width w.

   A. Give pseudo-code.

   B. As presented, the vertical-horizontal algorithm only required communication in one dimension (vertical or horizontal) at any given time. However, in a given dimension, communication in two directions (up-down or left-right) is required. Modify your pseudo-code of Part A to only communicate in one direction at a given time.

   C. Optimize your pseudo-code for Part B for the case that h≠w to do a vertical-horizontal or a horizontal-vertical routing and analyze the complexity of your algorithm and the sizes of the queues required.

384. The vertical-horizontal routing algorithm presented for a mesh of height h and width w uses queues of size $O(h)$ (or size $MIN(h,w)$ assuming that a vertical-horizontal or horizontal-vertical routing is used depending on whether h or w is smaller). Discuss and analyze methods that are limited to queues of size $O(1)$ but may pass packets in both the vertical and horizontal directions during all phases of the algorithm.

385. Recall the dynamic programming algorithm that was presented in the chapter on strings to compute the edit distance between two strings.

   A. Describe a parallel algorithm to compute the edit distance between two strings of length n that operates on a mesh of height 1 and width n in $O(n)$ time.

   Hint: Compute diagonal cross sections as described in the discussion of relaxed ordering that followed the presentation of the algorithm.

   B. Generalize your solution to Part A for strings of different lengths.

386. Define a *Boolean function* as a function of n, $n \geq 1$, Boolean inputs (0 or 1) to a single Boolean output. (An arbitrary Boolean function from n inputs to m outputs can be viewed as a collection of yes-no Boolean functions, one for each output).

A. *Boolean expressions* can be used to denote Boolean functions. They are formed from variable names, the AND ($\wedge$), OR ($\vee$), and NOT ($\neg$) operators, and parentheses. When parentheses are omitted, $\neg$ has precedence over $\wedge$ which has precedence over $\vee$. Also, we write variable names next to each other to denote $\wedge$; for example, ab is the same as $a \wedge b$. A Boolean expression is in *disjunctive normal form* if it is the sum of a number of *terms*, where a *term* is the AND of a number of input variables, each of which may be complemented or uncomplemented. For example, the expression $(a \vee b) \wedge (c \vee \neg d)$ is not in disjunctive normal form, but the equivalent expression

$$(ac) \vee (bc) \vee (a \neg d) \vee (b \neg d)$$

is in disjunctive normal form. Show that any Boolean expression can be put in disjunctive normal form.

B. Explain why any Boolean function can be represented by a Boolean expression. *Hint:* Think of the correspondence between the rows of a truth table and the terms of an expression in disjunctive normal form.

C. Explain why Parts A and B imply that any Boolean function can be constructed from the AND, OR, and NOT functions.

D. Define the NAND function to be $\neg (a \wedge b)$. Show that any Boolean function can be constructed from just NAND operations. Show that the function NOR defined as $\neg (a \vee b)$ also has this property.

387. In the same style of drawing as used in the example for AND, OR, and NOT circuits for cMOS:

A. Draw a cMOS circuit for NAND.

B. Draw a cMOS circuit for NOR.

388. In the same style of drawing as used in the example for NAND and NOR circuits for nMOS:

A. Draw a nMOS circuit for AND.

B. Draw a nMOS circuit for OR.

C. Draw a nMOS circuit for NOT.

389. Consider the 1-bit nMOS adder example:

A. Fill in the details of the explanation for how it works given in the example.

B. Explain how to layout a k-bit nMOS adder, $k \geq 1$, by using copies of the 1-bit nMOS adder presented in the example.

C. Draw a cMOS 1-bit adder that works just like the 1-bit nMOS adder example, except that the pull-ups are replaced by complementary circuitry.

390. Consider the cMOS circuit for RAM memory that was presented.

A. Assuming that a k-bit register is formed by replicating horizontally the 1-bit addressable register presented to form a circuit that is $O(1)$ high by $O(k)$ wide, explain why the RAM memory layout presented is $O(\sqrt{n} + \log(n))$ high by $O(k\sqrt{n} + \log(n))$ wide and hence uses a total area of $O(kn)$.

B. Explain how an $O(1)$ high by $O(k)$ wide k-bit register as used in Part A can be folded into an area that is $O(\sqrt{k})$ by $O(\sqrt{k})$. That is, assuming that $\sqrt{k}$ 1-bit registers are placed in each row, explain why the width has to be increased by only $O(1)$ to accommodate the wires to connect adjacent rows.

C. Assuming that a k-bit register uses an area that is $O(\sqrt{k})$ by $O(\sqrt{k})$ as described in Part B, explain why the RAM memory layout presented is $O(\sqrt{kn} + \log(n))$ high by $O(\sqrt{kn} + \log(n))$ wide and hence uses a total area of $O(kn)$.

391. Consider the proof of the (non-restricted) area-time tradeoff $AT^2 = \Omega(n^2)$.

A. We relied on the fact that no matter what LS output bits are in Y, the values of input bits not in S can be chosen to force $Y = X$. Explain why this is true.

B. We placed the partitioning line so that at least $n/3$ of the outputs were produced on each side of the line. Explain why for any $k \geq 2$, at least n/k outputs on each side of the line still give $AT^2 = \Omega(n^2)$.

C. The $AT^2 = \Omega(n^2)$ bound had a constant of 1/9. Under the assumption that at most one least significant output bit is produced at each pin, explain why the constant can be improved to 1/4. Discuss what kinds of assumptions allow the constant to be further improved.

Chapter Notes

For further reading on parallel algorithms, the reader may refer to a host of texts, including Akl [1997], Cosnard and Trystram [1995], Jaja [1992], Gibbons and Rytter [1988], Gibbons and Spirakis [1993], Lakshmivarahan and Dhall [1990], Leighton [1992], Quinn [1987], Ranka and Sahni [1990], and Reif [1993]. Because "parallel algorithms" is a rather broad term, these books can have very different emphasis; for example, Leighton [1992] focuses more on networks and Jaja [1992] focuses more on the PRAM model (both are an excellent source of further references).

The class of problems that can be solved in poly-log time ($O(\log(n)^k)$ for some $k \geq 1$) with a polynomial number of processors and memory is often called NC, for "Nick's class", due to the work of N. Pippenger [1979].

For all the algorithms presented here, we have assumed that enough processors are available to match the problem size. In practice, most algorithms can be scaled so that as much parallel computing as possible can be employed with the number of processors available and then serial computing can make up the difference. For example, with recursive merge sorting on a PRAM, if there are less than n processors available to sort a list of n elements, we can process recursive calls one at time in the normal serial fashion until the problem size gets down to the number of processors. For a discussion of this issue, see the book of JaJa [1992].

Many other networks with properties similar to the HC/CCC/Butterfly have been proposed; e.g., the *shuffle-exchange network* (e.g., see the presentation in the book of Ullman [1984]) and the *Mobius Cubes* (Cull [1995]. Bhatt and Leiserson [1982] address the assembly of tree machines. Fat trees are presented by Leiserson [1985]).

The classic *Thinking Machines CM*-1 *and CM*-2 parallel computer (built by Thinking Machines Inc. in the 1980's — see the book of Hillis [1982]), that had up to $65,536 = 2^{16}$ processors, was based on a modified hypercube of dimension 12 where each of the $4,096 = 2^{12}$ vertices were 4 by 4 meshes of 16 processors. The later CM-5 machine was based on fat trees.

Our general characterization of chip technology as wires on a number of layers where when wires on special layers cross transistors are formed suffices for the theoretical considerations here. However, actual fabrication processes are quite complex; for further reading, see for example the book of Weste and Eshraghian [1993]. As discussed in the exercises, the single-metal CMOS technology is not the simplest technology that has been used. CMOS has both n-type (a 0 opens the switch) and p-type transistors (a 1 opens the switch), but it is possible to only have only one type of transistor and instead of using complementary circuitry a thing sort of like a big resistor (called a "pull-up") can be fabricated. Unfortunately, the constant power drain through the pull-ups makes nMOS technology consume more power (and make chips run hotter), and this technology was more or less abandoned in favor of CMOS. See the book of Mead and Conway [1980] for a presentation of nMOS technology.

The book of Akl [1985] is devoted to parallel sorting algorithms. Olariu, Pinotti, and Zheng [2000] consider fixed-size parallel sorting devices.

The book of Karpinski and Rytter [1998] is devoted to parallel algorithms for graph matching.

Grenlaw and Petreschi [2000] and Grenlaw, Halldorsson and Petreschi [2000] consider parallel algorithms for Prufer codes, which can be used to represent trees.

Gonzalez and Storer [1985], Storer and Reif [1991], DeAgostino and Storer [1992], Storer [1992], Royals, Markas, Kanopoulos, Reif, and Storer [1993], and Storer, Belinskaya, and DeAgostino [1995] consider parallel algorithms for data compression.

Appendix: Common Sums

Introduction: Sums (often called a "series") arise often in algorithm analysis. Although we can usually bound sums with simple algebra or with a proof by induction, integrals can sometimes provide more exact bounds or more easily lead to a closed form.

A. Approximating Sums with Integrals

For two integers $a<b$, if $f(x)$ is a continuous non-negative *non-decreasing* function between a and b (for $a \leq x \leq b$, $f(x) \geq 0$ and if $x<y$ then $f(x) \leq f(y)$) for which it is possible to compute its integral, then we can bound

$$\sum_{i=a}^{b} f(i)$$

by thinking of this sum as an approximation to the area under the curve defined by $f(x)$. If the sum is represented by unit wide bars that are positioned to the left of each point, then:

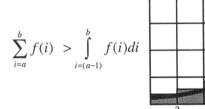

$$\sum_{i=a}^{b} f(i) > \int_{i=(a-1)}^{b} f(i)di$$

However, if we think of the bars as being positioned to the right of each point, then:

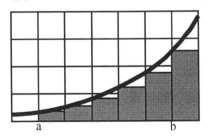

$$\sum_{i=a}^{b} f(i) < \int_{i=a}^{b+1} f(i)di$$

Non-increasing functions: Using similar reasoning, when $f(x)$ is a continuous *non-increasing* (if $x<y$ then $f(x) \geq f(y)$) function we have:

$$\int_{i=a}^{b+1} f(i)di < \sum_{i=a}^{b} f(i) < \int_{i=(a-1)}^{b} f(i)di$$

Example: Bounding a Hill-Shaped Sum

The techniques for non-increasing and non-decreasing functions can be "pieced" together to bound more complicated functions. To illustrate this idea, for $a \leq M \leq b$, where a and b are integers and M is a real number, consider a continuous non-negative "hill-shaped" function $f(i)$ that is non-decreasing from $i=a$ to $i=M$ and then non-increasing from $i=M$ to $i=b$. Note that although M is a maximum value of the function between a and b, it may not be unique; it could be, for example, that $f(\lfloor M \rfloor) = f(M) = f(\lceil M \rceil)$.

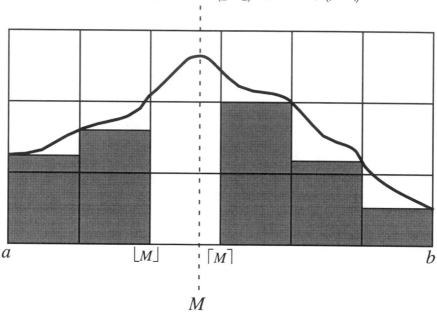

By bounding the non-decreasing portion from a to $\lfloor M \rfloor$ and the non-increasing portion from $\lceil M \rceil$ to b, we leave out only $f(\lfloor M \rfloor)$ and $f(\lceil M \rceil)$; the minimum of these is bounded by the integral, and then we can add on the maximum of these, which is $\leq f(M)$:

$$\sum_{i=a}^{b} f(i) = \sum_{i=a}^{\lfloor M \rfloor - 1} f(i) + f(\lfloor M \rfloor) + f(\lceil M \rceil) + \sum_{i=\lceil M \rceil + 1}^{b} f(i)$$

$$\leq \text{maximum}\{f(\lfloor M \rfloor), f(\lceil M \rceil)\} + \int_{i=a}^{\lfloor M \rfloor} f(i)di + \int_{i=\lfloor M \rfloor}^{\lceil M \rceil} f(i)di + \int_{i=\lceil M \rceil}^{b} f(i)di$$

$$\leq f(M) + \int_{i=a}^{b} f(i)di$$

Note: When $\lfloor M \rfloor = a$ or $\lceil M \rceil = b$, this bound essentially includes the upper bounds on non-increasing and non-decreasing functions as special cases, where the first or the last term of the sum is explicitly left out and the integral bounds the rest.

B. Arithmetic Sum

Idea: Even though the terms vary, the average value is equivalent to the largest value.

Theorem: For any integer $n \geq 0$

$$\sum_{i=0}^{n} i = \tfrac{1}{2}n^2 + \tfrac{1}{2}n$$

$$\sum_{i=0}^{n} i^2 = \tfrac{1}{3}n^3 + \tfrac{1}{2}n^2 + \tfrac{1}{6}n$$

$$\sum_{i=0}^{n} i^3 = \tfrac{1}{4}n^4 + \tfrac{1}{2}n^3 + \tfrac{1}{4}n^2$$

and in general, for $k > 1$:

$$\sum_{i=0}^{n} i^k = \tfrac{1}{k+1}n^{k+1} + \tfrac{1}{2}n^k + \Theta\!\left(n^{k-1}\right)$$

Proof: The sum is $O(n^{k+1})$ because each term is $O(n^k)$ and the sample exercises to the first chapter showed how to crudely argue it is $O(n^{k+1})$. The sample exercises for induction and recursion verified the exact coefficients for cases $k=1,2,3$. For a more precise general analysis, we can bound the sum, which is *non-decreasing*, by an integral and observe that the first two terms of the binomial sum $(n+1)^{k+1}$ are n^{k+1} and $(k+1)n^k$:

$$\sum_{i=0}^{n} i^k < \int_{i=0}^{n+1} i^k \, di = \left[\frac{1}{k+1} i^{k+1} \right]_{i=0}^{i=(n+1)} = \frac{1}{k+1}(n+1)^{k+1} = \frac{1}{k+1}n^{k+1} + n^k + O\!\left(n^{k-1}\right)$$

For the second coefficient, the integral overestimates the ith term of the sum by an area *greater* than a little triangle of width 1, height ki^{k-1}, and area $ki^{k-1}/2$ (i.e., the slope of the top of the triangle is defined by $\dfrac{d(i^k)}{di} = ki^{k-1}$ — see the figure below), for a total of:

$$\sum_{i=0}^{n} \frac{k}{2} i^{k-1} = \sum_{i=1}^{n} \frac{k}{2} i^{k-1} > \frac{k}{2} \int_{i=0}^{n} i^{k-1} \, di = \frac{k}{2}\left[\frac{1}{k} i^k \right]_{i=0}^{i=n} = \frac{n^k}{2}$$

But the overestimate is *less* than sum of the triangles defined by the dashed lines:

$$\frac{\left(1^k - 0^k\right)}{2} + \frac{\left(2^k - 1^k\right)}{2} + \cdots + \frac{\left((n+1)^k - n^k\right)}{2} = \frac{(n+1)^k}{2} = \frac{n^k}{2} + O\!\left(n^{k-1}\right)$$

Hence the correct value for the second coefficient is exactly 1/2.

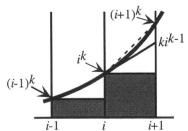

C. Simple Geometric Sum (unweighted, $k=0$)

Idea: Summing powers of r when $r<1$ cost no more asymptotically than the first term (its like when the family picks at the leftover birthday cake — if each person takes only half of what is left, you will never run out of cake). On the other hand, when $r>1$, just the last term by itself is large.

Theorem: For an integer $n \geq 0$ and a real r such that $r>0$ and $r \neq 1$:

$$\sum_{i=0}^{n} r^i = \frac{1-r^{n+1}}{1-r} < \begin{cases} 0<r<1: & \dfrac{1}{1-r} = O(1) \\[2ex] r>1: & \left(\dfrac{r}{r-1}\right)r^n = \Theta(r^n) \end{cases}$$

Note: When $r>1$, it "looks better" to write this sum equivalently as:

$$\sum_{i=0}^{n} r^i = \frac{r^{n+1}-1}{r-1}$$

Proof:

Let S denote this sum; then:

$$
\begin{array}{rcllllll}
S & = & 1 & +r & +r^2 & +... & +r^n & \\
rS & = & & r & +r^2 & +... & +r^n & +r^{n+1} \\
S-rS & = & 1 & +0 & +0 & +... & +0 & -r^{n+1}
\end{array}
$$

Since $r \neq 1$, we now can divide by $1-r$ to get:

$$S = \frac{1-r^{n+1}}{1-r}$$

Corollary: For $n>0$ a power of 2:

$$\sum_{i=0}^{\log_2(n)} \frac{1}{2^i} = 2 - \frac{1}{n}$$

$$\sum_{i=0}^{\log_2(n)} 2^i = 2n - 1$$

D. Linear Weighted Geometric Sum (k=1)

Idea: Multiplying each term of a geometric sum by a polynomial in i does not affect the asymptotic complexity when $r<1$ because the exponential decay far out-weighs the contribution of the polynomial terms. When $r>1$, again the last term dominates. Here we begin by considering the simple case that the polynomial is just i.

Theorem: For an integer $n \geq 0$ and a real r such that $r>0$ and $r \neq 1$:

$$\sum_{i=0}^{n} ir^i = \frac{r-r^{n+1}}{(1-r)^2} - \left(\frac{r}{1-r}\right)nr^n < \begin{cases} 0<r<1: \ \dfrac{r}{(1-r)^2} = O(1) \\[2ex] r>1: \ \left(\dfrac{r}{r-1}\right)\left(nr^m\right) = \Theta\left(nr^n\right) \end{cases}$$

Proof:

If S denotes this sum, then:

$$
\begin{aligned}
S &= r &+2r^2 &+3r^3 &+\dots && +nr^n \\
rS &= && r^2 &+2r^3 &+\dots & +(n-1)r^n & +nr^{n+1} \\
S-rS &= r && +r^2 &+r^3 &+\dots & +r^n & -nr^{n+1}
\end{aligned}
$$

Hence, by dividing by $1-r$ and then factoring out r:

$$
\begin{aligned}
S &= \left(\tfrac{r}{1-r}\right)\sum_{i=0}^{n-1} r^i - \left(\tfrac{r}{1-r}\right)nr^n \\[1ex]
&= \left(\tfrac{r}{1-r}\right)(\text{geometric sum to } n-1) - \left(\tfrac{r}{1-r}\right)nr^n \\[1ex]
&= \left(\tfrac{r}{1-r}\right)\frac{1-r^n}{1-r} - \left(\tfrac{r}{1-r}\right)nr^n \\[1ex]
&= \frac{r-r^{n+1}}{(1-r)^2} - \left(\tfrac{r}{1-r}\right)nr^n
\end{aligned}
$$

If $r>1$, then the first term is negative, the second term positive, and the expression only gets smaller by throwing out the first term, from which the theorem follows.

If $r<1$, then the first term is positive, the second term is negative, and the expression only gets larger by throwing out the second term and then throwing out the $-r^{n+1}$, from which the theorem follows.

Corollary: For $n \geq 1$ a power of 2:

$$\sum_{i=0}^{\log_2(n)} \frac{i}{2^i} = 2 - \log_2(n)/n - 2/n < 2$$

$$\sum_{i=0}^{\log_2(n)} i2^i = 2n\log_2(n) - 2n + 2 < 2n\log_2(n)$$

E. Quadratic Weighted Geometric Sum ($k=2$)

Idea: If we think of each term as i^k times r^i, for $k=1$, we reduced the sum for the k=1 case to the $k=0$ case. For $k=2$, we reduce the sum to a $k=1$ sum plus a $k=0$ sum.

Theorem: For an integer $n \geq 0$ and a real r such that $r>0$ and $r \neq 1$:

$$\sum_{i=0}^{n} i^2 r^i < \begin{cases} 0<r<1: & \dfrac{r+r^2}{(1-r)^3} = O(1) \\ \\ r>1: & \left(\dfrac{r}{r-1}\right)\left(n^2 r^m\right) = \Theta\left(n^2 r^n\right) \end{cases}$$

Proof:

If S denotes this sum, then:

$$\begin{aligned} S &= r &&+2^2 r^2 &&+3^2 r^3 &&+... &&+n^2 r^n \\ rS &= &&\quad r^2 &&+2^2 r^3 &&+... &&+(n-1)^2 r^n &&+n^2 r^{n+1} \\ S-rS &= r &&+\left(2^2-1\right)r^2 &&+\left(3^2-2^2\right)r^3 &&+... &&+\left(n^2-(n-1)^2\right)r^n &&-n^2 r^{n+1} \end{aligned}$$

Hence, by dividing by $1-r$ and then factoring out r:

$$S = \left(\tfrac{r}{1-r}\right)\sum_{i=0}^{n}\left((i+1)^2 - i^2\right)r^i - \left(\tfrac{r}{1-r}\right)n^2 r^n$$

$$= \left(\tfrac{r}{1-r}\right)\sum_{i=0}^{n}(2i+1)r^i - \left(\tfrac{r}{1-r}\right)n^2 r^n$$

$$= \left(\tfrac{r}{1-r}\right)\left(2\sum_{i=0}^{n-1}ir^i + \sum_{i=0}^{n-1}r^i\right) - \left(\tfrac{r}{1-r}\right)n^2 r^n$$

$$= \left(\tfrac{r}{1-r}\right)\left(2(\text{linear weighted sum}) + (\text{simple geometric sum})\right) - \left(\tfrac{r}{1-r}\right)n^2 r^n$$

If $r>1$, then the first term is negative, the second term positive, and the expression only gets smaller by throwing out the first term, from which the theorem follows:

$$S = \left(\tfrac{r}{r-1}\right)\left(n^2 r^m\right)$$

If $r<1$, then the first term is positive, the second term is negative, and the expression only gets larger by throwing out the second term and then substituting the $r<1$ bounds for the linear weighted and simple geometric sums (from 0 to $n-1$) to get:

$$S < \left(\tfrac{r}{1-r}\right)\left(2\left(\frac{r}{(1-r)^2}\right) + \left(\frac{1}{1-r}\right)\right) = \frac{r+r^2}{(1-r)^3}$$

Note: Even for $k=2$, we can see a pattern emerging; $(i+1)^2$ gives us the corresponding binomial coefficients (the 2-row of Pascal's triangle — see the first chapter) and subtracting i^2 removes the first one to leave the coefficients 2 and 1, which are the coefficients of smaller order sums.

F. Cubic Weighted Geometric Sum ($k=3$)

Idea: For $k=3$, we reduce the sum to a $k=2$ sum plus a $k=1$ sum plus a $k=0$ sum.

Theorem: For an integer $n \geq 0$ and a real r such that $r>0$ and $r \neq 1$:

$$\sum_{i=0}^{n} i^3 r^i < \begin{cases} 0<r<1: \quad \dfrac{r+4r^2+r^3}{(1-r)^4} = O(1) \\[4mm] r>1: \quad \left(\dfrac{r}{r-1}\right)\left(n^3 r^m\right) = \Theta\left(n^3 r^n\right) \end{cases}$$

Proof:

If S denotes this sum, then:

$$\begin{aligned} S &= r &&+2^3 r^2 &&+3^3 r^3 &&+... &&+n^3 r^n \\ rS &= &&r^2 &&+2^3 r^3 &&+... &&+(n-1)^3 r^n &&+n^3 r^{n+1} \\ S-rS &= r &&+(2^3-1)r^2 &&+(3^3-2^3)r^3 &&+... &&+(n^3-(n-1)^3)r^n &&-n^3 r^{n+1} \end{aligned}$$

Hence, by dividing by $1-r$ and then factoring out r:

$$S = \left(\tfrac{r}{1-r}\right)\sum_{i=0}^{n-1}\left((i+1)^3 - i^3\right)r^i \; - \left(\tfrac{r}{1-r}\right)n^3 r^n$$

$$= \left(\tfrac{r}{1-r}\right)\sum_{i=0}^{n-1}\left(3i^2 + 3i + 1\right)r^i \; - \left(\tfrac{r}{1-r}\right)n^3 r^n$$

$$= \left(\tfrac{r}{1-r}\right)\left(3\sum_{i=0}^{n-1}i^2 r^i + 3\sum_{i=0}^{n-1}ir^i + \sum_{i=0}^{n-1}r^i\right) - \left(\tfrac{r}{1-r}\right)n^3 r^n$$

$$= \left(\tfrac{r}{1-r}\right)\left(3(\text{quadratic w. sum}) + 3(\text{linear w. sum}) + (\text{simple geo. sum})\right) - \left(\tfrac{r}{1-r}\right)n^3 r^n$$

If $r>1$, then the first term is negative, the second term positive, and the expression only gets smaller by throwing out the first term, from which the theorem follows:

$$S = \left(\tfrac{r}{r-1}\right)\left(n^3 r^m\right)$$

If $r<1$, then the first term is positive, the second term is negative, and the expression only gets larger by throwing out the second term and then substituting the $r<1$ bounds for the quadratic weighted, linear weighted, and simple geometric sums (from 0 to $n-1$) to get:

$$S < \left(\tfrac{r}{1-r}\right)\left(3\left(\frac{r+r^2}{(1-r)^3}\right) + 3\left(\frac{r}{(1-r)^2}\right) + \left(\frac{1}{1-r}\right)\right) = \frac{r+4r^2+r^3}{(1-r)^4}$$

Note: For $k=3$, we see the pattern continuing; $(i+1)^3$ gives us the corresponding binomial coefficients (the 3-row of Pascal's triangle) and subtracting i^3 removes the first one to leave the coefficients 3, 3, and 1, which are the coefficients of smaller order sums.

G. Weighted Geometric Sum (for any non-negative integer k)

Idea: Motivated by $k=0$ through $k=3$, we add some number of each of the sums for $k=0$ through $k-1$, given by binomial coefficients.

Theorem: For integers $k,n \geq 0$, and a real r such that $r>0$ and $r \neq 1$:

$$\sum_{i=0}^{n} i^k r^i = \begin{cases} 0 < r < 1: & O(1) \\ r > 1: & \Theta\left(n^k r^n\right) \end{cases}$$

Proof: If S denotes this sum, then:

$$
\begin{aligned}
S &= r &&+2^k r^2 &&+3^k r^3 &&+... &&+n^k r^n \\
rS &= &&r^2 &&+2^k r^3 &&+... &&+(n-1)^k r^n &&+n^k r^{n+1} \\
S - rS &= r &&+\left(2^k-1\right)r^2 &&+\left(3^k-2^k\right)r^3 &&+... &&+\left(n^k-(n-1)^k\right)r^n &&-n^k r^{n+1}
\end{aligned}
$$

Hence, by dividing by $1-r$ and then factoring out r:

$$S = \left(\tfrac{r}{1-r}\right)\sum_{i=0}^{n-1}\left((i+1)^k - i^k\right)r^i \; - \left(\tfrac{r}{1-r}\right)n^k r^n$$

$$= \left(\tfrac{r}{1-r}\right)\sum_{i=0}^{n-1}\left(\sum_{m=0}^{k-1}\binom{k}{m}i^m r^i\right) - \left(\tfrac{r}{1-r}\right)n^k r^n \quad \text{—put in expression for binomial sum less first term}$$

$$= \left(\tfrac{r}{1-r}\right)\sum_{m=0}^{k-1}\left(\binom{k}{m}\sum_{i=0}^{n-1} i^m r^i\right) - \left(\tfrac{r}{1-r}\right)n^k r^n \quad \text{—change order of summation}$$

$$= \left(\tfrac{r}{1-r}\right)\sum_{m=0}^{k-1}\left(\binom{k}{m}(\text{weight } m \text{ geometric sum from 0 to } n-1)\right) - \left(\tfrac{r}{1-r}\right)n^k r^n$$

If $r>1$, then the first term is negative, the second term positive, and the theorem follows:

If $r<1$, the second term is negative and since the first term is a function only of r and k, it follows that its is $O(1)$. We can crudely bound the constant by observing that since

$$\binom{k}{m} = \frac{k!}{(k-m)!m!} = \frac{k(k-1)!}{(k-m)(k-1-m)!m!} = \frac{k}{k-m}\binom{k-1}{m} < k\binom{k-1}{m}$$

it follows inductively, using $k=2$ as the base case, that:

$$S < \left(\tfrac{r}{1-r}\right)\left(\binom{k}{k-1}\sum_{i=0}^{n-1}i^{k-1}r^i + \sum_{m=0}^{k-2}\left(\binom{k}{m}\sum_{i=0}^{n-1}i^m r^i\right)\right)$$

$$< \left(\tfrac{r}{1-r}\right)\left(k\sum_{i=0}^{n-1}i^{k-1}r^i + k\sum_{m=0}^{k-2}\left(\binom{k-1}{m}\sum_{i=0}^{n-1}i^m r^i\right)\right)$$

$$= \left(\tfrac{r}{1-r}\right)\left(2k(\text{weight }(k-1)\text{ geometric sum from 0 to } n-1)\right)$$

$$< \left(\tfrac{2kr}{1-r}\right)^{k-2}\left(\frac{r+r^2}{(1-r)^3}\right) = \frac{(2kr)^{k-2}\left(r+r^2\right)}{(1-r)^{k+1}}$$

A Second Proof of the Weighted Geometric Sum, Using Calculus

Idea: We integrate $x^i r^i$ with a simple function plus a "smaller" integral. Similar to the first proof, for $r>1$ we can throw out the smaller integral, but for $r<1$ we "unwind" the smaller integrals to produce a sum (this proof applies when $k \geq 0$ is any real number). For simplicity, we focus just on O (i.e., upper bounding the sum), rather than Θ.

For $r>1$, we have:

$$\sum_{i=0}^{n} i^k r^i = \sum_{i=1}^{n} i^k r^i = n^k r^n + \sum_{i=1}^{n-1} i^k r^i$$

$$< n^k r^n + \int_{i=1}^{n} i^k r^i di = n^k r^n + \left[\frac{i^k r^i}{\ln(r)} \right]_{r=1}^{r=n} - \frac{k}{\ln(r)} \int_{i=1}^{n} i^{k-1} r^i$$

$$< \left(1 + \tfrac{1}{\ln(r)}\right) n^k r^n = O\left(n^k r^n\right)$$

For $r<1$, we have a hill shaped function (see the example in Appendix A) with a derivative of zero at $k/\ln(1/r)$, and since we only care about the range 1 to n, we set:

$M := 1/(\ln(1/r))$; **if** $M<1$ **then** $M:=1$; **if** $M>n$ **then** $M:=n$

Then we have:

$$\sum_{i=0}^{n} i^k r^i = \sum_{i=1}^{n} i^k r^i < M^k r^M + \int_{i=1}^{n} i^k r^i di$$

$$= M^k r^M - \left[\frac{i^k r^i}{\ln(1/r)} \right]_{i=1}^{i=n} + \tfrac{k}{\ln(1/r)} \int_{i=1}^{n} i^{k-1} r^i di$$

$$< M^k r^M + \frac{r}{\ln(1/r)} + \frac{k}{\ln(1/r)} \int_{i=1}^{n} i^{k-1} r^i di$$

$$< M^k r^M + \frac{r}{\ln(1/r)} + \frac{kr}{\ln(1/r)^2} + \left(\frac{k}{\ln(1/r)} \right)^2 \int_{i=1}^{n} i^{k-2} r^i di$$

$$\vdots$$

$$< M^k r^M + \frac{r}{\ln(1/r)} \sum_{j=0}^{k} \left(\frac{k}{\ln(1/r)} \right)^j$$

This expression is $O(1)$ since it is a function only of k and r. We can bound the constant by computing the geometric sum for each of the three possibilities:

$$\sum_{i=0}^{n} i^k r^i < \begin{cases} \dfrac{k}{\ln(1/r)}<1: \ r\left(1+\dfrac{1}{\ln(1/r)-k}\right) = O(1) \\[3mm] \dfrac{k}{\ln(1/r)}=1: \ r\left(2+\dfrac{1}{k}\right) = O(1) \\[3mm] M=\dfrac{k}{\ln(1/r)}>1: \ M^k r^M + \dfrac{r}{k}\left(\dfrac{M^{k+2}}{M-1}\right) = O(1) \end{cases}$$

Examples of the weighted geometric sum $y = x^k r^x$ when $r < 1$

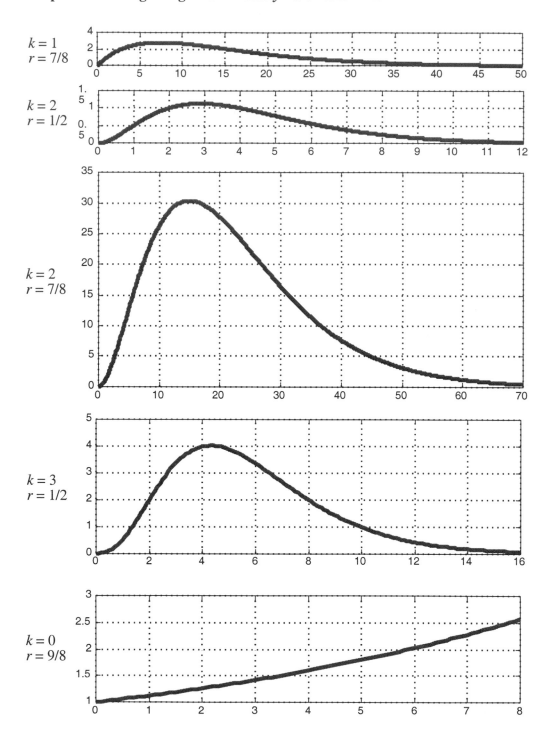

$k = 1$
$r = 7/8$

$k = 2$
$r = 1/2$

$k = 2$
$r = 7/8$

$k = 3$
$r = 1/2$

$k = 0$
$r = 9/8$

I. Harmonic Sum

Idea: This sum might appear at first to be one that converges, but it really just a discrete computation of the logarithm function.

Theorem: For an integer $n \geq 1$:

$$\sum_{i=1}^{n} \frac{1}{i} = \ln(n) + O(1) = O(\log(n))$$

Proof: We can group terms together to get upper and lower bounds:

$$\sum_{i=1}^{n} \frac{1}{i} = 1 + \frac{1}{2} + \frac{1}{3} \cdots \frac{1}{n}$$

$$\leq 1 + (\frac{1}{2} + \frac{1}{3}) + (\frac{1}{4} + \frac{1}{5} + \frac{1}{6} + \frac{1}{7}) + \ldots$$

$$\leq \sum_{i=0}^{\lfloor \log_2(n) \rfloor} \left(2^i \cdot \frac{1}{2^i} \right) = \lfloor \log_2(n) \rfloor + 1 = O(\log(n))$$

$$\sum_{i=1}^{n} \frac{1}{i} = 1 + \frac{1}{2} + \frac{1}{3} \cdots \frac{1}{n}$$

$$\geq 1 + \frac{1}{2} + (\frac{1}{3} + \frac{1}{4}) + (\frac{1}{5} + \frac{1}{6} + \frac{1}{7} + \frac{1}{8}) + \ldots$$

$$\geq 1 + \sum_{i=1}^{\lfloor \log_2(n) \rfloor} \left(2^{i-1} \cdot \frac{1}{2^i} \right) = \frac{1}{2} \lfloor \log_2(n) \rfloor + 1 = \Omega(\log(n))$$

For more precise bounds, we bound the sum, which is *non-increasing*, by integrals:

$$\sum_{i=1}^{n} \frac{1}{i} = 1 + \sum_{i=2}^{n} \frac{1}{i} \leq 1 + \int_{i=1}^{n} \frac{1}{i} = 1 + \left[\ln(i) \right]_{i=1}^{i=n} = \ln(n) + 1 = O(\log(n))$$

$$\sum_{i=1}^{n} \frac{1}{i} = \frac{1}{n} + \sum_{i=1}^{n-1} \frac{1}{i} \geq \frac{1}{n} + \int_{i=1}^{n} \frac{1}{i} = \frac{1}{n} + \left[\ln(i) \right]_{i=1}^{i=n} = \ln(n) + \frac{1}{n} = \Omega(\log(n))$$

Note 1: Since $\ln(10) < 2.31$ and $\log_2(10) > 3.32$, for $n \geq 10$: $\ln(n) + 1 < \log_2(n)$

Note 2: It is possible to show that

$$\sum_{i=1}^{n} \frac{1}{i} = \ln(n) + \gamma + \varepsilon$$

where $\gamma = 0.57721...$ is "Euler's constant" and $\varepsilon < \frac{1}{2n}$.

J. Sums of Inverse Powers

Theorem: For an integer $n \geq 1$ and a real number $r > 1$:

$$\sum_{i=1}^{n} \frac{1}{i^r} \leq \frac{r}{r-1} = O(1)$$

Proof: For $n=1$, the theorem is true since $r>1$; for $n \geq 2$ we bound the sum, which is *non-increasing*, by an integral:

$$\sum_{i=1}^{n} \frac{1}{i^r} = 1 + \sum_{i=2}^{n} \frac{1}{i^r}$$

$$< 1 + \int_{i=1}^{n} \frac{1}{i^r} di$$

$$= 1 + \left[\frac{i^{-r+1}}{-r+1} \right]_{i=1}^{i=n}$$

$$= 1 + \left(\frac{1}{r-1} - \frac{1}{(r-1)i^{n-1}} \right)$$

$$< 1 + \frac{1}{r-1}$$

$$= \frac{r}{r-1}$$

Corollary:

$$\sum_{i=1}^{n} \frac{1}{i^2} \leq 2$$

Note: If we set $r=1$, we have the harmonic sum, which diverges. For the "break-even" point where weighting makes the sum converge, it takes more than $10^{10^{87}}$ terms for

$$\sum_{i=4}^{\infty} \frac{1}{i \log_2(i) \left(\log_2 \log_2(i) \right)^2}$$

to converge to 38.43... with two-decimal accuracy, and the sum

$$\sum_{i=4}^{\infty} \frac{1}{i \log_2(i) \log_2 \log_2(i)}$$

is less than 10 for all practical purposes even though it does diverge (Zwillinger [1996]).

Bibliography

K. Abramson [1963]. *Information Theory and Coding*, McGraw Hill.

K. Abrahamson [1987]. "Generalized String Matching", *SIAM Journal of Computing 16:6*, 1039–1051.

I. T. Adamson [1996]. *Data Structures and Algorithms: A First Course*, Springer.

G. M. Adel'son-Vel'skii and Y. M. Landis [1962]. "An Algorithm for the Organization of Information", *Doklady Akad. Nauk SSR 146*, 263–266 (Russian); English translation in *Soviet Math. Doklady 3* (1962), 1259–1262.

A. Aho and M. Corasick [1975]. "Efficient String Matching: An Aid to Bibliographic Search", *Communications of the ACM 18*, 333–340.

A. V. Aho, J. E. Hopcroft, and J. D. Ullman [1974]. *The Design and Analysis of Computer Algorithms*, Addison–Wesley.

A. V. Aho, J. E. Hopcroft, and J. D. Ullman [1983]. *Data Structures and Algorithms*, Addison–Wesley.

A. V. Aho, R. Sethi, and J. D. Ullman [1986]. *Compilers: Principles, Techniques, and Tools*, Addison–Wesley.

E. K. Ahuja and J. B. Orlin [1989]. "A Fast and Simple Algorithm for the Maximum Flow Problem", *Operations Research 37*, 748–759.

E. K. Ahuja, J. B. Orlin [1989], and R. E. Tarjan [1989]. "Improved Time Bounds for the Maximum Flow Problem", *SIAM Journal of Computing 18*, 939–954.

A. N. Akansu and R. A. Haddad [1992]. *Multiresolution Signal Decomposition: Transforms, Subbands, and Wavelets*, Academic Press.

S. G. Akl [1997]. *Parallel Computation: Models and Methods*, Prentice Hall.

S. G. Akl [1985]. *Parallel Sorting Algorithms*, Academic Press.

N. Alon [1990]. "Generating Pseudo-Random Permutations and the Maximum Flow Problem", *Information Processing Letters 35*, 201–204.

A. Andersson, T. Hagerup, J. Hastad, and O. Petersson [1994]. "The Complexity of Searching a Sorted Array of Strings", *Proceedings Annual ACM Symposium on the Theory of Computing*, 317–325.

A. Andersson, T. Hagerup, S. Nilsson, and R. Raman [1995]. "Sorting in Linear Time?", Proceedings 27th Annual ACM Symposium on the Theory of Computing, 427–436.

A. Apostolico [1984]. "On Context-Constrained Squares and Repetitions in a String", *R.A.I.R.O. Theoretical Informatics 18*, 147–159.

A. Apostolico [1985]. "The Myrad Virtues of Subword Trees", in A. Apostolico and Z. Galid, eds., *Combinatorial Algorithms on Words*, NATO ASI Series, Computer and Systems Sciences Vol. 12, Springer, 85–96.

A. Apostolico, ed. [1997]. *Pattern Matching Algorithms*, Oxford U. Press.

A. Apostolico [1992]. 'Self-Alignments and Their Applications", *Journal of Algorithms 13*, 446–467.

A. Apostolico, M. E. Bock, and S. Lonardi [1999]. "Linear Global Detectors of Redundant and Rare Substrings", *Proceedings Data Compression Conference*, IEEE Computer Society Press, 168–177.

A. Apostolico, M. Crochemore, Z. Galil, and U. Manber [1992–2000]. *Combinatorial Pattern Matching* (an annual symposium), Springer, (the first two symposiums had no proceedings; the editors listed here are for the third symposium).

A. Apostolico and R. Giancarlo [1986]. "The Boyer-Moore String Searching Strategies Revisited", *SIAM Journal of Computing 15*, 98–105.

A. Apostolico and Z. Gaili [1997]. *Pattern Matching Algorithms*, Oxford U. Press.

A. Apostolico, C. Iliopoulos, G. M. Landau, B. Schieber, and U. Vishkin [1988]. "Parallel Construction of a Suffix Tree with Applications", *Algorithmica 3*, 347–365.

A. Apostolico, and S. Lonardi [1998]. "Some Theory and Practice of Greedy Off-Line Textual Substitution", *Proceedings Data Compression Conference*, IEEE Computer Society Press, 119–128.

A. Apostolico, and S. Lonardi [2000]. "Compression of Biological Sequences by Greedy Off-Line Textual Substitution", *Proceedings Data Compression Conference*, IEEE Computer Society Press, 143–152.

A. Apostolico and F. P. Preparata [1982]. "A Structure for the Statistics of all Substrings of a Textstring with or without Overlap", *Proceedings 2nd World Conference on Mathematics*.

V. L. Arlazarov, A. Dinic, M. A. Kronrod, and I. A. Faradzev [1970]. "On Ecnomical Construction of the Transitive Closure of a Directed graph", *Doklady Acad. Nauk SSR 194*, 487–488 (Russian); English translation in *Soviet Math. Doklady 11:5*, 1209–1210.

R. Armoni, A. Ta-Shma, A. Wigderson, and S. Zhou [2000]. "An $O(\log(n)^{4/3})$ Space Algorithm for (s,t) Connectivity", *Journal of the ACM 47:2*, 294–311.

Z. Arnavut, D. Leavitt, and M. Abdulazizoglu [1998]. "Block Sorting of Multset Permutations and Compression", Technical Report, Department of Mathematics and Computer Science, State U. New York at Fredonia; an abstract appears in the Proceedings of the 1998 IEEE Data Compression Conference, 524.

M. J. Atallah, ed. [1999]. *Algorithms and Theory of Computation Handbook*, CRC Press.

L. Auslander and S. V. Parter [1961]. "On Embedding Graphs in the Plane", *Journal of Math. and Mech. 1:3*, 517–523

Y. Azar, A. Broder, A. Karlin, and E. Upfal [2000]. "Balanced Allocations", *SIAM Journal of Computing 29:1*, 180–200; an extended abstract appeared in *Proceedings 26th Anual ACM Symposium on the Theory Computing*, 1994.

S. Baase [1988]. *Computer Algorithms*, Addison–Wesley.

L. Banachowski, A. Kreczmar, and W. Rytter [1991]. *Analysis of Algorithms and Data Structures*, Addison–Wesley.

B. Balkenhol and S. Kurtz [2000]. "Universal Data Compression Based on the Burrows-Wheeler Transform: Theory and Practice", *IEEE Transactions on Computers 49:10*, 1043–1053.

R. A. Baeza–Yates and G. H. Gonnet [1992]. "A New Approach to text Searching", *Communicaions of the ACM 35:10*, 74–82.

D. Baron and A. C. Singer [2001]. "On the Cost of Worst-Case Coding Length", to appear IEEE Transactions on Information Theory; also, Technical Report, Dept. Electrical Engineering, U. Illinois.

D. Basu [1991]. "Text Compression Using Several Huffman Trees", *Proceedings Data Compression Conference*, IEEE Computer Society Press, 452.

G. Battista and R. Tamassia [1989]. "Incremental Planarity Testing", *Proceedings 30th Annual IEEE Symposium on the Foundations of Computer Science*, 436–441.

R. Bayer [1972]. "Symmetric Binary B-Trees: Data Structure and Maintenance Algorithms", *Acta Informatica 1:3*, 173–189.

R. Bayer and E. M. McCreight [1972]. "Organization and Maintenance of Large Ordered Indices", *Acta Informatica 1*.

J. Beidler [1995]. *Data Structures and Algorithms: An Object-Oriented Approach Using Ada 95*, Springer.

T. Bell, J. Cleary, and I. Witten [1990]. *Text Compression*, Prentice Hall.

T. Bell, I. Witten, and J. Cleary [1989]. "Modelling for Text Compression", *ACM Computing Surveys 21:4*, 557–591.

R. Bellman [1957]. *Dynamic Programming*, Princeton U. Press.

J. Bentley [1975]. "Multidimensional binary search trees used for associative searching", *Communications of the ACM 18:9*, 509–517.

J. Bentley [1999]. Private communication.

J. Bentley, D. Sleator, R. Tarjan, and V. Wei [1986]. "A Locally Adaptive Data Compression Scheme", *Communications of the ACM 29:4*, 320–330.

P. Berenbrink, A. Czumaj, A. Steger, and B. Vöcking [2000]. "Balanced Applications: The Heavily Loaded Case", *Proceedings 32nd Annual ACM Symposium on the Theory Of Computing*, 745–754.

C. Berge [1957]. "Two Theorems in Graph Theory", *Proceedings of the National Academy of Sciences 43*, 842–844.

C. Berge [1962]. *The Theory of Graphs and Its Applications*, Wiley.

V. Bhaskaran and K. Konstantinides [1995]. *Image Compression Standards: Algorithms and Architecture*, Kluwer.

S. N. Bhatt and C. E. Leiserson [1982]. "How to Assemble Tree Machines", *Proceedings ACM Symposium on the Theory of Computing*, 77–84.

N. Biggs [1993]. *Algebraic Graph Theory*, Cambridge U. Press.

D. Bini and V. Pan [1994]. *Polynomial and Matrix Computations*, Birkhäuser.

B. Bloom [1970]. Space / Time Tradeoffs in Hash Coding with Allowable Errors, *Communications of the ACM 13:7*, 422-426.

M. Blum, R. W. Floyd, V. R. Pratt, R. L. Rivest, and R. E. Tarjan [1972]. "Time Bounds for Selection", *Journal of Computer and System Sciences 7:4*, 448–461.

A. Blumer and R. McEliece [1988]. "The Renyi redundancy of generalized Huffman codes", *IEEE Trans. Inform. Theory IT-34*, 1242–1249.

A. Blumer, A. Efrenfeucht, and D. Haussler [1987]. "Average Sizes of Suffix Trees and DWAGS", *Proceedings 1ˢᵗ Conference of Combinatorics*, U. Montreal, Canada.

A. Blumer, J. Blumer, and D. Haussler [1985]. "The Smallest Automaton recognizing the Subwords of a Text", *Theoretical Computer Science 40*, 31–55.

A. Blumer, J. Blumer, D. Haussler, and R. McConnell [1984]. "Building a Complete Inverted File for a Set of Text Files", *Proceedings of the Annual ACM Symposium on the Theory of Computing*, 349–358.

A. Blumer, J. Blumer, D. Haussler, and R. McConnell [1987]. "Complete Inverted Files for Efficient Text retrieval and Analysis", *Journal of the ACM 34:3*, 578–595.

K. P. Bogart [1983]. *Introductory Combinatorics*, Pitman Publishing Co.

J. A. Bondy and U. S. R. Murty [1976]. *Graph Theory with Applications*, Elsevier.

R. S. Boyer and J. S Moore [1977]. "A Fast String Searching Algorithm", *Communications of the ACM 20*, 762–772.

R. N. Bracewell [1986]. *The Fourier Transform and its Applications*, McGraw-Hill.

E. Orn Brigham [1974]. *The Fast Fourier Transform*, Prentice Hall.

R. A. Brualdi [1977]. *Introductory Combinatorics*, North-Holland.

R. G. Busacker and T. L. Saaty [1965]. *Finite Graphs and Networks: An Introduction with Applications*, McGraw-Hil.

G. Brassard and P. Bratley [1988]. *Algorithmics: Theory and Practice*, Prentice Hall.

G. Brassard and P. Bratley [1996]. *Fundamentals of Algorithmics*, Prentice Hall.

P. Bratley and J. Millo [1972]. "Computer Recreations; Self-Reproducing Automata", *Software Practice & Experience 2*, 397–400.

J. Burger, D. Brill, and F. Machi 1980]. "Self-Reproducing Programs", *Byte*, 74–75.

M. Burrows and D. J. Wheeler [1994]. "A Block-Sorting Lossless Data Compression Algorithm", SRC Research Report 124, Digital Systems Research Center, 130 Lytton Avenue, Palo Alto, CA 94301.

R. Capocelli and A. De Santis [1988]. "Tight upper bounds on the redundancy of Huffman codes", *Proceedings IEEE International Symposium on Information Theory*, Kobe, Japan; also in *IEEE Trans. Inform. Theory IT-35:5* (1989).

R. Capocelli and A. De Santis [1991]. "A note on D-ary Huffman codes", *IEEE Trans. Inform. Theory IT-37:1.*

R. Capocelli and A. De Santis [1991]. "New Bounds on the Redundancy of Huffman Code", *IEEE Trans. Inform. Theory IT-37*, 1095–1104.

R. Capocelli and A. De Santis [1992]. "Variations on a Theme by Gallager", *Image and Text Compression*, Kluwer Academic Press.

R. Capocelli, R. Giancarlo, and I. Taneja [1986]. "Bounds on the redundancy of Huffman codes", *IEEE Trans. Inform. Theory IT-32:6*, 854–857.

B. Carpentieri and J. A. Storer [1994]. "Split-Merge Video Displacement Estimation", *Proceedings of the IEEE 82:6*, 940–947.

B. Carpentieri and J. A. Storer [1994]. "Optimal Inter-Frame Alignment for Video Compression", *International Journal of Foundations of Computer Science 5:2*, 65–177.

B. Carpentieri and J. A. Storer [1996]. "A Video Coder Based on Split-Merge Displacement Estimation", *Journal of Visual Communication and Visual Representation 7:2*, 137–143.

J. L. Carter and M. N. Wegman [1979]. "Universal Classes of Hash Functions", *Journal of Computer and System Sciences 18*, 143–154.

B. Chapin [2000]. "Switching Between Two List Update Algorithms for Higher Compression of Burrows-Wheeler Transformed Data", *Proceedings Data Compression Conference, IEEE Computer Society Press*, 183–192.

M. Charikar, S. Guha, E. Tardos, and D. Shmoys [1999]. "A Constant-Factor Approximation Algorithm for the k-Median Problem", *Proceedings 31st ACM Symposium on the Theory of Computing*, 1–10.

B. Chazelle [2001]. "The Soft Heap: An Approximate Priority Queue with Optimal Error Rate", *Journal of the ACM 47:6*, 1012–1027.

B. Chazelle [2001]. "A Minimum Spanning Tree Algorithm with Inverse-Ackermann Type Complexity", *Journal of the ACM 47:6*, 1028–1047.

W.H. Chen, C. H. Smith, and S. C. Fralick [1977]. "A Fast Computational Algorithm for the Discrete Cosine Transform", *IEEE Transactions on Communications 25:9*, 1004–1009.

M. T. Chen and J. Sieferas [1985]. "Efficient and Elegant Subword-Tree Construction", in A. Apostolico and Z. Galid, eds., *Combinatorial Algorithms on Words*, NATO ASI Series, Computer and Systems Sciences Vol. 12, Springer, 97–107.

J. Cheng and G. Langdon [1992]. "QM-AYA Adaptive Arithmetic Coder", *Proceedings Data Compression Conference*, IEEE Computer Society Press, 428.

J. Cheriyan, T. Hagerup, and K. Mehlhorn [1996]. "$O(n^3)$-time Maximum-Flow Algorithm", *SIAM Journal of Computing 45*, 1144–1170.

B. V. Chernasky [1977]. "Algorithm for the Construcion of Maximal Flow in Networks with Complexity of $O(V^2E^{1/2})$ Operations", *Mathematical Methods of Solution of Echnomical Problems 7*, 112–125 (Russian).

B. V. Cherassky and A. V. Goldberg [1997]. "On Implementing the Push-Relabel Method for the Maximum Flow Problem", *Algorithmica 19*.

N. Christofides [1975]. *Graph Theory an Algormithic Approach*, Academic Press.

C. K. Chui [1992]. *An Introduction to Wavelets*, Academic Press.

C. K. Chui [1992b]. *Wavelets: A Tutorial in Theory and Applications*, Academic Press.

C. K. Chui [1995]. *Wavelets: Theory, Algorithms, and Applications*, Academic Press.

C. K. Chui [1998]. *Wavelet Tool Box*, Academic Press.

C. K. Chui [1998b]. *Wavelets in a Box*, Academic Press.

R. J. Clarke [1985]. *Transform Coding of Images*, Academic Press.

J. Cleary and I. Witten [1984]. "Data Compression Using Adaptive Coding and Partial String Matching", *IEEE Transactions on Communications 32:4*, 396–402.

J. Cleary and I. Witten [1984b]. "A Comparison of Enumerative and Adaptive Codes", *IEEE Trans. Information Theory IT-30:2*, 306–315.

J. Cleary, W. Teahan, and I. Witten [1995]. "Unbounded Length Contexts for PPM", *Proceedings Data Compression Conference,* IEEE Computer Society Press, 52–61.

R. Cleve and J. Watrous [2000]. "Fast Parallel Circuits for the Quantum Fourier Transform", *Proceedings 41st IEEE Annual Symposium on Foundations of Computer Science*, 526–535.

R. Cole [1991]. "Tight Upper Bounds on the Complexity of the Boyrer-Moore Pattern Matching Algorithm", *Proceedings Second Annual Symposium on Discrete Algorithms*, 224–233.

R. Cole and R. Hariharan [2000]. "Faster Suffix Tree Construction with Missing Suffix Links", *Proceedings ACM Annual Symposium on the Theory of Computing*, 407–415.

E. Von Collani and K. Drager [2001]. *Binomial Distribution Handbook*, Birkhäuser.

J. Combes and A. Grossman [1989]. *Wavelets: Time-Frequency Methods and Phase Space*, Springer.

C. Constantinescu and J. A. Storer [1994]. "On-Line Adaptive Vector Quantization with Variable Size Codebook Entries", *Information Processing and Management 30:6*, 745–758.

C. Constantinescu and J. A. Storer [1994b]. "Improved Techniques for Single-Pass Vector Quantization", *Proceedings of the IEEE 82:6*, 933–939.

J. W. Cooley and J. Tukey [1965]. "An Algorithm for the Machine Calculation of Complex Fourier Series", *Mathematics of Computation 19*, 297–301.

D. Coppersmith and S. Winograd [1987]. "Matrix Multiplication Via Arithmetic Progressions", *Proceedings 19th Annual ACM Symposium on the Theory of Computing*, 1–6.

T. H. Cormen, C. E. Leiserson, R. L. Rivest, and C. Stein [2001]. *Introduction to Algorithms*, second edition, McGraw-Hill; the first edition, by the first three authors, was printed in 1990.

M. Cosnard and D. Trystram [1995]. *Parallel Algorithms and Architectures*, Thomson Computer Press.

D. J. Cooke and H. E. Bez [1984]. *Computer Mathematics*, Cambridge U. Press.

G. Cormack and R. Horspool [1984]. "Algorithms for Adaptive Huffman Codes", *Information Processing Letters 18*, 159–165.

M. Crochermore, A. Czumaj, L. Gasieniec, S. Jarominek, T. Lecroq, W. Plandowski, and W. Rytter [1992]. "Speeding up Two String Matching Algorithms", *Proceedings 9th Annual Symposium on the Theoretical Aspects of Computer Science*, Springer, A. Finkel and M. Jantzen Eds., 589–600.

P. Cull [1995]. The Mobius Cubes, *IEEE Transactions on Computers 44:5*, 647–659.

W. H. Cutler [1977]. "The Six Piece Burr", *J. Recreational Math. 10:4*, 241–250.

W. H. Cutler [1986]. "Holey 6-Piece Burr", manuscript.

W. H. Cutler [1994]. "A Computer Analysis of All 6-Piece Burrs", manuscript.

G. Dahlquist and A. Bjorck [1974]. *Numerical Methods*, Prentice Hall.

G. B. Dantzig [1951]. "Application of the Simplex Method to a Transportation Problem", In *Activity Analysis and Production and Allocation*, Wiley, 359–373.

I. Daubechies [1992]. *Ten Lectures on Wavelets*, Society for Industrial and Applied Mathematics Press, Philadelphia, PA, Capital City Press, Montpelier, VT.

S. DeAgostino and J. A. Storer [1992]. Parallel Algorithms for Optimal Compression Using Dictionaries with the Prefix Property", *Proceedings Data Compression Conference*, IEEE Computer Society Press, 152–161.

S. DeAgostino and J. A. Storer [1992]. "On-Line versus Off-Line Computation in Dynamic Text Compression", *Information Processing Letters 59*, 169–174.

L. Debnath [2000]. *Wavelet Transforms and Time-Frequency Signal Analysis*, Birkhäuser.

A. K. Dewdney [1985]. "Bill's Baffling Burrs...", *Sci. American (Oct.)*, 16–27.

M. Dietzfelbinger, A. Karlin, K. Mehlhorn, F. M. Meyer Auf Der Heide, H. Ronhert, and R. E. Tarjan [1994]. "Dynamic Perfect Hashing: Upper and Lower Bounds", *SIAM Journal of Computing 23:4*, 738–761.

Dijkstra [1959]. "A Note on Two Problems in Connection with Graphs", *Numerische Mathematik 1*, 269–271.

E. A. Dinic [1970]. "Algorithm for Solution of a Problem of Maximum Flow in a Network with Power Estimation", *Soviet Math. Doklady 11*, 1277–1280.

E. A. Dinic [1973]. "Metod Porazyadnogo Sokrashcheniya Nevyazok I Transportnye Zadachi", *Issledovaniya Po Diskretnoi Matematike*, Nauka, Moskva, 46–57 (Russian).

A. J. W. Duijvestijn [1972]. "Correctness Proof of an In-Place Permutation", *BIT 12:3*, 318–324.

J. Earley [1970]. "An Efficient Context-Free Parsing Algorithm", *Communications of the ACM 13:2*, 94–102.

J. Edmonds and R. M. Karp [1972]. "Theoretical Improvements in Algorithmic Efficiency for Network Flow Problems", *Journal of the ACM 19*, 248–264.

M. Effros [1999]. "Universal Lossless Source Coding with the Burrows-Wheeler Transform", *Proceedings Data Compression Conference,* IEEE Computer Society Press, 178–187.

S. Even [1979]. *Graph Algorithms*, Computer Science Press / W. H. Freeman Press.

S. Even and O. Kariv [1975]. "An $O(n^{2.5})$ Algorithm for Matching in General Graphs", *Proceedings 16th Annual IEEE Symposium on the Foundations of Computer Science*, 100–112.

N. Faller [1973]. "An Adaptive System for Data Compression", *7th Asilomar Conference on Circuits and Systems*, 593–597.

L. Fan, P. Cao, J. Almedia, and A. Broder [1998]. "Summary Cache: A Scalable Wide-Area Web Cache Sharing Protocol", *Proceedings SIGCOMM*; full paper is contained in Technical Report 1361, Computer Science Dept., U. Wisconsin.

M. Farach-Colton, P. Ferragina, and S. Muthukrishnan [2001]. "On the Sorting-Complexity of Suffix Tree Construction", *Journal of the ACM 47:6*, 987–1011.

E. Feig and E. Linzer [1990]. "Discrete Cosine Transform Algorithms for Image Data Compression", *Proceedings Electronic Imaging*, 84–87.

P. M. Fenwick [1996]. "The Burrows-Wheeler Transform for Block Sorting Text Compression: Principles and Improvements", *The Computer Journal 39:9*, 731–740.

E. R. Fiala and D. H. Greene [1989]. "Data Compression with Finite Windows", *Communications of the ACM 32:4*, 490–505.

M. J. Fischer and A. R. Meyer [1971]. "Boolean Matrix Multiplication and Transitive Closure", *Proceedings 12th Annual Symposium on Switching and Automata*, 129–131.

R. W. Floyd [1962]. "Algorithm 97: Shortest Path", *Communications of the ACM 5:6*, 345.

R. W. Floyd and R. L. Rivest [1975]. "Expected Time Bounds for Selection", *Communications of the ACM 18:3*, 165–172.

L. R. Ford and D. R. Fulkerson [1956]. "Maximal Flow Through a Network", *Canadian Journal of Mathematics 8*, 399–404.

L. R. Ford and D. R. Fulkerson [1962]. *Flows in Networks*, Princeton U. Press.

W. D. Frazer and A. C. McKellar [1970]. "Samplesort: A Sampling Approach to Minimal Storage Tree Sorting", *Journal of the ACM 17:3*, 496–507.

E. Fredkin [1960]. "Trie Memory", *Communications of the ACM 7:12*, page 101.

M. L. Fredman, J. Komlos, and E. Szemeredi [1984]. "Storing a Sparse Table with O(1) Worst-Case Acess Time", *Journal of the ACM 31:3*, 538–544.

M. L. Fredman and D. E. Willard [1990]. "BLASTING Through the Information Theoretic Barrier with FUSION TREES", *Proceedings 22nd Annual ACM Symposium on the Theory of Computing*, 1–7.

H. N. Gabow [1985]. "Scaling Algorithms for Network Problems", *Journal of Computer and System Sciences 31*, 148–168.

H. N. Gabow, H. Kaplan, and R. E. Targan [1999]. "Unique Maximum Matching Algorithms", *Proceedings 31st ACM Annual Symposium on the Theory of Computing*, 70–78.

D. Gale and L. S. Shapely [1962]. "College Admissions and the Stability of Marriage", *American Mathematical Monthly 69*, 9–15.

Z. Galil [1978]. "An $O(V^{5/3}E^{2/3})$ Algorithm for the Maximal Flow Problem", *Acta Informatica 14*, 221–242.

Z. Galil [1979]. "On Improving the Worst-Case Running Time of the Boyer-Moore String-Searching Algorithm", *Communications of the ACM 22*, 505–508.

Z. Galil and A. Naamad [1980]. "An $O(EV\log^2 V)$ Algorithm for the Maximal Flow Problem", *Journal of Computer and Systems Sciences 21*, 203–217.

Z. Galil [1985]. "Optimal Parallel Algorithms for String Matching", *Information and Control 67*, 144–157.

Z. Galil and R. Giancarlo [1986]. "Improved String Matching with k Mis-matches", *SIGACT News 17*, 52–54.

Z. Galil and J. Sieferas [1983]. "Time-Space Optimal String Matching", *Journal of Computer and System Sciences 26*, 280–294.

R. Gallager [1978]. "Variations on a Theme by Huffman", *IEEE Transactions on Information Theory 24:6*, 668–674.

J. Gallant, D. Maier, and J. A. Storer [1980]. "On Finding Minimal Length Superstrings", *Journal of Computer and System Sciences 20*, 50–58.

M. Gardner [1978]. "Mathematical Games", *Sci. American (Jan.)*, 14–25.

M. R. Garey and D. S. Johnson [1979]. *Computers and Intractability: A Guide to the Theory of NP-Completeness*, W. H. Freeman and Company, San Francisco.

C. F. Gauss [1866]. "Theoria Interpolationis Methodo Nova Tractata", in *Werke III, Nachlass*, Konigliche Gesellschaft der Wissenschaften, 265–330.

R. Giancarlo [1995]. "A Generalization of the Suffix Trie Construction to Square Matrices, with Applications", *SIAM J. of Computing 24:3*, 520–562.

R. Giancarlo and R. Grossi [1995]. "On the Construction of Classes of Suffix Trees for Square Matrices: Algorithms and Applications", *Proceedings ICALP*.

R. Giancarlo and D. Guaiana [1996]. "When Suffix Trees Bloom or Fully Dynamic Data Structures for Image Data Compression", Technical Report, Dipartimento di Matematica ed Applicazioni, U. Palermo, Palermo, Italy.

R. Giancarlo and D. Guaiana [1997]. "On-Line Construction of Two-Dimensional Suffix Tries", Technical Report, Dipartimento di Matematica ed Applicazioni, U. Palermo, Palermo, Italy.

A. Gibbons [1985]. *Algorithmic Graph Theory*, Cambridge U. Press.

A. Gibbons and P. Spirakis, eds. [1993]. Lectures on Parallel Computation, Cambridge U. Press.

A. Gibbons and W. Rytter [1988]. *Efficient Parallel Algorithms*, Cambridge U. Press.

R. Giegerich [2000]. "Explaining and Controlling Ambiguity in Dynamic Programming", *Proceedings Combinatorial Pattern Matching (CPM)*, Montreal, Canada, 46–59.

E. Gilbert [1971]. "Codes Based on Inaccurate Source Probabilities", *IEEE Transactions on Information Theory 17:3*, 304–314.

E. N. Gilbert and E. F. Moore [1959]. "Variable Length Encodings", *Bell System Technical Journal 38:4*, 933–968.

A. V. Goldberg [1987]. "Efficient Graph Algorithms for Sequential and Parallel Computers", Ph.D. Thesis, Department of Electrical Engineering and Computer Science, MIT.

A. V. Goldberg and R. E. Tarjan [1986]. "A New Approach to the Maximum Flow Problem", *Proceedings 18th Annual ACM Symposium on the Theory of Computing*, 136–146.

A. V. Goldberg and S. Rao [1998]. "Beyond the Flow Decomposition Barrier", *Journal of the ACM 45:5*, 1–15.

A. J. Goldstein [1963]. "An Efficient and Constructive Algorithm for Testing Whether a Graph Can be Embedded in the Plane", Technical Report, Department of Mathematics, Princeton U.

M. J. Golin and H. S. Na [2001]. "Optimal Prefix-Free Codes That End in a Specified Pattern and Similar Problems", *Proceedings Data Compression Conference*, IEEE Computer Society Press, 143–152.

S. Golomb [1980]. "Sources which Maximize the Choice of a Huffman Coding Tree", *Information and Control 45*, 263–272.

G. H. Golub and C. F. Van Loan [1983]. *Matrix Computations*, Johns Hopkins U. Press.

M. C. Golumbic [1980]. *Algorithmic Graph Theory and Perfect Graphs*, Academic Press.

G. H. Gonnet [1981]. Expected Length of the Longest Probe in Hash Code Searching", *Journal of the ACM 28*, 289–304.

G. H. Gonnet and R. Baeza-Yates [1991]. *Handbook of Algorithms and Data Structures*, Addison–Wesley.

M. Gonzalez and J. A. Storer [1985]. "Parallel Algorithms for Data Compression", *Journal of the ACM 32:2*, 344–373.

R. C. Gonzalez and Paul Wintz [1987]. *Digital Image Processing*, Addison–Wesley.

R. C. Gonzalez and R. E. Woods [1992]. *Digital Image Processing*, Addison–Wesley.

M. Goodrich and R. Tamassia [1995]. *Data Structures and Algorithms in Java*, Wiley.

M. Goodrich and R. Tamassia [2001]. *Algorithm Engineering: Design, Analysis, and Java Case Studies*, Wiley.

G. C. Gotlieb and L. R. Gotlieb [1978]. *Data Types and Structures*, Prentice Hall.

R. M. Gray and J. W. Goodman [1995]. *Fourier Transforms*, Kluwer.

R. Greenlaw, M. Halldorsson, and R. Petreschi [2000]. "Computing Prufer Codes Efficiently in Parallel", Journal of Discrete Applied Mathematics; see also, Technical Report, Dept. Computer Science, Armstrong Atlantic State U.

R. Greenlaw and R. Petreschi [2000]. "On Computing Prufer Codes and Their Corresponding Trees Optimally in Parallel", Proceedings JIM; see also, Technical Report, Dept. Computer Science, Armstrong Atlantic State U.

K. Grochenig [2000]. *Foundations of Time-Frequency Analysis*, Birkhäuser.

S. L. Graham, M. A. Harrison, and W. L. Ruzzo [1976]. "On-Line Context-Free Language Recognition in Less than Cubic Time", *Proceedings 8th Annual Symposium on the Theory of Computing*, 112–120.

R. L. Graham and P. Hell [1985]. "On the History of the Minimum Spaning Tree Problem", *Annals of the History of Computing 7*, 43–57.

R. L. Graham, D. E. Knuth, and O. Patashnik [1989]. Concrete Mathematics: A Foundation for Computer Science, Addison–Wesley.

W. K. Grassmann and J. P. Tremblay [1996]. *Logic and Discrete Mathematics*, Prentice Hall.

D. H. Greene and D. E. Knuth [1988], *Mathematics for the Analysis of Algorithms*, Birkhäuser.

D. Gries and J. Xue [1988]. "The Hopcroft-Tarjan Planarity Algorithm, Presentations and Improvements", Technical Report 88–906, Department of Computer Science, Cornell U.

R. Grossi and J. S. Vitter [2000]. "Compressed Suffix Arrays and Suffix Trees with Applications to Text Indexing and String matching", *Proceedings ACM Annual Symposium on the Theory of Computing*, 397–406.

Gthompso [2001]. The "Quine" web page: www.nyx.net/~gthompso/quine.htm

L. J. Guibas and R. Sedgewick [1978]. "A Diochromatic Framework for Balanced Trees", *Proceedings 19th Annual Sumposium on the Foundations of Computer Science*, 8–21.

L. J. Guibas and A. M. Odlyzko [1980]. "Periods in Strings", *Journal of Combinatorial Theory 30*, 19–42.

L. Guivarch, J. Carlach, and P. Siohan [2000]. "Joint Source-Channel Soft Decoding of Huffman Codes with Turbo-Codes", *Proceedings Data Compression Conference*, IEEE Computer Society Press, 83–92.

D. Gusfield [1989]. *The Stable Marriage Problem: Structure and Algorithms*, MIT Press.

D. Gusfield [1997]. *Algorithms on Strings, Trees, and Sequences: Computer Science and Computational Biology*, Cambridge U. Press.

D. Gusfield, G. M. Landau, and B. Schieber [1991]. "An Efficient Algorithm for the All-Pairs Suffix-Prefix Problem", Technical Report, Computer Science Division, U. of California, Davis.

M. Habib, C. McDiarmid, J. Ramirez-Alfonsin, and B. Reed [1998]. *Probabilistic Methods for Algorithmic Discrete Mathematics*, Springer.

T. Hagerup [1994]. "Optimal Parallel String Algorithms: Merging, Sorting, and Computing the Minimum", *Proceedings Annual ACM Symposium on the Theory of Computing*, 382–391.

T. Hagerup and C. Rub [1989]. "A Guided Tour of Chernoff Bounds", *Information Processing Letters 33*, 305–308.

L. Hales and S. Hallgren [2000]. "An Improved Quantum Fourier Transform Algorithm and Applications", *Proceedings 41st IEEE Annual Symposium on Foundations of Computer Science*, 515–524.

M. Hall [1948]. "Distinct Representatives of Subsets", *Bulletin of the AMS 54*, 922–926.

M. Hankamer [1979]. "A modified huffman procedure with reduced memory requirements", *IEEE Transactions on Communications 27:6*, 930–932.

R. M. Haralick [1976]. "A Storage Efficient Way to Implement the Discrete Cosine Transform", *IEEE Transactions on Computers C-25*, 764–765.

F. Harary [1969]. *Graph Theory*, Addison–Wesley.

R. Hariharan [1994]. "Optimal Parallel Suffix Tree Construction", *Proceedings Annual ACM Symposium on the Theory of Computing*, 290–299.

B. G. Haskell, A. Puri, and A. N. Netravali [1997]. Digital Video: *An Introduction to MPEG-2*, Chapman and Hall.

L. Hay [1980]. "Self-Reproducing Programs", *Creative Computing*, 134–136.

H. Helfgott [1997]. "Sort-Based Compression", manuscript, Computer Science Department, Brandeis U.

F. Hennie [1977]. *Introduction to Computability*, Addison–Wesley.

P. Henrici [1964]. *Elements of Numerical Analysis*, Wiley.

T. Hickey [1997]. Course notes, Department of Computer Science, Brandeis University.

W. D. Hillis [1982]. *The Connection Machine*, MIT Press.

C. A. R. Hoare [1961]. "Algorithm 63 (Partition) and Algorithm 65 (Find)", *Communications of the ACM 4:7*, 321–322.

C. A. R. Hoare [1962]. "Quicksort", *Computer Journal 5:1*, 10–15.

D. S. Hochbaum [1997], ed. *Approximation Algorithms for NP-Hard Problems*, PWS Publishing Company.

K. Hoffman and R. Kunze [1971]. *Linear Algebra*, Prentice Hall.

C. M. Hoffman and M. J. O'Donnell [1982]. "Pattern Matching in Trees", *Journal of the ACM 29:1*, 68–95.

M. Hofri [1987]. *Probabilistic Analysis of Algorithms*, Springer.

D. Hofstadter [1980]. *Godel, Escher, and Bach*, Basic Books, Chapter 16.

J. E. Hopcroft [1974]. Lecture notes, Cornell U.

J. E. Hopcroft and R. M. Karp [1973]. "An $n^{5/2}$ Algorithm for Maximum Matching in Bipartite Graphs", *Siam Journal of Computing 2*, 225–231.

J. E. Hopcroft and R. E. Tarjan [1974]. "Efficient Planarity Testing", *Journal of the ACM 21:4*, 549–568.

J. E. Hopcroft and J. D. Ullman [1979]. *Introduction to Automata Theory, Languages, and Computation*, Addison–Wesley.

J. E. Hopcroft, R. Motwani, and J. D. Ullman [2001]. *Introduction to Automata Theory, Languages, and Computation*, Addison–Wesley.

E. Horowitz and S. Sahni [1998]. *Computer Algorithms*, Computer Science Press / W. H. Freeman Press.

T. C. Hu [1968]. "A Decomposition Algorithm for Shortest Paths in a Network", Operations Research 16, 91–102.

T. C. Hu and A. C. Tucker [1971]. "Optimum Binary Search Trees", *SIAM Journal of Applied Mathematics 21:4*, 514–532.

D. Huffman [1952]. "A Method for the Construction of Minimum-Redundancy Codes", *Proceedings of the IRE* 40, 1098–1101.

R. Y. K. Isal and A. Moffat [2001]. "Parsing Strategies for BWT Compression", *Proceedings Data Compression Conference*, IEEE Computer Society Press.

J. JaJa [1992]. *An Introduction to Parallel Algorithms*, Addison–Wesley.

G. Jacobson [1992]. "Random Access in Huffman-coded Files", *Proceedings Data Compression Conference*, IEEE Computer Society Press, 368–377.

M. Jacobsson [1978]. "Huffman coding in Bit-Vector Compression", *Information Processing Letters* 7:6, 304–307.

P. Jacquet and W. Szpankowski [1994]. "Autocorrelation on Words and Its Applications", *Journal of Combinatorial Theory, Series A 66:2*, 237–269.

N. S. Jayant and P. Noll [1984]. *Digital Coding of Waveforms: Principles and Applications to Speech and Video*, Prentice Hall.

T. Jiang, M. Li, and P.Vitanyi [2000]. "A Lower Bound on the Average-Case Complexity of Shellsort", *Journal of the ACM 47:5*, 883–904.

O. Johnsen [1980]. "On the Redundancy of Binary Huffman Codes", *IEEE Transactions on Information Theory* 26:2, 220–222.

D. S. Johnson [1985]. Technical Report, Bell Laboratories, Murray Hill, NJ.

D. S. Johnson, A. Demers, J. D. Ullman, M. R. Garey, and R. L. Graham [1974]. "Worst-Case Performance Bounds for Simple One-Dimensional Packing Algorithms", *SIAM Journal of Computing 3*, 299–325.

D. S. Johnson [1973]. *Near-Optimal Bin Packing Algorithms*, Ph.D. Thesis, Department of Mathematics, MIT.

D. S. Johnson and C. Mcgeoch, eds. [1992]. *Network Flow and Matching, Volume 12 of DIMACS Series in Discrete Mathematics and Theoretical Computer Science*, Center for Discrete Mathematics and Theoretical Computer Science (DIMACS), Rutgers U., published by the American Mathematical Society.

A. Karatsuba and Y. Ofman [1962]. "Mulitplication of Multidigit Numbers on Automata", *Doklady Akad. Nauk SSSR 145*, 293–294 (Russian).

D. Karger, P. Klein, and R. Tarjan [1995]. "A Randomized Linear-Time Algorithm to Find Minimum Spanning Trees", *Journal of the ACM 42:2*, 321–328.

J. Karkkainen, G. Navarro, and E. Ukkonen [2000]. "Approximate String Matching over Ziv-Lempel Compressed Text", *Proceedings Combinatorial Pattern Matching (CPM)*, Montreal, Canada (Springer), 195–209.

R. Karp [1961]. "Minimum-Redundancy Coding for the Discrete Noisless Channel", *IRE Transactions on Information Theory 7*, 27–39.

R. M. Karp [1972]. "Reducibility among Combinatorial Problems", in R. E. Miller and J. W. Thatcher (eds.), *Complexity of Computer Computations*, Plenum Press, 85–103.

R. M. Karp and M. O. Rabin [1987]. "Efficient Randomized Pattern Matching", *IBM Journal of Research and Development 31:2*, 249–260.

M. Karpinski and W. Rytter [1998]. *Fast Parallel Algorithms for Graph Matching Problems*, Oxford Science Publications.

T. Kasami [1965]. "An Efficient Recognition and Syntax Algorithm for Context-Free Languages", Scientific Report AFCRL-65-758, Air Force Cambridge Research Lab, Bedford, MA.

T. Kasami and K. Torii [1969]. "An Efficient Recognition and Syntax Algorithm for Unambiguous Context-Free Grammars", *Journal of the ACM 16:3*, 423–431.

A. V. Karzanov [1974]. "Determining the Maximal Flow in a Network by a Method of Preflows, *Soviet Math. Doklady 15*, 434–437.

T. Kida, M. Takeda, A. Shinohara, and S. Arikawa [1999]. "Shift-And Approach to Pattern Matching in LZ Compressed Text", *Proceedings Combinatorial Pattern Matching (CPM)*, Warwick, UK (Springer), 1–13.

S. R. Kim and K. Parl [2000]. "A Dynamic Edit Distance Table", *Proceedings Combinatorial Pattern Matching (CPM)*, Springer, 60–68.

V. King, S. Rao, and R. E. Tarjan [1994]. "A Faster Deterministic Maximum Flow Algorithm", *Journal of Algorithms 17*, 447–474; an extended abstract appeared in *Proceedings of the 3rd Annual ACM-SIAM Symposium on Discrete Algorithms*, 157–164.

J. H. Kingston [1990]. *Algorithms and Data Structures*, Addison–Wesley.

S. C. Kleene [1952]. *Introduction to Mathematics*, Van Nostrand.

S. C. Kleene [1956]. "Representation of Events in Nerve Nets and Finite Automata", *Automata Studies*, Shannon and McCarthy eds., Princeton U. Press, 3–40.

S. T. Klein and D. Shapira [2000]. "A New Compression Method for Compressed Matching", *Proceedings Data Compression Conference*, IEEE Computer Society Press, 400–409.

S. T. Klein and D. Shapira [2001]. "Pattern Matching in Huffman Encoded Texts", *Proceedings Data Compression Conference*, IEEE Computer Society Press, 449–458.

G. D. Knott [2000]. *Interpolating Cubic Splines*, Birkhäuser.

R. D. Knott [2001]. Fibonacci Numbers Web Page, Department of Computing, University of Surrey, UK, www.ee.surrey.ac.uk/Personal/R.Knott/Fibonacci/fib.html.

D. Knuth [1969]. *The Art of Computer Programming, Volumes I, II, III*, Addison–Wesley.

D. Knuth [1971]. "Optimal Binary Search Trees", *Acta Informatica 1*, 14–25.

D. Knuth [1982]. "Huffman's Algorithm via Algebra", *Journal Combinatorial Theory Series A* 32, 216–224.

D. Knuth [1985]. "Dynamic Huffman Coding", *Journal of Algorithms* 6, 163–180.

D. E. Knuth, J. H. Morris Jr., and V. R. Pratt [1977]. "Fast Pattern Matching in Strings", *SIAM Journal of Computing* 6, 323–350.

B. Kolman and R. C. Busby [1984]. *Discrete Mathematical Structures for Computer Science*, Prentice Hall.

S. R. Kosaraju [1989]. "Efficient Tree Pattern Matching", Proceedings 30th Annual IEEE Symposium on the Foundations of Computer Science, Research Triangle Park, NC, 178–183.

S. R. Kosaraju [1994]. "Real-Time Pattern Matching and Quasi-Real-Time Construction of Suffix Tries", Proceedings Annual ACM Symposium on the Theory of Computing, 310–316.

D. C. Kozen [1992]. *The Design and Analysis of Algorithms*, Springer.

R. Krause [1962]. "Channels which Transmit Letters of Unequal Duration", *Information and Control* 5:1, 13–24.

L. I. Kronsjo [1979]. *Algorithms: Their Complexity and Efficiency*, Wiley.

J. B. Kruskal [1956]. "On the Shortest Spanning Tree of a Graph and the Travelling Salesman Problem", *Proceedings Americn Mathematical Society* 7, 48–50.

C. Kuratowski [1930]. "Sur Le Problème des Courbes Gauches en Topologie", *Fundamenta Mathematicae 15*, 271–238.

Laarhaven and Aarts [1987]. *Simulated Annealing*, D. Reidel Publishing Co.

N. J. Laarson [1998]. "The Context Trees of Block Sorting Compression", *Proceedings Data Compression Conference*, IEEE Computer Society Press, 189–198.

S. Lakshmivarahan and S. K. Dhall [1990]. *Analysis and Design of Parallel Algorithms: Arithmetic and Matrix Problems*, McGraw-Hill.

G. M. Landau, B. Schieber, and U. Vishkin [1987]. "Parallel Construction of a Suffix Tree", *Proceedings 14th ICALP*, Lecture Notes in Computer Science 267, Springer, 314–325.

G. Landau and U. Vishkin [1986]. "Efficient String Matching with k Errors", *Theoretical Computer Science 43*, 239–249.

G. Langdon [1981]. "Tutorial on Arithmetic Coding", *Technical Report RJ3128*, IBM Research Lab., San Jose, CA.

G. Langdon [1984]. "An Introduction to Arithmetic Coding", *IBM Journal of Research and Development 28:2*, 135–149.

G. Langdon [1991]. "Probabilistic and Q-Coder Algorithms for Binary Source Adaption", *Proceedings Data Compression Conference*, IEEE Computer Society Press, 13–22.

G. Langdon and J. Rissanen [1981]. "Compression of black-white images with arithmetic coding", *IEEE Transactions on Communications COM-29:6*, 858–867.

N. J. Larsson [1999]. *Structures of String Matching and Data Compression*, Ph.D. Dissertation, Department of Computer Science, Lund U., Sweden.

L. Larmore and D. Hirschberg [1987]. "A Fast Algorithm for Optimal Length-Limited Codes", *Technical Report*, Dept. of Information and Computer Science, University of California, Irvine, CA.

E. Lawler [1976]. *Combinatorial Oprimization: Networks and Matroids*, Holt, Rinehart, and Winston.

B. G. Lee [1984]. "A New Algorithm to Compute the Discrete Cosine Transform", *IEEE Transactions on Acoustics, Speech, and Signal Processing 32:6*, 1243–1245.

T. Lengauer [1989]. "Hierarchical Planarity Testing Algorithms", *Journal of the ACM 36:3*, 474–509.

F. T. Leighton [1992]. *Introduction to Parallel Algorithms and Architectures: Arrays, Trees, Hypercubes*, Morgan Kaufmann Publishers, San Mateo, CA.

F. T. Lieghton and S. Rao [1999]. "Multicommodity Max-Flow Min-Cut Theorems and Their use in Approximation Algorithms", *Journal of the ACM 46:6*, 787–832.

C. E. Leiserson [1985]. "Fat-Trees: Universal Networks for Hardware-Efficient Supercomputing", *IEEE Transactions on Computers 34:10*, 892–901.

A. M. Lesk [1994]. "Computational Mocular Biology", in *Encyclopedia of Computer Science and Technology 31*, A. Kent and J. G. Williams eds., 101–165.

L. S. Levy [1980]. *Discrete Structures of Computer Science*, Wiley.

H. R. Lewis and L. Denenberg [1991]. *Data Structures and Their Algorithms*, Harper Collins.

M. Liddell and A. Moffat [2001]. "Length-Restricted Coding in Static and Dynamic Frameworks", *Proceedings Data Compression Conference*, IEEE Computer Society Press, 133–142.

A. Ligtenberg and M. Vetterli [1986]. "A Discrete Fourier-Cosine Transform Chip", *IEEE Journal on Selected Areas in Communications SAC-4*, 49–61.

J. Lin and J. A. Storer [1991]. Processor-Efficient Algorithms for the Knapsack Problem", *Journal of Parallel and Distributed Computing 11*, 332–337.

J. Lin and J. A. Storer [1994]. Design and Performance of Tree-Structured Vector Quantizers", *Information Processing and Management 30:6*, 851–862.

J. Lin, J. A. Storer, and M. Cohn [1992]. Optimal Pruning for Tree-Structured Vector Quantization", *Information Processing and Management 28:6*, 723–733.

R. Lipton and D. Lopresti [1985]. "A Systolic Array for Rapid String Comparison", Chapel Hill Conference on VLSI; also, Technical Report, Department of Computer Science, Princeton U.

C. L. Liu [1968]. *Introduction to Combinatorial Mathematics*, McGraw-Hill.

C. L. Liu [1985]. *Elements of Discrete Structures*, McGraw-Hill.

R. Lowarance and Wagner [1975]. "An Extension of the String-to-String Correction Problem", *Journal of the ACM 22:2*, 177–183.

L. Lovasz [1979]. *Combinatorial Problems and Exercises*, North Holland.

M. G. Maab [2000]. "Linear Bidirectional On-Line Construction of Affix Trees", *Proceedings Combinatorial Pattern Matching (CPM)*, Springer, 320-334.

D. Maier and J. A. Storer [1978]. "A Note Concerning the Superstring Problem", *Proceedings Twelfth Annual Conference on Information Sciences and Systems*, Johns Hopkins U.

D. Maier [1990]. Course notes, Oregon Graduate Center.

M. E. Majster and A. Reiser [1980]. "Efficient On-Line Construction and Correction of Position Trees", *SIAM Journal of Computing 9:4*, 785–807.

J. Makhoul [1980]. "A Fast Cosine Transform in One and Two Dimensions", *IEEE Transactions on Acoustics, Speech, and Signal Processing 28:1*, 27–34.

V. Makinen [2000]. "Compact Suffix Array", *Proceedings Combinatorial Pattern Matching (CPM)*, Montreal, Canada (Springer), 305–319.

V. M. Malhotra, M. P. Kumar, and S. N. Maheshwari [1978]. "An $O(|V|^3)$ Algorithm for Finding Maximum Flows in a Network", *Information Processing Letters 7*, 277–278.

S. Mallat [1998]. *A Wavelet Tour of Signal Processing*, Academic Press.

U. Manber [1989]. *Introduction to Algorithms: A Creative Approach*, Addison–Wesley.

Z. Manna [1974]. *Mathematical Theory of Computation*, McGraw-Hill.

D. Manstetten [1992]. "Tight bounds on the redundancy of Huffman codes", *IEEE Trans. Inform. Theory IT-38:1*, 144–151.

G. Manzini [2001]. "An Analysis of the Burrows–Wheeler Transform", *Journal of the ACM 48:3*, 407-330.

M. W. Marcellin, M. J. Gormish, and M. P. Boliek [2000]. "An Overview of JPEG2000", *Proceedings Data Compression Conference*, IEEE Computer Society Press, 523–541.

G. Martin, G. Langdon, and S. Todd [1983]. "Arithmetic Codes for Constrained Channels", *IBM Journal of Research and Development 27:2*, 94–106.

M. A. McBeth [1990]. "The Turing Machine of Ackermann's Function", Manuscript, Computer Centre, Goldsmith's College, London.

E. M. McCreight [1976]. "A Space-Economical Suffix Tree Construction Algorithm", *Journal of the ACM 23:2*, 262–272.

J. H. McClellan, R. W. Schafer, and M. A. Yoder [1999]. *DSP First: A Multimedia Approach*, Prentice Hall.

D. McIntyre and M. Pechura [1985]. "Data Compression Using Static Huffman Code-Decode Tables", *Journal of the ACM 28:6*, 612–616.

R. McNaughton and H. Yamada [1960]. "Regular Expressions and State Graphs for Automata", *IEEE Transactions on Electronic Computers 9:1*, 39–47.

C. Mead and L. Conway [1980]. *Introduction to VLSI Systems*, Addison–Wesley.

K. Mehlhorn [1980]. "An Efficient Algorithm for Constructing Nearly Optimal Prefix Codes", *IEEE Transactions on Information Theory 26:5*, 513–517.

K. Mehlhorn [1984]. *Data Structures and Algorithms, Volumes I, II, II*, Springer.

K. Mehlhorn and U. Vishkin [1984]. "Randomized and Deterministic Simulations of PRAMs by Parallel Machines with Restricted Granularity of Parallel Memory", *Acta Informatica 21*, 339–374.

S. Michali and V. V. Vazirani [1980]. "An $O(\sqrt{V} \cdot E)$ Algorithm for Finding Maximum matching in General Graphs", *Proceedings 21st Annual IEEE Symposium on the Foundations of Computer Science*, 17–27.

R. Miller and L. Boxer. *Algorithms: Sequential and Parallel, A Unified Approach*, Prentice Hall.

R. Milidiu, E. Laber, and A. Pessoa [1999]. "A Work Efficient Parallel Algorithm for Constructing Huffman Codes", *Proceedings Data Compression Conference*, IEEE Computer Society Press, 277–286.

S. Mitarai, M. Hirao, T. Matsumoto, A. Shinohara, M. Takeda, and S. Arikawa [2001]. "Compressed Pattern Matching for SEQUITUR", *Proceedings Data Compression Conference*, IEEE Computer Society Press, 469–478.

M. Mitzenmacher [1996]. *The Power of Two Choices in Randomized Load Balancing*, Ph.D. Thesis, Department of Computer Science, University of California at Berkeley.

M. Mitzenmacher [2001]. "On the Hardness of Finding Multiple Preset Dictionaries", *Proceedings Data Compression Conference,* IEEE Computer Society Press, 411–418.

M. Mitzenmacher [2001b]. Compressed Bloom Filters, *Proceedings 20th ACM Symposium on Principles of Distributed Computing*, 144-150.

M. Mitzenmacher [2001c]. "Compressed Bloom filters for Distributed Web Caches", Technical Report, Computer Science Department, Harvard U..

M. Mitzenmacher, A. W. Richa, and R. Sitaraman [2001]. "The Power of Two Random Choices: A Survey of Techniques and Results", *Handbook of Randomized Computing*, Volume 1, S. Rajasekaran, P. M. Pardalos, J. H. Reif, and J. Rolim editors, Kluwer Academic Publishers, 255-312.

S. L. Mitchell [1979]. "Linear Algorithms to Recognize Outerplanar and Maximal Outerplanar Graphs", *Information Processing Letters 9:5*, 229–232.

J. Mitchell, W. B. Pennebaker, C. E. Fogg, and D. J. LeGall [1996]. *MPEG Video Compression Standard*, Chapman and Hall.

A. Moffat [1989]. "Word-based text compression", *Software-practice and experience 19*, 185–198.

A. Moffat [1990]. "Implementing the PPM Data Compression Scheme", *IEEE Transactions on Communications 38:11*, 1917–1921.

A. Moffat [2002]. *Compression Algorithms: Fast and Effective Source Coding*, Manuscript, Dept. of Computer Science and Software Engineering, The University of Melbourne, Australia.

A. Mofat, R. Neal, and I. Witten [1995]. "Arithmetic Coding Revisited", *Proceedings Data Compression Conference, IEEE Computer Society Press*, 202–211.

A. Moffat and A. Turpin [1996]. "On the Implementation of Minimum-Redundancy Prefix Codes", *Proceedings Data Compression Conference,* IEEE Computer Society Press, 170-179.

A. Moat, R. Neal, and I. Witten [1995]. "Arithmetic Coding Revisited", *Proceedings Data Compression Conference,* IEEE Computer Society Press, 202–211.

A. Moffat, A. Turpin, and J. Katajainen [1995]. "Space-Efficient Construction of Optimal Prefix Codes", *Proceedings Data Compression Conference,* IEEE Computer Society Press, 192–201.

A. Moffat and A. Wirth [2001]. "Can We Do Without Ranks in Burrows Wheeler Transform Compression", *Proceedings Data Compression Conference*, IEEE Computer Society Press.

B. M. E. Moret and H. D. Shapiro [1991]. *Algorithms from P to NP,* Benjamin Cummings Publishing Co.

R. Morris [1968]. "Scatter Storage Techniques", *Communications of the ACM 11:1*, 35–44.

J. H. Morris and V. R. Pratt [1970]. "A Linear Pattern Matching Algorithm", Technical Report 40, Computing Center, U. of California, Berkeley.

D. R. Morrison [1968]. "PATRICIA – Practical Algorithm To Retrieve Information Coded in Alphanumeric", *Journal of the ACM 15:4*, 514–534.

J. L. Mott, A. Kandel, and T. P. Baker [1986]. *Discrete Mathematics for Computer Scientists and Mathematicians*, Prentice Hall.

G. Motta [2000]. Private communication.

G. Motta, J. A. Storer, and B. Carpentieri [1999]. "Adaptive Linear Prediction Lossless Image Coding", *Proceedings Data Compression Conference*, IEEE Computer Society Press, 491–500.

G. Motta, J. A. Storer, and B. Carpentieri [2000]. "Improving Scene Cut Quality for Real-Time Video Decoding", *Proceedings Data Compression Conference*, IEEE Computer Society Press, 470-479.

R. Motwani and P. Raghavan [1995]. *Randomized Algorithms*, Cambridge U. Press.

G. Navarro, T. Kida, M. Takeda, A. Shinohara, and S. Arikawa [2001]. "Faster Approximate String Matching over Compressed Text", *Proceedings Data Compression Conference*, IEEE Computer Society Press, 459–468.

G. Navarro and M. Raffinot [1999]. "A General Practical Approach to Pattern Matching over Lempel-Ziv Compressed Text", *Proceedings Combinatorial Pattern Matching (CPM)*, Springer, 1–13.

G. Navarro and J. Tarhio [2000]. "Boyer-Moore String Matching over Ziv-Lempel Compressed Text", *Proceedings Combinatorial Pattern Matching (CPM)*, Springer, 166–180.

R. Neapolitan and K. Naimipour [1996]. *Foundations of Algorithms*, Jones and Bartlett Publishers.

Y. Nekritch [2000]. "Decoding of Canonical Huffman Codes with Look-Up Tables", *Proceedings Data Compression Conference*, IEEE Computer Society Press, 566.

M. R. Nelson [1996]. "Data Compressin with the Burrows-Wheeler Transform", *Dr. Dobb's Journal*, September, 46–50.

Y. Nievergelt [1999]. *Wavelets Made Easy*, Birkhäuser.

J. Nievergelt and K. H. Hinrichs [1993]. *Algorithms and Data Structures*, Prentice Hall.

J. Nievergelt and E. M. Reingold [1973]. "Binary Search Trees of Bounded Balance", *SIAM Journal of Computing 2:1*, 33–43.

B. Noble and J. W. Daniel [1977]. *Applied Linear Algebra*, Prentice Hall.

H. Nussbaumer [1981]. *Fast Fourier Transform and Convolution Algorithms*, Springer.

S. Olariu, M. C. Pinotti, and S. Q. Zheng [2000]. "An Optimal Hardware-Algorithm for Sorting Using a Fixed-Size Parallel Sorting Device", *IEEE Transactions on Computers 49:12*, 1310-1324.

M. O'Nan [1976]. *Linear Algebra*, Harcourt Brace Jovanovich.

A. V. Oppenheim, R. W. Schafer, J. R. Buck [1999]. *Discrete-Time Signal Processing*, Prentice Hall.

O. Ore 1963]. *Graphs and Their Uses*, Random House.

C. H. Papadimitriou and K. Steiglitz [1982], *Combinatorial Optimization*, Prentice Hall.

I. Parberry [1998]. "Everything You Wanted to Know About the Running Time of Mergesort But Were Afraid to Ask", Technical Report, Department of Computer Sciences, U. North Texas.

H. Park, J. Son and S. Cho [1995]. Area Efficient fast Huffman Decoder for Multimedia Applications", *Proceedings IEEE International Conference on Acoustics, Speech, and Signal Processing*, 3279–3282.

T. Park and V. Prasanna [1993]. "Area Efficient VLSI Architectures for Huffman Coding", *Proceedings IEEE International Conference on Acoustics, Speech, and Signal Processing*, I: 437–440.

W. B. Pennebaker and J. L. Mitchell [1993]. *JPEG Still Image Data Compression Standard*, Van Nostrand Reinhold.

W. Pennebaker, J. Mitchell, G. Langdon, and R. Arps [1988]. "An overview of the basic principles of the Q-coder", *IBM Journal of Research and Development 32:6*, 717–726.

P. Pevzner and M. S. Waterman [1995]. "Open Problems in Computational Molecular Biology", *Proceedings of the Third Israel Symposium on Computers and Systems*, IEEE Computer Society Press.

N. Pippenger [1979]. "On Simultaneous Resource Bounds", *Proceedings 20th Annual IEEE Symposium on the Foundations of Computer Science*, 307–311.

S. Phillips and J. Westbrook [1993]. "Online Load Balancing and Network Flow", *Proceedings 25th Annual ACM Symposium on the Theory of Computing*, 402–411.

E. Polak [1997]. *Optimization Algorithms and Consistent Approximations*, Springer.

G. Polya, R. E. Tarjan, and D. R. Woods [1983]. *Notes on Introductory Combinatorics*, Birkhäuser.

W. K. Pratt [1991]. *Digital Image Processing*, Wiley.

F. Preparata and R. T. Yeh [1973]. *Introduction to Discrete Structures for Computer Science and Engineering*, Addison–Wesley.

W. H. Press, S. A. Teukolsky, W. T. Vetterling, and B. P. Flannery [1999]. *Numerical Recipes in C: The Art of Scientific Computing*, Cambridge U. Press.

R. C. Prim [1957]. "Shortest Connection Networks and Some Generalizations", *Bell System Technical Journal 36*, 1389–1401.

J. G. Proakis and D. G. Manolakis [1996]. *Digital Signal Processing: Principles, Algorithms, and Applications*, Prentice Hall.

P. Purdom and C. Brown [1985]. *The Analysis of Algorithms*, Holt, Rinehart, and Winston.

V. R. Pratt [1979]. Shellsort and Sorting Statistics, Garland Publishing.

M. J. Quinn [1987]. *Designing Efficient Algorithms for Parallel Computers*, McGraw-Hill.

S. Rajasekaran, P. M. Pardalos, J. H. Reif, and J. Rolim, editors [2001]. *Handbook of Randomized Computing*, Kluwer Academic Publishers, 255-312.

V. Ramachandran and J. H. Reif [1989]. "An Optimal Parallel Algorithm for Graph Planarity", *Proceedings 30th Annual IEEE Symposium on the Foundations of Computer Science*, 282–287.

S. Ranka and S. Sahni [1990]. *Hypercube Algorithms*, Springer.

K. R. Rao and R. Yip [1990]. *Discrete Cosince Transform: Algorithms, Advantages, Applications*, Academic Press.

J. H. Reif, ed. [1993]. *Synthesis of Parallel Algorithms*, Morgan Kaufmann.

J. H. Reif and J. A. Storer [1987]. Minimizing Turns for Discrete Movement in the Interior of a Polygon", *IEEE Journal of Robotics and Automation 3:3*, 182–193.

J. H. Reif and J. A. Storer [1994]. "A Single-Exponential Upper Bound for Finding Shortest Paths in Three Dimensions", *Journal of the ACM 41:5,* 1013–1019.

J. H. Reif and J. A. Storer [2001]. "Optimal Encoding of Non-Stationary Sources", *Information Sciences*, 87–105; se also "Optimal Lossless Compression of a Class of Dynamic Sources", *Proceedings IEEE Data Compression Conference*, 501–510, 1998.

E. M. Reingold and W. J. Hansen [1983]. *Data Structures*, Little Brown and Co.

E. M. Reingold, H. Nievergelt, and N. Deo [1977]. *Combinatorial Algorithms: Theory and Practice*, Prentice Hall.

J. Rissanen and G. Langdon [1979]. "Arithmetic coding", *IBM J. Res. Dev. 23*, 149–162.

J. Rissanen and K. Mouhiuddin [1989]. "A Multiplication-Free Multialphabet Arithmetic Code", *IEEE Transactions on Communications 37:2*, 93–98.

F. Rizzo, B. Carpentieri, and J. A. Storer [1999]. "Experiments with Single-Pass Adaptive Vector Quantization", *Proceedings Data Compression Conference*, IEEE Computer Society Press, 546; see also "Improving Single-Pass Adaptive VQ", *International Conference on Acoustics, Speech, and Signal Processing (ICASSP)*, IMDSP2.10.

F. Rizzo, B. Carpentieri, and J. A. Storer [2001]. LZ-Based Image Compression, *Information Sciences 135*, 107–122.

F. Rizzo and J. A. Storer [2001]. "Overlap in Adaptive Vector Quantization", *Proceedings Data Compression Conference*, IEEE Computer Society Press, 401–410.

F. S. Roberts [1984]. *Applied Combinatorics*, Prentice Hall.

E. Rocke [2000]. "Using Suffix Trees for Gapped Motif Discovery", *Proceedings Combinatorial Pattern Matching (CPM)*, Montreal, Canada (Springer), 335–349.

M. Rodeh, V. R. Pratt, and S. Even [1981]. "Linear Algorithm for Data Compression vis String Matching", *Journal of the ACM 28:1*, 16–24.

K. H. Rosen [2000]. *Elementary Number Theory and Its Applications*, Addison–Wesley.

K. A. Ross and C. R. B. Wright [1999]. *Discrete Mathematics*, Prentice Hall.

D. M. Royals, T. Markas, N. Kanopoulos, J. H. Reif, and J. A. Storer [1993]. "On the Design and Implementation of a Lossless Data Compression and Decompression Chip", *IEEE Journal of Solid-State Circuits*, 948–953.

W. Rytter [1980]. "A Correct Preprocessing Algorithm for the Boyer-Moore String Matching Algorithm", *SIAM Journal of Computing 9*, 509–512.

K. Sadakane [1999]. "A Modified Burrows-Wheeler Transform for Case-Insensitive Search with Application to Suffix Array Compression", Technical Report, Department of Computer Science, U. Tokoyo.

S. Sahni [1981]. *Concepts in Discrete Mathematics*, Camelot Publishing Co.

D. Sankoff and J. B. Kruskal [1983]. *Time Warps, String Edits, and Macromolecules: The Theory and Practice of Sequence Comparison*, Addison–Wesley.

W. J. Savitch [1970]. "Relationship Between Nondeterministic and Deterministic Tape Complexities", *Journal of Computer and System Sciences 4:2*, 177–192.

L. L. Schumaker and G. Webb, eds. [1993]. *Recent Advances in Wavelet Analysis*, Academic Press, Dan Diego, CA.

M. Sipser [1997]. *Introduction to the Theory of Computation*, PWS Publishing Co.

K. Sadakane [1998]. "A Fast Algorithm for Making Suffix Arrays and for Burrows-Wheeler Transform", *Proceedings Data Compression Conference*, IEEE Computer Society Press, 129–138.

S. C. Sahinalp [1994]. "Symmetry Breaking for Suffix Tree Construction", *Proceedings Annual ACM Symposium on the Theory of Computing*, 300-309.

M. Schindler [1997]. ""A Fast Block-Sorting Algorithm for Lossless Data Compression", *Proceedings Data Compression Conference*, IEEE Computer Society Press, 469.

R. Sedgewick and P. Flajolet [1998]. *Analysis of Algorithms, Volume I and II*, Addison–Wesley.

J. Setubal and J. Meidanis [1997]. *Introduction to Computational Mocular Biology*, PWS.

J. Seward [2001]. "Space-Time Tradeoffs in the Inverse B-W Transform, *Proceedings Data Compression Conference*, IEEE Computer Society Press.

J. Sieferas [1997]. Lecture notes, Dept. of Computer Science, U. Rochester, Rochester, NY.

A. Siegel [1989]. "On Universal Classes of Fast Hash Functions, Their Time-Space Tradeoff and Their Applications", *Proceedings 30th IEEE Symposium on the Foundations of Computer Science*, 612–621.

Y. Shibata, T. Matsumoto, M. Takeda, A. Shinohara, and S. Arikawa [2000]. "A Boyer-Moore Type Algorithm for Compressed Pattern Matching", *Proceedings Combinatorial Pattern Matching (CPM)*, Springer, 181–194.

Y. Shibata, M. Takeda, A. Shinohara, and S. Arikawa [1999]. "Pattern Matching in Compressed Text using Antidictionaries", *Proceedings Combinatorial Pattern Matching (CPM)*, Springer, 1–13.

Y. Shiloach [1978]. " An $O(nI\log^2 I)$ Maximum Flow Algorithm, Technical Report STAN-CS-78-802, Computer Science Department, Stanford U.

R. C. Singleton [1969]. "Algorithm 347: An Algorithm for Sorting with Minimal Storage", *Communications of the ACM 12:3*, 185–187.

S. S. Skiena [1998]. *The Algorithm Design Manual*, Springer.

D. D. Sleator and R. E. Tarjan [1980]. "An $O(nm\log(n))$ Algorithm for Maximum Network Flow", Technical Report STAN-CS-80-831, Computer Science Department, Stanford U.

D. D. Sleator and R. E. Tarjan [1983]. "Self-Adjusting Binary Search Trees", *Proceedings 15th Annual ACM Symposium on the Theory of Computing*, 235–245.

A. O. Slisenko [1977]. "String-Matching in Real Time", Preprint P-7–77, The Steklov Institute of Mathematics, Leningrad Branch (Russian).

A. O. Slisenko [1978]. "String-Matching in Real Time: Some Properties of the Data Structure", Mathematical Foundations of Computer Science, *Proceedings 7th Symposium*, Zakopane, Poland, Springer, 493–496.

A. O. Slisenko [1980]. "Determination in Real Time of All Perodicities in a Word", Soviet Mathematics *Doklady 21:2*, 392–395.

A. O. Slisenko [1981]. "Detection of Periodicities and String-Matching in Real Time", *Soviet Journal of Mathematics 22:3*, 1316–1387; translated from Zapiska Nauchnykh Seminarov Leningradskogo Otdeleniya Matematicheskogo Instituta im. V. A. Steklova AN SSR 105 (1980), 62–183 (Russian).

G. Smith [1984]. Class notes, Computer Science Department, Brandeis U.

D. F. Stanat and D. F. McAllister [1977]. *Discrete Mathematics in Computer Science*, Prentice Hall.

T. Standish [1980]. *Data Structures*, Addison–Wesley.

G. A. Stephen [1994]. *String Searching Algorithms*, World Scientific Press, Singapore.

J. A. Storer [1979]. "A Note on the Complexity of Chess", *Proceedings Thirteenth Annual CISS* , 160-166.; a full version of this paper later appeared as: "On the Complexity of Chess", *Journal of Computer and System Sciences 27:1*, 77–100.

J. A. Storer [1980]. "The Node Cost Measure for Embedding Graphs on the Planar Grid", *Proceedings Twelfth Annual ACM Symposium on the Theory of Computing*, Los Angeles, CA, 201–210; a full version of this paper later appeared as: "On Minimal Node Cost Planar Embeddings", *Networks 14:2*, 181–212, 1984.

J. A. Storer [1981]. Constructing Full Spanning Trees for Cubic Graphs", *Information Processing Letters 13:1*, 8–11, October, 1981.

J. A. Storer [1983]. "An Abstract Theory of Data Compression", *Theoretical Computer Science 24* , 221–237; a preliminary version of this paper appeared as "Toward an Abstract Theory of Data Compression", *Proceedings Twelfth Annual Conference on Information Sciences and Systems* 391–399, 1978.

J. A. Storer [1985]. "Textual Substitution Techniques for Data Compression", *Combinatorial Algorithms on Words*, Springer, 111–129.

J. A. Storer [1988]. *Data Compression: Methods and Complexity Issues*, Computer Science Press (a subsidiary of W. H. Freeman Press).

J. A. Storer [1991]. The Worst-Case Performance of Adaptive Huffman Codes", *Proceedings IEEE Symposium on Information Theory*, Budapest, Hungary.

J. A. Storer [1992], ed. Image and Text Compression, Kluwer.

J. A. Storer, D. Belinskaya, and S. DeAgostino [1995]. Near Optimal Compression with Respect to a Static Dictionary on a Practical Massively Parallel Architecture", *Proceedings Data Compression Conference*, IEEE Computer Society Press, 172–181.

J. A. Storer [2000]. "Error Resilient Dictionary Based Compression", *Proceedings of the IEEE 88:11*, 1713–1721; earlier work upon which this paper is based appeared as: "The Prevention of Error Propagation in Dictionary Compression with Update and Deletion", *Proceedings Data Compression Conference*, IEEE Computer Society Press, 199–208; "Error Resilient Data Compression with Adaptive Deletion", in *Compression and Complexity of Sequences,* IEEE Press, 285–294, 1998.

J. A. Storer and H. Helfgott [1997]. Lossless Image Compression by Block Matching", *The Computer Journal 40:2/3*, 137–145.

J. A. Storer and R. Keller [1974]. "A Proof of Manna, Prob. 3–22", unpublished manuscript, completed as part of the course EE 503, taught by R. Keller, Dept. of Electrical Engineering and Computer Science, Princeton U.

J. A. Storer, A. Nicas, and J. Becker [1985]. "Uniform Circuit Placement", *VLSI Algorithms and Architectures 1*, 255–273.

J. A. Storer and J. H. Reif [1991]. A Parallel Architecture for High Speed Data Compression", *Journal of Parallel and Distributed Computing 13*, 222–227.

J. A. Storer and J. H. Reif [1994]. "Shortest Paths in the Plane with Polygonal Obstacles", *Journal of the ACM 41:5,* 982–1012.

J. A. Storer and J. H. Reif [1995]. "Error Resilient Optimal Data Compression", *SIAM Journal of Computing 26:4*, 934–939.

J. A. Storer and J. H. Reif [1997]. "Low-Cost Prevention of Error-Propagation for Data Compression with Dynamic Dictionaries", *Proceedings Data Compression Conference*, IEEE Computer Society Press, 171–180.

J. A. Storer and T. G. Szymanski [1978]. "The Macro Model for Data Compression", *Proceedings Tenth Annual ACM Symposium on the Theory of Computing*, 30-39; a full version of this paper later appeared as " Data Compression Via Textual Substitution", *Journal of the ACM 29:4*, 928–951.

G. Strang amd T. Nguyen [1996]. *Wavelets and Filter Banks*, Wellsley-Cambridge Press.

V. Strassen [1969]. "Gaussian Elimination is Not Optimal", *Numerische Mathematik 13*, 354–356.

D. M. Sunday [1990]. "A Very Fast Substring Search Algorithm", *Communications of the ACM 33:8*, 143.

B. W. Suter [1997]. *Multirate & Wavelet Signal Processing*, Academic Press.

M. N. S. Swamy and K. Thulasiraman [1981]. *Graphs, Networks, and Algorithms*, Wiley.

W. Szpankowski [1990]. "Patricia Tries Again Revisited", *Journal of the ACM 37:4*, 691–711.

W. Szpankowski [1993]. "A Generalized Suffix Tree and Its (Un)Expected Asymptotic Behaviours", *SIAM Journal of Computing 22:6*, 1176–1198.

G. B. Thomas [1968]. *Calculus and Analytic Geometry*, Addison–Wesley.

R. E. Tarjan [1975]. "Efficiency of a Good But Not Linear Set Union Algorithm", *Journal of the ACM 22:2*, 215–225.

R. E. Tarjan [1983]. *Data Structures and Network Algorithms*, Society for Industrial and Applied Mathematics (SIAM) Press.

F. Taylor and J. Mellott [1998]. *Hands On Digital Signal Processing*, McGraw-Hill.

K. Thompson [1968]. "Regular Expression Search Algorithm", *Communications of the ACM 11:6*, 419–422.

T. Tietelbaum [1974]. Lecture notes for Computer Science 611, Cornell U.

P. Topiwala [1998]. *Wavelet Image and Video Compression*, Kluwer.

J. P. Tremblay and P. G. Sorenson [1984]. *An Introduction to Data Structures with Applications*, McGraw-Hill.

B. D. Tseng and W. C. Miller [1978]. "On Computing the Discrete Cosine Transform", *IEEE Transactions on Computers C-27*, 966–968.

A. Tucker [1980]. *Applied Combinatorics*, Wiley.

J. Turner [1989]. "Approximation Algorithms for the Shortest Common Superstring Problem", *Information and Computation 83*, 1–20.

W. T. Tutte, ed. [1984]. *Graph Theory*, Encyclopedia of Mathematics and its Applications, Addison–Wesley.

E. Ukkonen [1985]. "Approximate String Matching", *Information and Control 64*, 100-118.

E. Ukkonen [1993]. "Algorithms for Approximate String Matching Over Suffix Trees", Proceedings 4th Symposium on Combinatorial Pattern Matching, Padova, Italy, 228–242.

E. Ukkonen [1995]. "On-Line Construction of Suffix-Tries", Algorithmica 14, 249–260.

J. D. Ullman [1984]. *Computational Aspects of VLSI*, Computer Science Press / W. H. Freeman.

L. G. Valiant [1975]. "General Context-Free Language Recognition is Less Than Cubic Time", *Journal of Computer and System Sciences 10:2*, 308–315.

S. Vempala [1999]. "On the Approximability of the Traveling Salesman Problem", Technical Report, Laboratory for Computer Science, MIT.

A. Vetterli and J. Kovacevic [1995]. *Wavelets and Subband Coding*, Prentice Hall.

U. Vishkin [1990]. "Deterministic Sampling – A New Technique for Fast Pattern Matching, *Proceedings Annual ACM Symposium on the Theory of Computing*, Baltimore, MD, 170-180.

J. Vitter [1987]. "Design and analysis of dynamic Huffman codes", *Journal of the ACM 34*, 825–845.

J. Vitter [1989]. "Dynamic Huffman coding", *ACM Trans. Math. Software 15*, 158–167.

B. Vöcking [1999]. "How Asymetry helps Load Balancing", *Proceedings IEEE 40th Annual Symposium on Foundations of Computer Science*, 131-140.

P. A. J. Volf and F. M. J. Williams [1998]. "Switching Between Two Universal Source Coding Algorithms", *Proceedings Data Compression Conference*, IEEE Computer Society Press, 491–500.

P. A. J. Volf and F. M. J. Williams [1998b]. "The Switching Method: Elaborations", *Proceedings 19th Symposium on Information Theory*, 13–20.

R. A. Wagner and Fischer [1974]. "The String to String Correction Problem", *Journal of the ACM 21:1*, 168–173.

W. D. Wallis [2000]. *A Beginner's Guide to Graph Theory,* Birkhäuser.

S. Warshall [1962]. "A Theorem on Boolean Matrices", *Journal of the ACM 9:1*, 11–12.

M. S. Waterman [1995]. *Introduction to Computational Biology: Maps, Sequences, Genomes*, Chapman and Hall, London.

P. Weiner [1973]. "Linear Pattern Matching Algorithms", *Proceedings 14th Annual Symposium on the Foundations of Computer Science*, 1–11.

E. W. Weisstein [1999]. *CRC Concise Encyclopedia of Mathematics*, CRC Press.

N. H. E. Weste and K. Eshraghian [1993]. *Principles of CMOS VLSI Design*, Addison–Wesley.

S. G. Williamson [1980]. "Finding a Kuratowski Subgraph in Linear Time", Technical Report, Department of Mathematics, Institut de Recherche Mathematique Avance, Universite Louis Pasteur, Strasbourg.

S. G. Williamson [1984]. "Depth-First Search and Kuratowski Subgraphs", Journal of the ACM 31:4, 681–693.

S. Winograd [1970]. "On the Algebraic Complexity of Functions", in *Actes du Congres International Des Mathemacticiens*, Volume 3, 283–288.

S. Winograd [1973]. "Some Remarks on Fast Multiplication of Polynomials", in *Complexity of Sequential and Parallel Numerical Algorithms*, J. F. Traub, ed., Academic Press, 181–196.

I. Witten, A. Moffat, and T. Bell [1994]. *Managing Gigabytes*, Van Nostrand Reinhold.

I. Witten, R. Neal, and J. Cleary [1987]. "Arithmetic Coding for Data Compression", *Communications of the ACM 30:6*, 520-540.

S. Wu and U. Manber [1992]. "Fast Text Searching Allowing Errors", *Communications of the ACM 35:10*, 83–90.

R. Yeung [1991]. "Local redundancy and progressive bounds on the redundancy of a Huffman code", *IEEE Trans. Inform. Theory IT-37:3*, 687–691.

D. H. Younger [1967]. "Recognition and Parsing of Context-Free Languages in Time n^3", *Information and Control 10:2*, 189–208.

R. Zitowolf [1991]. "A Systolic Architecture for Sliding Window Data Compression", *Proceedings VLSI Signal Processing IV*, IEEE Press., 339–351.

U. Zwick [1999]. "All Pairs Shortest Paths", *Proceedings 31st ACM Annual Symposium on the Theory of Computing,* Alanta, GA, 61–69.

D. Zwillinger, ed. [1996]. *CRC Standard Mathematical Tables*, 30th edition, CRC Press.

Notation

- $O()$ = "Big O" notation for asymptotic complexity
  $\Omega()$ = "Big Ω" notation for asymptotic complexity
  $\Theta()$ = both O and Ω

- $\lceil i \rceil$ = The least integer greater than or equal to i.
  $\lfloor i \rfloor$ = The greatest integer less than or equal to i.

- $x\ MOD\ y = x - \lfloor x/y \rfloor y$

    Note: This notation is usually used when x and y are positive integers (so $x\ MOD$ y is the remainder after x is divided by y), but it is defined for any two real numbers x and y with the convention that $x\ MOD\ 0 = x$.

- $A \cup B$ = the union of sets A and B; the elements that are in A or B (or both).
  $A \cap B$ = the intersection of sets A and B; the elements that are in both A and B.
  $A\text{-}B$ = the difference between sets A and B; the elements in A but not in B.
  $|A|$ = the number of elements in the set A.

- $n! = 1 * 2 * 3 \ldots * n$

- $n^{1/i} = \sqrt[i]{n}$, $i>1$
  $n^{-i} = 1/n^i$, $i>0$

- $\log_b(n)$ = starting with 1, the number of times we must multiply by b to get n
  e = 2.7182... (the "natural logarithm base")
  $\ln(n) = \log_e(n)$
  $\log(n)$ = a notation to indicate that the logarithm base doesn't matter

- $a^{b^c} = a^{\left(b^c\right)}$

- $\displaystyle\sum_{i=1}^{n} f(i)$ = the sum from $i=1$ to n of $f(i)$

  $\displaystyle\prod_{i=1}^{n} f(i)$ = the product from $i=1$ to n of $f(i)$

- ∞ = infinity; a quantity larger than any positive integer
  $-\infty$ = minus infinity, a quantity smaller than any negative integer

Index

C

central processing unit (*see* CPU)
CFL, 174, 191–192, 201
character bit vectors, 370–372
Chernoff bound (*see* hashing)
child
    in a tree (*see* tree)
    array (*see* tree)
    number (*see* tree)
chip, 520
Chomsky normal form (*see* CFL)
clause (*see* Boolean expression)
clique (*see* graph)
clique problem, 325, 344-345
closed hashing (*see* hashing)
cMOS, 516, 518–519, 538–539
CNF (*see* Boolean expression)
coloring a tree red-black, 261
CombineBits, 366–372
common subsequence (*see* longest
    common subsequence)
common sums (*see* sum)
commutative ring (*see* ring, DFT)
compact trie, 384
complete binary tree (*see* tree)
complete *k*-ary trees (*see* heap)
complex number or exponential (*see*
    DFT)
compression
    arithmetic code, 397–402, 419
    binary prefix code algorithm, 381
    binary run-length code, 9, 44
    Burrows–Wheeler transform,
        403–404, 420
    dynamic dictionary, 382–383,
        416–417
    Huffman tree / trie / code, 181, 195,
        194-195, 381, 397, 415–416
    JPEG, 452, 468–469
    MPEG, 453, 468–469
    MTF / move-to-front, 405–406, 420
    of trees, 411
    prefix code, 381
    run-length code, 9, 44–45
    sliding window, 395

uniquely decodable, 381, 411
update method, 382-393
using superstrings, 411
variable length code, 381, 394, 397-
    402, 405, 414-420, 452-453
computer chip (*see* chip)
conjunctive normal form (*see* Boolean
    expression)
connected component or graph (*see*
    graph)
constant *e* (*see e*)
context-free language or grammar (*see*
    CFL)
converting a string to a number, 92,
    108–110
convolution (*see* polynomial
    multiplication, DFT)
cos / cosine (*see* trigonometric
    functions)
cosine transform (*see* DCT)
CPU, 1, 471
CRCW
    array max, 482–483
    PRAM, 472–473, 528–531
CREATE (*see* list)
credit (*see* amortized complexity)
CREW PRAM, 472–473
cube and peg puzzle (*see* puzzle)
cube connected cycles (*see* hypercube)
cubic weighted geometric sum (*see*
    geometric sum)
cut (*see* maximum flow)
CUT (*see* list)
cycle (*see* graph)
CYK algorithm (*see* CFL)

D

data compression (*see* compression)
data distribution, 488, 491, 502–503,
    512, 536–537
DATA (*see* binary search tree, list, tree)
data in the leaves, 259

used for polynomial multiplication, 168

used for post-order traversal, 130

used for prefix-sum, 479-480

used for pre-order traversal, 130

used for quick sort, 166, 177, 486

used for Strassen's algorithm, 169

used for summing an array, 478, 489

used for switching networks, 532-533

used for the Fourier transform, 431-437

DIVIDE, 330

dominating set, 325, 344

don't care position, 365

dot product, 465

double butterfly network, 533–534

double exponential tree, 483, 529

doubly ended queue, 81

dynamic dictionary (*see* compression)

dynamic programming

comparison with divide and conquer, 192

examples, 171–174

introduction, 171

saving sub-problems, 192

sums, 175, 191–194

used for CFL recognition, 174

used for edit distance, 395

used for knapsack, 171

used for longest common subsequence, 306

used for matrix multiplication, 173

used for optimal binary search trees, 158-159

used for paragraphing, 172

used for Pascal's triangle, 12

used for path finding, 307

used for shortest paths, 305

used for transitive closure, 306

E

e, 20, 207, 217–218

ED (*see* edit distance)

edge

cover, 343

list, 291

edit distance, 395, 413, 419, 537

Edmonds–Karp algorithm (*see* maximum flow)

elimination of recursion (*see* recursion)

empty

buckets (*see* hashing)

set, 374

string, 374

tree (*see* tree)

EMPTY (*see* list)

end statement, 4

ENQUEUE (*see* queue)

equivalence of red-black and 2-3 trees (*see* red-black tree)

equivalence relation or class, 333

EREW PRAM, 472–473, 489–495

erewListSuffixSum, 479

ESTACK (*see* union-find)

Euler path, 302, 335–336

Euler's formula for planar graphs, 329

evaluating an arithmetic expression (*see* arithmetic expression)

exclusive OR (*see* XOR)

expected number of empty buckets (*see* hashing)

expected time, 6

exponential, 19–21

expression tree, 137

F

face, 329-330

factorial

assembly language, 5, 27

correctness, 101

definition, 5, 583

elimination of recursion, 100

pseudo-code, 5

M

N

O

O notation (*see* asymptotic complexity)
$O(1)$, 7
odd-even bubble sort (*see* sorting)
off-line heap operations (*see* union-find)
omega (Ω) notation (*see* asymptotic complexity)
on the fly array initialization, 271, 282
OneCount, 368
open hashing (*see* hashing)
operand, 63, 93
optimal binary search tree (*see* binary search tree)
optimal ordering of matrices (*see* matrix multiplication)
OR, 37, 365–372, 515–516
ordered tree (*see* tree)
ordering lemma (*see* binary search tree)
out-degree (*see* graph)
outer-planar graph, 330, 352
overlap lemma, 182

P

PACK, 365
packet routing, 506, 514, 535
paragraphing, 172, 190–191
parent array (*see* tree)
PARENT (*see* binary search tree, heap, tree)
parent (*see* tree)
parity, 35
partition problem, 190
PARTITION, 166, 177, 184–187
Pascal's triangle (*see* binomial coefficient)
path
    compression (*see* union-find)
    definition(*see* graph)
    finding, 307, 337–338
PATRICIA trees, 417
pattern

diagram, 375–378
    matching, 374–378, 410, 415
PERCDOWN (*see* heap)
percolate update, 392–393
PERCUP (*see* heap)
perfect hashing (*see* hashing)
perfect integer square (*see* puzzle)
permutation, 272, 282–283, 434
planar
    graph, 329–331, 352
    grid, 330
    separator, 330
pointer, 68–70
polynomial
    evaluation, 10
    hash function (*see* hashing)
    multiplication, 169, 421, 440
    polynomial-time reduction, 326
POP (*see* stack)
POP(i) (*see* stack)
postfix expression, 63, 76–77, 82–83, 94, 112–114
post-order traversal (*see* tree)
POWER (*see* integer exponents)
PRAM
    model, 471–487
    simulation, 488
Pratt–Rodeh–Even algorithm, 417–418
prefix expression, 93, 94, 110–112, 112–114, 377
prefix property (*see* compression)
prefix-sum, 479–481, 500, 509, 525–526
pre-order traversal (*see* tree)
PREV (*see* list)
Prim's algorithm, 298–299
principal root of unity (*see* DFT)
procedure statement, 4
production in a CFL (*see* CFL)
proof by induction, 87–88, 95
pseudo-code, 4
PSPACE-complete or hard, 351
PUSH (*see* stack)
puzzle

burr, 347–348, 352–353
cube and peg, 52
Kev's cubes, 53
perfect square, 47
Towers of Hanoi, 98, 102–105, 115
Traffic Jam, 347

Q

quadratic weighted geometric sum (*see* geometric sum)
queue
    definition, 64
    DEQUEUE, 64
    ENQUEUE, 64
    FRONT, 64
    REAR, 64
quick sort (*see* sorting)

R

RAM, 1, 471, 519
random access memory (*see* RAM)
randomized algorithm
    introduction, 176
    used for duplicate search, 474
    used for hashing, 204-205, 215-216
    used for k^{th} largest, 178
    used for quick sort, 177
    used for routing, 506
    used for statistical smapling, 176
    used for string matching, 364
RB leaf (*see* heap)
Rcharsize, 369
RCHILD (*see* binary search tree)
REAR (*see* queue)
recurrence relation (*see* divide and conquer)
recursion, 87–88, 99
red-black tree
    black height, 248
    definition 248,

equivalence to 2-3 trees, 250
facts, 249
height, 253
height-balanced, 253
INSERT, 251–252
relaxed, 263–265
virtual leaf, 248
virtual min-height, 248
reduction to flow (*see* maximum matching)
reflexive relation (*see* equivalence relation)
REFINE, 398
register, 1, 5, 27, 518
regular expression, 374, 410
relaxed red-black tree (*see* red-black tree)
repeat statement, 4
representing arrays (*see* array)
RESCAN, 387–389
residual graph (*see* maximum flow)
resizing a hash table (*see* hashing)
reverse bubble sort (*see* sorting)
reversing a list (*see* list)
revisedPACK, 367
RIGHT, 397
right-augmented butterfly, 502
right-heavy (*see* AVL tree)
right-to-leftMATCH, 356
ring, 460–461
RL (*see* binary search tree)
Rones, 369
root (*see* tree)
root of unity (*see* DFT)
rotate left or right (*see* binary search tree)
rotation (*see* binary search tree)
routing (*see* packet routing)
RR (*see* binary search tree)
RSIB (*see* tree, heap)
rule in a CFL (*see* CFL)
run-length code (*see* data compression)

S

sa+dc+abMATCH, 367
sacbv+dc+ab+mmMATCH, 371
sacbvMATCH, 370
sacbvPACK, 370
saMATCH, 365
satisfiable Boolean expression (*see* Boolean expression)
Savitch's theorem (*see* PSPACE-complete)
SCAN, 386
scLOW, 297
self-adjusting binary search tree (*see* binary search tree)
self-printing program, 38–41, 52, 55–56
self-reproducing (*see* self-printing)
SendRight, 512
separator (*see* planar separator)
series (*see* sum)
sets over a small universe, 269
shift-and string matching, 365–373, 408–409, 415
ShiftLeft, 37–38, 365–372
ShiftRight, 35–38
shortest common superstring (*see* superstring)
shortest path (*see* minimum cost path)
shrinking a blossom (*see* maximum matching)
ShuffleFix, 504
sibling (*see* tree)
simulated annealing, 183, 196–197, 346
sin / sine (*see* trigonometric functions)
sine transform (*see* discrete sine transform)
single-source minimum cost path, 303–304, 337
sink (*see* maximum flow)
six piece rectangular burr puzzle (*see* puzzle)
size of the largest bucket (*see* hashing)
SIZE (*see* list)
sliding dictionary (*see* compression)

sliding suffix trie, 390–393, 412
sorting
    area-time tradeoff, 522–524
    better than $O(n\log(n))$ time (*see* heap)
    bubble sort, 8
    bucket sorting, 273, 283, 357
    displacement sort, 44, 526
    heap sort (*see* heap)
    introduction, 6
    lexicographic sorting, 357, 379
    lower bounds (*see* heap)
    merge sort, 78–80, 164–165, 184, 485, 492–495, 504–505, 513, 531
    odd-even bubble sort, 43
    position sort, 44
    quick sort, 166, 177, 184–187, 486
    reverse bubble sort, 43
    string sorting, 357, 379
    topological sort, 301
    tree sort, 149
    vector sorting, 273, 357
source (*see* maximum flow)
Spack, 365–372
spanning tree or forest (*see* graph)
splay tree (*see* binary search tree)
SPLAY-STEP (*see* binary search tree)
SPLICE (*see* list)
SPLIT (*see* binary search tree, 2-3 tree)
square root computation, 124-125
stable marriage, 324, 342, 350
stable sorting, 184
stack
    definition, 61
    POP, 61
    POP(i), 75, 296
    PUSH, 61
    TOP, 61
standard matrix multiplication algorithm (*see* matrix multiplication)
statistical sampling, 176
Strassen's algorithm (*see* matrix multiplication)

W

Warshall's algorithm, 306
weighted geometric sum (*see* geometric
    sum)
weighted hash function (*see* hashing)
while statement, 4
worst case time, 6

X

XOR, 35–36, 366

Y

yes-no problem, 325

Z

zeroth order Huffman code, 381, 411
zero term, 17